Recent Advances in the Diagnosis and Management of Plant Diseases

L.P. Awasthi
Editor

Recent Advances in the Diagnosis and Management of Plant Diseases

Springer

Editor
L.P. Awasthi
Department of Plant Pathology
N.D. University of Agriculture
and Technology
Kumarganj, Faizabad, Uttar Pradesh,
India

ISBN 978-81-322-3420-3 ISBN 978-81-322-2571-3 (eBook)
DOI 10.1007/978-81-322-2571-3

Springer New Delhi Heidelberg New York Dordrecht London

Softcover re-print of the Hardcover 1st edition 2015

Printed on acid-free paper

Springer (India) Pvt. Ltd. is part of Springer Science+Business Media (www.springer.com)

Everyone who remembers his or her own educational experience remembers teachers, not methods and techniques.

Sidney Hook

Prof. H.N. Verma
Hon'ble Vice-Chancellor, Jaipur National University, Jaipur.

I am proud to dedicate this book to my respected teacher, Professor (Dr.) H.N. Verma. Without whose love, guidance, continued support, and encouragement, the task so demanding, would have been almost insurmountable. Sir, you are my best guide, mentor, and Guru.

Prof. L.P. Awasthi

Foreword

AMITY UNIVERSITY
UTTAR PRADESH
AMITY INSTITUTE OF VIROLOGY AND IMMUNOLOGY

Prof. Narayan Rishi
Advisor

FOREWORD

"Recent Advances in Diagnosis and Management of Plant Diseases" is an edited book on commercially important plant diseases, their diagnosis and management. The book contains 21 chapters dedicated to various important topics such as Fungal, Bacterial, Viral and Nematode diseases of crop plants, vegetables, fruits (Pomegranate and Citrus), fiber crop (Mesta), eco-friendly management of pests and diseases through microbial biopesticides, integrated disease management, management of diseases in organic cultivation, management of viral diseases through antiviral proteins isolated from higher plants and nutritional disorders of wheat and citrus and their management. In addition there is a chapter on recent techniques of diagnosis of pathogens. Efficient and reliable diagnosis of a pathogen is a prerequisite for disease management strategy. Each chapter has been contributed by well known scientists on the topic. Use of microbial bio- pesticides is an important eco-friendly tool for management of pests and diseases and has become very popular. A chapter therefore, on commercialization of microbial bio pesticides is most appropriate. Fusarium wilt diseases, damping-off diseases of seedlings in nursery and downy mildews are serious problems for long. A chapter each on the current status of these three problems with their recent management strategies have been added in the book. Four chapters have been devoted on novel techniques of detection and diagnosis of fungal, bacterial and viral pathogens and their applications in management. Two chapters have been devoted on applications of antiviral proteins isolated from higher plants such as *Boerhavia diffusa* in management of viral diseases. Soybean mosaic virus disease is a widely prevalent and serious problem in Soybean. In this chapter integrated disease management strategies using available information on biology of the pathogen, epidemiology and genetics of resistance have been highlighted.

Prof. L.P. Awasthi, has a long standing of teaching and research in Plant Pathology at N.D. University of Agriculture and Technology, Faizabad, U.P. This book is a very good attempt towards addressing several serious problems in Plant Pathology for long. It is a must in all libraries of higher education in Agriculture, Biotechnology, Microbiology and Molecular Biology. It will be helpful to postgraduate and research students and faculty in the areas of Life Sciences.

Narayan Rishi

(Narayan Rishi)

J-3 Block, Amity University Campus, Sector - 125, Noida - 201 303, Gautam Buddha Nagar, U.P. (INDIA)
Tel.: +91(0)-120-4392961, **Fax** : +91(0)-120-4735699, Mob.: 9560097725
E-mail : aivi@amity.edu, nrishi@amity.edu; **Website** : www.amity.edu

Preface

The modern system of agriculture is based on a substantial use of microorganisms. It aims at mutually reinforcing relationship between agricultural productivity and conservation of nature. Sustainable agriculture for the developing countries is to maintain food production or, more realistically, to maintain an increasing trend in food grain production while preserving the underlying resources base. The changed agricultural scenario has focused our attention on ways to manage plant diseases, reduce crop losses, avoid wide fluctuations in production, and sustain the higher levels of productivity. The global change in climatic conditions may also influence infection, development, perpetuation, and severity of many plant diseases, which may directly affect crop production. The multiplicity of crops, largely of tropical and subtropical climate, overlapping growing seasons, a large number of varieties, production systems, and various cultural practices favor the occurrence of a number of plant diseases, a few of them are quite serious and cause great economic losses.

A large number of diagnostic tools have been developed and used for the identification of many viral, bacterial, fungal, nematode, and nonparasitic disorders in crops. The management of different plant diseases is also very important which is presently done through chemicals, bioagents, plant products, and cultural practices.

Many diagnostic kits and sophisticated equipments are being used for the exact diagnosis of causal organisms. Of the various strategies available for disease management, the chemical-based strategies have been widely acceptable. The agrochemicals used in plant protection cause environmental pollution and human health hazards and affect microbial life, besides their very high cost. On the other hand bioagents and many substances isolated from a large number of plants have exhibited a significant role in the management of different crop diseases. These phytochemicals are easily biodegradable and do not have any residual effect on the environment, flora, and fauna. To avoid or minimize the use of chemical pesticides, the most logical approach is the integrated disease management (IDM).

This book *Recent Advances in the Diagnosis and Management of Plant Diseases* incorporates critical reviews and research articles on important diseases of different crops and their suitable management strategies. This volume contains 21 chapters covering important diseases of crops, caused by bacteria, fungi, viruses, viroids, phytoplasma, nematode, and nonparasitic diseases of commercial field crops and their management. Chapters cover diseases of wheat, barley, rice, pulses, cucurbits, solanaceous vegetables, chickpea, okra, soybean, mesta, pomegranate, and citrus, techniques for detection and diagnosis of plant pathogens, and the management of various diseases through biopesticides and antiviral agents/virus inhibitors from higher plants.

Faizabad L.P. Awasthi

Acknowledgments

I am highly indebted to Prof. H.N.Verma, Hon'ble vice-chancellor, Jaipur National University, Jaipur – my best teacher, my mentor, and Guru for his love, guidance, support, encouragement, and advice throughout my whole life. I owe him everything and it is impossible for me to express thanks to him in words.

I am extremely grateful to all learned contributors for their cooperation in compiling useful information on different aspects of disease diagnosis and their management strategies. I am sure, detailed accounts on different aspects will be a great help to students, researchers, and teachers of plant pathology and for planning future strategies for the management of crop diseases using latest and modern technologies. Thanks are due also to all contributors each of whom had endeavored to present an update of their specialized aspect. While dealing with such a voluminous work, errors may likely to occur, despite my best efforts.

I record my sincere gratitude to my wife Mrs. Poornima Awasthi, daughters Er. Girishma Awasthi Mishra and Kopal Awasthi, son Bhuvnesh Krishna Awasthi, sons-in-law Er. Vineet Mishra and Er. Akshat Dixit, grandson Atharv, and my Ph.D. students Samir and Aditya who extended full cooperation in many invisible ways during the preparation of this manuscript.

I am confident that the book will be widely accepted by students, teachers, and researchers in the field of plant pathology and life sciences.

Faizabad L.P. Awasthi

Contents

About the Editor

Prof L.P. Awasthi is a distinguished scientist who has made significant contributions to the field of plant pathology in general and plant virology in particular. He did his Ph.D. in botany (plant pathology/virology) from Lucknow University, Lucknow. Prof. Awasthi's professional experience includes more than 40 years of teaching and research in plant pathology. He has guided 77 M.Sc. (Ag.) and 35 Ph.D. students. The major areas of his research interest include:

- The diagnosis and characterization of viral diseases of important crop plants
- Molecular characterization and the mode of action of resistance inducing defensive antiviral proteins from plants

The work has demonstrated for the first time that the inducible plant defense system(s) against viruses can be switched on after treatment with certain highly specific basic phytoproteins and has opened a new field of "plant immunology." This more recently developed a novel strategy of immunizing plants using the phytoproteins that shows great promise as it is versatile, nonspecific and entirely risk free, and a meaningful virus disease control technology. Immunization by plant products that stimulate the plant's natural disease resistance mechanism may provide a new strategy for crop protection against viruses. Thus, a modest beginning toward successful virus control has been made. This strategy of defense in plant, in outcome, but not mechanism, is comparable to the inducible defense system in animals.

Prof. Awasthi has published more than 350 research papers in foreign and Indian journals of repute and many popular articles in Hindi and English in agriculture journals/bulletins and laboratory manuals and has a lot of contributory chapters in edited books.

Prof. Awasthi visited Karl-Marx University, Leipzig; University of Berlin; Institute of Plant Pathology, Aschersleben; and Halle University, Germany. Prof. Awasthi is working as an editor/referee in the editorial boards of many Indian and foreign journals. Apart from this he is a member of several national, state-level committees and many professional societies. He has been conferred the Plant Pathology Leadership Award by the Indian Phytopathological Society, India, for his outstanding contributions, scientific excellence, and distinguished services for the cause of plant pathological research, education, and technology dissemination, which has impacted the science of plant pathology in the country.

1 Commercialisation of Microbial Biopesticides for the Management of Pests and Diseases

Shripad Kulkarni

So far, use of synthetic chemical pesticides had been the widely used approach for reducing the estimated 45 % gross crop loss due to pests and diseases, amounting to around Rs 290 billion per annum. Fortunately, realisation of the negative effects of these chemicals on nature and natural resources like pollution, pesticide residue, pesticide resistance, etc., have forced many to shift focus on to more reliable, sustainable and environment-friendly agents of pest control, the biopesticides. In spite of the claimed efficacy, their use, however, has remained very low due to a number of socio-economic, technological and institutional constraints. Nonetheless, rise in income levels due to a growing economy coupled with increasing awareness of health-related effects of chemical pesticides has increased the demand of organic food. The striking feature of biopesticides is environment friendliness and easy biodegradability, thereby resulting in lower pesticide residues and largely avoiding pollution problems associated with chemical pesticides. Further, use of biopesticides as a component of Integrated Pest Management (IPM) programmes can greatly decrease the use of conventional (chemical) pesticides, while achieving almost the same level of crop yield.

The term "biopesticides" is usually referred to all biological materials and organisms that can be formulated and used as pesticide to control obnoxious pests and diseases threatening the productivity of crops and animals. These include microscopic organisms like virus, bacteria, fungi, protozoa, nematodes, antagonistic fungi, bacteria and macroscopic animals like insects (parasitoid predators), mites, plants (botanicals) and semiochemicals that alter behaviour of insects. The modes of action of biopesticides involve competition, antagonism/inhibition, toxication, infection, infestation and predation (Hajeck 2004).

Trichoderma spp. have been widely used as antagonistic fungal agents against several pests as well as plant growth enhancers. Faster metabolic rates, antimicrobial metabolites and physiological conformation are key factors which chiefly contribute to antagonism of these fungi. Mycoparasitism, spatial and nutrient competition, antibiosis by enzymes and secondary metabolites and induction of plant defence system are typical biocontrol actions of these fungi. On the other hand, *Trichoderma* spp. have also been used in a wide range of commercial enzyme productions, namely, cellulases, hemicellulases, proteases and -1,3-glucanase (Chet et al. 1990; Harman 2005).

Pseudomonas fluorescens bacteria effectively control wilts and root rot diseases of groundnut, cotton, banana, soya bean, tomato, pigeon pea,

S. Kulkarni (✉)
Department of Plant Pathology, Institute of Organic Farming UAS, Dharwad 580005, India
e-mail: shripadkulkarni@rocketmail.com

L.P. Awasthi (ed.), *Recent Advances in the Diagnosis and Management of Plant Diseases*,
DOI 10.1007/978-81-322-2571-3_1

etc. It also controls the rice blast and sheath blight of paddy. This bacterium enters the plant system and acts as a systemic biocontrol agent against diseases. It is suitable for all kinds of crops.

Biopesticides are an important group of pesticides that can reduce pesticide risks. Among more than 900 species of entomopathogenic fungi belonging to 100 genera that are recorded, only 10 species have been commercially exploited, *viz. Beauveria bassina, Metarhizium anisopliae, Nomuraea rileyi, Verticillium lecanii, Hirsutella thompsonii, Paecilomyces* sp. Fungi are the only insect pathogens that are capable of invading insect by penetrating the cuticle. This is the most useful means of infection. Therefore, those insects that feed by sucking, such as aphids, scales, etc., are attacked only by fungal pathogens. This mode of infection means that fungi are very dependent on environmental conditions, in particular, high humidity, in order to achieve infection.

Most successes with fungi have been with Deuteromycetes which cause epizootic on foliage-feeding insects in tropical environments only. *Metarhizium anisopliae* (Metchnikoff) can infect more than 200 insect hosts. The potential is proved beyond doubt in the management of different species of grasshoppers, termites, root grub, pyrilla of sugarcane BPH in paddy, coconut rhinoceros beetle, coffee berry borer, etc. *Verticillium lecanii* has been reported to be highly potent against aphids, thrips, scales, hoppers, whiteflies, mites, etc. This fungus is being used in the management of soft-sucking pests in greenhouses. Epizootics created by the fungus could reduce the greenbug infestation considerably in coffee plantations in South India. *Beauveria bassiana* has been exploited to manage the pests like coconut rhinoceros beetle, root grubs, coffee stem and berry borers, *Helicoverpa*, etc.

Nomuraea rileyi is an important natural mortality factor of some of the lepidopteran hosts like *Spodoptera, Helicoverpa*, semiloopers, cutworms, etc. in crop ecosystem under transitional high humid climates. In the low-cost mass-multiplication technique developed by UAS, Dharwad made this fungus available for farmers at a cheaper cost of Rs 200/kg which is recommended at the rate of 250–500 g/ha in groundnut, soya bean, crops, etc. *Hirsutella thompsonii* is another pathogenic fungi which infects many of the mite pests on fruits, crops, plantation crops, etc. These entomopathogens being facultative can be easily and cheaply mass multiplied on any carbon-rich substrate especially sorghum, rice, etc.

1.1 Factors Affecting Growth of Biopesticides

However, some of the factors which have restricted the growth of biopesticides are (Cook 1985):

- Low reliability because of low stability in effect
- Target specificity which distracts farmers
- Slow in action compared to synthetics
- Shorter shelf life
- Erratic availability of biopesticides in the market
- Already established and strong market of chemical pesticides
- Regulatory system favourable to chemical pesticides
- The gradual disappearance of multiple or mixed cropping, which is known to keep away the magic bullet-chemical pesticide

1.2 Usage of Biopesticides

The global weighted average consumption level of biopesticides is approximately 1 kg/ha. With the global organic farming area comprising about 24 million hectares, global biopesticide consumption is thus estimated at about 24 million kg.

1.3 Industry Overview

The biopesticide market is growing very rapidly. In 2005, biopesticides accounted for about 2.5 % of the total pesticide market, which was merely 0.2 % during 2000. This share is expected to grow to about 4.2 % by 2010 while the market value is estimated to reach more than US$ 1 billion (source: BCC research). However, the overall growth rate of biopesticides is estimated to be about 10 % per annum for the next 5 years.

In terms of use, orchards claim the largest share (55 %) of the total biopesticides used. Region wise, North America consumes the largest share (40 %) of the global biopesticide production followed by Europe and Oceanic countries accounting for 20 % each.

1.4 Biopesticide in India

Biopesticides represent only 2.89 % (as on 2005) of the overall pesticide market in India and is expected to exhibit an annual growth rate of about 2.3 % in the coming years (Thakore 2006). In India, so far only 12 types of biopesticides have been registered under the Insecticide Act (1968). Neem-based pesticides (Schmutterer 1990), *Bacillus thuringiensis*, NPV and *Trichoderma* are the major biopesticides produced and used in India. Consumption of biopesticides has increased from 219 metric tons in 1996–1997 to 683 metric tons in 2000–2001, and about 85 % of the biopesticides used are neem-based products. Consumption of chemical pesticides has significantly fallen from 56,114 MT to 43,584 MT during the same period.

The major advantages of the use of biopesticides are the following:

- Economical, once method is developed
- Selective and hence no side effects
- Self-propagating and self-perpetuating as they are biological entitles
- No resistance development
- Safe to nontarget organisms
- Virtually permanent unless ecosystem is disturbed
- Effective against pests that are not accessible to other approaches
- Compatible with other pest management tools except with broad-spectrum toxicants

1.5 General Characters of Biopesticides

It is necessary for biopesticides to have some useful characters as mentioned below (Thakur and Sandhu 2010):

- Have a narrow target range and a very specific mode of action
- Are slow acting
- Have relatively critical application times
- Suppress, rather than eliminate, a pest population
- Have limited field persistence and a short shelf life
- Are safer to humans and the environment than conventional pesticides
- Present no residue problems

Microbial insecticides are another kind of biopesticides. They come from naturally occurring or genetically altered bacteria, fungi, algae, viruses or protozoans (Vega et al. 2009).

They suppress pests by mechanism:

- Producing a toxin specific to the pest
- Causing a disease on pest
- Preventing establishment of other microorganisms through competition

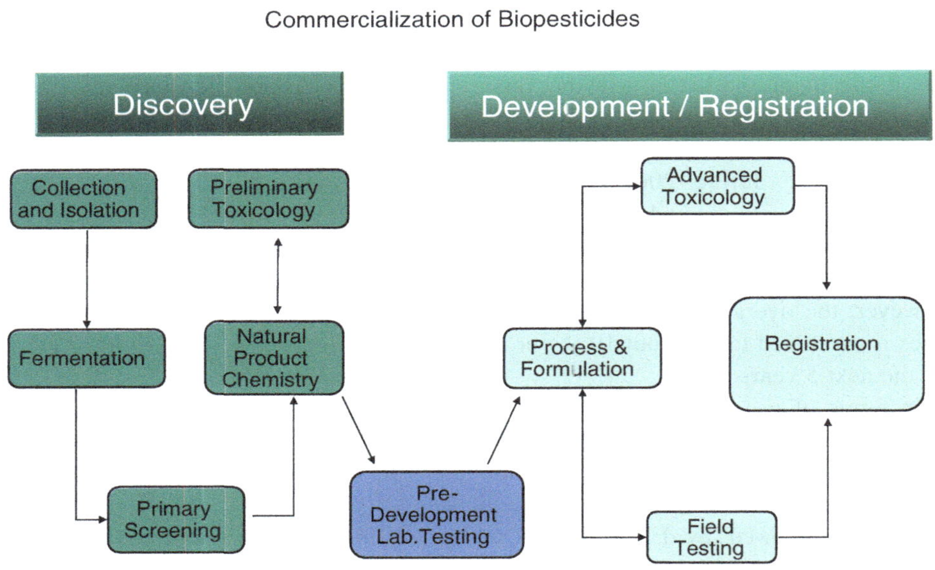

1.6 Different Microbial Pesticides for Pest Management

There are specific entomopathogenic fungi for the management of different insect pests. Their mechanism slightly differs, even site of action and their requirement for multiplication and invasion varies considerably (Bhattacharyya et al. 2004).

1.6.1 *Beauveria bassiana*

It is an entomopathogenic fungus for protecting root and stem from destroying insect pests. It directly invades the insect body, when conidia become attached to the insect cuticle, and after germination, the hyphae penetrate the cuticle and proliferate in the insect's body. High humidity or free water is essential for conidial germination and infection establishes between 24 and 48 h. The infected insect may live for 3–5 days after hyphal penetration, and after death, the conidiophores bearing conidia are produced on cadaver (Li et al. 2001; Roy et al. 2010).

Crops: cotton, *Helicoverpa*; cabbage, DBM; sugarcane, root grub; coffee, berry borer; black pepper, pollu beetle; palms, red weevil; and paddy, BPH

Target pests: caterpillars, weevils, leafhoppers, bugs, grubs and leaf-feeding insects

1.6.1.1 Method of Application

Foliar application (borer and cutworm)

The product should be sprayed on growing plants using hand, ground or aerial equipment.

Soil application (for root grubs)

Beauveria bassiana (4–5 Kg/Ac) can be applied around the root zone and incorporated into the soil along with FYM (500 kg) during June–July.

1.6.1.2 Frequency of Application

Frequency of application depends on the pest and the crop. For greenhouse pest problems, application once in every 15–20 days are recommended. All applications should be based on monitoring of pest populations.

1.6.1.3 Dosage

Foliar spray: 400 g/acre in 200 L of water, i.e. 2 g per litre of water. The spray volume depends on the crop canopy.

Soil application: 5 kg/acre.

The products may be used alone or tank mixed with other products such as sticking agents, insecticidal soaps, emulsifiable oils and insecticides or used with beneficial insects. Do not use with fungicides and wait 48 h after application before applying fungicides.

1.6.2 Nomuraea rileyi

Nomuraea rileyi is a biological formulation based on friendly fungus *Nomuraea rileyi*. It acts on harmful caterpillars by gradually paralysing and thus killing them. It targets larvae and immature stages of leaf rollers, cutworms, leaf-eating caterpillars and other lepidopteran pests. It is available in powder form at our institute.

Nomuraea rileyi (Farlow) Samson is an entomopathogenic fungus found in several countries, including India. This fungus attacks important caterpillar pests of soya bean, groundnut and maize such as *Spodoptera* and *Helicoverpa* causing epizootics. Environmental conditions (solar radiation, temperature, humidity, etc) greatly affect this microorganism and bring down their field viability and persistence.

1.6.2.1 Mode of Action

The spores of this fungus when come in contact with the cuticle (skin) of susceptible insects germinate and grow directly through the cuticle to the inner body of their host. The fungus proliferates throughout the insect's body and drains the insect of nutrients, eventually killing it (Ignoffo 1981; Mathew et al. 1998).

Crops: soya bean, groundnut, chilli, tomato, potato, cauliflower, paddy, lucerne and corn

Target pests: leaf-eating caterpillars, borers

1.6.2.2 Application Methods

Foliar Spray: The product should be sprayed on the growing plants using hand-spraying equipment.

1.6.2.3 Frequency of Application

Applications should be repeated at least once in 10–15 days. All applications should be based on monitoring of pest populations.

1.6.2.4 Dosage

Foliar spray: 400 g/acre in 200 L of water, i.e. 2 g per litre of water. The spray volume depends on the crop canopy.

The products may be used alone or tank mixed with other products such as sticking agents, insecticidal soaps, emulsifiable oils and insecticides or used with beneficial insects. Do not use with fungicides and wait 48 h after application before applying fungicides.

1.6.3 *Metarhizium anisopliae*

It is based on formulation of a friendly fungus *Metarhizium anisopliae* which effectively controls white grubs, brown leafhoppers of paddy, pyrilla of sugarcane, green semilooper, termites, spotted pod borer of pigeon pea, coconut rhinoceros beetle, mango hopper, rhizome weevil of banana and DBM of cauliflower.

It infects insects that come in contact with it. Once the fungus spores attach to the surface of the insect, germinate and begin to grow, they then penetrate the exoskeleton of the insect and grow very rapidly inside the insect causing the insect to die. Other insects that come in contact with infected insects also become infected with the fungus.

1.6.3.1 Mode of Action

The spores of this fungus when come in contact with the cuticle (skin) of susceptible insects germinate and grow directly through the cuticle to the inner body of their host. The fungus proliferates throughout the insect's body and drains the insect of nutrients, eventually killing it (Hasan et al. 2002).

Crops: white grubs, brown leafhoppers of paddy, pyrilla of sugarcane, green semilooper, termites, spotted pod borer of pigeon pea, coconut rhinoceros beetle, mango hopper, rhizome weevil of banana and DBM of cauliflower

Target pests: white grubs, brown leafhoppers, pyrilla, semilooper, termites, spotted pod borer, rhinoceros beetle, rhizome weevil and diamond-back moth, root weevils and mango hoppers

1.6.3.2 Application Methods

Foliar Spray: The product should be sprayed on the growing plants using hand-spraying equipment.

Soil application (root grubs and weevils): It (4–5 kg/acre) can be applied around the root zone and incorporated into the soil along with FYM (500 kg) during June–July.

1.6.3.3 Frequency of Application

Frequency of applications depends on the pest and the crop.

1.6.3.4 Dosage

Foliar spray: 400 g/acre in 200 L of water, i.e. 2 g per litre of water. The spray volume depends on the crop canopy.

Soil application: 5 kg/acre

1.6.4 Verticillium lecanii

Verticillium lecanii is another friendly fungus with talc-based formulation. The fungus inserts its mycelium through the body wall of all stages of insects and kills the insects by infecting it.

The death of the insect occurs because of release of certain toxic elements like destraksin desmethyl, destraksin, etc. which kill the hardy stage-like pupae of the insects.

Verticillium lecanii (formerly known as *Cephalosporium lecanii*) was first described in 1861 and is a cosmopolitan fungus found on insects. It is a common pathogen of scale insects in tropical and subtropical climates. *V. lecanii* is known as a "white-halo" fungus because of the white mycelial growth on the edges of infected scale insects. The conidia (spores) of *V. lecanii* are slimy and attach to the cuticle of insects. The fungus infects insects by producing hyphae from germinating spores that penetrate the insect's integument; the fungus then destroys the internal contents and the insect dies.

The fungus eventually grows out through the cuticle and sporulates on the outside of the body. Infected insects appear as white to yellowish cottony particles. Diseased insects usually appear in 7 days. However, due to environmental conditions, there may be some considerable lag time from infection to death of insects. *V. lecanii* works best at temperatures of 15–25 °C and a relative humidity of 85–90 %. The fungus needs high humidity for at least 10–12 h. This can be a problem as many plant pathogenic fungi (e.g. *Botrytis*) favour these same environmental conditions. *V. lecanii* spores are damaged by ultraviolet radiation. In greenhouses, heating pipes may reduce the effectiveness of the fungus, because this creates a microclimate where the air is drier and humidity is lower. In addition, *V. lecanii* is generally not useful in interiorscapes due the low-humidity conditions in these environments.

The fungal mycelium of *V. lecanii* produces a cyclodepsipeptide toxin called bassianolide, which has been shown to kill silkworm. The fungus produces other insecticidal toxins such as dipicolinic acid. The activity of *V. lecanii* depends on the strain of the fungus. *V. lecanii* strains with small spores infect aphids, whereas fungal strains with large spores infect whiteflies. Certain strains have also been reported to be pathogenic on rust fungi. Higher doses of the fungus result in faster kill. Virulence depends on the density of spores and rate of sporulation, which is dependent on environmental conditions. Virulence varies with the method of conidial production. Less virulent conidia are obtained from fermented media as compared to shaken liquid or solid media. Formulated products from conidial production can last up to 1 year. These products are easy to wet and dilute for spraying. Also, the fungus can stick to the surface of leaves and host insects. It is effective against hardy insects like scales, aphids, thrips, hoppers, etc.

1.6.4.1 Mode of Action

The spores of this fungus when come in contact with the cuticle (skin) of target insects germinate and grow directly through the cuticle to the inner body of their host. The fungus proliferates throughout the insect's body, draining the insect of nutrients and eventually killing it in around a week's time.

Crops: ornamentals and vegetables in greenhouses, nurseries, lawns, landscape perimeters. Vegetables, field crops and other agricultural crops

Target pests: whiteflies, thrips, aphids, scales, mites and mealy bugs

1.6.4.2 Application Methods

It should be sprayed on foliage of plants using hand, ground or aerial spray equipment. It is advisable to provide a good coverage on the undersides of the leaves.

1.6.4.3 Frequency of Application

Applications should be repeated at least once in 15–20 days for two times. For greenhouse pest problems, applications in every 10–15 days are recommended. All applications should be based on monitoring of pest populations.

1.6.4.4 Dosage

Foliar spray: 400 g/acre in 200 L of water, i.e. 2 g per litre of water. The spray volume depends on the crop canopy.

The products may be used alone or tank mixed with other products such as sticking agents, insecticidal soaps, emulsifiable oils and insecticides or used with beneficial insects. Do not use with fungicides and wait 48 h after application before applying fungicides.

1.7 Different Microbial Formulations for Disease Management

1.7.1 *Pseudomonas fluorescens*

It is a formulation of *Pseudomonas fluorescens* – a very potent microbe that not only cures serious plant diseases like damping off, scab, root and stem rot and blights but also controls some species of nematodes. It also helps plants to absorb available phosphorous and works as catalyst for *Trichoderma viride*. Bacteria and fungal propagules on the leaves of crops often serve as nucleation sites for ice formation, and ice crystals often form when they are present and the temperature falls below freezing, with resulting damage to the leaf. If these are replaced on plant leaves with competitive antagonists that lack the ice-nucleating protein, frost is prevented, even at temperatures as low as −5 °C. Other strains of *Pseudomonas fluorescens* are antagonistic to foliar or rhizosphere bacteria and fungi through the production of siderophores and antibiotics (Haas and Défago 2005).

Crops: paddy, pomegranate, potato, eggplant, tomatoes, chilli, cut flowers, orchards, etc.

Target diseases: *Fusarium*, *Xanthomonas*, *Pythium* spp., *Phytophthora* spp., *Rhizoctonia solani*, *Botrytis cinerea*, *Sclerotium* spp.

1.7.1.1 Method and Frequency of Application

Suspend *P. fluorescens* in sufficient water (500 g/100 L) to achieve uniform application. Two to three applications in vegetables and ornamentals and 4–5 applications in lawn and landscape crops are recommended. Applications during early stages of plant growth protect the plant during critical stages of development.

1.7.1.2 Dosage

Soil application: 2 kg/acre along with dried FYM.

Seed treatment: at 4–5 g per kg of seeds as per standard wet treatment.

Seedling root dipping: at 10 g/L prior to planting. Overdosing does not cause any harmful side effects.

1.7.2 *Trichoderma harzianum*

They are the front-rank killer of harmful fungal diseases like *Fusarium, Rhizoctonia* and *Sclerotium* which cause great havoc to all important crops like soya bean, cotton, sugarcane, cereals and many more. Formulation of potent antagonistic fungi, *Trichoderma harzianum* successfully controls soil-borne fungal pathogens that cause diseases in crops. *Trichoderma* parasitise pathogenic fungi and limit their growth and activity. They also produce toxic metabolites and form a protective coating on seeds against soil-borne pathogenic fungi. The hidden fungal diseases spring a surprise with their sudden attack before a farmer can do anything. It can be effectively used for seed treatment as well as soil application (Mukhopadhyay et al. 1992; Verma et al. 2007).

Crops: cereals, oilseeds, eggplant, potato, chilli, tomatoes, cucumbers, cut and pot flowers, orchards, vineyard ornamentals in greenhouses, lawn nurseries, etc.

Target diseases: *Pythium* spp., *Ganoderma* spp., *Rhizoctonia solani, Fusarium* spp., *Botrytis cinerea, Sclerotium* spp., *Sclerotinia* sp.

1.7.2.1 Method of Application

Suspend *Trichoderma harzianum* in sufficient water (500 g/100 L) to achieve uniform application. Apply at the rate of 100–200 g per cubic metre (loose) of greenhouse potting mix, soil or planting beds.

Trichoderma harzianum can be applied through low-pressure watering nozzles. Agitate to maintain suspension. For best effect, treat potting mix several days before use for seeding or transplants. For bulbs and ornamentals: dip bulbs in *Trichoderma harzianum* suspension (10 g/L) prior to planting.

1.7.2.2 Frequency of Application

Two to three applications in vegetable ornamentals and 4–5 applications in lawns and landscape crops are recommended. Applications during early stages of plant growth protect the plant during critical stages of development.

1.7.2.3 Dosage

Soil application: 2 kg/acre along with 200 kg FYM

Seed treatment: at 4–6 g per kg of seeds as per standard wet treatment

Seedling treatment: at 10 g/L prior to planting

1.7.3 *Bacillus subtilis*

It is an antagonistic bacterial biocontrol agent, which controls many soil- and airborne diseases of paddy, groundnut, cotton, vegetables, soya bean, etc. Foliar application of *Bacillus subtilis* along with *Pseudomonas fluorescens* controls leaf diseases of many crops. The bacterium colonises the developing leaf and root system of the plant and thus competes with and thereby suppresses plant diseases. The plant growth-promoting rhizobacteria (PGPR) having an antagonistic interaction with various soil-borne plant pathogens protects plants against seed and root diseases (Saleem and Kandasamy 2002).

Crops: chilli, grapes, potato, cucumbers, eggplant, tomatoes, cut flowers, orchards, vineyard ornamentals in greenhouses, lawn nurseries, etc.

Target diseases: *Phytophtora* spp., *Iodium* spp., *Alternaria, Aspergillus* spp., *Pythium* spp., *Rhizoctonia solani, Botrytis cinerea, Sclerotium rolfsii*, *Sclerotinia* spp. and many powdery and downy mildew-causing organisms

1.7.3.1 Method of Application

Suspend *Bacillus subtilis* in sufficient water (500 g/100 L) to achieve uniform application. It can be applied through low-pressure watering nozzles as a spray.

1.7.3.2 Frequency of Application

Two to three applications in vegetable ornamentals and 4–5 applications in lawn and landscape crops are recommended. Applications during early stages of plant growth protect the plant during critical stages of development.

1.7.3.3 Dosage

Spray: 5 g/L
Seed treatment: at 4–5 g per kg of seeds as per standard wet treatment
Seedling root dipping: at 10 g/L prior to planting

1.7.4 *Paecilomyces lilacinus*

Paecilomyces are soil-borne hyphomycetous fungi belonging to class Deuteromycetes and occur in the soil rhizosphere. *Paecilomyces* is an egg parasite and referred to as opportunistic fungus. Many enzymes produced by *P. lilacinus* have been studied, and serine protease with biological activity is against *Meloidogyne hapla* eggs. Strains of *P. lilacinus* have shown to produce proteases and a chitinase, enzymes that could weaken a nematode eggshell.

Before infecting a nematode egg, *P. lilacinus* flattens against the egg surface and becomes closely appressed to it. *P. lilacinus* produces simple appressoria anywhere on the nematode eggshell either after a few hyphae grow along the egg surface or after a network of hyphae form on the egg. The presence of appressoria appears to indicate that the egg is, or is about to be, infected. In either case, the appressorium appears the same, as a simple swelling at the end of a hypha, closely appressed to the eggshell. Adhesion between the appressorium and nematode egg surface must be strong enough to withstand the opposing force produced by the extending tip of a penetration hypha. When the hypha has penetrated the egg, it rapidly destroys the juvenile within, before growing out of the now empty eggshell to produce conidiophores and to grow towards adjacent eggs (Mendoza et al. 2007; Moore et al. 2008).

Crops: eggplant, potato, chilli, tomatoes, cucumbers, flowers, orchards, vineyard ornamentals in greenhouses, lawn nurseries and landscape
Target Pests: plant parasitic nematodes in soil. Examples include *Meloidogyne* spp. (root knot nematodes), *Radopholus similis* (burrowing nematode), *Heterodera* spp. and *Globodera* spp. (cyst nematodes), *Pratylenchus* spp. (root lesion nematodes), *Rotylenchulus reniformis* (reniform nematode), *Nacobbus* spp. (false root-knot nematodes)

1.7.4.1 Method of Application

Suspend *Paecilomyces lilacinus* in sufficient water (500 g/100 L) to achieve uniform application. Apply at the rate of 100–200 g per cubic metre (loose) of greenhouse potting mix, soil or planting beds.

For best effect, treat potting mix several days before use for seeding or transplants.

1.7.4.2 Frequency and Dosage of Application

Soil application: 5 kg/ha along with Farm Yard Manure
Seed treatment: at 4–5 g per kg of seeds as per standard wet treatment
Seedling dip treatment: at 10g/L prior to planting

References

Bhattacharyya A, Samal AC, Kar S (2004) Entomophagous fungus in pest management. News Lett 5:12
Chet I, Ordentlich A, Shapira Oppenheim A (1990) Mechanism of biocontrol of soil borne plant pathogens by rhizobacteria. Plant Soil 129:85–92
Cook RJ (1985) Biological control of plant pathogens: theory to application. Phytopathology 75:25–29
Haas D, Défago G (2005) Biological control of soil-borne pathogens by fluorescent pseudomonads. Nat Rev Microbiol 3:307–319
Hajek A (2004) Natural enemies: an introduction to biological control. Cambridge University Press, Cambridge
Harman GE (2005) Overview of mechanisms and uses of *Trichoderma* spp. Phytopathology 96:190–194

Hasan S, Bhamra AK, Sil K, Rajak RC, Sandhu SS (2002) Spore production of *Metarhizium anisopliae* (ENT-12) by solid state fermentation. J Indian Bot Soc 8:85–88

Ignoffo CM (1981) The fungus *Nomuraea rileyi* as a microbial insecticide: fungi. In: Burges HD (ed) Microbial control of pests and plant diseases. Academic Press, London, pp 513–538

Li ZZ, Li CR, Huang B, Meizhen MZ (2001) Discovery and demonstration of the teleomorph of *Beauveria bassiana* (Bals.) Vuill., an important entomogenous fungus. Chin Sci Bull 46(9):751–753

Mathew SO, Sandhu SS, Rajak RC (1998) Bioactivity of *Nomuraea rileyi* against *Spilosoma obliqua*: effect of dosage, temperature and relative humidity. J Indian Bot Soc 77:23–25

Mendoza AR, Sikora RA, Kiewnick S (2007) Influence of *Paecilomyces lilacinus* strain 251 on the biological control of burrowing nematode *Radopholus similis* in banana. Nematropica 37:203–213

Moore SR, Castillo JD, Lawrence KS (2008) NemOut for *Meloidogyne arenaria* management on peanuts. Plant Dis Manag Rep 2:N028

Mukhopadhyay AN, Shresta SM, Mukherjee PK (1992) Biological seed treatments for control of soil-borne plant pathogens. FAO Plant Prot Bull 40:21–30

Roy H, Vega F, Chandler D, Goettel M, Pell J, Wajnberg E (2010) The ecology of fungal entomopathogens. Springer, Dordrecht

Saleem B, Kandasamy U (2002) Antagonism of Bacillus species (strain BC121) towards Curvularia lunata. Curr Sci 82(12):1457–1463

Schmutterer H (1990) Properties and potentials of natural pesticides from neem tree. Annu Rev Entomol 35:271–298

Thakur R, Sandhu SS (2010) Distribution, occurrence and natural invertebrate hosts of indigenous entomopathogenic fungi of Central India. Indian J Microbiol 50(1):89–96

Vega FE et al (2009) Fungal entomopathogens: new insights into their ecology. Fungal Ecol 2:149–159

Verma M, Brar SK, Tyagi RD, Surampalli RY, Valero JR (2007) Antagonistic fungi, *Trichoderma* spp.: panoply of biological control. Biochem Eng J 37:1–20

Management of Plant Diseases Under Organic Cultivation

2

Shripad Kulkarni and S. Lingraju

Indian agriculture shifted from a state of food deficiency to food sufficiency through introduction of fast growing, high-yielding hybrid varieties and the use of high dose of chemical fertilizers, pesticides, and herbicides. No doubt green revolution yielded rich dividends but at the same time gifted serious pest and disease problems and degradation of environment followed by deleterious effects on environment (Singh 2005).

Most of horticultural crops are eaten fresh or used for health care; hence, any contamination in the form of pesticide/chemical residue may lead to health hazards; hence, organic horticulture offers a better possibility of producing healthy food. Plant diseases caused by different groups of organisms including fungi, bacteria, viruses, rickettsia, spiroplasma, nematodes, and few others have remained important in causing significant losses in different crops indicating the urgent need of their integrated management. The continuous and indiscriminate use of chemical pesticides has posed several serious problems such as pesticide residue, development of resistant strains, environmental pollution, and adverse effect on beneficial microorganisms and created a greater concern over global food safety and security.

S. Kulkarni (✉)
Department of Plant Pathology, Institute of Organic Farming UAS, Dharwad 580005, India
e-mail: shripadkulkarni@rocketmail.com

S. Lingraju
University of Agriculture Sciences, Institute of Organic Farming

Organic farming relies on crop rotation, crop residues, animal manures, legumes, green manures, off-farm organic wastes, cultural practices, mineral-bearing rocks, and aspects of biological pest control to maintain soil productivity and to supply plant nutrients and to control diseases, insects, weeds, and other pests (Fernando et al. 2004).

2.1 Cultural Methods

By practicing the following methods, we can regulate/modify the pest and disease incidence effectively.

2.1.1 Selection of Adopted and Resistant Varieties

By choosing varieties which are well adopted to the local environmental conditions (such as temperature, nutrient supply, pests, and disease resistance) crop is allowed to grow healthier and stronger against attack of pests and pathogens.

2.1.1.1 Defense Mechanisms Influencing Diseases

1. Phenols: Host enzymes like polyphenol oxidase and peroxidase oxidize phenolics to quinines, and the quinines are more fungitoxic than phenolics.

L.P. Awasthi (ed.), *Recent Advances in the Diagnosis and Management of Plant Diseases*,
DOI 10.1007/978-81-322-2571-3_2

2. Phytoalexins: Phytoalexins are mostly isoflavonoids, terpenoids, and polyacetylene compounds and synthesized de novo on infection by the pathogens. Phenylalanine and acetic acid may be involved in the biosynthesis of phytoalexins, and phenylalanine ammonia lyse (PAL) has been considered to be the key enzyme.
3. Lignin: Phenylalanine and cinnamic acid are the important precursors. Lignin may act as physical barrier to the pathogens.
4. Callose: It is the substance found in the sieve tubes and may prevent the leakage of sieve tube sap or water in the cell walls. Penetration of incompatible pathogens into the host tissues results in the production of papillae, and the papillae may mostly contain callose.
5. Sugars: Sugars are precursors of synthesis of phenolics, phytoalexins, lignin, and callose.
6. Amino acids: Amino acids are the cornerstones for synthesis of proteins, and some of them are essential for the synthesis of phenolics, phytoalexins, and lignin.

2.1.2 Cropping System

Cropping system in a particular agricultural ecosystem plays a major role. By adopting a suitable cropping system in right time, we can avoid most of the harmful pests and diseases. Some of the practical practices are as follows:

1. Activity in the soil will enhance the presence of beneficial organisms. For instance, plants colonized by arbuscular mycorrhiza may increase pests also. So one has to be careful when selecting proper green-manuring crop.
2. Crop rotation: Cauliflower–paddy–cauliflower rotation is highly effective in controlling stalk rot of cauliflower disease caused by *Sclerotinia sclerotiorum*, and it reduces infection by >60 % and >161 % increase in seed yield has been observed (Gupta and Thind 2006).
3. Adopting mixed cropping system pest and disease incidence can be minimized since pest has less host plants to feed on.
4. By following crop rotation practices, we can increase the soil fertility and reduce the chances of soilborne diseases.
5. By cultivating green manure cover crops like horse gram, cowpea, sunn hemp, *Sesbania*, dhaincha, and *Gliricidia*, we can increase the biological activity in the soil and enhance the presence of beneficial organism.

2.1.3 Selection of Clean Seed and Planting Materials

Seeds and planting materials are the primary sources of diseases, and hence, selection and use of disease-free seeds after inspection for pathogens and weeds are very much essential. Further, it is advised to get seeds and planting materials from the reliable safe sources only. Use healthy seeds and its hot water treatment at 50 °C for 25 min, or dipping in sodium phosphate solution (90 g/L) for 20 min is effective for tomato mosaic which is the most dangerous disease of tomato.

2.1.4 Selection of Optimum Planting/Sowing Time and Spacing

Most of the pests or diseases attack the crops only in a certain life stage. By adopting sufficient spacing between plants, we can reduce the spread of a disease as well as allow good sunlight to the plants which facilitates less moisture on the leaves leading to hinder and of pathogen development and infection. In the same way, more sunlight allows plants to do more photosynthesis. This practice not only avoids disease and pests in cropping system but also increases the crop productivity.

2.1.5 Balanced Organic Nutrition

Gradual and steady growth makes plants less vulnerable to infection. So this steady growth could be achieved by applying organic fertilizers timely and moderately because excess and indiscriminate use of fertilizers often results in damaging the roots. This damage facilitates to secondary infection. To overcome this problem, **farmers** can adopt integrated nutrient management system with organic manures like FYM, compost, nutrients slowly when the plant needs. Further, by using liquid bio-fertilizers like potash mobilizers, *Frateuria aurantia* along with organic manures provides balanced potassium and contributes to the prevention of fungi and bacterial infections.

2.1.6 Addition of More and More Organic Matter

Organic matter content of the soil is directly related to density and activities of microorganisms in the soil; therefore pathogenic and soilborne fungal population can be reduced. Besides this, organic matter provides:

- All the nutrients that are required by the plants
- Correct C:N ratio in the soil
- Good physical chemical and biological support to soil
- More water-holding capacity to the soil
- Cover from evaporation losses of the moisture from the soil

Ultimately, organic matter supplies substances which strengthen plants with their own protection mechanisms.

2.1.7 Soil Amendments

The decomposition of organic matter helps in alteration of the physical, chemical, and biological conditions of the soil, and the altered conditions reduce the inoculum potential of a soilborne pathogens. In addition, the practice improves soil structures, which promotes root growth of the host. Various biochemical substances like antibiotics and phenols are released during decomposition, which in turn induce resistance in the root system. Soil amendments like sunflower, rapeseed cakes, mustard cake, gypsum, and bean straw can be used.

Rhizoctonia solani is a causal organism of damping-off in nursery stage and wire stem, bottom rot and head rot after transplanting plants in crucifiers. Soil amended with neem cake 1 kg/m^2 and solarized by covering with white polythene sheet for 2 weeks and treatment of seeds with *Trichoderma harzianum* proved to be effective in the management of damping-off of vegetables crops. Soil amended with cellulose powder was also found effective in reducing the disease incidence, which may be due to increase in C:N ratio in soil. This increase results in the decrease of fungal population, but actinomycetes and bacterial populations increased.

Organic amendment is used here to mean organic material incorporation into the soil that comes from external sources such as processing residues or industrial waste products. Organic material added as fresh crop residue and grown in the field in rotation – break, cover, trap, antagonistic, or green manure crops – is discussed below. Incorporation into the soil of large amounts of any organic material will reduce nematode densities. Oil cakes, coffee husks, paper wastes, crustacean skeletons, sawdust, and chicken manure have been used with some success. Nematode control may be due to any one or more of the following mechanisms:

- Toxic and nontoxic compounds present in the organic material
- Toxic metabolites produced during microbial degradation
- Enhancement of the soil antagonistic potential

Chitin amendments have received much interest in the past as an organic amendment in that they stimulate the antagonistic potential in soil toward nematodes. Organic amendments have also been combined with various biocontrol agents with reports of enhanced levels of control. The use of organic amendments is often limited by availability and, in some cases, by the large quantities needed. In addition to their effects on nematode density, organic amendments also improve soil structure and water-holding capacity, reduce diseases, and limit weed growth, all of which ultimately lead to a stronger plant and improved tolerance to nematode attack.

2.1.8 Biofumigation

This term normally refers to suppression of soil-borne pests and pathogens by biocidal compounds, principally isothiocyanates, released in soil, when glucosinolates in cruciferous crop residues are hydrolyzed. Soil amended with fresh or dried cruciferous residues at 38 °C day and 27 °C night temperatures reduced *Meloidogyne incognita* galling by 95–100 % after 7 days' incubation in controlled environment tests. It should be noted here, however, that many cruciferous plants are good hosts of some important species of *Meloidogyne.*

The term biofumigation is now used more freely whenever volatile substances are produced through microbial degradation of organic amendments that result in significant toxic activity toward a nematode or disease. The release of toxic compounds already present in antagonistic plants used as amendments, e.g., neem, marigold, and castor, or the production of toxic compounds due to microbial fermentation of nutrient-rich organic amendments, e.g., velvet bean, sunn hemp, or elephant grass, leads to significant levels of nematode control.

Biofumigation using fresh marigold as an amendment is used effectively in root-knot management in protected cultivation in Morocco. *Tagetes* is grown in the raised beds prior to the planting of susceptible horticultural crops. The crop is then incorporated into the soil after 2–3 months. The beds are fitted with drip irrigation and covered with plastic mulch. The soil in the bed is then biofumigated under conditions of the high temperature and optimum soil moisture.

Control due to any form of biofumigation is probably the result of multifaceted mechanisms (Maloney 1995) including:

1. Nonhost or trap cropping depending on the host status of the plant used
2. Lethal temperature due to solarization
3. Nematicidal action of toxic by-products produced during organic matter degradation
4. Stimulation of antagonists in the soil after biofumigation

2.1.9 Water Management

By practicing good water management, water logging in the field and stress on plants can be avoided; otherwise, pathogens take chance and infect the crop. Further, sprinkling water on foliage shall be avoided as it increases the diseases by giving chance to pathogenic fungal spores to germinate.

2.1.10 Use of Proper Sanitation Measures

"Pull and Burn" is the best method to control disease and removal of infected plant parts (leaves, fruits) from the ground to prevent the disease from spreading. It eliminates residues of infected plants after harvesting.

2.1.11 Soil Solarization

Soil solarization for 4 weeks during summer months coupled with application of neem cake at 400 g/m^2 proved effective against damping-off (76.9 %) in nursery and resulted in significant, higher number of healthy transplants.

2.2 Mechanical Methods

(a) Removing and burning and clipping of lower leaves up to 20 cm and weeding to reduce the Alternaria blight in tomato.
(b) This is the best method, before reaching the loss beyond economic level. To save crop from fungal and bacterial diseases, pull and burn method is most effective.

2.3 Botanicals

Plants during their long evolution have synthesized a diverse array of chemicals to prevent the colonization of pathogens. They produce secondary metobolites like terpenoid alkaloids, flavonoids, and phenolic compounds. These secondary metabolites are having disease-suppressing properties. In India, several plants such as nicotinoids, natural pyrethrins, rotenoids, and neem products have been used in the past for suppression of diseases. Among the botanical pesticides, neem occupies very important place in the pest and disease control. Different parts of neem tree can affect more than 200 insects and diseases; some of them are effective against nematodes, fungi, bacteria, and viruses (Ray et al. 2004; Varma and Dubey 1999).

2.4 Biological Control

Management of soilborne pathogens is difficult because of nonavailability of desired level of resistance against major soilborne diseases caused by species of *Fusarium*, Sclerotium, *Macrophomina*, *Pythium*, *Phytophthora*, and few others forced to search or resorting to new approaches to manage the diseases (Krishna Chandra and Srivatha 2005). Therefore, biocontrol agents or antagonists are means of plant disease control that has gained importance in the recent years. The biocontrol agents multiply in soil and remain near root zone of the plants and offer protection even at later stages of crop growth.

2.4.1 Advantages of Biocontrol Agents

Following are the advantages of biocontrol agents (Kishore and Pande 2005):

1. They avoid environmental pollution of soil, air, and water as is being experienced in chemical control.
2. They avoid adverse effect on the beneficial microbes including antagonists in the soil, whereas chemicals are lethal.
3. They are less expensive compared to chemical control.
4. Continuous uses of bioagents avoid the development of resistant strains of pathogens.
5. Bioagent application is usually once and does not need repeated application.
6. Biological control becomes part of modern large-scale agriculture and helps in increasing crop production within existing resources maintaining biological balance.

2.4.2 Mechanisms of Disease Control by Biocontrol Agents

For managing the diseases bioagents act in a different way, even single bioagent has got different mechanism or some time senergetic effect of these mechanisms results in disease suppression (Khan and Gangopadhyay 2008):

1. Antibiosis
2. Competition
3. Hyperparasitism
4. Mycorrhizae
5. Plant growth-promoting rhizobacteria (PGPR)
6. Cross protection
7. Induced resistance

2.4.3 Bioagents Used in the Management of Plant Diseases

1. *Trichoderma viride*
2. *T. harzianum*
3. *T. hamatum*
4. *T. koningii*
5. *Gliocladium virens*
6. *Pseudomonas fluorescens*

7. *Paecilomyces lilacinus*
8. *Bacillus subtilis*
9. *Glomus fasciculatum*
10. *Agrobacterium radiobacter* K-84 and K-1026
11. Mild strains of viruses
12. *Cladosporium herbarum*

2.4.4 Types of Formulations

Based on type of disease, crop and environment farmers need different formulations for the effective management of plant diseases (Zaidi and Singh 2004).

1. Powder formulation
2. Encapsulations in organic polymer like sodium alginate
3. Pelleting biomass and bran with sodium alginate
4. Wheat bran sawdust water for soil application
5. Molasses-enriched (Kaolin) clay granules
6. Liquid-coating formulation bioprotectant as powder on which suspension of aqueous binder is sprayed on seeds for 0.1-mm-thick layer
7. Emulsifiable concentrate with 1–10 spores/ml

2.4.5 Methods of Application of Bioagents

Farmers need to adopt different application strategy based on type of disease, extent of occurrence and its life cycle (Weller 1988; Paulitz and Belanger 2001).

1. Broadcast
2. Furrow application
3. Root-zone application
4. Seed treatment
5. Wound application
6. Spraying

References

Fernando WGD, Nakkeeran S, Zhang Y (2004) Ecofriendly methods in combating Sclerotinia sclerotiorum (Lib.) De Berry. Recent Res Dev Envion Biol 1:339–347

Gupta SK, Thind TS (2006) Disease problems in vegetables production. Scientific Publishers (India), Jodhpur

Khan M, Gangopadhyay S (2008) Eficacy of Pseudomonas fluorescens in controlling root rot of chickpea caused by Macrophomina phaseolina. J Mycol Plant Pathol 38(3):580–587

Kishore GK, Pande S (2005) Natural fungicides for management of phyto-pathogenic fungi. Annu Rev Plant Pathol 3:331–356

Krishna Chandra GS, Srivatha RSH (2005) Bio control agents and bio pesticides. Published by Regional center of organic farming, Hebbal, Bangalore

Maloney A (1995) Three sources for non-chemical management of plant disease: towards an ecological framework. In: Andrews JH, Tommerup IC (eds) Advances in plant pathology, vol. II. Academic, London, pp 106–129

Paulitz TC, Belanger RR (2001) Biological control in greenhouse system. Annu Rev Phytopathol 39:103–133

Ray AB, Sarma BK, Singh UP (2004) Medicinal properties of plants-antifungal, antibacterial and antiviral activities. International Book Distributing Co, Lucknow

Singh RS (2005) Plant diseases, 8th edn. Oxford and IBH publishing Co. Pvt. Ltd, New Delhi

Varma J, Dubey NK (1999) Perspective of botanical and microbial products as pesticides of tomorrow. Curr Sci 76:172–179

Weller DM (1988) Biological control of soil borne plant pathogens in the rhizosphere with bacteria. Annu Rev Phytopathol 26:379–407

Zaidi NW, Singh US (2004) Mass multiplication and delivery of Trichoderma and Pseudomonas. J Mycol Plant Pathol 34(3):732–741

Combined Approach of Morphological and Molecular Diagnosis of *Fusaria* spp. Causing Diseases in Crop Plants

3

Prameeladevi Thokala, Deeba Kamil, Pankaj Pandey, Prabhakaran Narayanasamy, and Nita Mathur

Most plant pathologists, at some time in their career, must identify a culture of a *Fusarium* species. The complexity of the problem varies, depending on the host from which the culture originated and the degree of resolution required in the identification. *Fusarium* species cause a huge range of diseases on an extraordinary range of host plants. The fungus can be soilborne, airborne, or carried in plant residue and can be recovered from any part of a plant from the deepest root to the highest flower. In addition, *Fusarium* taxonomy has been plagued by changing species concepts, with as few as nine or well over 1,000 species being recognized by various taxonomists during the past 100 years, depending on the species concept employed. The taxonomy of *Fusarium* spp. is based primarily on the morphology and development of conidia and conidiophores and, to a lesser degree, on host plant association and colony morphology. However, if symptoms are novel or unusual for the diseased host, as the characters are continuously variable and there is no way of knowing the degree of variation tolerable within the individual species, the information provided by this morphological approach alone may not be particularly valuable.

Recently, molecular approaches have been used to study the phylogeny of a wide range of phytopathogenic fungi (Paavanen-Huhtala et al. 1999; Jasalavich et al. 2000; Yli-Mattila et al. 2002). Phylogenetic species recognition, based on DNA sequence data from multiple loci, allows greater numbers of species to be distinguished than in the exclusive use of morphological features (Taylor et al. 2000). The nuclear ribosomal repeat includes both highly conserved genes and more variable spacer regions (Taylor et al. 2000). The intergenic spacer region (IGS), which separates rDNA repeat units, appears to be more rapidly evolving than any region in the rDNA repeat units (Hillis and Dixon 1991). Closely related species may show considerable divergence in IGS, often reflecting both length and sequence variation (Hillis and Dixon 1991); length and restriction site variation may even occur within the rDNA of an individual in some fungi (Martin 1990). However, the multiple copies of IGS do not evolve independently (Martin 1990). Therefore, sequences of the IGS and internal transcribed spacer (ITS1 and ITS2) regions have been used for RFLP analysis of *Fusarium oxysporum* (Appel and Gordon 1995;

P. Thokala (✉) • D. Kamil • P. Pandey
P. Narayanasamy • N. Mathur
Division of Plant Pathology, Indian Agricultural Research Institute, New Delhi 110012, India
e-mail: prameelancha@yahoo.co.in

L.P. Awasthi (ed.), *Recent Advances in the Diagnosis and Management of Plant Diseases*,
DOI 10.1007/978-81-322-2571-3_3

Paavanen-Huhtala et al. 1999), *Fusarium avenaceum* (Paavanen-Huhtala 2000), and different species of *Fusarium* section *Sporotrichiella* (Bulat et al. 1997).

Our objective in this chapter is to outline a practical approach for authentic identification of *Fusarium* species by using the combination of morphological and molecular methods.

3.1 Overall Identification Strategy

The first step in the identification process is to clearly describe the plant disease and the symptoms observed on the diseased plant and to note the weather conditions under which the disease occurred. By using the following methods, the isolation and purification of causal agent are done and identified by using morphological and molecular criteria. These steps are evaluated in more detail in the text that follows.

3.2 Diseases Symptoms and Distribution

Strains of *Fusarium* species can cause an extraordinarily broad range of plant diseases. The most important are the crown and root rots, stalk rots, head and grain blights, and vascular wilt diseases that are well known to most pathologists (Nelson et al. 1981; Summerell et al. 2001), but lesser known diseases such as malformation disease in mango (Ploetz 2001) and bakane disease in rice can have important local economic impact. The nature of the disease provides important clues as to the species that will be recovered and often limits the range of species that need to be distinguished (Table 3.1).

Fusarium species recovered from both natural and agricultural ecosystems have distinct climatic preferences (Backhouse and Burgess 2002; Backhouse et al. 2001; Burgess and Summerell 1992). The climate, and even local variations in weather, can limit the range of species observed, even if several are present, and influence their relative frequency. In broad terms, there are species that prefer tropical climates, hot arid climates, or temperate climates. Some *Fusarium* spp. have a cosmopolitan range (Table 3.2).

Table 3.1 *Fusarium* species commonly recovered as causes of diseases from economically important host plants

Plant species	*Fusarium* species
Banana	*F. oxysporum* f. sp. *cubense*
Cotton	*F. oxysporum* f. sp. *vasinfectum*
Legumes	*F. avenaceum* and *formae speciales* of *F. oxysporum* and *F. solani*
Maize	*F. graminearum*, *F. proliferatum*, *F. oxysporum subglutinans*, *F. verticillioides*
Rice	*F. fujikuroi*
Sorghum	*F. andiyazi*, *F. proliferatum*, and *F. thapsinum*
Vegetables	*Formae speciales* of *F. oxysporum* and *F. solani*
Wheat/barley/oats	*F. culmorum*, *F. graminearum*, and *F. pseudograminearum*

Table 3.2 Species of *Fusarium* regularly recovered from various parts of diseased plants as saprophytes

Plant part	*Fusarium* species
Roots and stem bases	*F. acuminatum*, *F. avenaceum*, *F. compactum*, *F. equiseti*, *F. proliferatum*, *F. oxysporum*, *F. solani*
Leaves and aerial parts	*F. proliferatum*, *F. semitectum*
Flowers	*F. semitectum*
Seed and grain	*F. chlamydosporum*, *F. equiseti*, *F. poae*, *F. semitectum*

Symptoms caused by *Fusaria* on different hosts

3.2.1 Symptoms caused by *Fusaria* on different hosts

Banana wilt

wilt of cotton

Tomato wilt

Mango Malformation

Cob rot of maize

Guava wilt

Bakanae disease of rice

3.3 Isolation and Preservation

3.3.1 Isolation Techniques

There are many techniques to isolate soil fungi. The soil dilution plate technique was first developed for the isolation of bacteria, but it has been successfully applied on soil fungi which give quantitative results (Warcup 1955; Garrett 1981). Similarly, suspension-plating method is used for estimation of *F. oxysporum f. melonis* population in soils (Wensly and McKeen 1962). The screened immersion-plate technique gives a wider range and variety of fungal species isolated from soils (Chesters and Thornton 1956). On the other hand, direct soil plating method gives an advantage of detecting low fungal population in soils (Reinking and Wollenweber 1927; Warcup 1950). Moreover, *Fusarium* species could also be isolated by using living root or sterile straw baiting techniques, e.g., peas, flax, grass, banana tissue, and wheat straw (Burgess et al. 1994).

However, plating of soil dilutions or individual soil particles spread onto nutrient agar is performed by many researchers in general (McMullen and Stack 1983). Comparatively, debris isolation technique gives a higher diversity of *Fusarium* species recovered (McMullen and Stack 1983). The use of modified Nash and Snyder's medium (MNSM = PPA) is effective to determine the population of *F. solani* f. sp. *glycines* in soybean soils, while Komada's medium is selective for *F. oxysporum* (Komada 1975).

3.3.2 Preservation

There are several techniques to preserve *Fusarium* cultures into a collection. Sterilized carnation leaf pieces are good substrates for long-term preserving cultures of *Fusarium* species that was kept at −30 °C (Fisher et al. 1982). A spore suspension in sterilized 15 % glycerol kept in deep freezer at −70 °C has also been used for preservation (Leslie and Summerell 2006). The isolates that are preserved by using this method could remain viable up to 10 years. However, lyophilization preservation technique could maintain the viable cells for an extended period of time for more than 20 years. Lyophilization preservation technique is done by freeze-drying the culture with a colonized leaf piece (Tio et al. 1977). Another method used to preserve the cultures is soil preservation (Leslie and Summerell 2006). The soil must be sterilized completely in order to preserve the *Fusarium* species. This method is also considered as a long-term preservation technique.

3.4 Morphological Identification

Fusarium species are of frequent occurrence in most soils and because of their competitive ability can be easily isolated in the presence of phycomycetes, dry-spored molds, actinomycetes, or bacteria.*Fusarium sporodochia* found on woody stems or other plant tissue should be moistened, and a dilute spore suspension should be made from the mass of spores. *Fusarium* species are identified on the basis of cultural characters *viz.*, growth rate, culture pigmentation; microscopic characters *viz.*, phialides, conidia (microconidia and macroconidia) and chlamydospores.

1. *Fusarium oxysporum* **Schlecht**

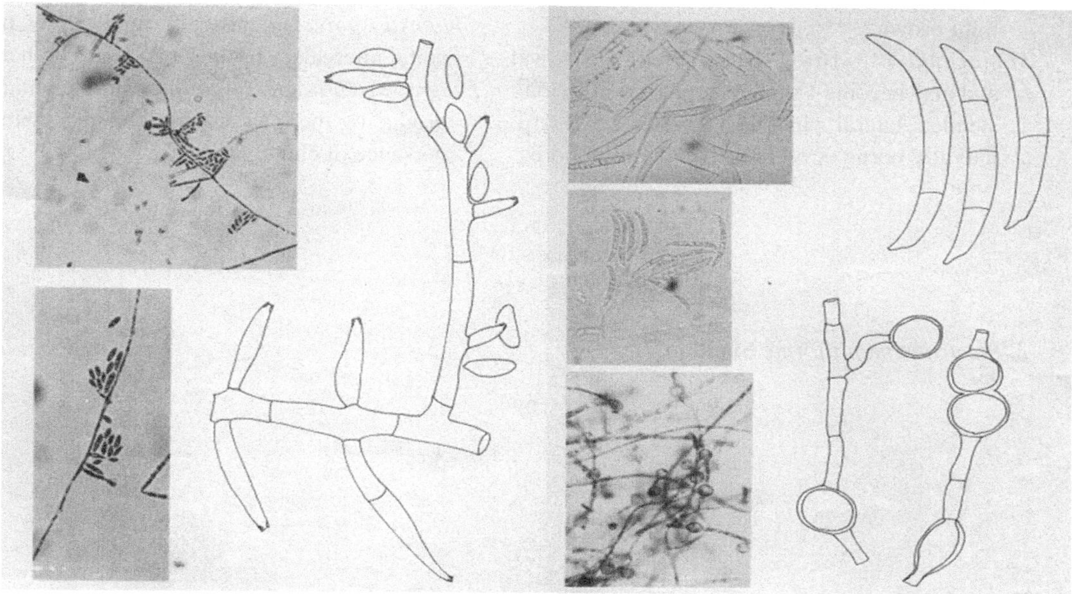

Growth rate: 4.5 cm

Culture pigmentation: white, peach, salmon, vinaceous gray to purple, violet

Microconidia: oval–ellipsoidal, cylindrical, straight or curved, 5–12 × 2.2–3.5 μ, produced from simple, short, lateral phialides often grouped forming *Tubercularia*-like sporodochia

Macroconidia: generally 3–5 septate, 27–60 × 3–5 μ, thin walled, fusoid

Chlamydospores: globose, formed singly or in pairs, intercalary or on short lateral branches

Diagnostic characters: the short simple phialides producing the microconidia together with the presence of chlamydospores

2. *Fusarium solani* **(Mart.) Sacc**

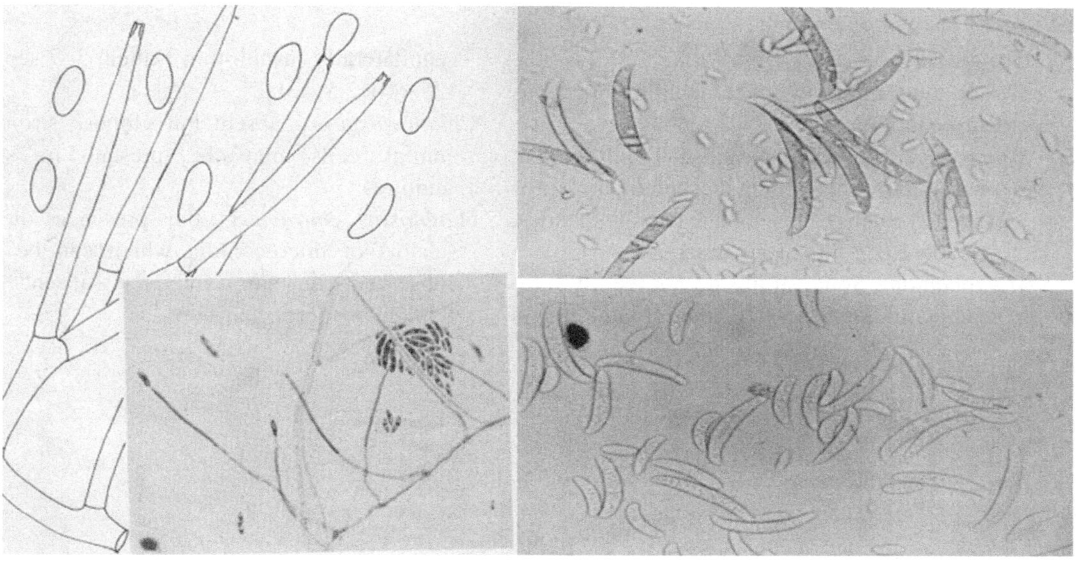

Growth rate: 3.2 cm

Culture pigmentation: grayish-white to white, light brown

Microconidia: 8–16 × 2–4 μ, cylindrical to oval and may become 1 septate, produced from long slender, lateral phialides 45–80 × 2.5–3.0 μ, laterally borne or on branched conidiophores

Macroconidia: generally 3–5 septate, 27–60 × 3–5 μ

Chlamydospores: globose, formed singly or in pairs, intercalary or on short lateral branches

Diagnostic characters: the short simple phialides producing the microconidia together with the presence of chlamydospores

3. *Fusarium moniliforme* **Sheldon**

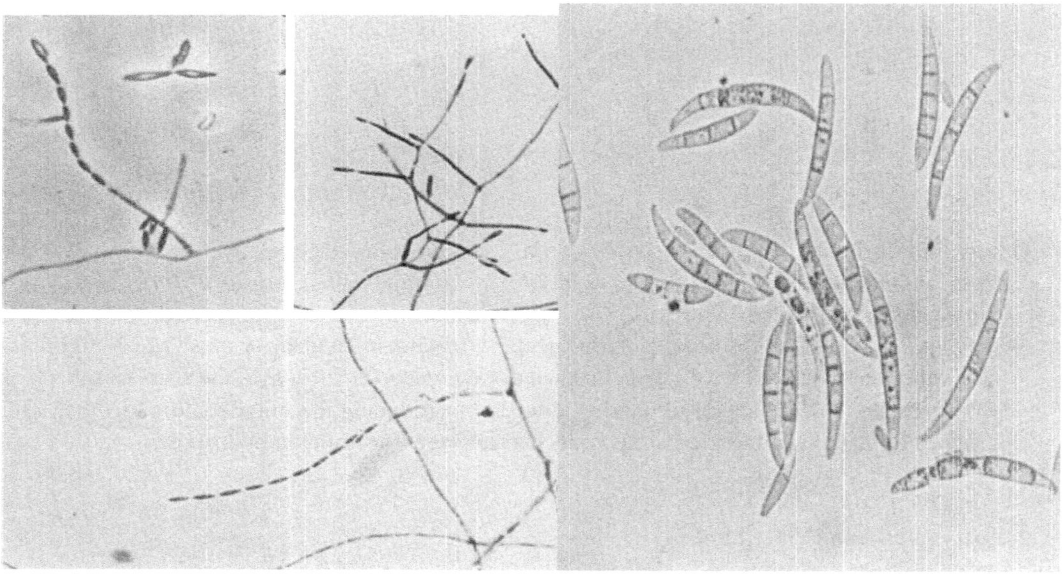

Growth rate: 4.6 cm.

Culture pigmentation: peach salmon, vinaceous purple to violet.

Microconidia: fusoid to clavate, 5–12 × 1.5–2.5 μ, occasionally becoming 1 septate and produced in chains from subulate lateral phialides, 20–30 × 2.0–3.0 μ, at the base.

Macroconidia: Some strains do not readily form macroconidia, but when present they are in equilaterally fusoid, thin walled, 3–7 septate, 25–60 × 2.5–4.0 μ.

Chlamydospores: absent but globose stromatic initial cells may be present in some cultures.

Diagnostic characters: the presence of the chains of microconidia which can be best observed in situ and the absence of chlamydospores.

4. ***Fusarium decemcellulare*** **Brick**

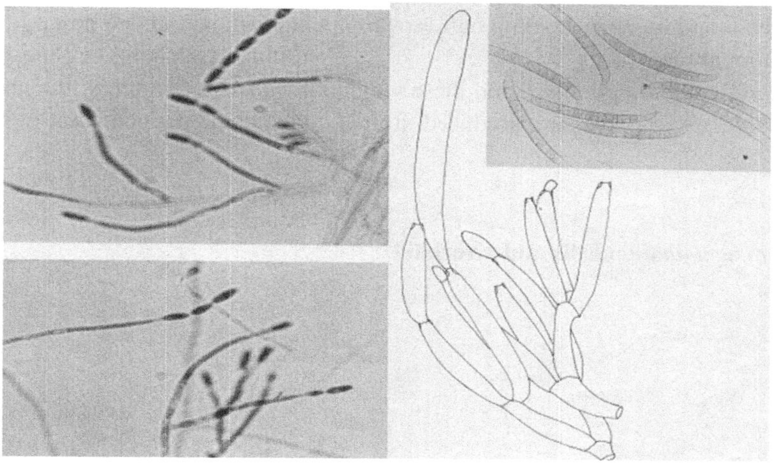

Growth rate: 3.2 cm.

Culture pigmentation: rose darkening to red, aerial mycelium white but pustules of macroconidia cream to yellow.

Microconidia: formed in chains from well-developed phialides. They are oval, aseptate, to 1 septate, 10–15 × 3.0–5.0 μ.

Macroconidia: formed on sporodochia from well-developed phialides. They are 7–10 septate, 55–130 × 6–10 μ.

Chlamydospores: absent.

Diagnostic characters: The presence of the chains of microconidia, distinct spore shape and size, pigmentation.

5. ***Fusarium equiseti*** **(Corda) Sacc**

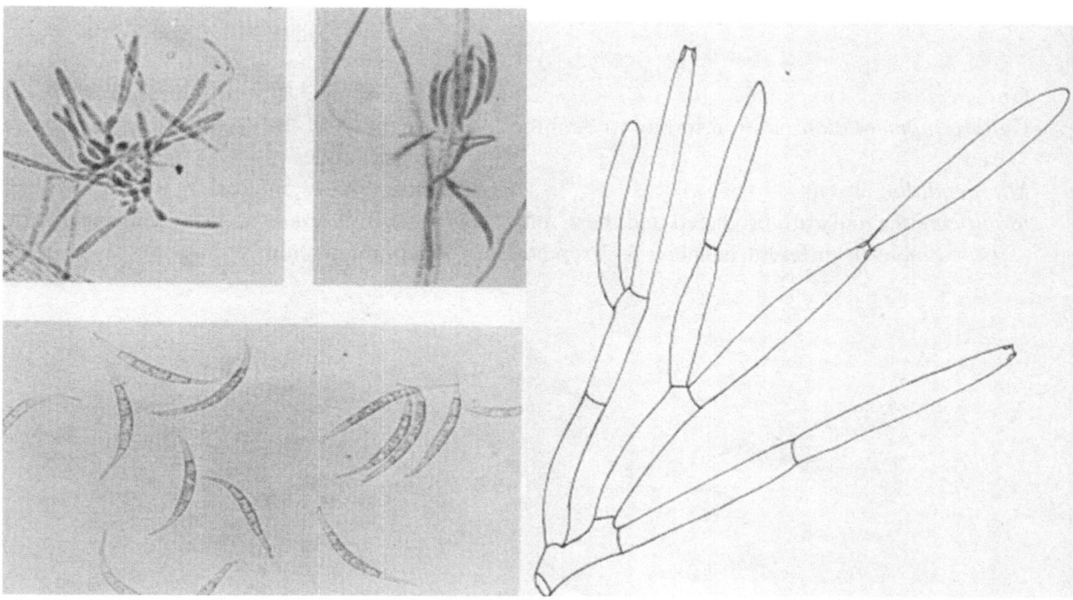

Growth rate: 5.9 cm

Culture pigmentation: peach usually changing to avellaneous and finally becoming buff brown

Microconidia: absent

Macroconidia: only are produced and these may be variable in size and are produced from single solitary or grouped phialides; conidia 4–7 septate, 22–60 × 3.5–9.0 μ

Chlamydospores: globose, 7–9 μ d, intercalary, solitary, in chains or clumps

Diagnostic characters: the absence of microconidia and pigmentation

6. ***Fusarium acuminatum*** **Ellis and Everhart**

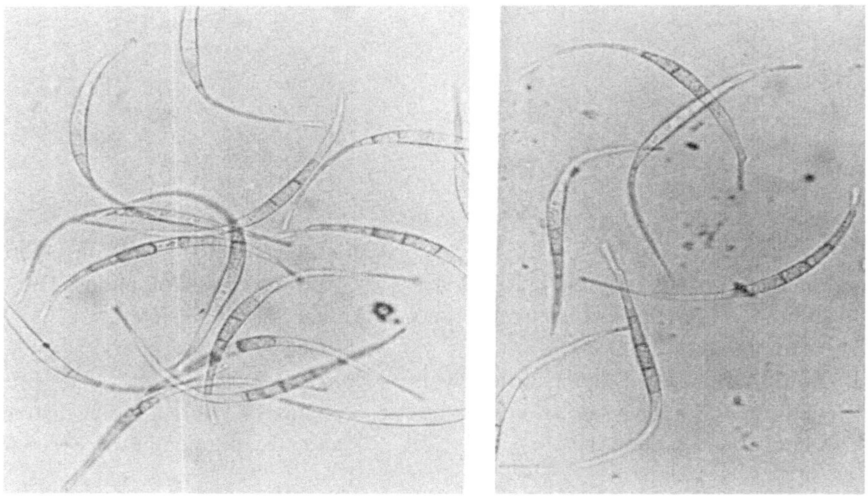

Growth rate: 4.5 cm.

Culture pigmentation: saffron to bay to carmine red

Microconidia: absent

Macroconidia: only are produced and these may be variable in different isolates, 3–7 septate, 30–70 × 3.5–5.0 μ often with an incurved elongation of the apical cell and are produced from phialides

Chlamydospores: intercalary in knots or in chains

Diagnostic characters: spore shape and carmine-red pigmentation

7. *Fusarium udum* **Butler**

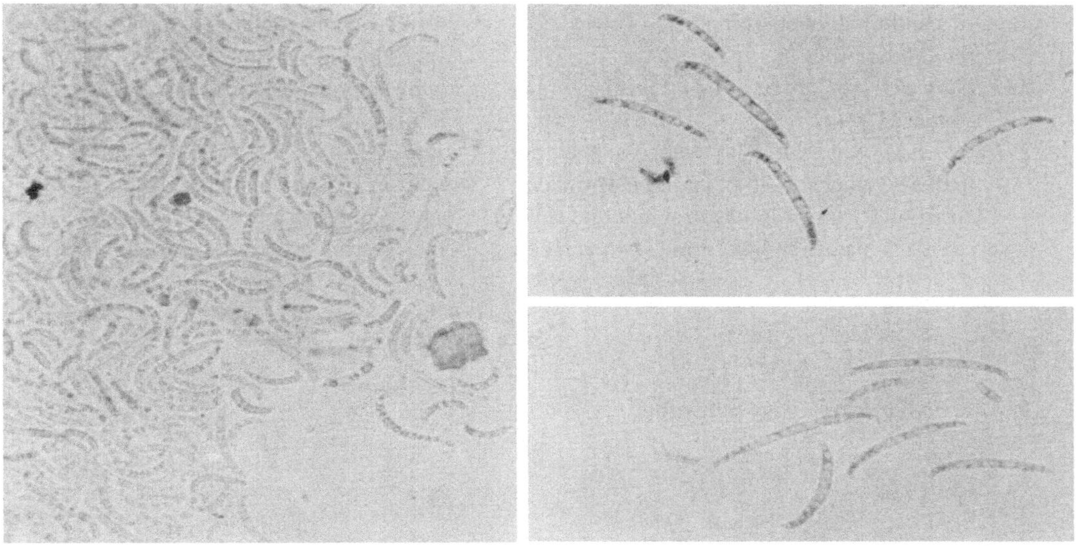

Growth rate: 4.2 cm

Culture pigmentation: Pale sulfurous to rose-buff becoming salmon-orange with production of conidia, occasional strains have purple pigmentation

Conidia: no clear distinction between microconidia and macroconidia. Conidia variable with a strongly curved or hooked apex, 6–8 × 3–3.5 μ and 30–40 × 3–3.5 μ

Chlamydospores: sparse, oval to globose, 8–11 × 8–12 μ

Diagnostic characters: extremely variable conidia with strongly curved apex and limited host range on *Cajanus* and *Crotalaria*

8. *Fusarium pallidoroseum* **Berk and Rav**

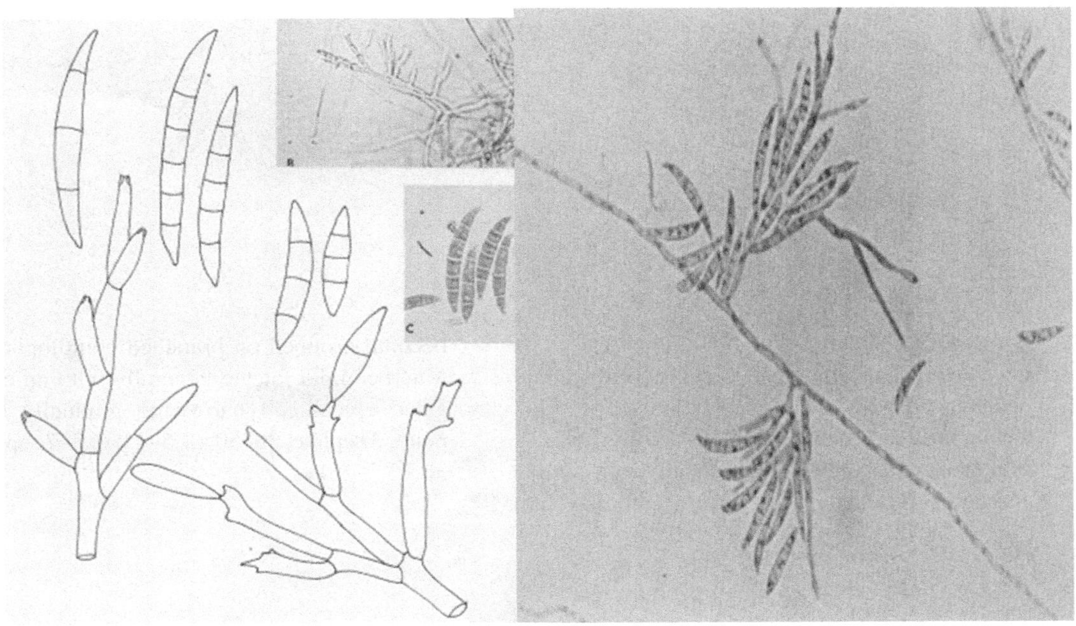

Growth rate: 6.1 cm

Culture pigmentation: peach changing to avellaneous and finally becoming buff-brown

Microconidia: absent

Macroconidia: of two types, primary and secondary

Primary macroconidia: with wedge-shape foot cell, 0–5 septate, 7.5–35 × 2.5–4.0 μ, formed as blastospores from polyblastic sympodial cells, up to five separate spores formed by each cell

Secondary macroconidia: with typical heeled foot cell, 3–7 septate, 20–46 × 3.0–5.5 μ, formed from phialides usually grouped in sporodochia

Chlamydospores: often sparse, globose, 10–12 μ d, becoming brown, intercalary, single, or in chains

Diagnostic characters: the presence of primary and secondary macroconidia, pigmentation, and spore form and presence of chlamydospores

9. *Fusarium graminearum* **Schwabe**

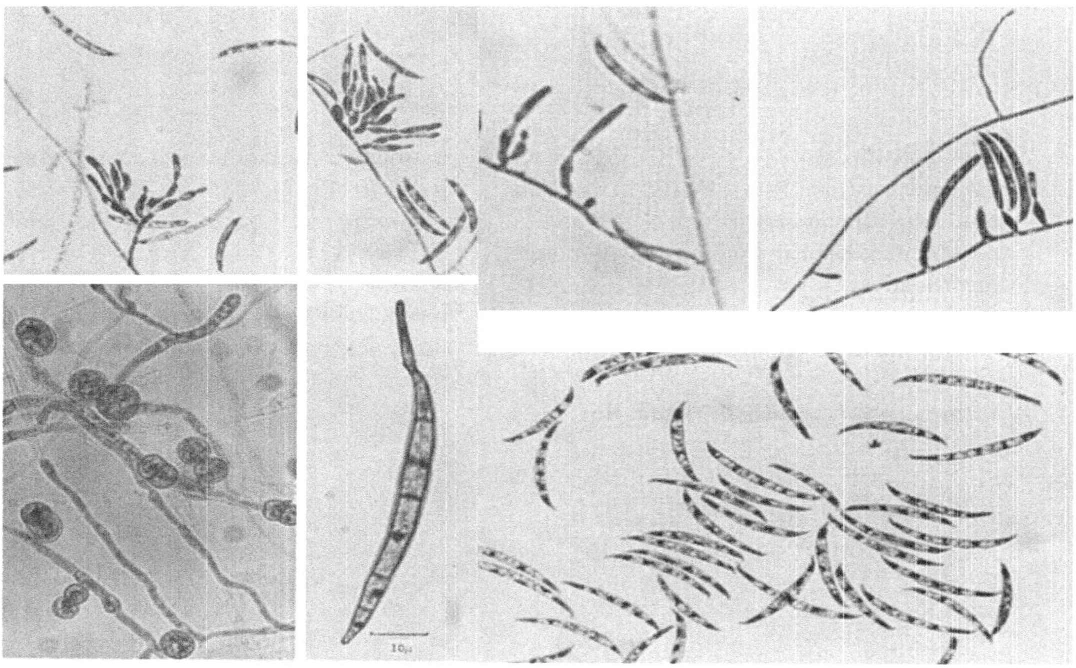

Growth rate: 8.9 cm.

Culture pigmentation: rose, coral becoming vinaceous with a brown tinge.

Microconidia: absent.

Macroconidia: only are produced from simple lateral phialides which may or may not become grouped on branched conidiophores. Macroconidia falcate generally with an elongated epical cell narrowing gradually to a point, 3 septate, 30–50 × 3.5–4.0 μ, 5–7 septate, 36 × 3.5–5.0 μ.

Chlamydospores: absent or rare. If present, intercalary, 10–12 μ d in knots or in chains.

Diagnostic characters: The long falcate macroconidia often formed sparsely in many strains are characteristic. Many isolates of this species with floccose aerial mycelium and rose to coral pigmentation produce neither macroconidia nor chlamydospores until surface of colony is washed clean of mycelium and culture reincubated.

10. *Fusarium dimerum* **Penz**

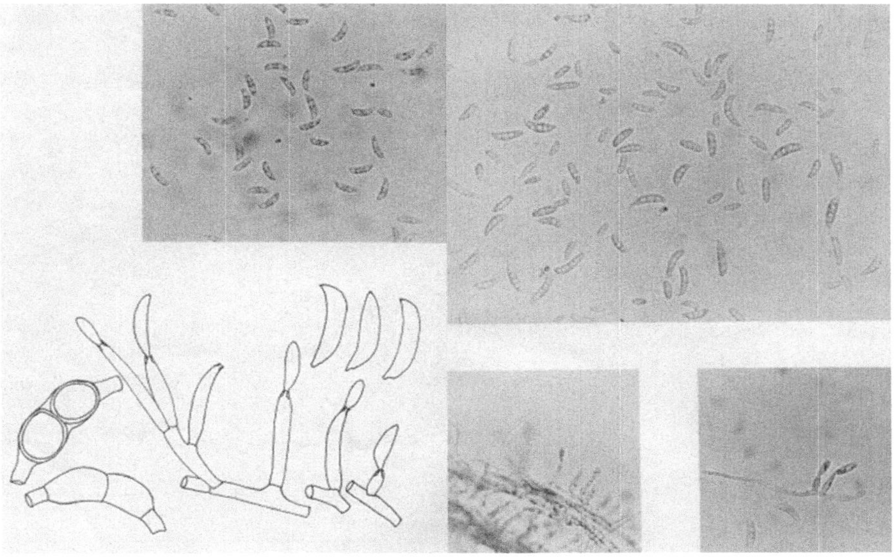

Growth rate: 2.7 cm.

Culture pigmentation: orange-beige to apricot.

Conidia: Somewhat heterogeneous probably representing primary and secondary conidia as occasional phialides develop, 0 septate, 6.5–10.5 × 2.3–2.5 μ, 1–2 septate, 10–12 × 3.0–3.5 μ.

Chlamydospores: globose, oval to smooth, 8–12 μ d, intercalary, formed singly or in chains.

Diagnostic characters: Conidial form and presence of chlamydospores separate it from the related species.

11. *Fusarium aquaeductuum* **Lagerh**

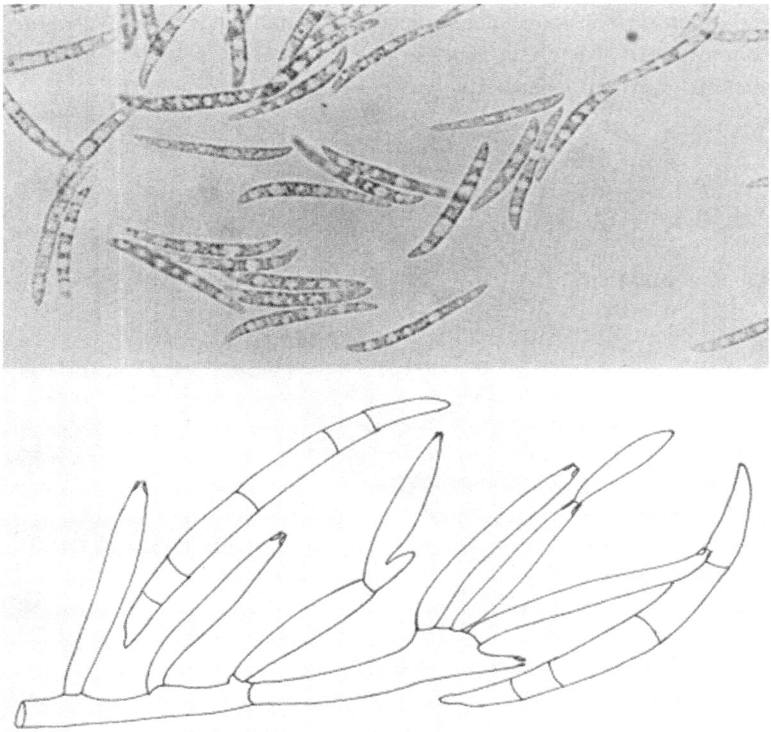

Growth rate: 0.5 cm.

Culture pigmentation: pale cream becoming orange or salmon-pink with convoluted, merismoid or fibrillose surface appearance.

Microconidia: Absent.

Macroconidia: only are formed from pionnotes sporodochia. indistinctly 1–5 septate, variable in length, 15–65 × 2.5–4.0 μ.

Chlamydospores: absent.

Diagnostic characters: slow growth, macroconidia shape and size, absence of chlamydospores. Most occur in sewage or polluted water and also occur as parasites on sphaeriaceous fungi.

3.5 Molecular Identification

Conventional characterization of toxigenic *Fusarium* species has been based mainly on morphological methods (Leslie et al. 2001), which are the most routinely performed. Nevertheless, recognition by morphological characters sometimes is not enough for accurate identification of fungal isolates at the species level. Furthermore, morphological characterization is time-consuming and requires considerable expertise in *Fusarium* taxonomy and physiology (Leslie and Summerell 2006). As identification of

Fusarium species is critical to predict the potential mycotoxigenic risk of the isolates, there is a need for accurate and complementary tools which permit a rapid, sensitive, and reliable specific diagnosis of *Fusarium* species.

The concerns indicated above may be overcome by appropriate DNA sequencing and species-specific PCR assays (Jurado et al. 2006a, b). Various PCR assays have been developed for the identification of toxigenic species of *Fusarium*. Some of them are based on single-copy genes directly involved in mycotoxin biosynthesis while others are species specific (Gonzalez Jaén et al. 2004; Mulé et al. 2005). The last ones often amplify multicopy target sequences, such as IGS or ITS regions (intergenic spacer and internal transcribed spacer of rDNA units, respectively), which increase the sensitivity of the assay in comparison with PCR assays based on single-copy sequences. The use of these PCR approaches has been already useful in epidemiological analyses (Jurado et al. 2004, 2006a; b; Sreenivasa et al., 2008). Nevertheless, the worldwide distribution of toxigenic *Fusarium* species is a challenge to the universality of any species-specific PCR assay, and therefore, specificity should be confirmed in strains from various crops and/or geographic locations. Regarding DNA sequencing, the translation elongation factor 1-α (*TEF1*-α) gene appears to occur consistently as a single copy in *Fusarium* and shows a high level of sequence polymorphism among closely related species (Geiser et al. 2004), even when compared with the intron-rich portions of protein-coding genes such as calmodulin, β-tubulin, and histone H3 (Rahjoo et al. 2008). For these reasons, *TEF1*-α has become the marker of choice as a single-locus identification tool in *Fusarium*.

3.5.1 Different Techniques

3.5.1.1 RAPD–PCR Technique

El-Fadly et al. (2008) explored the possible utilization of random amplified polymorphic DNA (RAPD–PCR) technique for identifying *Fusarium* spp., either alternatively or complementary to those based upon morphological and pathological characteristics.

Random amplified polymorphic DNA (RAPD–PCR) was used to identify some *Fusarium* isolates. Seven *Fusarium* isolates which were identified by their morphological and pathological characteristics as *F. semitectum*, *F. culmorum*, *F. moniliforme*, *F. solani*, *F. graminearum*, *F. oxysporum* f. sp. *lycopersici*, and *F. oxysporum* f. sp. *vasinfectum* were used in this study. RAPD analysis was carried out using eight random primers; each of them consisted of ten base pairs. Genetic variability among such species and *formae speciales* under study were recorded. Six out of the eight primers were differentiated between some of the tested *Fusarium* species, since 100 % similarity was recorded between two or three different species of the fungus, while the rest of the two primers clearly distinguished each of all studied *Fusarium* spp. including the two *formae speciales*.

In conclusion, RAPD–PCR technique is a useful tool for differentiating between species and *formae speciales* of the genus *Fusarium* either alternatively or complementary to methods based upon morphological and pathological characteristics.

3.5.1.2 Amplified Translation Elongation Factor 1-α Gene Fragment

Nitschke et al. (2009) used this tool for reliable identification based on sequence information of the translation elongation factor 1-α (*TEF1*-α) gene for the numerous *Fusarium* spp. being isolated from sugar beets. In all, 65 isolates from different species (*Fusarium avenaceum*, *F. cerealis*, *F. culmorum*, *F. equiseti*, *F. graminearum*, *F. oxysporum*, *F. proliferatum*, *F. redolens*, *F. solani*, *F. tricinctum*, and *F. venenatum*) were obtained from sugar beet at different developmental stages from locations worldwide. Database sequences for additional species (*F. sporotrichioides*, *F. poae*, *F. torulosum*, *F. hostae*, *F. sambucinum*, *F. subglutinans*, and *F. verticillioides*), isolated from sugar beets in previous studies, were included in the analysis. Molecular sequence analysis of the partial *TEF1*-α gene fragment revealed sufficient variability to differentiate between the *Fusarium* spp., resulting in species-dependent separation of the isolates analyzed. This interspecific divergence could be translated into a polymerase chain reaction restriction fragment length polymorphism assay using only two subsequent restriction digests for the differentiation of 17 of 18 species.

3.5.1.3 IGS–RFLP Analysis

The intergenic spacer (IGS) regions of the rDNA of several *Fusarium* spp. strains obtained from the collaborative researchers were amplified by polymerase chain reaction (PCR), and an IGS–RFLP analysis was performed by Konstantinova and Yli-Mattila (2004). Restriction digestion with AluI, MspI, and PstI allowed differentiation between the related *Fusarium poae* and *Fusarium kyushuense* species. *Fusarium langsethiae* was also separated from *Fusarium sporotrichioides* (including var. minus) on the basis of the banding patterns after MspI digestion, while specific XhoI, AluI, and MspI restriction patterns were found in the IGS amplicons of *F. sporotrichioides* var. minus. According to the phylogenetic analysis of IGS–RFLP patterns, *F. langsethiae* (except for one strain), *F. sporotrichioides*, *F. poae*, and *F. kyushuense* strains formed four well-supported clades with high-bootstrap values. Based on the sequence differences in the IGS region, species-specific primers were designed for the *F. langsethiae*/*F. sporotrichioides* group and for *F. poae*. The specificity and sensitivity of the primers were tested on various *Fusarium* species and isolates and on several other important fungal genera associated with cereals. The *F. poae-specific* primers, designed in this study, showed the same specificity as primers Fp82f/Fp82r developed previously. The two phylogenetic subgroups *of F. langsethiae*, found by IGS sequencing analysis, were separated on the basis of size differences of the amplification products with primers CNL12/PulvIGSr specific for the *F. langsethiae*/*F. sporotrichioides* group.

RFLP analysis of the amplified IGS region is a useful molecular assay for characterization and a phylogenetic study of several related *Fusarium* species—*F. langsethiae*, *F. sporotrichioides*, *F. sporotrichioides* var. *minus*, *F. poae*, and *F. kyushuense*. The primers designed in this study were highly specific and allowed identification of *F. poae* and the *F. langsethiae*/*F. sporotrichioides* group.

3.5.1.4 Real-Time PCR Assay

Real-time PCR assays allow species-specific quantification of *Fusarium* biomass. A real-time PCR technique was applied for the quantification of trichothecene-producing *Fusarium* species as well as the highly toxigenic *Fusarium graminearum* present in barley grain and malt (Sarlin et al. 2006). PCR results were compared to the amounts of trichothecenes detected in the samples to find out if the PCR assays can be used for trichothecene screening instead of expensive and laborious chemical analyses.

3.5.1.5 Analysis of the ITS rRNA Region

Accurate morphological identification of *Fusarium* spp. beyond the genus is time-consuming and insensitive. Young Mi et al. (2000) examined the usefulness of the nuclear ribosomal RNA (rRNA) internal transcribed spacer regions (ITS1 and ITS4) to detect and differentiate *Fusarium* spp.

To investigate the genetic relationship among 12 species belonging to the *Fusarium* section *Martiella*, *Dlaminia*, *Gibbosum*, *Arthrosporiella*, *Liseola*, and *Elegans*, the internal transcribed spacer (ITS) regions of ribosomal DNA (rDNA) were amplified with primer pITS1 and pITS4 using the polymerase chain reaction (PCR). After the amplified products were digested with seven restriction enzymes, restriction fragment length polymorphism (RFLP) patterns were analyzed. The partial nucleotide sequences of the ITS region were determined and compared. Little variation was observed in the size of the amplified product having sizes of 550 bp or 570 bp. Based on the RFLP analysis, the 12 species studied were divided into five RFLP types. In particular, strains belonging to the section *Martiella* were separated into three RFLP types. Interestingly, the RFLP type of *F. solani* f. sp. *piperis* was identical with that of isolates belonging to the section *Elegans*. In the dendrogram derived from RFLP analysis of the ITS region, the *Fusarium* spp. examined were divided into two major groups. In general, section *Martiella* excluding *F. solani* f. sp. *piperis* showed relatively low similarity with the other section. The dendrogram based on the sequencing analysis of the ITS2 region also gave the same results as that of the RFLP analysis. As expected, 5.8S, a coding region, was highly conserved, whereas the ITS2 region was more variable and informative. The difference in the ITS2 region between the length of *F. solani* and its *formae speciales* excluding *F. solani* f. sp. *piperis* and that of other species was caused by the insertion/deletion of nucleotides in positions 143–148 and 179–192.

3.5.1.6 Combined Approach

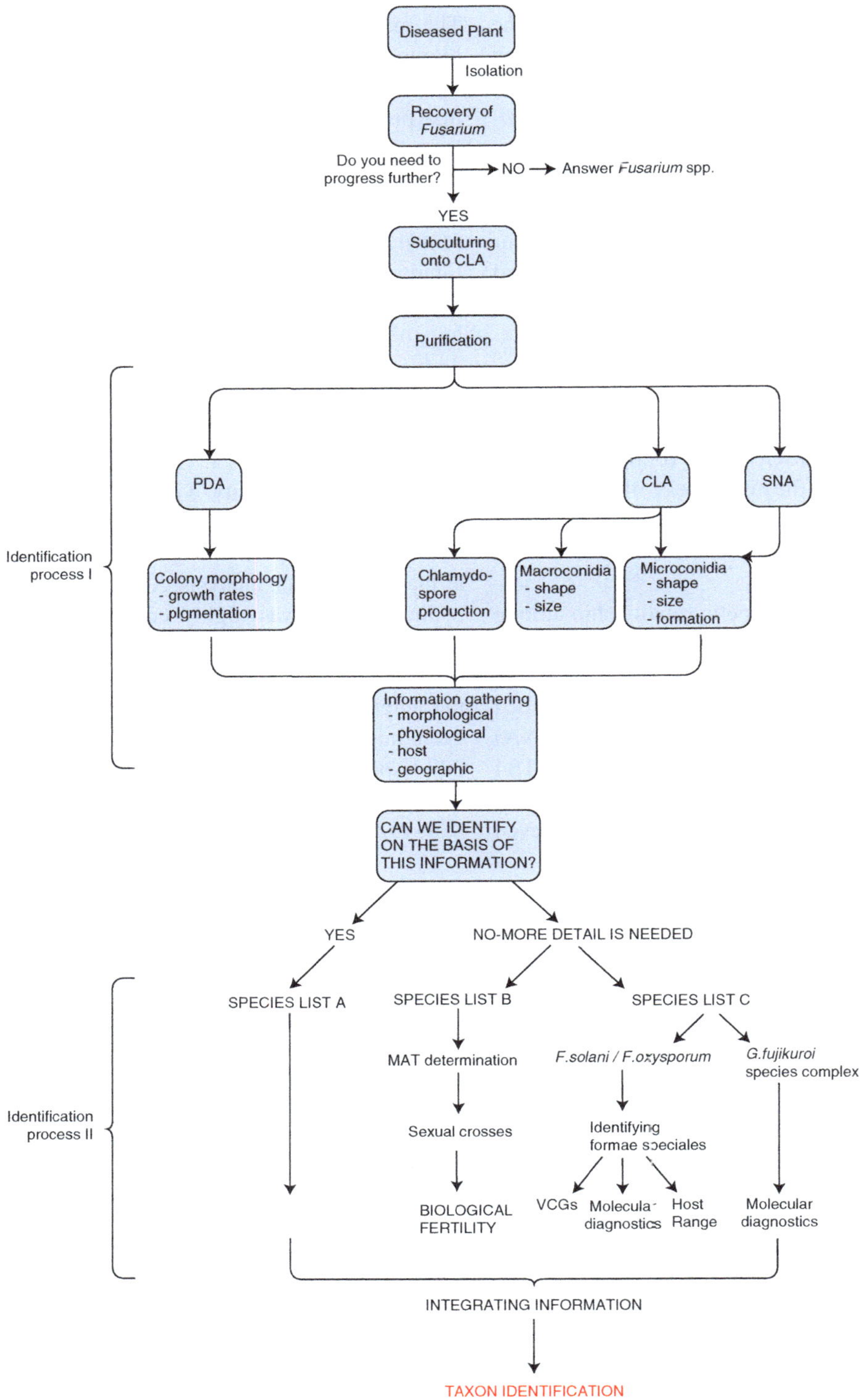

3.6 Conclusion

Different concepts have been used to define the fungal species. Morphological identification of plant pathogenic fungi is the first and the most difficult step in the identification process. This is especially true for *Fusarium* species. Although morphological observations may not suffice for complete identification, a great deal of information is usually obtained on the culture at this stage. But these approaches for identifying fungi are laborious and time-consuming and provide insufficient taxonomic resolution. The major disadvantages are that all the assays based on phenotypes are too sensitive to growth conditions and depend on gene expression. For species that cannot be reliably identified in this way, especially for members of the *G. fujikuro* complex, additional analysis such as DNA sequencing and species-specific PCR assays must be conducted. The integration of morphology and molecular-based techniques gave fair solution in easing some of the complex problems that are associated with either morphology or molecular techniques. Therefore, in this chapter, the combined morphological and molecular approaches are described and incorporated for authentic identification of *Fusarium spp.*

References

Appel DJ, Gordon TR (1995) Intraspecific variation within populations of *Fusarium oxysporum* based on RFLP analysis of the intergenic spacer region of the rDNA. Exp Mycol 19:120–128

Backhouse D, Burgess LW (2002) Climatic analysis of the distribution of *Fusarium graminearum, F. pseudograminearum and F. culmorum* on cereals in Australia. Australas Plant Pathol 31:321–327

Backhouse D, Burgess LW, Summerell BA (2001) Biogeography of *Fusarium*. In: Summerell BA, Leslie JF, Backhouse D, Bryden WL, Burgess LW (eds) Fusarium: Paul E. Nelson memorial symposium. American Phytopathological Society, St. Paul, pp 124–139

Bulat SA, Alekhina IA, Kozlova ED (1997) Molecular taxonomy of *Fusarium* fungi by means of ribotyping rDNA sequencing and UP–PCR analysis. A case study of taxa in *Sporotrichiella* section. Cereal Res Commun 25:565–570

Burgess LW, Summerell BA (1992) Mycogeography of *Fusarium*: survey of *Fusarium* species in subtropical and semi-arid grasslands soils in Queensland. Mycol Res 96:780–784

Burgess LW, Summerell BA, Bullock S, Gott KP, Backhouse D (1994) Laboratoly manual for Fusarium research, 3rd edn. The University of Sydney Royal Botanic Gardens, Sydney

Chesters CGC, Thornton RH (1956) A comparison of techniques for isolating soil fungi. Trans Br Mycol Soc 39:301

El-Fadly GB, El-Kazzaz MK, Hassan MAA, El-Kot GAN (2008) Identification of some *Fusarium* spp. using RAPD-PCR technique. Egypt J Phytopathol 36(1–2):71–80

Fisher NL, Burgess LW, Toussoun TA, Nelson PE (1982) Carnation leaves as a substrate and for preserving cultures of *Fusarium* species. Phytopathology 72:151–153

Garrett SD (1981) Soil fungi and soil fertility, 2nd edn. Pergamon Press, Oxford

Geiser DM, Jimenez Gasco MM, Kang S, Mkalowska I, Veeraraghavan N, Ward TJ, Zhang N, Kuldau GA, O'Donnell K (2004) FUSARIUM-IDv.1.0: a DNA sequence database for identifying *Fusarium*. Eur J Plant Pathol 110:473–479

Gonzalez Jaén MT, Mirete S, Patiño B, López-Errasquín E, Vázquez C (2004) Genetic markers for the análysis of variability and for production of specific diagnostic sequences in fumonisin-producing strains of *Fusarium verticillioides*. Eur J Plant Pathol 110:525–532

Hillis DM, Dixon MT (1991) Ribosomal DNA: molecular evolution and phylogenetic inference. Q Rev Biol 66:411–453

Jasalavich CA, Ostrofsky A, Jellison J (2000) Detection and identification of decay fungi in spruce wood by Restriction Fragment Length Polymorphism analysis of amplified genes encoding rRNA. Appl Environ Microbiol 66:4725–4734

Jurado M, Vázquez C, López-Errasquin E, Patiño B, González-Jaén MT (2004) Analysis of the occurrence of *Fusarium* species in Spanish cereals by PCR assays. In: Proceedings of the 2nd international symposium on Fusarium head blight and 8th European Fusarium seminar, vol 2, pp 460–464

Jurado M, Vázquez C, Callejas C, González-Jaén MT (2006a) Occurrence and variability of mycotoxigenic *Fusarium* species associated to wheat and maize in the South West of Spain. Mycotoxin Res 22:87–91

Jurado M, Vázquez C, Marín S, Sanchis V, González-Jaén MT (2006b) PCR-based strategy to detect contamination with mycotoxigenic *Fusarium* species in maize. Syst Appl Microbiol 29:681–689

Komada H (1975) Development of a selective medium for quantitative isolation of *Fusarium oxysporum* from natural soil. Rev Plant Prot Res 8:114–125

Konstantinovaa P, Yli-Mattila T (2004) IGS–RFLP analysis and development of molecular markers for identification of Fusarium poae, Fusarium langsethiae, Fusarium sporotrichioides and Fusarium kyushuense. Int J Food Microbiol 95(3):321–331

Leslie JF, Summerell BA (2006) The Fusarium laboratory manual. Blackwell Publishing, Ames, p 388

Leslie JF, Zeller KA, Summerell BA (2001) Icebergs and species in populations of *Fusarium*. Physiol Mol Plant Pathol 59:107–117

Martin FN (1990) Variation in the ribosomal DNA repeat unit within single-oospore isolates of the genus *Pythium*. Genome 33:585–591

McMullen MP, Stack RW (1983) Effects of isolation techniques and media on the differential isolation of *Fusarium* species. Phytopathology 73:458–462

Mulé G, González Jaén MT, Hornok L, Nicholson P, Waalwijk C (2005) Advances in molecular diagnosis of toxigenic *Fusarium* species: a review. Food Addit Contam 22:316–323

Nelson PE, Toussoun TA, Cook RJ (eds) (1981) *Fusarium*: diseases, biology and taxonomy. Pennsylvania State University, University Park

Nitschke E, Nihlgard M, Varrelmann M (2009) Differentiation of eleven *Fusarium* spp. Isolated from sugar beet, using restriction fragment analysis of a polymerase chain reaction–amplified translation elongation factor 1α gene fragment. Mycology 99(8):921–929

O'Donnell K (1996) Progress towards a phylogenetic classification of *Fusarium*. Sydowia 48:57–70

Paavanen-Huhtala S (2000) Molecular-based assays for the determination of diversity and identification of *Fusarium* and *Gliocladium* fungi. PhD thesis, Annales Universitatis Turkvensis AII 139, University of Turku, Finland, pp 198

Paavanen-Huhtala S, Hyvonen J, Bulat SA, Yli-Mattila T (1999) RAPD–PCR, isozyme, rDNA RFLP and rDNA sequence analyses in identification of Finnish *Fusarium oxysporum* isolates. Mycol Res 103:625–634

Ploetz RC (2001) Malformation: a unique and important disease of mango, *Magnifera indica* L. In: Summerell BA, Leslie JF, Backhouse D, Bryden WL, Burgess LW (eds) Fusarium: Paul E. Nelson memorial symposium. American Phytopathological Society, St. Paul, pp 233–247

Rahjoo V, Zad J, Javan-Nikkhah M, Mirzadi Gohari A, Okhovvat SM, Bihamta MR, Razzaghian J, Klemsdal SS (2008) Morphological and molecular identification of *Fusarium* isolated from maize ears in Iran. J Plant Pathol 90:463–468

Reinking OA, Wollenweber HW (1927) "Tropical Fusaria" in Philippine. J Sci 32(2):103–252

Sarlin T, Yli-Mattila T, Jestoi M, Rizzo A, Paavanen-Huhtala S, Haikara A (2006) Real-time PCR for quantification of toxigenic *Fusarium* species in Barley and Malt. Eur J Plant Pathol 114:371–380

Sreenivasa MY, González-Jaén MT, Dass RS, Rharith Raj AP, Janardhana GR (2008) A PCR-based assay for the detection and differentiation of potential fumonisin-producing *Fusarium verticillioides* isolated from Indian maize Kernels. Food Biotechnol 22:160–170

Summerell BA, Leslie JF, Backhouse D, Bryden WL, Burgess LW (eds) (2001) Fusarium: Paul E. Nelson memorial symposium. American Phytopathological Society, St. Paul

Taylor JW, Jacobson DJ, Kroken S, Kasuga T, Geiser DM, Hibbett DS, Fisher MC (2000) Phylogenetic species recognition and species concepts in fungi. Fungal Genet Biol 31:21–32

Tio M, Burgess LW, Nelson PE, Toussoun TA (1977) Techniques for the isolation, culture and preservation of the Fusaria. Aust Plant Pathol Soc Newslett 6:11–13

Warcup JH (1950) The soil-plate method for isolation of fungi from soil. Nature (London) 146:117

Warcup JH (1955) On the origin of colonies of fungi developing on soil-dilution plates. Trans Br Mycol Soc 38:298–301

Wensly RN, McKeen CD (1962) Rapid test for pathogenicity of soil isolates of *Fusarium oxysporum* of sp. *melonis*. Can J Microbiol 8:818–819

Yli-Mattila T, Paavanen-Huhtala S, Bulat SA, Alekhina IA, Nirenberg HI (2002) Molecular, morphological and phylogenetic analysis of the *Fusarium avenaceum/F. arthrosporioides/F. tricinctum* species complex—a polyphasic approach. Mycol Res 106:655–669

Young-Mi L, Choi Y, Min B (2000) PCR-RFLP and sequence analysis of the rDNA ITS region in the *Fusarium* spp. J Microbiol 38(2):66–73

Damping-Off Disease of Seedlings in Solanaceous Vegetables: Current Status and Disease Management

4

Kanika Pagoch, J.N. Srivastava, and Ashok Kumar Singh

Vegetables play an important role in a balanced diet by providing not only energy but also supplying vital protective nutrients like minerals and vitamins. In addition to their role in nutrition, vegetables increase attractiveness and palatability of a diet by providing sensory appeal through their test and flavours. Vegetables are a major and very important constituent of human diet (Thamburaj and Singh 2005).

The vegetable crops propagated by seeds, like cucurbits, beans, radish, turnip, leafy vegetable and okra, are required to be sown directly in the field, whereas some crops like, tomato, brinjal, chilli, etc. are first sown in nurseries for raising seedlings and then transplanting. The disease is common in nursery beds and young seedlings by several fungi such as *Pythium*, *Phytopthora*, *Fusarium*, *Rhizoctonia* and *Colletotrichum*.

Damping-off disease of seedlings is widely distributed all over the world. It was first studied by Hensin Germany. Damping off is a seedling disease common to most of solanaceous vegetable, *viz.* tomato, brinjal and chilli. The disease is of common in nursery beds and young seedlings. Several seed- and soil-borne fungi can kill before the tender radicle and plumule are established in the nursery bed (Sohi 1982; Fageria et al. 2003).

The pathogen attracts the seed and seedling roots during germination either before or after emergence. Within days, a number of seedlings are destroyed by pathogens, and also later after several weeks, damping-off seedling may develop root rot or stem canker (Atkinson 1895). The pathogen attracts underground soil line or crown roots of seedlings. In some damping-off fungi, foliar blight may also occur.

Depending upon host variety and environmental factors, 25–75 % losses are caused due to this disease (Gupta and Paul 2001). Damping off of chilli (*Pythium aphanidermatum* (Edson) Fitz) was responsible for 90 % mortality of seedlings both in the nursery and main field (Sowmini 1961). The fungus has a wide host range and attracts the plants belonging to families Cruciferae, Leguminosae and Chenopodiaceae (Alexander 1931).

The amount of damage the disease causes to seedlings depends on the fungus, soil moisture and soil temperature and other factors rather than upon the particular species of plant concerned. Normally, however, cool wet soils favour the development of the disease (Alexander 1931).

K. Pagoch (✉)
Division of Plant Pathology, SKUAST-Jammu, Jammu, India
e-mail: apsinghagron@gmail.com

J.N. Srivastava
Department of Plant Pathology, Bihar Agricultural University, Sabour, Bhagalpur, Bihar, India

A.K. Singh
Division of Plant Pathology, Sher-e-Kashmir University of Agricultural Sciences & Technology, Jammu, J & K, India

L.P. Awasthi (ed.), *Recent Advances in the Diagnosis and Management of Plant Diseases*,
DOI 10.1007/978-81-322-2571-3_4

The disease is responsible for poor germination and stand of seedling in the nursery bed and often the infected seedling carry the pathogen to the main field where transplanting is done. Older plants are seldom killed by damping-off fungi mainly because the development of secondary stem tissue forms roots, and stems still can be attacked, resulting in poor growth and reduced yields (Singh 1995).

Therefore, vegetables require more attention by the farmers and scientists' at field level; correct diagnosis of malady with suitable control measures need to be explored for better production of vegetables (Fageria et al. 2003).

4.1 Symptoms

Damping-off disease in vegetables occurs in two phases based on the time of infection.

4.1.1 Pre-emergence Damping Off

In this phase, the infection takes place before the hypocotyl has broken the seed coat or as soon as the radicle and plumule emerge out of the seeds, the seedling disintegrate before they come out of soil surface (Singh 1995). In fact, the seeds may rot or seedlings are killed before they emerge through the soil surface. This is referred to as pre-emergence damping off which results in poor field emergence/poor seed germination. If germination has occurred the hypocotyl (emerging shoot) shows water-soaked lesion (Singh 2000). The disease is often not recognized by the farmer who attributes the failure of emergence to poor quality of the seed.

4.1.2 Postemergence Damping Off

Postemergence damping off is characterized by development of disease after the seedlings have emerged out of the soil surface but before the stems are lignified (Atkinson 1895). The infection results as lesion formation on the collar region giving a pinched appearance. The infection point at the stem becomes hard and thin and such symptoms are commonly called "wire stem" appearance at the base of the stem. Infection usually occurs at the ground level or through roots. The infected tissues appear soft and water soaked. As the disease advances the stem becomes constricted at the base and plants collapse. Seedlings that appear healthy one day may have collapsed by the next morning (Singh 1995). The top of the plant may appear healthy when it falls over but quickly wilts and dies. The roots may or may not be decayed. Generally, the cotyledons and leaves wilt slightly before the seedlings are prostrated, although sometimes they remain green and turgid until collapse of the seedlings occurs (Brien and Chamberlain 1937). Transplanting of seedlings on infested soil, early infection of damping-off resulting soon dies of seedlings. In fields and nurseries the disease often occurs in a roughly circular pattern, or radiates from initial infection points, causing large spots or areas in which nearly all the seedlings are killed. This is because of the tendency of fungi to grow radially from the point of origin thereby causing large spots or areas in which nearly all the seedlings are killed (Wick 1998). Other above-ground symptoms of root rot include stunting, low vigour or wilting on a warm day. Foliage of such plants may yellow and fall prematurely starting with the oldest leaves.

Besides this, damping off is often confused with plant injury caused by insect feeding (severe mite, aphid, scale infestation), excessive fertilization, high levels of soluble salts, excessive heat or cold injury, excessive or insufficient soil moisture, insufficient light or nitrogen, root feeding by nematodes or insect larvae, or chemical toxicity in air or soil (Brien and Chamberlain 1937) (Fig. 4.1).

4.2 Causal Organisms

The fungi responsible for damping off include species of *Pythium*, *Phytopthora*, *Fusarium*, *Rhizoctonia* and *Colletotrichum*. Other fungi that occasionally cause this disease include *Glomerella*, *Alternaria*, *Phoma* and *Botrytis*. These fungi are not host specific and more or less

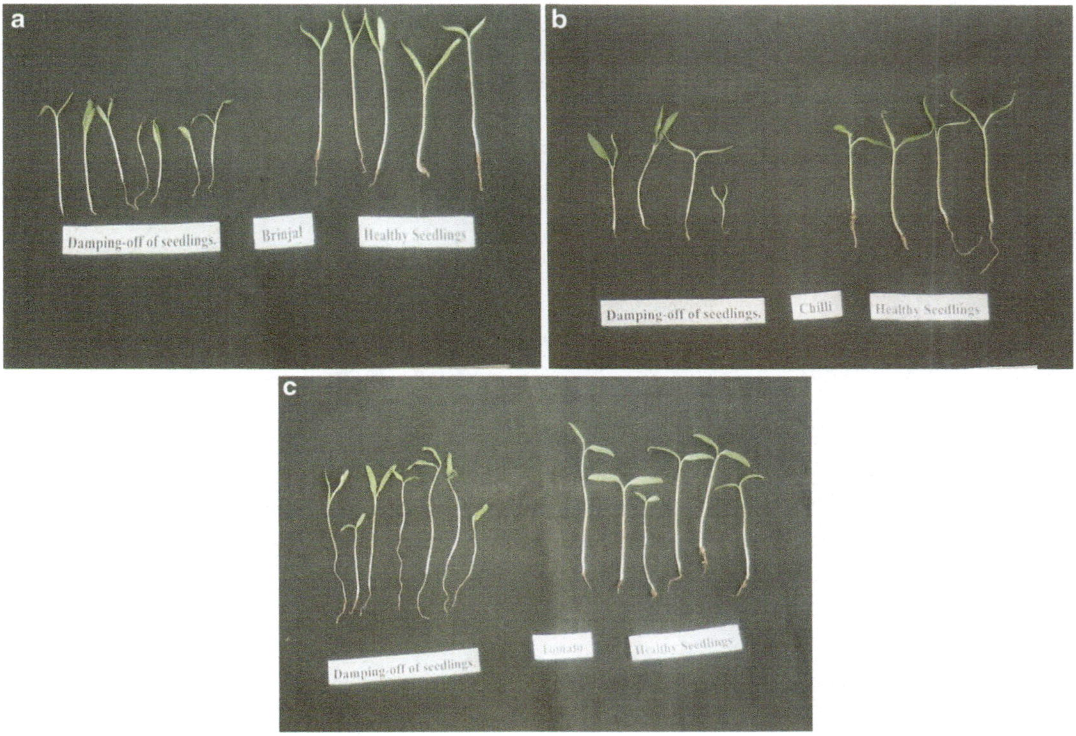

Fig. 4.1 Damping off symptoms in brinjal (**a**), chilli (**b**) and tomato (**c**) seedlings

associated with vegetable crops. But *Pythium* spp. are generally known as damping-off fungi (Chupp and Sherf 1960). These include *P. aphanidermatum* (Eds.) Fitz., *P. debaryanum* (Hesse), *P. butleri* (Subram), *P. ultimum* (Trow) and *P. arrhenomanes* (Drechsler) in which *P. aphanidermatum* is the most common fungus responsible for damping-off disease.

4.3 Disease Cycle and Environmental Relations

Pythium species is soil-borne pathogen and also weak saprophyte and poor parasite. The pathogen perennates in soil through oospores present in plant debris or most commonly through its mycelium (Rangaswami 2002). The mycelial stage of the fungus is capable of infecting the host plant and multiplying very rapidly. The species of *Pythium* enter to the soil through the pre-colonized host residue carrying oospores and sporangia which are the survival structures. When proper host and proper growing condition become available, the pathogen infects the seed and seedlings causing damping-off disease (Singh 1995). Germination of these leads to primary infection of seedlings and the asexual spores later formed carry on the secondary infection and rapid spread of disease (Fig. 4.2).

High soil moisture and relatively high soil temperatures favour rapid development of damping off. High soil moisture makes soil nutrients available to the oospores which germinate and produce zoospores. In the presence of high soil moisture, there is rapid dissemination of the zoospores which attack the seeds and germinating seedlings (Gattani and Kaul 1951). With the help of hydrolytic enzymes, the fungus causes rapid breakdown of the host tissues prior to actual colonization of the tissues (killing in advance).

After invading the host tissues, the fungus rapidly forms oospores. The oospores persist in

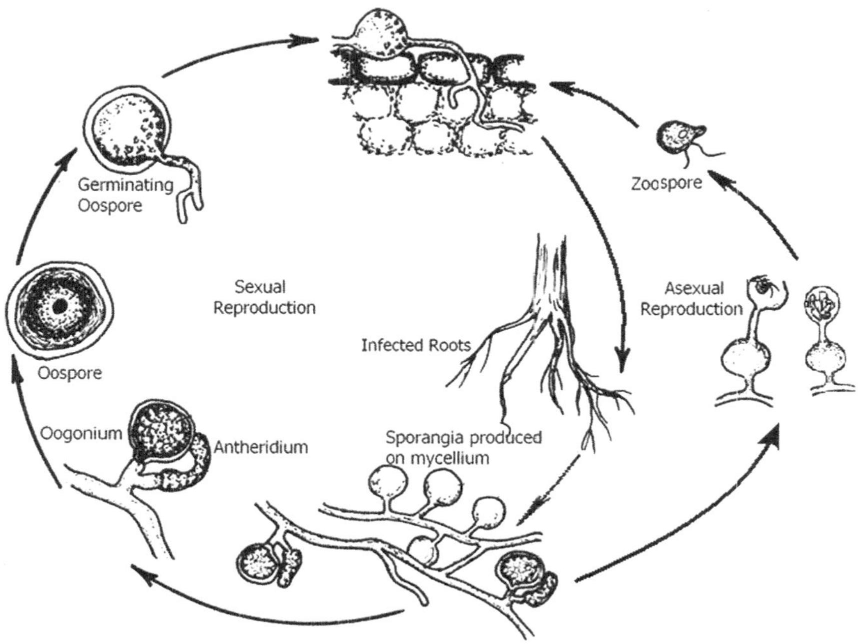

Fig. 4.2 The disease cycle of Pythium damping off in solanaceous vegetable

the soil, resisting adverse condition. When there is sufficient moisture, they germinate and produce the mycelium which later forms the asexual stage of reproduction (Chamount 1979). The zoospores formed in the vesicle of sporangia are commonly released in soil water and spread from place to place. Often, when the climatic conditions are unfavourable for the asexual reproduction of the fungus, sexual reproduction starts, resulting in formation of oospores, which help to survive adverse environmental conditions thus the fungus capable of living for many years in soil, completing its life cycle both saprophytically and as a facultative parasite (Clinton 1920).

4.4 Epidemiology

High soil moisture (90–100 %) with temperatures between 24–30 °C favours the development of the disease. But the disease is most severe in ill-aerated, ill-drained soil. Such conditions are common in compact, heavy soil. Loose soil, having a good proportion of sand, exhibits less loss from the disease. These types of soils are not suitable for the pathogen (Gupta and Paul 2001). The infection is being favoured by poor aeration, narrow spacing, prolongation of juvenile stage of the seedlings and general weakening of the plants under such condition. On the whole, the conditions which predispose the seedlings to damping off are overcrowding of seeds of seedlings, growth under too damp condition, excess of water/moisture in soil and the occurrence of too much decaying organic matter (Linderman 1989).

Nitrogen applications made too early promote damping off. Germinating seed and new seedlings do not need much supplemental nutrition; the endosperm contains sufficient food required by the seedlings to grow nitrogenous fertilizer (Rajan and Singh 1974). An ideal nursery should be located on light, well-drained soil conditions (Singh 2000).

Pythium debaryanum and P. ultimum prefer cool (20 °C) conditions, while *P. aphanidermatum* is more severe at temperatures above 20

°C. Presence of seed or root exudates is known to encourage sporangial germination in *P. aphanidermatum* and *P. irregulare* (Baker and Cook 1974).

4.5 Management Strategies for Disease

4.5.1 Prevention Practices for Disease

The best way to control damping-off disease is to prevent it. There are many prevention techniques, and a combination of them is most effective. Certain cultural and sanitation practices can help in reducing inoculum load of the disease-causing pathogens. It is essential to adopt timely control measures to avoid losses due to pathological problems during emergence (Hewson et al. 1998). The following steps are necessary for successful integrated management to avoid, exclude or eradicate the pathogen from the place of activity.

4.5.1.1 Avoidance of the Pathogen Can Be Done by Adopting following Practices

- Selected area for planting a crop and raising a nursery should be free from pathogenic fungi responsible for pre- and postemergence damping-off disease.
- During summer season it is necessary to plough the soil deep in the open sun to expose and check the growth of propagules of pathogens existing in the soil at high temperatures (Palti 1981).
- One of the most important aspects in the control of damping off is to ensure high-quality seed. Avoid sowing of cracked or injured seeds (Suryanarayana 1978).
- Planting should be done at the right time to discourage the pathogen activity and enhance seed germination and emergence (Cook et al. 1978).
- Persistence of the pathogen can be checked by 2–3 years of crop rotation involving non-host crops. Long crop rotations are helpful in managing the pathogens. Inter- and mixed cropping check the severity of pathogens in soil which attack the crop even at early stages of emergence (Curl 1963).

4.5.1.2 Exclusion of the Pathogen Can Be Done by Seed Treatment, Soil Treatment and Rouging of Diseased Plants

- *Seed treatment:* Systematic fungicides like Vitavax, carbendazim and benomyl may be used as seed treatment when the infection is within the seed (Evans 1971). Contact/nonsystemic fungicides like thiram, captan, mancozeb, organomercurials, etc. are also used as seed treatment for damping-off pathogen (Tripathi and Grover 1978; Baylis 1941; Chupp and Sherf 1960; Harman 1991).
- The pathogens eradicate by physical treatments like hot water treatment at 52 °C, for 30 min. Solar energy treatments are also useful pathogen eradication (Pandey et al. 2002).
- The diseased seed and seedlings should be removed and destroyed as soon as observed (Cook et al. 1978).
- Proper cleaning of the seed is necessary before planting.

4.5.1.3 Eradication of the Pathogen

The pathogens can be eradicated by several methods.

1. *Soil treatment:* Suitable soil treatment can be easily done in limited areas in nurseries which may control some specific pathogens infesting a particular crop. Such treatment will also eradicate inoculum carried on the seed coat (Chaube and Varshney 2003).
2. *Soil fumigation:* Application of formalin (formalin + charcoal ash at 15:85) at 30 g/sq. ft. about 3″ deep in soil is effective.
3. *Soil solarization:* Soil-borne pathogens are eradicated by solarization (Chen and Katan 1980). Soil disinfestations by use of white

transparent polythene in hot summer are one of the most effective approach for management of soil-borne diseases particularly in nursery bed. Nursery bed soil can be mulched by white transparent polythene for 14–30 days. The main objective of soil solarization is to eliminate pathogen, insect, biotic agent and also weeds. Dominant soil pathogens and also damping off causing pathogens *Pythium*, *Phytopthora*, *Fusarium*, *Rhizoctonia* and *Colletotrichum* spp. are effectively reduced in nursery beds (Pandey and Pandey 2004).

Solanaceous vegetable nursery bed soil requires mulching only 14 days by white transparent polythene (Manomohan and Sivaprakasam 1994). The percentage germination of seed was also enhanced (Katan 1980). Soil solarization of nursery beds has been found very effective in reducing the damping off in tomato, chilli and brinjal (Pandey and Pandey 2005).

4.5.2 Management Practices

4.5.2.1 Cultural Practices

Cultural practices, such as thin sowing to avoid overcrowding, use of light soil in the nursery beds, light but frequent watering of the nursery, use of well-decomposed farm yard manure, avoidance of excessive use of nitrate forms of nitrogenous fertilizers, proper drainage and no repetition of the same crop in the same field do not favour pathogen development and thus reduce the chances of damping off (Mehrotra et Aneja 2001; Palti 1981).

Avoidance of nursery sowing in the same bed year after year and also apply crop rotation for 2–3 years with non-host crop (Mukhopadhyay 1994).

Burning of 12 in. thick stack of farm trash over the nursery bed provides partial sterilization of soil and is most common and effective in reducing the pathogen population.

4.5.2.2 Chemical Control

Seed Treatment The incidence of damping off can be reduced by treating the seed with a fungicide prior to sowing. Fungicides will protect the seed from the soil-borne pathogens, but have little effect against the seed-borne fungi. Treating of seed with suitable seed protectants is one of the most effective control measure against pre-emergence stage of damping off. The protectant is applied in dry or wet form to the seed and a layer around the seed coat which keeps the pathogen away until the seed has emerged (Person and Chilton 1942; Pandey et al. 2002).

Seed protectants, *viz.* Agrosan GN (An organo-mercurial), captan, Difolatan, thiram, Bavistin, etc. recommended as dry seed treatment at 2.5–3 g/kg seed to check damping off (Tripathi and Grover 1978; Leach and Smith 1945).

Copper sulphate (10 %) solution as seed soaking treatment also reduced the disease (Shyam 1991).

Ceresan and Semesan recommended as dry seed treatment at 2.5/kg seed to check damping off (Clinton 1920).

Thiram and captan and organomercurial fungicides have proved to be highly effective in controlling seed-borne pathogens (seed rots, damping off of seedlings due to *Pythium*, *Fusarium* spp. etc.) in different vegetables resulting in improved seed germination and crop stand (McCallans 1948).

The fungicide metalaxyl has systemic properties and may be used prior to sowing to reduce populations of *Pythium* and *Phytophthora* in the soil (Sahani et al. 1967).

4.5.2.3 Soil Treatment

1. Soil fumigation:

Periodical soil treatment of the nursery can be done by using formalin dust (15 parts of formalin and 85 parts of charcoal ash) at 30 g/sq. foot (3 in. deep) to check damping off (Gupta and Paul 2001).

2. Drenching of soil:

To control postemergence stage of damping-off disease, soil drenching with fungicides is very effective.

Soil drench with captan, captafol (Difolatan), Blitox-50 and thiram at 0.2–0.5 % gave better control of the disease. But these may prove costly (Tripathi and Grover 1978).

To provide protection from postemergence damping off, nursery should be drenched with captan at 0.2 % or carbendazim 50WP at 0.1 % or Mancozeb 75 WP at 0.25 % (Joseph 1997).

A good all-purpose preventive treatment for damping off is a 50–50 mixture of captan and benomyl which may be applied as a drench. Drenching of nursery soil with 0.3 % captan at 5 l/m^2 before sowing treated seeds provides excellent control of damping off.

4.5.2.4 Biological Control

Bioagents Against Damping of Disease

Fungicides are available to control damping-off disease, but chemical control by seed treatment or soil drenching has several limitations besides being environmentally hazardous (Richardson 1991). Therefore, biological control as a mechanism to reduce soil-borne plant pathogens is gaining importance in recent years due to chemicals that are used to control the diseases are expensive but also contribute to soil pollutants and adversely affect nontarget species (Richardson 1991). Several microorganisms antagonistic to pathogens are available which can be used as seed or seed bed treatment (Fig. 4.3a, b). *Trichoderma viride*, *T. harzianum*, *T. hamatum*, *T. reesei* and *T. koningii* have been reported to control species of *Pythium*, *Rhizoctonia*, *Fusarium*, *Sclerotium rolfsii* (Bagyaraj and Govindan 1996; Papavizas and Lumsden 1980). Damping off due to *Pythium indicum* in tomato was controlled by the application of *T. viride* (Krishnamurthy and Bhaskaran 1994a, b).

Chilli damping off due to *P. aphanidermatum* was controlled by seed treatment with conidia of *T. viride* and *T. hamatum* in nursery (Ramanathan and Sivaprakasam 1994; Hazarika et al. 2000).

Seed and soil treatment of chilli and nursery soil with *T. harzianum* and *T. reesei* effectively controlled damping off and enhanced germination and emergence (Krishnamurthy and Bhaskaran 1994a, b). Vesicular arbuscular mycorrhizae also control soil-borne pathogens and check seed rot and seedling damage problem (Jalali and Thaereja 1981).

Biological seed treatment can be done by priming on seed, coating the seeds, seedling dipping and dry powder treatment depending upon the nature of biocontrol agents. Generally 6–10 g *Trichoderma* for one kg of seed is used for seed treatment but spore concentration should be in between 10^6 and 10^9 ml. Similarly 10–25 g powder should be applied in per m^2 area depending upon the soil type and organic matter. Seed treatment by *Trichoderma* should be along with soil application at 10 g/m^2 for nursery diseases. Some *Trichoderma* spp. are insensitive to fungicides at lower doses of pencycuron, copper hydroxide and captan that can be incorporated while application (Pandey et al. 2002) (Fig. 4.3).

Leaf Extracts Against Damping Off

Plant extracts have also been successfully used to control emergence problems due to pathogens including *P. aphanidermatum* (Jacob et al. 1989). Due to the presence of phenolic substances and resins, gummy and non-volatile substances the plant extracts are effective against *Pythium* spp.

Narayana and Shukla (2001) evaluated the antifungal activity of 37 plants against *P. aphanidermatum* and reported that maximum inhibition (94.4 %) and least postemergence damping off was recorded by *O. paniculata* extracts among all the plants. Tomato seeds soaked in 20 % leaf extract of *Bougainvillea glabra* or *Piper betle* for 6 h before sowing increased germination by 75 % and damping off due to *P. aphanidermatum*, and *S. rolfsii* was also inhibited by the leaf extract of the two plants (Muthuswami 1972). Drenching of soil after sowing, with

Fig. 4.3 Seed and soil treatment with *Trichoderma harzianum* (**a**) and seed and soil treatment with *Trichoderma viride* (**b**) in brinjal

extracts of *Tamarindus indica* and *Leucaena leucocephala*, was also found very effective against damping off due to *P. indicum* (Johenson 1914; Mukhopadhyay et al. 1992).

4.6 Integrated Disease Management

Integrated disease management provides a combination of cultural, biological and chemical tools to control and/or manage crop diseases effectively (Satija and Hooda 1987). Cultural controls keep *Pythium* spp. from reaching the roots, while biological and chemical controls inhibit or suppress *Pythium* spp. in the root zone.

4.6.1 Integration of Biocontrol Agents and Fumigants

Fumigation with dazomet, methyl isothiocyanate/1,3-dichloropropene and mixtures of methyl bromide and chloropicrin effectively decreases the populations of *Pythium*. Moreover, several biocontrol agents have also been used successfully for the control of damping off (Urech et al. 1977). Strashnow et al. (1985) reported that under greenhouse condition the combined treatment of *T. harzianum* (equivalent to 200 kg/ha) with lower dose of methyl bromide completely controlled disease incidence of *R. solani* in bean seedlings. Under field condition the combination of *T. harzianum* (200 kg/ha) and methyl bromide gave significant synergistic effect on damping off of carrot seedlings caused by *R. solani*.

4.6.2 Integration of Biocontrol Agents and Fungicides

It has been observed that in certain cases the biocontrol agents or the chemicals alone could not provide satisfactory result for the management of a particular soil-borne disease (Sokhi and Thind 1996a, b). However, the integration of biocontrol agents with certain compatible chemicals may give synergistic effect and provide better disease control than either treatments alone (Sokhi and Thind 1996a, b). In radish, the conidial suspension of *T. harzianum* and benodanil was found effective to minimize pre-emergence damping off caused by *R. solani*. The control of damping off by both seed treatment with *T. harzianum* and soil mix of benodanil was additive but not interactive (Lifshitz et al. 1985). Bacterial species, *viz. Pseudomonas cepacia*, *P. fluorescens* and *Corynebacterium* species as seed dressing in combination with captan provided effective control of the damping off and root rot of peas caused by *Pythium* and Aphanomyces (Parke et al. 1991). The control of damping off of tomato seedlings by both seed treatment with *T. harzianum* and drenching soil with fungicides is suitable control measure (Rakesh and Indra 2007).

4.6.3 AM Fungi and *Azospirillum* in Suppression of Damping Off

AM fungi are known to colonize a number of tropical plants including vegetables. AM association is known to help in the growth of various crops like carrot, tomato, etc. (Sasal 1991). Reddy et al. (2006), reported that G. *fasciculatum* proved as the benefactor and enhanced the plant growth, nutrition status, yield and reduction in disease severity. The dual inoculations (mycorrhiza with fungal pathogen) showed significant suppression in the progression of the pathogen and consequently reduction in the severity of the damping off. Mosse (1973) indicated that mycorrhizal colonization induce chemical, physiological and morphological alterations in the host plant which may result in an increase in host resistance. The potential in biological suppression of soil-borne pathogens gives a wider vision of AM fungi, in that they act as an alternative strategy for the host plant in conditions that are deleterious to root growth. Therefore, the introduction and consequent management of such symbiotic colonization could be employed for the advantage of the crop.

Due to the imposition of competition with the pathogen for space, nutrition and host photosynthates (Harley and Smith 1983) or the alterations of the physiology of the host which induces host defence mechanisms (Schenck and Kellam 1978), the arbuscular mycorrhizal fungi (AMF) are able to suppress the damping-off disease. Suppression of damping off by AMF has already been reported in cucumber and ginger (Joseph 1997).

Chilli seedlings pre-inoculated with native AMF recorded the least percent of disease incidence of 22.3, and this was significantly superior as against the control which recorded 65.1 % disease incidence (Kavitha et al. 2003). Dual inoculation of AMF along with *Azospirillum* also reduced damping off by 72.7 % over control. This may be due to the effect of *Azospirillum* and AMF interaction which makes the plant healthier by way of enhanced uptake of nutrients and trigger the host defence mechanism.

4.6.4 Chemical Control

Although fungicides are slowly taking the back seat in our fight against plant pathogens in view of some associated adverse effects on the environment, their role in managing several devastating plant diseases cannot be overlooked. Prevent *Pythium* diseases by practicing integrated disease management strategies based on cultural and biological controls. Use fungicides as a last resort at the onset of disease (Thind 2007). Fungicides have been commonly used for the control of plant diseases all over the world since the nineteenth century and in many cases have become an integral component of our crop production system (Gupta and Bilgrami 1970). Among different methods of disease control, host resistance is still the most preferred choice, although lack of its durability has been a persistent drawback. Fungicides, despite certain drawbacks, are considered to play a significant role in containing losses due to plant diseases in the coming years (Mukhopadhyay 1994).

Effectiveness of fungicides to control damping off is highly variable (Govindappa and Grewal 1965). Several fungicides are registered for use in vegetable nurseries to control soil-borne diseases. Thiram and captan and organo-mercurial fungicides have proved to be highly effective in controlling seed-borne pathogens (seed rots, damping off of seedlings due to *Pythium*, *Fusarium* spp., etc.) in different vegetables resulting in improved seed germination and crop stand. Phenylamide fungicides have a unique potential of curbing plant pathogens belonging to the class oomycetes such as *Pythium*, *Phytopthora*, etc. Metalaxyl came as a landmark discovery and a breakthrough was achieved in the effective control of diseases caused by oomycetes. The fungicide metalaxyl has systemic properties and may be used prior to sowing to reduce populations of *Pythium* and *Phytophthora* in the soil (Urech et al. 1977).

Other fungicides having different chemistry but similar anti-oomycete activity spectrum such as prothiocarb and propamocarb (carbamates), hymexazol (isoxazoles), etc. Hymexazol is used as seed dressing or soil drench to control soil-borne *Pythium* spp. and *Aphanomyces* spp. in various vegetables. Pre-emergence damping off of cabbage and cauliflower can be controlled by seed treatment with apron-70 (White et al. 1984) or thiram (Sandhu and Gill 1983; Grewal and Singh 1965).

The first post-plant fungicide application should be made when most seedlings have emerged and the seeds begin to drop from cotyledon leaves. If frequent applications of fungicides are planned, alternation of the captan-benomyl mix with other fungicides is advised to minimize the build-up of resistant pathogens. But in view of the resistance risk associated with most of the systemic, site-specific compounds, there is need to develop and employ an ideal IDM module ensuring integration of cultural, chemical and biological means of disease management options in a holistic manner.

References

Alexander LJ (1931) The causes and control of damping-off of tomato, brinjal and chilli seedlings. Ohio Agric Exp Stat Bull 496:57–61

Atkinson GF (1895) Damping-off. N Y (Cornell) Agric Exp Stat Bull 94:233–272

Bagyaraj DJ, Govindan M (1996) Microbial control of fungal root pathogens. In: Mukerji KG (ed) Advances in botany. APH Publishing Corporation, New Delhi, pp 293–321

Baker KF, Cook RJ (1974) Biological control of plant pathogens. W. H. Freeman, San Francisco, pp 433–437

Baylis GTS (1941) Fungi which control pre-emergence injury to garden peas. Ann Appl Biol 28:210–218

Brien RM, Chamberlain EE (1937) Seedling damping-off. N Z Agric J 54:321–327

Chamount JP (1979) Inhibitions of pathogens of field legumes of mimosine. Bulletunde la Societe Botaque de France 126:537–541

Chaube HS, Varshney S (2003) Management of seed rot and damping-off diseases. In: Compendium training programme of CAS in plant pathology. GBPUAT, Pantnagar, pp 281–288

Chen Y, Katan J (1980) Effect of solar heating of soils by transparent polythene mulching on their chemical properties. Soil Sci 130:271–277

Chupp C, Sherf AF (1960) Vegetable diseases and their control. Ronald Press, New York

Clinton GP (1920) Damping-off. N Y (Cornell) Agric Exp Stat Bull 222:475–476

Cook RJ, Boosalis MG, Doupnik B (1978) Influence of crop residues on plant diseases. In: Crop residue management system. Ann Soc Agron 31:147–163

Curl EA (1963) Control of plant diseases by crop rotation. Bot Rev 29:413–479

Evans E (1971) Systemic fungicides in practices. Pestic Sci 2:192–196

Fageria MS, Choudhary BR, Dhaka RS (2003) Vegetable crops. production technology. Kalyani Publishers, Ludhiana, pp 25–74

Gattani ML, Kaul TN (1951) Damping off of tomato seedlings, its cause and control. Indian Phytopathol 4:156–161

Govindappa MH, Grewal JS (1965) Efficiency of different fungicides in controlling damping off of tomato. Indian J Agric Sci 35:210–215

Grewal JS, Singh RP (1965) Chemical treatment of seed and nursery bed to control damping off of cabbage and chilli. Indian Phytopathol 18:225–228

Gupta SC, Bilgrami RS (1970) Proc Natl Acad Sci India 40:6–8

Gupta VK, Paul YS (2001) Fungal diseases of tomato, chilli and brinjal. In: Gupta VK, Paul YS (eds) Diseases of vegetable crops. Kalyani Publishers, Ludhiana,pp 87–101

Harley JL, Smith SE (1983) Mycorrhizal symbiosis. Academic, London

Harman GE (1991) Seed treatment for biological control of plant diseases. Crop Prot 10:166–171

Hazarika DK, Sharma P, Paramanick T, Hazarika K, Phookan AK (2000) Biological management of tomato damping–off caused by *Pythium aphanidermatum*. Indian J Plant Pathol 18:36–39

Hewson RT, Sagenmuller A, Scholz-Tonga EM, Fabretti JP, Lobo SDAG, Blanquat ADE, Schreuder R (1998) Global integrated crop management success stories. In: Brighton crop protection conference, pests and diseases. Proceeding of international conference, vol 1, Brighton, UK, pp 161–168

Jacob CK, Sivaprakasam K, Jayarajan R (1989) Evaluation of some plant extracts and fungal antagonists for the biological control of pre and post emergence damping off of brinjal. J Biol Control 23:18–23

Jalali BL, Thaereja ML (1981) Management of soil borne plant pathogens by vesicular-arbuscular mycorrhizal model system. Indian Phytopathol 34:115–116

Johenson J (1914) The control of damping–off disease in plant bed. Wisconsin Agric Exp Stat Bull 31:29–61

Joseph PJ (1997) Stidies on damping-off of vegetable. Ph.D. thesis, Kerala Agricultural University, Thrissur, p 192

Katan J (1980) Solar pasteurization of soils for disease control-status and prospects. Plant Dis 64:450–454

Kavitha A, Meenakumari KS, Sivaprasad P (2003) Effect of dual inoculation of native arbuscular mycorrhizal fungi and Azospirillum on suppression of damping off in chilli. Indian Phytopathol 56:112–113

Krishnamurthy AS, Bhaskaran R (1994a) Biological control of damping off disease of tomato caused by *Pythium indicum* Balakrishnan. J Biol Control 4:52–54

Krishnamurthy AS, Bhaskaran R (1994b) *Trichoderma viride* in the control of damping off disease of tomato caused by *Pythium indicum*. In: Sivaprakasam K, Seetharaman K (eds) Crop diseases-innovative techniques & management. Kalyani Publishers, Ludhiana, pp 199–203

Leach LD, Smith PG (1945) Effect of seed treatment on protection, rate of emergence, and growth of garden peas. Phytopathology 35:191–206

Lifshitz R, Lifshitz S, Baker R (1985) Decrease in incidence of *Rhizoctonia* pre-emergence damping off by the use of integrated chemical and biological controls. Plant Dis 69:431–434

Linderman RG (1989) Organic amendments and soil-borne disease. Can J Plant Pathol 11:180–183

Manomohan D, Sivaprakasam K (1994) Biological control of damping off disease in chilli nursery. In: Sivaprakasam K, Seethraman K (eds) Crop diseases. Innovative techniques and management. Kalyani Publishers, Ludhiana, pp 215–243

McCallans SEA (1948) Evaluation of chemical as seed protectants by greenhouse tests with peas and other seeds. Contrib Boyce Thompson Inst Bull 15:91–117

Mehrotra RS, Aneja KR (2001) An introduction to mycology. New Age Int. (P) Ltd., Publication, New Delhi, pp 107–194

Mosse B (1973) Plant growth responses of vesicular-arbuscular mycorrhizae. New Phytol 72:127–136

Mukhopadhyay AN (1994) Bio-control of soil borne fungal plant pathogens: current status, future prospect and potential limitations. Indian Phytopathol 47:119–126

Mukhopadhyay AN, Shreshta SM, Mukherge PK (1992) Biological seed treatments for control of soil borne plant pathogen. FAO Plant Prot Bull 40:110–115

Muthuswami M (1972) Studies on damping off of tomato incited by *Pythium aphanidermatum* (Edson) Fitz. M. Sc. (Ag) thesis TNAU Coimbatore, pp 110

Narayana Bhat M, Shukla BK (2001) Evaluation of some leaf extracts against *Pythium aphanidermatum in vitro* and pot culture. Indian Phytopathol 54(3):395–397

Palti J (1981) Cultural practices and infectious crop diseases. Springer, Berlin

Pandey KK, Pandey PK (2004) Effect of soil solarization on disease management in vegetable nursery. J Mycol Plant Pathol 34:398–401

Pandey KK, Pandey PK (2005) Differential response of biocontrol agents against soil pathogens on tomato, chilli, and brinjal. Indian Phytopathol 58(3):329–331

Pandey KK, Pandey PK, Satpathy S (2002) Integrated management of diseases and insects of tomato, chilli and cole crops. I.I.V.R., Varanasi. Technical Bulletin no. 9. pp 22

Papavizas GC, Lumsden RD (1980) Biological control of soil–borne fungal propagules. Annu Rev Phytopathol 18:389–412

Parke JL, Ramd RE, Joy AE, King EB (1991) Biological control of *Pythium* damping off and *Aphanomyces* root rot of peas by application of *Pseudomonas cepacia and P. fluorescens* to seed. Plant Dis 75:987–992

Person LH, Chilton JP (1942) Seed and soil treatment for the control of damping off. J Agric Res 75:161–179

Rajan KM, Singh RS (1974) Effect of fertilizers on population of *Pythium aphanidermatum,* associated microflora and seedling stand of tomato. Indian Phytopathol 27:62–69

Rakesh K, Indra H (2007) Integrated management of damping off of tomato caused by *Pythium aphanidermatum*. J Mycol Plant Pathol 37(2):259–264

Ramanathan A, Sivaprakasam K (1994) Effect of seed treatment with plant extract, antagonist and fungicide against damping off disease in chilli. In: Sivaprakasam K, Seetharaman K (eds) Crop diseases innovative techniques and management. Kalyani Publishers, Ludhiana, pp 221–226

Rangaswami G (2002) Diseases of crop plants in India. Prentice Hall of India Pvt. Ltd., New Delhi, pp 286–334

Reddy BN, Raghavender CR, Sreevani A (2006) Approach for enhancing mycorrhiza-mediated disease resistance of tomato damping-off. Indian Phytopathol 59(3):299–304

Richardson ML (1991) Chemistry agriculture and environment. The Royal Society of Chemistry, Cambridge, p 546

Sahani ML, Singh RP, Singh R (1967) Efficacy of some new fungicides in controlling damping off of chillies. Indian Phytopathol 20:114–117

Sandhu KS, Gill SPS (1983) Effect of seed and soil treatments on damping off of cauliflower. Pesticides 16(11):17–18

Sasal K (1991) Effect of phosphate application on infection of vesicular-arbuscular mycorrhizal fungi on some horticultural crops. Sci Rep Miyagi Agric Coll 39:1–9

Satija DV, Hooda I (1987) Greenhouse evaluation of seed treatment with fungicides for the control of tomato and chilli damping off. Indian Phytopathol 40:222–225

Schenck NC, Kellam MK (1978) The influence of vesicular-arbuscular mycorrhizae on disease development. Florida Agric Exp Stat Tech Bull 16:798

Shyam KR (1991) Important disease of summer and winter vegetable crops and their management. In: Jain YC et al (ed) Horticultural techniques and post harvest management. Directorate of Extension Education, UHF, Nauni, pp 62–69

Singh RP (1995) Plant pathology. Central book depot, Allahabad, pp 187–193

Singh RS (2000) Plant disease. Oxford & IBH publishing Co., New Delhi, pp 173–177

Sohi HPS (1982) Studies on fungal diseases of Solanaceous vegetables and their control. In: Advances in mycology and plant pathology. Oxford & IBH, Calcutta, pp 19–30

Sokhi SS, Thind TS (1996a) Application of fungicides in integrated disease management in vegetables. In:

Rational use of fungicides in plant disease control. NATIC, Ludhiana, pp 133–142
Sokhi SS, Thind TS (1996b) Integration of fungicides with biocontrol agents in plant disease control. In: Rational use of fungicides in plant disease control. NATIC, Ludhiana, pp 183–194
Sowmini R (1961) Damping-off of seedlings and its management. M.Sc. (Ag.) thesis, University of Madras
Strashnow Y, Elad Y, Sivan A, Chet I (1985) Integrated control of *Rhizoctonia solani* by methyl bromide and *Trichoderma harzianum*. Plant Pathol 34:146–151
Suryanarayana D (1978) Seed pathology. Vikas Publication, New Delhi
Thamburaj S, Singh N (2003) Vegetable, tubercrops and spices. Indian Council of Agriculture Research, New Delhi, pp 10–75
Thind TS (2007) Changing cover of fungicide umbrella in crop protection. Indian Phytopathol 60(4):421–433
Tripathi NN, Grover RK (1978) Comparison of fungicides for the control of *Pythium butleri* causing damping off of tomato in nursery beds. Indian Phytopathol 30:489–496
Urech PA, Schwinn F, Staub T (1977) A novel fungicide for the control of late blight, downy mildews and related soil-borne diseases. Proc Br Crop Prot Conf Pests Dis 42:623–631
White JG, Crute IR, Wynn EC (1984) A seed treatment for the control of *Pythium* damping off diseases and *Peronospora parasitica* on brassicas. Ann Appl Biol 104:241–247
Wick RL (1998) Damping-off of bedding plants and vegetables. Department of Plant Pathology, Fernald Hall, University of Massachusetts, Amherst

Downy Mildew of Cucurbits and Their Management

5

C.P. Khare, J.N. Srivastava, P.K. Tiwari, A. Kotesthane, and V.S. Thrimurthi

Downy mildew (*Pseudoperonospora cubensis*) is an extremely destructive disease of cucurbits, and it was first reported from Cuba in 1868. During 1985–1988, epidemics of downy mildew (*Pseudoperonospora cubensis*) were recorded on cucumbers in Poland grown under plastic and in the field condition. It has now been reported from Japan, England, Brazil, New Jersey, Africa, etc. In India it occurs on all the cultivated cucurbits (Rondomanski and Wozniak 1989).

It occurs practically on members of Cucurbitaceae and mostly those which are cultivated, although it has been observed on the wild cucumbers and few other weed hosts (Doran 1932). In India, it is present all over the country causing heavy damage on muskmelon, watermelon, cucumber, sponge gourd, and ridge gourd but less destructive on bottle gourd, pumpkin, vegetable marrow, etc. (Gangopadhyay 1984).

The disease is prevalent in warm temperate and tropical regions of the world with abundant moisture. About 61 % reduction in crop yield has been recorded in cucurbits due to early infection of downy mildew, late infection being less harmful. The disease is confined mostly to the leaves. The loss of foliage result of early infection precludes normal flower set and fruit development. The fruits of infected plants resulting from the loss of foliage may have poor quality, viz., fail to get proper color, are tasteless, and look sunburnt (Gupta et al. 2001).

5.1 Symptoms

The infected leaves have green to yellowish and remaining dark green areas reflecting a mosaic like pattern, and there are symptoms characterized as development of irregularly shaped yellowish spots on the upper leaf surface. The spots quickly turn distinctly angular bounded by veins, become yellow and then necrotic, and also increase in number and size (Gupta et al. 2001). If the leaf is examined on the opposite side when dew or rain is present, the brown lesion will be covered, or at least bordered, by a pale gray to purple fungus growth (sporangia and sporangiophore). Later, the severely infected leaves become chlorotic, turn brown, and shrivel. In rainy (humid) weather, the entire vein is killed and showed wilt symptoms and injury to plants; a whole may be great enough to cause severe stunting and death. The infection results in the reduction in number and size of fruits and prevents fruit maturation, causing the fruit to have poor flavor. Pathogens are overwintering as active mycelium on either cultivated or wild cucurbits (Bains and Jhooty 1976).

C.P. Khare • P.K. Tiwari • A. Kotesthane • V.S. Thrimurthi
Division of Plant Pathology, Indira Gandhi Agricultural University, Raipur, Chhattisgarh, India

J.N. Srivastava (✉)
Department of Plant Pathology, Bihar Agricultural University, Sabour, Bhagalpur, Bihar, India
e-mail: j.n.srivastava1971@gmail.com

L.P. Awasthi (ed.), *Recent Advances in the Diagnosis and Management of Plant Diseases*,
DOI 10.1007/978-81-322-2571-3_5

Symptoms appear on the upper surface of the leaves as angular yellow spots. During favorable condition, the underside of these spots is covered with a grayish moldy growth. The yellow spots on the upper portion of the leaves appear just like indefinite mosaic pattern lesion on the upper surface. As the spots enlarge, a general yellowing of the leaves occurs followed by the death of the tissue. The leaves subsequently wither and die (Thamburaj and Singh 2005).

5.2 Causal Organism

Pseudoperonospora cubensis (Berk. and Curt.) Rostow. is an obligate parasite causing downy mildew of cucurbits. The mycelium of *Pseudoperonospora cubensis* is hyaline, coenocytic, and intercellular with small, ovate haustoria (Charles 1998). The mycelium produces long, branched sporangiophores which come out through stomata on the lower surface of the leaf. Sporangiophore is dichotomously or monochotomously branched with acute angle at the tip and also arises in groups of 1–5. Sporangia are lemon shaped, grayish to purple in color, ovoid to ellipsoidal, and thin walled and have an apical papilla at distal end. They measure 21–39 × 14–23 μm in size (Singh and Thind 2005).

Badadoost (2001) stated that the sporangia are lemon shaped, colored, and borne on the gracefully curved and pointed tips of branched sporangiophore. These are wind-borne and can successfully disperse to long distances if the air is moist. Sporangia give rise to biflagellate zoospores which swim and germinate producing germ tubes to penetrate the host leaf. Zoospores are biflagellate and 12–13 μm in diameter, and oospores are globose, yellow or hyaline, and 23–25 mm in size, but production of oospores is extremely rare.

5.3 Epidemiology

Perpetuation of pathogens occurs through mycelium or sporangia from one season to next season, while oospores are not common. In India, agroclimatic conditions are not favorable throughout the year. But the fungus is able to survive on wild cucurbits. The fungus survives between cropping periods and sporangia survive cold weather (Sherif and Macnab 1986). Sporangia can survive below freezing temperature (−18 °C) for 3–4 months and may serve as resting structure in the absence of oospores (Lange et al. 1989). Sporangia surviving on the greenhouse crops may form the primary source of infection to the later sown crops in the field (Bains and Jhooty 1978; Jhooty et al. 1989). Sporangia can be disseminated by water, splashed by rain, or carried by cucumber beetles.

The pathogen produces a germ tube in the germination process, and that germ tube penetrates through the stomata of the host surface and causes infection. The pathogen can infect the plants at temperatures between 10 and 27 °C, with optimum day temperature of 25–30 °C and night temperature of 15–21 °C (Ullasa and Amin 1988). Sporulation and infection are arrested above 35 °C, but the fungus can survive for several days at that temperature. Relative humidity of more than 75 % is conducive for disease development (Cohen and Rotem 1970).

Initiation and further progress of the disease depend mainly on moisture, temperature having a second negative effect on infectivity of airborne sporangia (Mahrishi and Siradhana 1988a, b). A film of moisture is necessary on the leaf surface for the infection to occur. The environmental conditions triggering epidemics had been determined to be leaf wetness from 22.00 until 10.00 h and a temperature of 15 °C for at least 6 h (Lehmann 1991). Disease severity was positively correlated with rainfall at 7 and 8–14 days before disease occurrence, but negatively correlated with average RH (Tsai et al. 1992).

Disease progress was highest between mid-August and September when the maximum temperature was 32–35 °C and the minimum temperature was 21–25 °C and RH 75–93 % (Gandhi et al. 1996).

Major changes induced in the resistant cultivars include heavy deposition enrichment with lignin-like material which encases the penetrating haustoria. Containment of the host cells and haustoria by such materials interrupts the flow of nutrients from and into the invaded cells (Cohen et al. 1989).

Ma et al. (1990) reported that the pathogen, *Pseudoperonospora cubensis*, is inhibited by high temperature. The damage caused by downy mildew in crops grown in plastic houses can be mitigated by closing doors and ventilation openings to allow the temperature to rise to 40–47 °C for 2 h every other day.

According to Palti and Rotem (1973), downy mildew epidemics have been reported under semiarid condition because of agriculture practices such as irrigation, favorable microclimate, and abundant pathogen reproduction. Hot and dry weather had a significant influence on the spread of disease in the fields. Huang et al. (1989) reported that increasing rainfall caused downy mildew as epidemic. Epidemics over a large area can result on account of multiple infections appearing uniformly, all over the fields, under favorable weather condition. Proximity of a given field to a source of inoculum was an important factor in the outbreak of disease epidemic (Cohen and Rotem 1970).

5.4 Management Practices

5.4.1 Chemical Control

Fungicide sprays are recommended for all cucurbits. Spray programs for downy mildew on any cucurbit are most effective when initiated prior to the first sign of disease because once downy mildew occurs in a planting, it becomes increasingly difficult for fungicides to control downy mildew. Both systemic and protectant fungicides are used for control of downy mildew. These can also be easily managed by spray of conventional fungicides as recommended against downy mildew.

The losses caused by *Pseudoperonospora cubensis* depend on growth stage at infection, rate of foliage growth, and pathogen development (Palti and Cohen 1980).

Wu (1994) reported that seed treatment, reduced RH, high-temperature treatment, and fungicides were used to control downy mildew (*Pseudoperonospora cubensis)*, anthracnose (*Colletotrichum orbiculare*), and sclerotinia rot (*S. sclerotiorum*) of cucumbers in the greenhouse.

Firstly, collect seed from disease-free fruit. If disease appeared on the crop, spray the crop with mancozeb at 0.25 % (2.5 g/L water) or zineb at 0.25 % (2.5 g/L water), copper fungicide at 0.3 % (3 g/L water) or chlorothalonil at 0.25 % (2.5 g/L water), or metalaxyl + mancozeb at 0.25 % (2.5 g/L water); repeat at weekly interval keeping in view the wet weather condition. Thorough spacing is needed in ensuring coverage of the undersurface of leaves as well (Anonymous 2006).

Treat seeds with Agrosan GN or Emisan at 2.5 g/kg of seed before sowing (Saha 2002). Spray the plants with Dithane M-45 or Indofil M-45 at 0.25 % (2.5 g/L water) or Dithane Z-78 at 0.3 % (3 g/L water) or Daconil at 0.2 % (2 g/L water) at a 7–10-day interval starting from the first appearance of the disease (Saha 2002).

The most effective control of downy mildew (*Pseudoperonospora cubensis)* was achieved with Arcerid [metalaxyl + Polykarbacin (metiram)], zineb and mixtures of Ridomil [metalaxyl] with Cuprosan (pyrifenox), and copper oxychloride, applied just in the appearance of the first symptoms (Chaban et al. 1990), while sprays of Ridomil plus M-45 (copper oxychloride + metalaxyl) at 0.5 %, Ridomil plus 48 (copper oxychloride + metalaxyl) at 0.3 %, Ridomil MZ-72 (mancozeb + metalaxyl) at 0.25 %, Mikal Cu (fosetyl) at 0.6 %, and Sandofan (copper oxychloride + oxadixyl) at 0.25 % gave the best control against *Pseudoperonospora cubensis* (Manole et al. 1990).

Mah (1988) tested five fungicides (metalaxyl-mancozeb, triforine, carbendazim, cyperal, and diathionon-copper) for their effectiveness in controlling cucumber downy mildew. Metalaxyl-mancozeb gave very good control of the disease, followed next in order of effectiveness by cyperal.

Weit and Neuhaus (1990) found that Ridomil and zineb (metalaxyl + zineb) gave best control, and recommendation for the use of this fungicide is given in order to minimize the risk of occurrence of resistance of the fungus *P. cubensis* to metalaxyl. Mixtures of mancozeb (0.1 %) and metalaxyl (0.05 %) spray initiated at conductive weather and repeated at a 10-day interval are very effective. Metalaxyl-resistant strains exhibit cross-resistance to other acylalanine fungicides. Use of metalaxyl has been abandoned in some countries for the above reasons. Metalaxyl-sensitive or metalaxyl-tolerant strains of *P. cubensis* are controlled by dimethomorph on cucumber and melons.

Thind et al. (1991) reported that in laboratory pot house and field studies, mancozeb at 0.3 % provided good control of *Pseudoperonospora cubensis* when used as a protectant but failed to check established infections even when applied only 24 h after inoculation with a sporangial suspension. Formulations of acylalanines, e.g., Ridomil MZ (metalaxyl + mancozeb), Galben M8-65 (benalaxyl + mancozeb), and fosetyl-aluminum, showed good protectant and eradicant activity under artificial (laboratory) and natural (field) conditions. Acylon, Pulsan, and Captan were similarly effective. Even after a gap of 15 days between treatments, these fungicides checked the disease. Ridomil MZ at 0.25 % had the longest persistence and best eradicant action, no disease developing even when application was delayed for 48 h after inoculation.

Golyshin et al. (1994) and Gonzalez et al. (1992) tested Acrobat 50 % (dimethomorph) singly or in combination with contact fungicides against *Pseudoperonospora cubensis* on cucumber. Dimethomorph + mancozeb or a tank mixture of dimethomorph + Daconil (chlorothalonil) (3–5 applications) recommended for disease control and preventive sprays recommended for control of primary infection.

Two or three sprays of Ridomil MZ-72 (metalaxyl) resulted in less downy mildew disease intensity, whereas sprays of folpet, Bordeaux mixture, and Aliette (fosetyl) + mancozeb were also effective. Treatments also gave higher monetary result and increase yield (Gaikwad and Karkeli 1994).

High-percentage control of *P. cubensis* on cucurbitaceous crops was achieved by spray applications of fosetyl-aluminum + folpet and using a tank mixture of fosetyl-aluminum with mancozeb, propineb, and zineb (Yucel and Gncu 1994).

Ilkweon et al. (1996) found that when fosetyl was applied four times from the start of the disease (downy mildew) at a 10-day interval, a yield index of 161 % was obtained.

Aliette (fosetyl) provided good control of downy mildew (*Pseudoperonospora cubensis*) in field and greenhouse cucumbers (Merz et al. 1995).

Brunelli and Collina (1996) reported that among copper oxychloride, copper hydroxide, anilazine, chlorothalonil, and fosetyl, chlorothalonil gave the best control against *Pseudoperonospora cubensis*, followed by fosetyl, which was less persistent, while the other products gave mediocre control.

Egan et al. (1998) reported that RH-7281 is a new, high performance fungicide currently under development for foliar use to control downy mildew.

Fugro et al. (1997) found that of the six fungicide treatments tested, fosetyl (as Aliette) gave the best disease control. Mancozeb, chlorothalonil, copper oxychloride, carbendazim, and mancozeb + fosetyl all controlled the disease to different degrees.

Mercer et al. (1998) reported that RPA 407213 combination with fosetyl-Al was highly active against *P. viticola*, *Pseudoperonospora cubensis* (on Cucurbitaceae), and *Peronospora parasitica* (on Brassicaceae).

Santos et al. (2003) observed that downy mildew caused by *Pseudoperonospora cubensis* is the main disease affecting melon fruits. Different intensities of the disease were achieved by spraying the following fungicide mixtures: methyl thiophenate (thiophenate-methyl) + chlorothalonil or metalaxyl + mancozeb. There was a significant reduction in fruit yield when the disease started at 24 and 36 days after planting, but when the disease started at 47 days, no effect in production was observed.

The fungicidal mixture metalaxyl + mancozeb was highly effective in controlling downy mildew in both dry and rainy seasons, while chlorothalonil + methyl hyphenate was effective only during the dry season (Santos et al. 2004). ICI A5504, a beta-methoxyacrylate compound, is particularly effective on Cucurbitaceae, providing unique control to both *P. cubensis* and *Sphaerotheca fuliginea* causing powdery mildew.

Ullasa and Amin (1988) observed that the incidence of *P. cubensis* was most effectively reduced by Daconil (chlorothalonil) followed by

Dithane M-45 (mancozeb) in field; 3–5 sprays of tank mixtures of dimethomorph + mancozeb or dimethomorph + chlorothalonil are recommended for checking primary infection. Weekly sprays of chlorothalonil or mancozeb give good control as protectants but fail to check established infection.

Daconil (chlorothalonil) and copper oxychloride gave effective disease control (Ullasa and Amin 1988).

When epidemic of downy mildew developed in susceptible cultivar, chlorothalonil and mancozeb were most effective in lowering the infection rate and reducing disease severity (Summer et al. 1981).

Motte et al. (1988) found that the control of downy mildew includes cultural measures to prevent outbreaks and chemical control measures using mancozeb 80 or Ridomil (metalaxyl) + zineb.

Rondomanski and Zurek (1988) recorded that the fungicides tested for control of downy mildew, mancozeb + metalaxyl (as Ridomil MZ-58), were the most effective.

Mahrishi and Siradhana (1988a, b) stated that sprays of Dithane M-45 (mancozeb) at 0.2 % decreased disease intensity, with three and four sprays at 10-day intervals or five sprays at 7-day intervals.

Mancozeb, mancozeb + metalaxyl, mancozeb + oxadixyl, and chlorothalonil fungicides were effective for disease control (Rondomanski and Wozniak 1989).

Bedlan (1990) reported that Galben M 8–65 (benalaxyl + mancozeb) is specifically approved for cucumber downy mildew, but it is not very effective.

Metalaxyl + mancozeb and chlorothalonil are the most effective fungicides for treating seedlings before inoculation (Tsai et al. 1992).

Khalil et al. (1992) reported that all five fungicides tested gave adequate control of this disease, but the most effective treatments were Trimeltox forte (200 g/100 L), Vitigran Blue [copper oxychloride] (300 g), and Dithane M-45 [mancozeb] (150 g), reducing disease intensity to 8.62, 7.75, and 9.25 %, respectively, compared with 92.5 % in the untreated plot, and increasing yields to 12.67, 14.87, and 17.79 t/ha. Ramirezarredondo (1995) conducted an experiment for chemical control of downy mildew in squash and found that oxadixyl + mancozeb (0.3 + 1.6 g/L), metalaxyl + mancozeb (0.2 + 1.4 g/L), and mancozeb (2.4 g/L) gave the best disease control and yield (2223, 1997, and 1919 exportation boxes/ha, respectively) compared with control treatment (1706 boxes/ha).

A study was conducted to determine the most effective treatment for controlling downy and powdery mildew (caused by *Pseudoperonospora cubensis* and *Erysiphe cichoracearum*, respectively) on bitter gourd (*Momordica charantia*), and among all treatments, eight sprays of 0.3 % copper oxychloride + 0.3 % wettable sulfur at 10-day intervals from 30 days after crop sowing were the most effective for the control of downy and powdery mildew diseases of bitter gourd during the rainy season, recording the highest yields and economic returns (Memane and Khetmalas 2003).

5.4.2 Cultural Practices

Destroying of wild cucurbits from vegetable-growing areas also helps to minimize the disease. Wide spacing between plants and conditions which do not favor high humidity in the microclimate and expose plants to sunlight reduce the disease. Management practices including irrigation should be used to reduce the relative humidity and also the leaf wetness, thereby reducing the chances for downy mildew development and its further spread. Potassium enrichment reduces incidence of downy mildew in cucumber. Follow a 3-year crop rotation to reduce soilborne inoculum and field sanitation by burning the crop debris after harvest (Saha 2002).

5.4.3 Bower System

The bower system is superior both in terms of additional yield and return on investment. Different training systems, i.e., ground, bush (dry bamboo sticks along with thorny branches),

Kniffin, and bower (both prepared with iron angles and galvanized iron wire), are used for less disease incidence of bitter gourd (cucurbits). Except for the ground training system, vines were trained on the support. The bower system had significantly less incidence of diseases, i.e., anthracnose (27.72 %), powdery mildew (23.80 %), and downy mildew (30.80 %), as well as the highest number of fruits per plant (30.46 %), longer and dark green fruits, and maximum yield (78.25 q/ha). The percentage of increase in yield over the ground training system was 71.83, 200.65, and 243.80 % in the bush, Kniffin, and bower systems, respectively, which was attributed to low incidence of diseases (Bhokare and Ranpise 2004). Joshi et al. (1994) conducted an experiment to assess the economic feasibility of different training systems for bitter gourd and found that the bower system is superior both in terms of additional yield and return on investment.

Lin Anchio (1995) conducted a trial and compared the yield of the cucumber cv. Swallow obtained using traditional PNG methods with those obtained with pruning and the use of stand support poles (SSPs) and thus concluded that there is no need to prune the lateral branches after topping, but the crop should be supported with stand poles.

Jaiswal et al. (1997) carried out a staking trial on cucumber and found that Bhaktapur local performed well for off-season production in terms of its good fruit yield, good-quality fruit bearing (shape, size, color status), and early fruit bearing. The effect of the staking system on days to the first harvest was not significant at any site. Farmer's practice of staking (i.e., use of bamboo sticks or tree branches) produced 10.7 and 49 % more fruits than the use of plastic string or no staking, respectively. However, fruit yield did not differ between staking systems under low to medium management condition.

5.4.4 Use of Resistant Variety

Pusa Hybrid-1 and Arka Chandan of pumpkin; Punjab Chappan Kaddu-1 of summer squash; Pusa Hybrid-3 and Pusa Summer Prolific round and long of bottle gourd; BL-240, Hybrid BTH-7, BTH-165, Phule Green, RHR BGH 1, and Arka Sujat of bitter gourd; IIHR-8 of sponge gourd; Poinsette and Priya of cucumber; Arka Manik of watermelon; and Punjab Rasila and Pusa Madhuras of muskmelon cultivar are tolerant to downy mildew disease (Thamburaj and Singh 2005).

Spurling (1973) reported that downy mildew (*P. cubensis*) is a serious disease of introduced cultivars of cucurbits grown by smallholders, but local cultivars are rarely damaged.

Reddy et al. (1995) reported that the incidence of downy mildew varied significantly with cultivar and plant growth. Cultivar Kalhatti showed complete resistance to downy mildew during the early stages of growth, while other cultivars were susceptible. The local cultivars Chitradurga and Bellary exhibited less disease incidence at 40 days after sowing (40 DAS) as compared with the other cultivars. At later stages of growth 40 days after sowing (60 DAS), the local cultivars Siddavanahalli and Bellary exhibited the lowest incidence.

5.4.5 Biological Control

Aqueous extracts of horse and cow manure have provided good control in Germany to cucumbers. The extracts increase chlorophyll content and peroxidase activity of treated plants and also inhibit release of zoospores from sporangia. Winterscheidt et al. (1990) reported that watery extracts of composted cattle manure, composted sea algae, composted grape, and composted manure significantly reduced the infection of cucumber leaves (*Cucumis sativus*) by *Pseudoperonospora cubensis*. All efficient extracts inhibited the germination of sporangia with zoospores. No induced resistance of the host was observed. The extracts had no curative effects.

Ma-Liping et al. (1996) also reported that compost extracts from horse and cow manures gave good control of *Pseudoperonospora cubensis* under greenhouse conditions, with relative efficiencies of 67.33 and 66.1 % compared with untreated plants. Sheep and pig manures were less effective (46.5 and 57.3 %, respectively).

Abou-Hadid et al. (2003) reported that *Trichoderma harzianum* and *Trichoderma hamatum* were the most effective antagonists against the pathogens of powdery or downy mildew disease.

References

Abou-Hadid AF, Abed EL, Moneim ML, Tia MMM, Aly AZ, Tohamy MRA (2003) Biological control of some cucumber diseases under organic agriculture. Acta Hortic 608:227–236

Anonymous (2006) Package and practices for vegetable crops. Diretorate of Extension Education, SKUAST-J, Chatha, Jammu, p 107

Anchio L (1995) Effect of pruning and use of stand support poles on cucumber yield in Papua. Harvest (Port Moresby) 17(1/2):9–16

Badadoost M (2001) Downy mildew of cucurbits. Bull Univ Illinois, Ag Services 345

Bains SS, Jhooty JS (1976) Overwintering of *Pseudoperonospora cubensis* causing downy mildew of Muskmelon. Indian Phytopathol 29:213–214

Bains SS, Jhooty JS (1978) Epidemiological studies on downy mildew of Muskmelon caused by *Pseudoperonospora cubensis*. Indian Phytopathol 31:42–46

Bedlan G (1990) Problems in treatment of cucumber downy mildew under glass. Pflanzenschutz-Wien 3:5–7

Bhokare SS, Ranpise SA (2004) Effect of different training systems on incidence of pest, diseases and yield of bitter gourd (*Momordica charantia L.*) cv. Konkan tara under Konkan conditions of Maharashtra. Haryana J Horticult Sci 33:139–141

Brunelli A, Collina M (1996) The protection of melon from *Pseudoperonospora cubensis*. Colture Protette 25(12):107–108

Chaban VV, Kitsno LV, Nedobytkyn VA (1990) Testing fungicides against downy mildew on cucumber. Ukraine Zashchita-RastenMoskva 9:27–28

Charles C (1998) Manual of vegetable plant diseases. Discovery Publishing House, New Delhi, pp 186–189

Cohen Y, Rotem J (1970) The relationship of sporulation to photosynthesis in some obligatory and facultative parasites. Phytopathology 60:1600–1604

Cohen Y, Eyal H, Hanania J, Malik J (1989) Ultrastructure of *Pseudoperonospora cubensis* in Muskmelon genotypes susceptible and resistant to downy mildew. Physiol Mol Plant Pathol 34:27–40

Doran WL (1932) Downy mildew of cucurbits. Massachusetts Agric Exp Stat Bull 283, pp 22

dos Santos AA, Cardoso JE, Vidal JC, Vianaf MP, Rossetti AG (2003) Effect of the downy mildew initiation on the melon fruit production. Fitopatol Bras 28(5):548–551

dos Santos AA, Cardoso JE, Vidal JC, Silva MCL (2004) Evaluation of chemical products on the control of downy mildew and stem canker of melon. Rev Cienc Agron 35(2):390–393

Egan AR, Michelotti EL, Young DH, Wilson WJ, Mattioda H (1998) RH-7281: a novel fungicide for control of downy mildew and late blight. Brighton crop protection conference: pests and diseases. Proceedings of an international conference, Brighton, UK, vol 2, pp 335–342

Fugro PA, Rajput JC, Mandokhot AM (1997) Sources of resistance to downy mildew in ridge gourd and chemical control. Indian Phytopathol 50(1):125–126

Gaikwad AP, Karkeli MS (1994) Control of downy mildew of cucumber with new fungicide. J Maharashtra Agril Univ 19(3):445–446

Gandhi SK, Maheshwari SK, Mehta N (1996) Epidemiological relationship between ridge gourd downy mildew and meteorological factors. Plant Dis Res 11(1):62–66

Gangopadhyay S (1984) Advance in vegetable disease. Associated Publishing Company, New Delhi, pp 151–161

Golyshin NM, Maslova AA, Goncharova TF (1994) Fungicides against peronosporaceae. Zashchita Rastenii Moskva 4:18

Gonzalez M, Barrios, Fand Rodriguez F (1992) Evaluation of the effectiveness of different fungicides against the mildew of cucumber caused by *Pseudoperonospora cubensis*. Proteccion de Plantas 2(4):75–86

Gupta VK, Pal YS (2001) Fungal disease of cucurbits. In: Gupta VK, Pal YS (eds) Diseases of vegetable crops, Kalyani Publishers, Ludhiana pp 71–75

Huang XM, Yang Z, Lu YH (1989) Effects of climatical factors on the occurrence of cucumber downy mildew in the suburbs of Shanghai. Acta Agric Shanghai 5(1):83–88

Ilkweon Y, Hanwoo D, Yongseub S, Sugan B, Sungkuk C, Boosull C (1996) Timely application of fosetyl-Al for the control of downy mildew on oriental melon (*Cucumis melo.* L.). J Agric Sci Crop Prot 38(2):378–381

Jaiswal JP, Bhattarai SP, Subedi PP (1997) Effect of different staking systems in cucumber var. Bhaktapur local for off-season production. *Working paper Lumle Re Alternaria solani* in hydroponically grown tomato. Phytopathology 89:722–727

Jhooty JS, Bains SS, Parkash V (1989) Epidemiology and control of cucurbit downy mildew. In: V. P. Agnihotri et al., (eds) Perspective in Plant Pathology. Today and Tomorrow's Printers and Publishers, New Delhi, 301–314

Joshi VR, Lawande KE, Pol PS (1994) Studies on the economic feasibility of different training systems in Bitter gourd. J Maharashtra Agril Univ 19(2):238–240

Khalil MR, Khan NU, Younas M, Shah NH, Hassan G (1992) Control of downy mildew (*Pseudoperonospora cubensis* (Berk. & Curt.) Rostow) of melon (*Cucumis melo*) with different fungicides. Pak J Phytopathol 4(1–2):50–53

Lange L, Eden U, Olison LW (1989) Zoosporogenesis in *Pseudoperonospora cubensis* the causal agent of cucurbit downy mildew. Nord J Bot 8:497–504

Lehmann M (1991) Study on forecasting the beginning of epidemics of downy mildew (*Pseudoperonospora cubensis*) in field cucumbers. Pflanzenschutz Wien 2:4–6

Ma SQ, Liang HH, Ma JX (1990) Study on the ecological way of preventing cucumber downy mildew: a report of a control experiment on the environmental temperature. Chin J Appl Ecol 1(2):136–141

Mah SY (1988) A comparison of five fungicides and two-spore germination suppressing agents for control of cucumber downy mildew (*Pseudoperonospora cubensis* Berk. and Curt.). Teknologi Sayur Sayuran (Malaysia) 1:61–65

Mahrishi RP, Siradhana BS (1988a) Effect of nutrition on downy mildew disease caused by *Pseudoperonospora cubensis* (Berk. & Curt.) Rostow. on muskmelon. Ann Arid Zone 27(2):153–155

Mahrishi RP, Siradhana BS (1988b) Studies on downy mildew of cucurbits in Rajasthan: incidence, distribution, host range and yield losses in muskmelon. Ann Arid Zone 27(1):67–70

Ma-Liping, Gao-Fen, Wu-Yingpeng, Qiao-Xiong Wu (1996) The inhibitory effects of compost extracts on cucumber downy mildew and the possible mechanism. Acta Phytophylacica Sinica 23(1):56–60

Manole N, Costache M, Varadie P, Paraschiv G, Gogoci I, Szabo A. Mirghis R (1990) Epidemiology and integrated control of *Pseudoperonospora cubensis* (Berk. et Curt) Rostov in cucumber. Academia de Stiinte Agricole si Silvice 23:57–73

Memane SA, Khetmalas MB (2003) Chemical control of downy and powdery mildew diseases of bitter gourd during rainy season. J Maharashtra Agril Univ 28(3):283–284

Mercer RT, Lacroix G, Gouot JM, Latorse MP (1998) RPA 407213: a novel fungicide for the control of downy mildews, late blight and other diseases on a range of crops. Brighton crop protection conference: pests & diseases. Proceedings of an international conference, Brighton, UK, vol 2, pp 319–326

Merz F, Schrameyer K, Sell P (1995) Establishment of a standard treatment against downy mildew in cucumbers. Gemuse-Munchen 31(7):438–439

Motte G, Muller R, Auerswald H, Beer M (1988) Monitoring and control of downy mildew of cucumbers. Gartenbau 35(1):15–16

Palti J, Cohen Y (1980) Downy mildew of cucurbits (*Pseudoperonospora cubensis*) the fungus and its hosts, distribution, epidemiology and control. Phytoparasitica 8(2):109–147

Palti J, Rotem J (1973) Epidemiological limitations of the forecasting of downy mildew and late blight in Israel. Phytoparasitica 1:119–126

Ramirezarredondo JA (1995) Chemical control of downy mildew (*P. cubensis* (Bert. & Curt.) Rostow.) in squash in the Mayo valley. Mexcio Revista Mexicana de Fitopatologia 13(2):126–130

Reddy BS, Thammaiah N, Patil RV, Nandihalli BS (1995) Studies on the performance of bitter gourd genotypes. Adv Agric Res India 4:103–108

Rondomanski W, Wozniak J (1989) Distribution and chemical control of downy mildew on cucumber. Biuletyn-Warzywniczy, Polonia, 2:145–149

Rondomanski W, Zurek B (1988) Downy mildew of cucurbits-a new threat to cucumber in Poland. Roczniki Nauk Rolniczych, E Ochrona Roslin 17(1) 199–209

Saha LR (2002) A hand book of plant diseases. Kalyani Publisher, Ludhiana, pp 361–362

Sherif AF, Macnab AA (1986) Vegetable disease and their control. Willey, New York, p 728

Singh PP, Thind TS (2005) Diseases of cucurbits and their management. In: Thind TS (ed) Diseases of fruit and vegetable. Kalyani Publishers, Ludhiana, pp 290–305

Spurling AT (1973) A review of mildew problems on cucurbit vegetable crops in Malawi. PANS 19(1):42–45

Summer DR, Phathak SC, Smitte D, Jhonson A, Glaze NC (1981) Control of cucumber foliar diseases, fruit rot and nematodes by chemicals applied through overhead sprinkler irrigation. Plant Dis 65(5):401–404

Thamburaj S, Singh N (2005) Vegetable, tuber crops and spices. Indian Council of Agriculture Research, New Delhi, pp 10–75

Thind TS, Singh PP, Sokhi SS, Grewal RK (1991) Application timing and choice of fungicides for the control of downy mildew of muskmelon. Plant Dis Res 6(1):49–53

Tsai WH, Tu CC, Lo CT (1992) Ecology and control of downy mildew on cucurbits. Plant Prot Bull Taipei 34(2):149–161

Ullasa BA, Amin KS (1988) Ridgegourd downy mildew epidemics in relation to fungicidal sprays and yield loss. Mysore J Agric Sci 22(1):62–67

Weit B, Neuhaus W (1990) Biology and control of cucumber downy mildew (*Pseudoperonospora cubensis*). Nachrichtenblatt fur den Pflanzenschutz in der DDR 44(1):5–8

Winterscheidt H, Minassian V, Weltzien HC (1990) Studies on biological control of cucumber downy mildew (*Pseudoperonospora cubensis*) with compost extracts. Gesunde Pflanzen (Germany, F R) 42(7):235–238

Wu SQ (1994) Integrated management of cucumber diseases in greenhouse. Bull Agric Sci Technol 2:24

Yucel S, Gncu M (1994) Studies on chemical control of downy mildew *Pseudoperonospora cubensis* Berk. and Curt. on cucurbits in the Mediterranean region. Bitki Koruma Bulteni 31(1–4):109–118

Advances in Diseases of Lentil (*Lens culinaris* Medik)

6

H.S. Tripathi

Lentil (*Lens culinaris* Medik) is an important grain legume crop and occupies a prominent position on account of its manifold uses in daily life. It is one of the early domesticated crops and plays an important role in human and animal nutrition as a source of low-cost protein and other valuable nutrients. Besides being protein rich, its seed is a rich source of minerals and vitamins (Table 6.1) for human nutrition and its straw is a valued animal feed. Its ability to fix atmospheric nitrogen and sequester carbon results in improved fertility and nutrient status of soil, which in turn contributes to the sustainability of production system. Lentil also provides an opportunity to contribute to food supplies through diversification of agricultural systems in many countries.

Two subspecies of *Lens culinaris* are cultivated in the world: macrosperma, which is grown in the Mediterranean region, and the microsperma, which is grown in Indian subcontinent and parts of the East (Jeswani 1988). The area under lentil cultivation in the world is 3.6 m/ha and about 76 % of it is in Asia (2.8 m/ha). Bangladesh, India, Iran, Syria and Turkey are the major lentil-producing countries in Asia, whereas Pakistan, China, Iraq, Lebanon, Jordan, Palestine and Yemen are minor producers (Table 6.2). Average yield of lentil for the Asian countries is higher (0.85 t/ha) as compared to world productivity (0.81 t/ha). In India, lentil is the second most important *rabi* pulse after chickpea. The area under cultivation in India, during 2004–2005, was 1.40 m ha with a production of 1.03 m tonnes and productivity 741 kg/ha (Anonymous 2005).

Being a cool season crop, lentil production is mainly confined to Northern and Central India. Bihar, Haryana, Madhya Pradesh, Maharashtra, Rajasthan, Orissa, Uttar Pradesh, Uttaranchal and West Bengal are major lentil-producing states. Uttar Pradesh and Madhya Pradesh together contribute to 78 % of the total national area as well as production of lentil. The potential yield of lentil is 1.5–1.8 t/ha, but we harvest only 0.62 t/ha in a farmer's field (Ali and Kumar 2000). The main constraints responsible for hampering the production of lentil seem to be the non-availability of quality seeds of high-yielding varieties, farmers' preference to grow bold seeded lentil which are comparatively poor yielder, cultivation of lentil under rain-fed conditions on marginal lands with low fertility and most importantly its unstable performance due to frequent infestation of diseases and pests.

Lentil is known to be attacked by a number of diseases. The potentially damaging diseases of lentil are wilt, rust and *Ascochyta* blight.

H.S. Tripathi (✉)
Centre of Advanced Studies in Plant Pathology,
G.B. Pant University of Agriculture and Technology,
Pantnagar, Uttarakhand 263145, India
e-mail: tripathihs@gmail.com

L.P. Awasthi (ed.), *Recent Advances in the Diagnosis and Management of Plant Diseases*,
DOI 10.1007/978-81-322-2571-3_6

Table 6.1 Nutritional composition of lentil seeds

Component		References
Crude protein	19.5–35.5 %	Bhatty (1988)
Fat	1.8–3.9 %	Bhatty (1988)
Carbohydrates	60.8 %	Bhatty (1988)
Starch	35–55%	Gupta (1988)
Cellulose	2–3%	
Lignin	10 %	
Free sugar	2–3%	
Potassium	24.2–29.0 g/kg	Abushakra and Tannous (1981)
Phosphorus	5.4–14.4 g/kg	
Iron	54–505 mg/kg	Summerfield and Muehlbauer (1982)
Zinc	18–330 mg/kg	
Vitamin A	200 mg/kg	
Vitamin B_1	0.5 mg/100 g	Savage (1988)
Vitamin B_2	0.21 mg/100 g	
Vitamin Niacin	1.8 mg/100 g	
Calorific value	346 cal/100 g	Gupta (1988)

Table 6.2 Major lentil-producing countries in Asia during 2006

Country	Area (ha)	Production (t)	Productivity (kg/ha)
Bangladesh	157,085	115,000	732
China	90,000	135,000	1500
India	1,393,300	982,820	705
Iran	260,000	100,000	385
Nepal	180,210	148,348	823
Pakistan	44,800	26,200	585
Syria	121,156	132,805	1096
Turkey	500,00	480.400	960
Total (Asia)	2,746,551	2,120,553	848
World	3,623,380	2,938,037	811
Asia (% of world)	75.8	72.1	104.5

Note: Adopted from FAO Statistical Yearbook 2006

6.1 Wilt

Prissyajnyak is the first reported lentil wilt caused by *Fusarium* sp. from Russia (USSR) in 1931. After 3 years, the occurrence of such a disease was reported from undivided Bengal (India) and later in other parts of India (Padwick 1941). The disease is widespread in most of the lentil-growing countries of the world, particularly in America, Bangladesh, Argentina, Brazil, India, Italy, France, Czechoslovakia, Myanmar, Nepal, Pakistan and USSR.

Mortality of lentil plants due to wilt varies from 25 % in seedling stage (Kannaiyan and Nene 1975) to 50 % in flowering stage (Khare et al. 1971). Under natural conditions, the disease incidence has been reported as high as 50–78 % in Madhya Pradesh (Agarwal et al. 1991), 70 % in areas of Czechoslovakia (Bojdova and Sinsky 1990) and 13.2 % in south Syria (El-Ahmed and Mousell 1986).

6.1.1 Symptoms

The pathogen affects the plant at three stages, i.e. pre-emergence, postemergence and adult stage. Under early infection, pre- and early postemergence mortality is noticed in patches showing seedling wilt. The vascular wilt is observed in case of plants infected at an advanced stage, i.e. at blooming. When the growth of infected plant is checked, one may observe that its leaves shrink or get curled starting from the lower part of the plants and extending upwards. Epinasty may be visible in some cases. The root system is poorly developed and becomes brown. Secondary root sometimes proliferate above the affected level. The wall of xylem vessels in the root are discoloured brown and contain fungal hyphae particularly if the plant is infected at an advanced stage of its growth. Vasudeva and Srinivasan (1952) reported that in a broadcast crop, wilt occurs in isolated patches, more or less circular in outline which enlarges as the disease advances. When the crop is sown in rows, the disease appears to progress along the lines.

6.1.2 Pathogen

The disease is caused by *Fusarium oxysporum* f.sp. *lentis*. The fungus is host specific and pathogenic for lentil only. Vasudeva and Srinivasan (1952) described the fungus as thin plectenchymatous stroma, pale to purple in colour. The aerial mycelium is woolly white, sometimes turning pink, collapsing after 3–4 weeks, becoming gelatinous and tough. Pinnotes and sporodochia are absent.

The fungus produces three types of asexual spores: microconidia, macroconidia and chlamydospores. The microconidia have been reported to be single celled, hyaline ovoid, cylindrical, oblong or slightly curved measuring 6.7–3.7 μm (Khare et al. 1971). Macroconidia are nearly straight to fusiform, falcate, slender, thin walled with indistinct septa, mostly single celled, rarely 1–5 septate and 10–50×5.0–6.7 μm in size. These conidia are formed freely on mycelium at the ends of free conidiophores. The base of conidia is papillate and it is where foot cell formation appears (Vasudeva and Srinivasan 1952). Chlamydospores are terminal or intercalary, borne on mycelium. They are one celled, but occasionally may be two celled, smooth, hyaline and rich in protoplasmic contents.

Sharma and Agnihotri (1972) studied the variability in the fungus and identified three isolates 'A', 'B' and 'C' on the basis of their virulence on lentil. Mehrotra and Claudius (1973) isolated 14 cultures of *F. oxysporum* f.sp. *lentis* from Madhya Pradesh (India) and found them different from each other with respect to their virulence on two varieties of lentil. At Jabalpur, India, eight isolates of lentil wilt pathogen (L_1–L_8) were studied in detail and differentiated on the basis of morphological and cultural characteristics as well as their virulence on lentil (Kannaiyan and Nene 1978a, b).

6.1.3 Host Range

The lentil wilt pathogen is reported to be highly host specific in nature (Vasudeva and Srinivasan 1952; Khare 1980; Kannaiyan and Nene 1978a, b). However, Padwick (1941) reported that the fungus could infect lentil as well as chickpea, but the chickpea isolate did not infect lentil.

6.1.4 Disease Cycle

The lentil wilt pathogen survives mainly in soil on infected plant residue and has a very high competitive saprophytic ability (Mehrotra and Claudius 1973). Ujevic et al. (1965) isolated *F. oxysporum* f.sp. *lentis* from lentil seeds and concluded that the pathogen is transmitted mainly by seeds. Khare et al. (1979) also reported the fungus to be associated with the seeds of lentil. However, Erskine et al. (1990) did not find lentil wilt pathogen associated with seeds obtained from wilt-infected plants. They considered the seed-borne nature of the disease related to either an external contamination or via trash.

The fungus enters the host through natural openings, such as emergence of secondary roots, or through wounds, which expose the xylem

(Khare 1980), and ultimately it colonizes the metaxylem and stops the supply of water and mineral ions by plugging the vessels with mycelial growth or tylosis formation resulting in loss of turgidity and finally wilting (Saxena et al. 1992).

Bhalla et al. (1992) reported that *F. oxysporum* infects roots of lentil by entering through the juncture of epidermal cells within 8 h after inoculation with or without forming appressoria. The cortical cell wall is ruptured 10–12 h after inoculation. There the fungi proliferate extensively within 24 h, but the endodermis and vascular cells were free of hyphae for at least 72 h.

The fungus produces certain toxic metabolites which reduce the vigour of seedlings of lentil, producing very small and dark green leaves due to loss of turgidity. They developed epinasty but later their tips turned yellow (Agarwal and Khare 1975). Production of hydrolytic enzymes was also reported by Mehrotra and Claudius (1973). The pathogen is reported to produce pectolytic and cellulolytic enzymes also.

6.1.5 Epidemiology

The incidence of lentil wilt is influenced by several biotic and abiotic factors. Sharma and Agnihotri (1972) reported that susceptibility of lentil plants to wilt decreases with advancing age and the plants are highly susceptible to wilt pathogen at an age of 22 days.

Lentil wilt is favoured in a soil with 25 % water holding capacity (Saxena and Khare 1988). They also reported that the incidence of wilt decreases with increase in soil moisture, showing no wilt at 75 % or more moisture level. Along with moisture, temperature also plays an important role in the development of lentil wilt. Vasudeva and Srinivasan (1952) observed the wilt incidence of 7.5, 96.4 and 41.7 % at soil temperature range of 13–23, 17–31 and 13.38 °C. They also reported that the disease appeared in November, but with a drop in temperature it has almost ceased in December to January and again appeared in February, reaching up to its maximum in March with a rise in temperature. Kaushal and Sharma (1998) reported the highest wilt incidence (75.1 %) at temperature range of 24–27 °C and the lowest (13.0 %) at the temperature range of 13–16 °C.

6.1.6 Host Resistance

Resistance is a fundamental attribute of all living systems. Plant essentially offers active resistance that largely determines the outcome of host-parasite interaction. Lentil varieties and germplasms have been screened against wilt in many countries especially in India, Czechoslovakia, Uruguay, Russia, Bangladesh, Bulgaria, Nepal, Egypt and Syria.

Resistance to wilt in Pant L 234 is controlled by two independent dominant genes with a complementary gene detected in resistant lines JL 446 and LP 286. A third dominant gene complementary to JL 446 and LP 286 is present in susceptible lines JL 641 and L 9–12. The third gene complementary to two genes in resistant lines is nonallelic to the third complementary gene in susceptible lines.

Eujayal et al. (1998) investigated the inheritance of resistance to lentil in a cross between resistant (IIL 5588) and susceptible lines (L 692-16-1 (5)). They reported that resistance to wilt was conditioned by a single dominant gene (fw) in the population. The map location of the fw locus was also identified for the first time through linkage to a random amplified polymorphic DNA (RAPD) marker (OPK 15900) at 10.8 cm. Two other RAPD markers OPB 17800 and OPD 15500 were identified by bulk segregant analysis and were found to be associated in the coupling phase with the resistant trait.

6.1.7 Disease Management

6.1.7.1 Cultural Control

Cultural control includes the manipulation of environment favourable to the crop and to the detriment of pathogen. To avoid lentil wilt, sowing should be done at appropriate time. Manipulation of sowing date may prove to be a good method but its efficacy will differ from place to place and season to season. Myalova (1973) reported infection of lentil by *F. oxysporum* to be seven times more severe in early sowing than in late ones. Kannaiyan and Nene (1975) reported that less disease was observed when crop was sown on 15 December, while in adult stage wilt was less in crop sown on 15 November. Bhatt and Sharma (1992) also reported more seedling wilt in a crop sown up to the third week of October and lesser in late-sown crop. Similar results were reported by Izquierdo and Morse (1975). In the former USSR more severely in the spring-sown crop compared to that sown in summer. However, in Syria late sowing of lentil in February increased wilt incidence compared to normal sowing in September. Crop rotation with non-hosts and destruction of diseased plant residues also reduce the wilt incidence.

6.1.7.2 Chemical Control

Scientists have made attempts to control the wilt pathogen by employing chemicals. Kovacikova (1970), Vishunavat and Shukla (1982) first obtained effective control of lentil wilt by treating the seeds with Captan (0.2 %) or Thiram (0.2 %). Khare et al. (1971) reported considerable reduction in wilt incidence in field when seeds were treated with Benlate or Chloroneb + Thiram at 1:100 w/w. Sharma et al. (1971), Agrawal et al. (1976) and Saxena (1987) found that Benlate, Ceresin wax, Mercuric chloride, Thiride and mixture of Captan + Thiram and PCNB were highly inhibitory to *F. oxysporum* f.sp. *lentis*. Kannaiyan and Nene (1975) found effective control of lentil wilt by seed treatment with Trianmol (0.05 % a.i.) or benomyl (0.1 % a.i.) under field conditions. Seed treatment with 0.15 % boric acid and $KMnO_4$ reduced infection by 20–70 % and yield increased by 22–23 % (Myalova 1973).

Ahmed and Ahmed (2000) evaluated the effect of several pesticides on population dynamics of *F. oxysporum* f.sp. *lentis* and recorded the lowest population at 75–105 days after seed treatment with carbendazim.

6.1.7.3 Biological Control

Yuen et al. (2001) reported isolates of *Pseudomonas, Erwinia, Rhizobium, Penicillium expansum* and *Trichoderma lignorum* to be antagonistic to *F. oxysporum* on lentil. Mehrotra and Claudius (1973) found that soil inoculation with *Trichoderma viride* and *Streptomyces gougerotii* controls lentil wilt. Mukherjee et al. (1989) reported antagonistic activity of *T. harzianum* and *T. koningii* against *F. oxysporum* f.sp. *lentis*. Singh and Mukhopadhyay (2000) reported that *Gliocladium virens* and its gamma radiation-induced mutants M-3 and M-7 effectively controlled the wilt complex of lentil caused by *R. solani*, *S. rolfsii* and *F. oxysporum* f.sp. *lentis*.

6.1.7.4 Integrated Management

De and Chaudhary (1999) assessed the effect of seed treatment using fungicide alone or in combination with different biological control agents as well as kerosene oil to manage lentil wilt and reported that *Bacillus subtilis* in combination with carboxin (Vitavax) showed 66 % wilt control and 45 % increase in yield, whereas separate combinations with *T. virens, T. harzianum* and *T. viride* reduced wilt incidence by 79 % and increased yield by 140, 224 and 241 %, respectively. *P. fluorescens* in combination with *T. viride* and carboxin (Vitavax) decreased wilt incidence by 65 % and increased yield by 229 %.

Ahmed et al. (2002) integrated host resistance with planting date and fungicidal seed treatment to manage lentil wilt and reported that lentil genotype had a greater effect on the onset and duration of *Fusarium* wilt than planting date or fungicide seed treatment. The per cent terminal wilt and areas under disease progress curve were lowest during November planting for all genotypes.

6.2 Rust

Rust is an important disease and has drawn attention of research workers, particularly in the northern part of the Indian subcontinent, Canada and the Mediterranean basin. The earliest record of this disease comes from India when it was recorded by Butler in 1918 in the Indo-Gangetic plains of India. Since then, the disease has occurred in several countries including Cyprus, Morocco, Sicily, Palestine, Portugal, Bulgaria, Turkey, Iran, Israel, Chile, Hungary, Syria, Nepal and Bangladesh.

The first documented lentil rust epidemic occurred in Cyprus when the entire crop was destroyed at Larnaca district during 1931 crop season (Nattrass 1932). A severe outbreak of the disease was reported in Sicily, causing heavy losses to the crop (Canonaco 1937). In India the disease appeared in epiphytotic form at Delhi during 1945–1946 and 1946–1947 *rabi* seasons (Prasada and Verma 1948). In the Narmada valley of Madhya Pradesh, severe outbreak of lentil rust was recorded in 1978 when as much as 100 % losses were recorded in some fields of lentil.

6.2.1 Symptoms

The disease appears on all aboveground parts of the plants. The first sign of infection becomes evident by formation of yellowish white aecia on the lower surface of leaves. Aecia are borne singly or arranged in a circular manner as small groups on leaflets. Aecia gradually turn light brown in colour. Later, brown uredia develop on both surfaces of leaves. Pustules are powder and may coalesce to form to large pustules. Telia are produced late in season and are more frequent on stems and pods. The disease causes severe deformity in plants and often the plant dies. In severe cases, plant may shed leaves and dries completely even before seed formation. The disease starts mostly from low-lying patches and radiates towards the border. Seed size is appreciably reduced.

6.2.2 Causal Organism

Rust of lentil is caused by *Uromyces viciae-fabae* (Pers.) de Bary. It is an autoecious and heterothallic fungus forming all the spore forms on lentil only. Pycnia are small, flask shaped and produced on the upper surface as well as lower surface of leaves and possess flexuous hyphae and nectar drop at the mouth (Prasada and Singh 1975). Aecia are formed after fertilization of receptive hyphae through pycniospores of opposite mating type. Fresh pycnia are produced till diploidization or the death of the host retaining the receptivity of haploid pustules till diploidization or the commencement of adverse condition (Prasada and Singh 1975).

The aecia are cup shaped. The peridium of aecium is short and whitish in colour. Aeciospores are produced in aecial cups in chains. They are dicaryotic, round to angular or elliptical with fine warts, measuring 14–22 μm in diameter.

Uredia appear late in the season. Uredospores are single celled, round to oval in shape, light brown, spiny and possess 3–4 germ pores. The size of uredospores ranges between 20 and 18 × 30 and 26 μm. Telia appear late in the season at pod formation stage. Telia are initially covered by host epidermis which later on ruptures to release teleutospores. Teleutospores are single celled, pedicellate, thick walled and light brown in colour with papillate apex. The size ranges between 25 and 18 × 38 and 27 μm.

The teleutospores produce four-celled basidium after germination on which four haploid hyaline basidiospores are formed. On germination, basidiospores infect the host and produce pycnia of different sexes.

6.2.3 Host Range

The pathogen has a wide host range. *Uromyces viciae-fabae* infects broad beans (*Vicia faba*) and other species of *Vicia* (Hiratsuka 1933), lentil (Nattrass 1932; Patil 1933) and *Lathyrus* sp. (Hiratsuka 1933; Patil 1933). Several species of *Vicia* and *Lathyrus* grown as ornamentals or as forage crops as well as many wild species are also infected by this rust. Couner and Bernier (1982) studied host range of various isolates of *U. viciae-fabae* from several species of *Vicia, Lathyrus,* pea and lentil. He reported that the isolates have a very extensive host range. Isolates of

U. viciae-fabae share so many hosts in common that it was impossible to classify them into formae speciales. They also concluded that the wild species of *Vicia* and *Lathyrus* may play an important role in epidemiology of the disease.

6.2.4 Epidemiology

Lentil rust appears in India mostly in January–February when the crop is at flowering stage.

6.2.4.1 Primary Inoculums and Secondary Spread

Teleutospores of fungus on infected plant debris present as admixture with seed lot act as primary source of inoculums for initiation of infection in the next season and cause fresh rust outbreaks. Unlike wheat stem rust (*Puccinia graminis tritici*) and linseed rust (*M. lini*), the teleutospores of this rust withstand the summer heat in the plains of India. Prasada and Verma (1948) reported that teleutospores when exposed to natural conditions and at room temperature were able to retain their viability up to 10 and 12 months, respectively, while those stored in refrigerator retain their viability up to nearly 2 years. Uredia and aecia of this rust perish during the summer that follows the harvest. The teliospores can germinate soon after their formation at temperature ranging from 12 to 22 °C with 60–70 % relative humidity. Since in nature these conditions are together available during the following winter, the teleutospores retain their viability until then (Fig. 6.1).

Uromyces viciae-fabae is one of those few rust fungi in which the repeating stage is predominantly the aeciospores and to a lesser extent the urediniospores (Chauhan and Singh 1994). The formation of aeciospores is governed by prevailing temperature between 17 and 22 °C where secondary aecia are produced, whereas uredia are formed at 25 °C, i.e. late in the season, quickly followed by telia (Prasada and Verma 1948).

6.2.4.2 Environmental Factors

Khare (1980) reported that high humidity and cloudy to drizzling weather with 20–22 °C temperature are favourable for disease development, and plants are more susceptible at flowering stage in lentil. Less amount of rain for longer duration offers a congenial environment for development and spread of disease. Higher temperature and heavy rainfall disfavour rust spread (Mittal 1997; Bakr 1991). The disease generally starts from low-lying patches in the field and radiates towards the border (Khare and Agarwal 1978). The disease is initiated in areas with dense crop canopy and luxuriant vegetative growth (Accantino 1964). Chauhan and Singh (1994) reported that severity of infection by *Uromyces viciae-fabae* and pustules per plant increased progressively with an increase in the duration of leaf wetness up to 24 h but did not increase further significantly. Thus 24 h period of leaf wetness is optimum for disease development.

6.2.5 Disease Management

6.2.5.1 Cultural Control

Cultural practices largely affect the environmental conditions favourable for growth and buildup of inoculums. The pathogen survives with the stored seeds of lentil mostly on inert matter associated with the seeds. Hence, disease can be checked by proper processing of seed lot. Diseased crop refuse should be discarded and burnt after harvest (Prasada and Verma 1948; Canonaco 1937). Due to seed-transmitted nature of the disease, the import of *Lens* sp. from South American countries is prohibited in the USA (Anonymous 1959). Singh and Dhingra (1980) and Mittal (1997) reported that early sowing (November) causes high disease severity compared to late sowing (December). Shukla and Amin (1991) reported increase in rust intensity on susceptible variety *Lens* 830 with delay in sowing. The intensity of rust was less with more yield when sowing was done up to the first week of November. The plants were not infected up to 10–11 weeks after sowing.

The pathogen attacks several other hosts such as broad bean, pea, *Lathyrus*, etc., of which some are grown round the year and thus help in multiplication and perpetuation of the fungus acting as a collateral host. Destruction of collateral hosts is

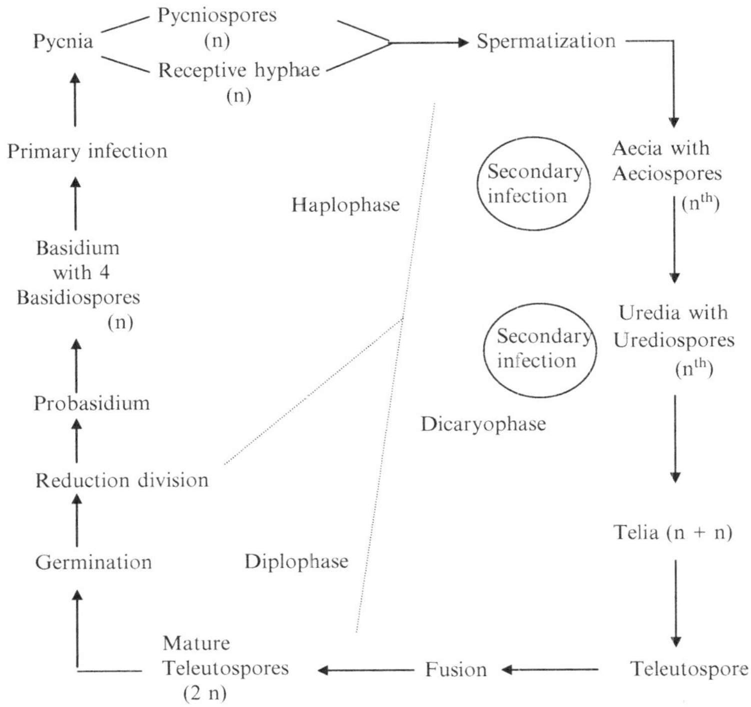

Fig. 6.1 Disease cycle of lentil rust caused by *U. viciae-fabae* (Pers.) de Bary

recommended to manage the disease (Kapooria and Sinha 1966; Couner and Bernier 1982). Controlled irrigation can eliminate the chances of heavy outbreak of disease (Agrawal and Prasad 1997).

6.2.5.2 Chemical Control

Several inorganic sulphur preparations are reported to give effective control of *U. viciae-fabae* (El-Helaly 1939; Accantino 1964). Organic sulphur fungicides like Ferbam, Ziram, Thiram and Zineb have been reported to give good control of *U. viciae-fabae* (Jackes and Webb 1956; Accantino 1964).

Prasada and Verma (1948) reported that treatment of lentil seeds with Agrosan GN results in complete control of disease. Singh (1985) found that the lentil crop grown from the seed treated with Vigil remains free from rust infection up to 60 days of sowing, while Bayleton protected the crop up to 50 days. Mohy et al. (1999) reported that treatment of lentil seeds with Mancozeb was highly effective against rust followed by seed treatment with carboxin and benomyl.

Accantino (1964) reported from Chile the control of lentil rust by nickel nitrate. Lentil plants sprayed with nickel nitrate solutions at flowering time remained free from the rust. Triadimefon, Calixin M and Calixin were found to be effective under field conditions for management of lentil rust (Upadhyay and Gupta 1994). Pande et al. (1995) reported best control of lentil rust by tridemorph (as Calixin) followed by Metiram (as Campogram) and benomyl (as Benlate). Ayub et al. (1996) evaluated six fungicides for their ability to control lentil rust. Among the fungicides tested, tilt gave the best control followed by Folicur and Calixin. Singh et al. (1985) evaluated various fungicides and foliar spray Dithane M-45 against rust of lentil was found to be the best followed by Sulfex and Karathane. The incidence of rust was least and yield was highest when Dithane M-45 was used.

Post-infection activity of ergosterol biosynthesis-inhibiting (EBI) fungicides was evaluated against *Uromyces viciae-fabae* (Gupta and Shyam 2000). All the EBI fungicides provided excellent post-infection activity in comparison to conventional nonsystemic fungicides like mancozeb and chlorothalonil. Fungicides like cyproconazole, flusilazole, penoconazole and hexaconazole were found to be the best followed by difenoconazole, triadimefon and fenarimol.

Sugha et al. (1994) reported that benomyl, carbendazim, thiabendazole and thiophanate methyl have very good potential for suppressing the early establishment of rust due to aeciospores, whereas benomyl, flutriafol and myclobutanil are effective in suppressing the late infections due to uredospores. Mancozeb when applied as foliar spray was the most effective fungicide reducing rust severity followed by oxycarboxin and benomyl. The spraying should be done twice, first after the first appearance of disease and then after two-day interval (Shukla and Amin 1991; Mohy et al. 1999).

6.2.5.3 Biological Control

Information in respect to biological control of lentil rust is lacking. Carrion et al. (1999) tried to control bean rust (*Uromyces appendiculatus*) by fungal antagonist (*Verticillium lecanii*). The antagonist proved to be more effective than the agrochemicals and significantly reduced the incidence and severity of rust. Singh (2003) evaluated some biocontrol agents and botanicals against pea rust. None of the bio agent was found to be effective, while neem oil was found to be better than other botanicals in respect to disease severity and grain yield of field pea.

6.2.5.4 Host Resistance

Resistance sources against lentil rust have been identified in several countries. *Uromyces viciae-fabae* is an autoecious rust, i.e. all the four stages of its life cycle are completed on lentil or its collateral host pea, broad bean and *Lathyrus*. Hence, evaluation of new matching virulence due to sexual recombination and selection pressure on pathogen is inevitable due to cultivation of resistant/free cultivars. Under such conditions hypersensitive resistance is likely to be short lived, and high-yielding slow-rusting cultivars can be successfully employed in disease-prone areas.

Basandrai et al. (2003) evaluated some promising genotypes against rust to identify slow-rusting resistance. Among various lines tested stocks LC 68-69-4-2 and L 4076 though showing disease reaction '7' and '5' yielded significantly better than check. It maybe due to the fact these varieties developed less TDS and AUDPC, thereby exhibiting slow rusting. Rating scale 1–9 was used for scanning disease severity.

6.3 *Ascochyta* Blight

The disease was observed in India (Sattar 1934). Since then it has been reported in 15 different countries including Australia, Canada, Chile, Greece, Ethiopia, Argentina, Italy, Pakistan, Cyprus, Spain, Turkey, Syria and the USA. It causes economic losses, especially in Argentina, Canada, Ethiopia, India and Pakistan (Erskine and Saxena 1991). In Canada, the disease occurs widely, especially in the humid regions of Saskatchewan and Manitoba. Severe epiphytotics of the disease were observed at Islamabad (Pakistan) during the year 1984–1985 and at least 4 times at Ludhiana (India) during a period of 20 years (Grewal 1988; Iqbal et al. 1992). Koul and Ambardar (1988) reported that *Ascochyta* blight along with *Alternaria* blight was responsible to cause heavy damage to the lentil crop in Kashmir even with 12–18 % disease severity.

The disease not only reduces the lentil yield but also affects the seed quality (Morrall and Sheppard 1981). The seeds from diseased plants become shrivelled and discoloured, with necrotic lesions and reduced germination and vigour (Cromey et al. 1987). Yield and seed quality reduction may lead to losses in potential income of more than 70 %. However, the magnitude of losses depends upon the cultivar used, the inoculum potential and prevailing weather conditions. Kaiser and Hannan (1982) reported that the disease may lead to growth and yield reduction to the extent of 24–59 % and more than 25 %, respectively, even though the environment may be unfavourable for its spread.

6.3.1 Symptoms

The symptoms appear as small, tan to dark brown circular lesions of 0.1–0.4 cm width with an indefinite or narrow brownish margin. All aboveground parts of the host are affected at different growth stages. The centre of the spot is light coloured and is speckled with tiny black fruiting bodies (pycnidia) arranged in concentric rings that are characteristics of the disease. The spots are more conspicuous on pods than on leaves and stems. Infected leaflets drop prematurely and the tips of the branches wilt, turn brown and die. Under severe infection, the leaves are blighted, pod size is reduced, and seeds from the affected pods are shrivelled and discoloured with necrotic lesions. The seeds become purplish brown and may have whitish patches of mycelia and tiny black fruiting bodies.

The lesions on stems are brown and elongated and may girdle the affected portion, causing rapid death of the portion of plant above the girdle (Davatzi-Helena 1980; Grewal 1988).

6.3.2 Causal Organism

The disease is caused by *Ascochyta lentis* Bond and Vassil. It produces gregarious, immersed and globose pycnidia, 175–300 μm in diameter, with a minute round ostiole and yellowish brown pseudoparenchymatous cortex. The conidia were cylindrical, straight or rarely curved, rounded at the ends with a median septum and 11.5–19.5×3.5–5.8 μ in size (Bondartzeva-Montevercles and Vassilievsky 1940).

The fungus grows well on PDA at a temperature ranging between 15 and 25 °C. Conidia are produced abundantly under continuous light, but some isolates show good linear growth in continuous light and some show zonation in alternate exposure to light and darkness (Kaiser et al. 1994).

6.3.3 Perpetuation

The disease is both stubble and seed borne (Ahmed and Beniwal 1988). The fungus is internally seed borne and has been detected in all parts of lentil seeds (Morall and Beauchamp 1988). It is reported to survive in infected lentil seeds for more than 30 years when stored at 4–8 °C. It survives for more than 3 years in infected pods and in seeds at 4–5 °C and 1–5 years on the soil surface (Kaiser and Hannan 1986). It loses its viability within 20 weeks at a soil depth of 16 cm. Kaiser et al. (1989) reported that the incidence of the pathogen on seed increases by 18–25 % during 4-year storage at 196–20 °C with no adverse effect on its pathogenicity. The frequency of transmission from infected seed to seedling is low, especially when the soil temperature is moderate to high and the main source of infection is the inoculum from the infected crop residues from previous crops (Bedi and Morrall 1990). Gossen and Morrall (1986) reported that the frequency of transmission of *A. lentis* from the seed to aerial parts of the seedling correlated significantly with the intensity of seed infection but was very low in warm and dry soil and high in cool soil with a moisture level near field capacity (Gossen and Morrall 1986).

6.3.4 Epidemiology

The disease spreads through the pycnidiospores produced at the foci of primary infection either through the infected seeds or the crop debris. Pedersen et al. (1994) found rain splash to be an effective source for short-range dispersal of conidia of *A. lentis*. Conidia in small airborne droplets and wind-drawn leaflets may be an important mechanism for the dispersal of conidia. Progress of foliar blight is rapid in the field, and epidemic levels can be reached under cool and wet weather conditions because spores are disseminated by rain splashes. The highest infection frequency is reported with a wetness period of 24–48 h and temperature of 10–15 °C. But the temperature had little effect on the lesion size and number of pycnidia per lesion (Pedersen et al. 1994).

The spores germinate within 6 h of inoculation, and germ tube grew along the leaf surface forming appressoria within 10 h. The fungus remains in the epidermal cell for 3–4 days, and then hyphae grow intercellularly in the mesophyll for 2–3 days after which it penetrates mesophyll cells. The pycnidia are formed after degeneration of infected host tissues, i.e. 10–14 days after inoculation (Roundhill et al. 1995).

6.3.5 Control

6.3.5.1 Cultural Control

Since diseased plant debris is the main source of primary inoculum, it should be destroyed after the harvest. Nene et al. (1988) suggested crop rotation, use of disease-free seed and early sowing to escape moist weather at harvest to minimize losses due to *Ascochyta* blight. Pederson and Morrall (1994) suggested the use of 10 cm wide barrier crop of lentil sprayed with chlorothalonil at early bloom or 14 m non-host crop furrier between the lentil crop and field with primary source of inoculum as a practical and economical method of disease control.

6.3.5.2 Chemical Control

Seed Treatment

A large number of fungicides have been evaluated to control the seed-borne infection of *A. lentis* in lentil seeds by seed treatment. Seed treatment with benomyl or thiabendazole increases the seedling emergence from infected lentil seeds (Kaiser and Hannan 1987). Russell et al. (1987)Thiabendazole and carbendazim protected the crop up to 143 days and both of them along with tridemorph reduced seed infection by the pathogen (Russell and Hagerty 1992). Iqbal et al. (1992) used a wet dip treatment of lentil seeds with Benlate, Calixin M, Tecto-60, Topsin-M, Antracol, Daconil and Dithane M-45. All the fungicides reduced the recovery of *A. lentis*, but Benlate, Calixin M, Tecto-60 and Topsin-M were more effective.

Beniwal et al. (1989) reported sun drying as an effective method to control the disease. The disease can be controlled to some extent by treating the seed with hot water at 55 °C for 25 min or by dry heat at 70 °C for 20 h (Ahmed and Beniwal 1991).

Foliar Sprays

Benomyl, Daconil, Captafol, Chlorothalonil, Folpet and Metiram are some of the fungicides reported to control the disease. Beauchamp et al. (1986) reported that Captafol, Chlorothalonil, Folpet and Metiram inhibit the conidial germination (100 %) at 10 μg/ml, while benomyl (32 μg/ml) prevented the mycelial growth. One spray of any of the above fungicides at early bloom to early pod stage protected the lentil crop from *Ascochyta* blight. Bashir et al. (1986) and Iqbal et al. (1989) reported that the disease was effectively controlled by three sprays of benomyl or Daconil (0.25). Malik et al. (1992) reported Daconil as the most effective crop spray treatment for reducing disease and increasing grain yield with cost benefit ratio of 1:3.54.

References

Accantino P (1964) Chemical control experiment on the lentil rust. Agric Technol 1963-63: 7–14

Agarwal SC, Khare MN (1975) Effect of fungicides on macroconidia formation of *Fusarium oxysporum* f.sp. *lentis*. Sci Cult 41:280–281

Agarwal SC, Singh K, Lal SS (1991) Plant protection of lentil in India. Paper presented in ICAR-ICARDA seminar "Lentils in South Asia", New Delhi, 11–15 March 1991, pp 1–37

Agrawal SC, Prasad KVV (1997) Diseases of lentil. Oxford and IBH Publishing Co. Pvt. Ltd., New Delhi, p 155

Agrawal SC, Khare MN, Agrawal PS (1976) Control of lentil rust (*Uromyces viciae fabae*) by use of fungicides. Indian Phytopathol 29:90–91

Ahmed S, Ahmed S (2000) Efficacy of pesticides on population dynamics of *F. oxysporum* f.sp. *lentis*. Ann Agric Res 21:527–529

Ahmed S, Beniwal SPS (1988) Seed borne nature of *Ascochyta lentis* in Ethiopia. LENS Newsl 15:30–31

Ahmed S, Beniwal SPS (1991) Ascochyta blight of lentil and its control in Ethiopia. Trop Pest Manag 37:368–373

Ahmed S, Akem C, Bayaa B, Erskine W (2002) Integrating host resistance with planting date and fungicide seed treatment to manage *Fusarium* wilt and so increase lentil yields. Int J Pest Manag 48(2):121–125

Ali M, Kumar S (2000) Problems and prospects of pulse research in India. Indian Farm 11:4–13

Anonymous (1959) Plant quarantine announcements. FAO Plant Prot Bull 7:149–150

Anonymous (2005) Project coordinators report. All India Coordinated Research Project on MULaRP. IIPR, Kanpur (UP), 20 p

Ayub A, Rahman MZ, Ali S, Khatun A (1996) Fungicidal spray to control leaf rust of lentil. Bangladesh J Plant Pathol 12:61–62

Bakr MA (1991) Plant protection of lentil in Bangladesh. Paper presented in ICAR-ICARDA seminar. Lentils in South Asia, New Delhi, 11–15 March, 1991, 20 p

Basandrai D, Basand Rai AK, Kalia V (2003) Rust (*Uromyces viciae fabae*) resistance and yield potential of some promising lentil genotypes. Indian J Mycol Plant Pathol 33:129–131

Bashir M, Alam SS, Malik BA (1986) *In vitro* evaluation of fungicides against *Ascochyta lentis*. LENS Newsl 13:26–27

Beauchamp CJ, Morrall RAA, Slinkard AE (1986) Effect of scheduling application of benomyl, captafol and chlorothalonil on ascochyta blight of lentil. Can J Plant Pathol 8:260–268

Bedi S, Morrall RAA (1990) Ascochyta blight in lentil crops and seed samples in Saskatchewan in 1988. Can Plant Dis Surv 70:125

Beniwal SPS, Ahmed S, Tadesse N (1989) Effect of sun drying of lentil seeds on the control of seed borne *Ascochyta lentis*. LENS Newsl 16:27–28

Bhalla MK, Nozolillo C, Schneider EF (1992) Observations on the responses of lentil root cells to hyphae of *Fusarium oxysporum*. J Phytopathol 35:335–341

Bhatt JC, Sharma AK (1992) Effect of sowing dates on disease incidence in lentil in Kumaon hills. Indian Phytopathol 45:LXIX (Abstr.)

Bojdova J, Sinsky T (1990) Species spectrum of the *Fusarium* genus on lentil in Czechoslovakia. Lens Newsl 17:29–31

Bondartzeva-Montevercles, Vassilievsky NI (1940) A contribution to the biology and morphology of some species of *Ascochyta* on leguminosae. Acta Inst Bot Acad Sci (USSR). 1938, Ser II:354–376

Canonaco A (1937) A heavy attack by rust on lentil. Rev Path Veg XXVII 9:281–285

Carrion G, Romero A, Rico Vary V (1999) Use of *Verticillium lecanii* as a biocontrol agent against bean rust (*Uromyces appendiculatus*). Fitopathologia 34:214–219

Chauhan RS, Singh BM (1994) Effect of different duration of leaf wetness on pea rust development. Plant Dis Res 9:200–201

Couner RL, Bernier CC (1982) Host range of *Uromyces viciae fabae*. Phytopathology 72:687–689

Cromey MG, Mulholland RI, Russell AC, Jenonyn WA (1987) *Ascochyta fabae* f.sp. *lentis* on lentil in New Zealand. N Z J Exp Agric 15:235–238

Davatzi-Helena K (1980) A disease of lentil caused by *Ascochyta lentis* Vassilievsky. Ann Inst Phytopathol Benaki 12:256–260

De RK, Chaudhary RG (1999) Biological and chemical seed treatment against lentil wilt. LENS Newsl 26:28–31

El-Ahmed M, Mousell N (1986) Lentil wilt in south Syria. Arab J Plant Prot 4:30

El-Helaly AF (1939) Preliminary studies on the control of bean rust. Bull Minst Agric Egypt 201:19

Erskine W, Saxena MC (1991) Breeding lentil at ICARDA for southern latitudes. Paper parented in ICAR-ACARDA seminar 'Lentils in South Asia' New Delhi, 11–15 March, 1991, pp 1–9

Erskine W, Bayaa B, Dholli M (1990) Effect of temperature and some media and biotic factors on the growth of *Fusarium oxysporum* f.sp. *lentis* and its mode of seed transmission. Arab J Plant Prot 8:34–37

Eujayal I, Erskine W, Dayaa B, Baum M, Penu E (1998) *Fusarium vascular* wilt in lentil: inheritance and identification of DNA markers for resistance. Plant Breed 117:497–499

Food and Agricultural Organization (2006) F.A.O. Yearbook. Food and Agricultural Organization, Rome

Gossen BD, Morrall RAA (1986) Transmission of *Ascochyta lentis* from infected lentil seeds and plant residue. Can J Plant Pathol 6:233–237

Grewal JS (1988) Diseases of pulse crops: an overview. Indian Phytopathol 41:1–14

Gupta SK, Shyam KR (2000) Post infection activity of ergosterol biosynthesis inhibiting fungicides against pea rust. J Mycol Plant Pathol 30:414–415

Hiratsuka N (1933) Studies on *Uromyces fabae* and its relative species. Jpn J Bot 6:32–37

Iqbal SM, Bakhsh A, Ahmed S, Bashir M (1989) Field evaluation of fungicides against ascochyta blight of lentil. Sarhad J Agric 5:307–309

Iqbal SM, Hussain S, Tahir M, Malik BA (1992) Effect of fungicidal seed treatment on lentil germination and recovery of seedborne *Ascochyta fabae* f.sp. *lentis*. LENS Newsl 19:53–55

Izquierdo JA, Morse R (1975) First selection for wilt resistance in lentil (*Lens culinaris* Medik) in Uruguay. LENS Newsl 2:20–28

Jackes H, Webb AJ (1956) Field tests for control of broad bean rust. N Z J Sci Tech Sect A 38:157–159

Jeswani LM (1988) Lentil. In: Baldev B, Ramanujam S, Jains HK (eds) Pulse crops. Oxford and IBH Publ. Co. Pvt. Ltd, New Delhi, pp 199–214

Kaiser WJ, Hannan RM (1982) *Ascochyta lentis*: incidence and transmission in imported lentil seeds. Phytopathology 72:944

Kaiser WJ, Hannan RM (1986) Incidence of seed borne *Ascochyta lentis* in infected lentil germplasm. Phytopathology 76:355–366

Kaiser WJ, Hannan RM (1987) Seed treatment fungicides for control of seed borne *Ascochyta lentis* on lentil. Plant Dis 71:58–62

Kaiser WJ, Stenwood PC, Hannan RM (1989) Survival and pathogenicity of *Ascochyta fabae* f.sp. *lentis* in lentil seeds after storage for 4 years at 20–196°C. Plant Dis 73:762–764

Kaiser WJ, Hannan RM, Rogers JD (1994) Factors affecting growth and sporulation of *Ascochyta fabae* f.sp. *lentis*. Plant Dis 78:374–379

Kannaiyan J, Nene YL (1975) Note on the effect of sowing dates on the reaction of 12 lentil varieties to wilt disease. Madras Agric J 62:240–242

Kannaiyan J, Nene YL (1976) Reaction of lentil germplasm and cultivars against three root pathogens. Indian J Agric Sci 46:165–167

Kannaiyan J, Nene YL (1978a) Host range of *Fusarium oxysporum* f.sp. *lentis, Sclerotium rolfsii* and *Rhizoctonia solani* isolated from lentil. LENS Newsl 5:11–12

Kannaiyan J, Nene YL (1978b) Strains of *Fusarium oxysporum* f.sp. *lentis* and their pathogenicity on some lentil lines. LENS Newsl 5:8–10

Kapooria RG, Sinha S (1966) Studies on the host range of *Uromyces fabae* (Pers.) de Bary. Indian Phytopathol 19:229–230

Kaushal RP, Sharma C (1998) Effect of temperature, soil pH and phosphorus level on lentil wilt development. LENS Newsl 245:65–67

Khare MN (1980) Wilt of lentil. JNKVV, Jabalpur, p 155

Khare MN, Agarwal SC (1978) Lentil rust survey in Madhya Pradesh. Proceedings of All India Pulse Workshop, Baroda, 3 p

Khare MN, Sharma HC, Agarwal SC (1971) Studies on wilt of lentil. Annual report 1970–1971. Department of Plant Pathology, JNKVV, Jabalpur, 59 p

Khare MN, Agarwal SC, Jain AC (1979) Lentil diseases and their control, Technical bulletin. JNKVV, Jabalpur, p 29

Koul ML, Ambardar VK (1988) Prevalence of some fungal diseases of pulses in Kashmir. Indian J Mycol Plant Pathol 18:76–77

Kovacikova E (1970) Seed treatment of lentil and pea against some fungal diseases. Ochr Rost 6:117–126

Malik MR, Iqbal SM, Hussain S, Khan MSA, Malik BA (1992) Economic efficiency of chemical control of lentil blight. Sarhad J Agric 8:671–678

Mehrotra RS, Claudius GR (1973) Role of metabolites and enzymes in the root rot and wilt diseases of *Lens culinaris*. Indian J Mycol Plant Pathol 3:8–16

Mittal RK (1997) Effect of sowing dates and disease development in lentil as sole and mixed crop with wheat. J Mycol Plant Pathol 27:203–209

Mohy UD, Din G, Khan MA, Khan SMC (1999) Evaluation of seed and foliar applied fungicides to control lentil rust. Pak J Phytopathol 11:77–80

Morall RAA, Beauchamp CJ (1988) Detection of *Ascochyta fabae* f.sp. *lentis* in lentil seeds. Seed Sci Technol 16:383–390

Morrall RAA, Sheppard JW (1981) Ascochyta blight of lentil in western Canada, 1978–1980. Can Plant Dis Surv 61:7–13

Mukherjee PK, Upadhyay JP, Mukhopadhyay AN (1989) Biological control of *Pythium* damping-off of cauliflower by *Trichoderma harzianum*. J Biol Control 3:119–124

Myalova LA (1973) *Fusarium* wilt of lentil. Zaschita Roslenii 4:53

Nattrass RM (1932) Annual report of the mycologist for 1931 annual report, Department of Agriculture, Cyprus, pp 56–64

Nene YL, Hanounik SB, Qureshi SH, Sen B (1988) Fungal and bacterial foliar diseases of pea, lentil, faba bean and chickpea. In: Summerfield RJ (ed) World crops: cool season food legumes, Current plant science and biotechnology in agriculture No. 5. Kluwer Academic Publishers, Dordrecht, pp 577–589

Nene YL, Kannaiyan J, Saxena GC (1975) Note on performance of lentil varieties and germplasm cultures against *Uromyces fabae* (Pers.) de Bary. Indian J Agric Sci 45:177–178

Padwick GW (1941) Report of the imperial mycologist. Sci Rep Agric Res Inst New Delhi 1939(40):94–101

Pande SK, Srivastava RK, Shahi HN (1995) Fungicidal evaluation against rust of lentil. Ann Plant Prot Sci 3:164

Patil MK (1933) India: disease in Bombay presidency. Int Bull Plant Prot VII 11:246

Pedersen EA, Morall RA, McCartney HA, Fitt BDL (1994) Dispersal of conidia of *Ascochyta fabae* f.sp. *lentis* from infected lentil plants by simulated wind and rains. Plant Pathol 143:50–55

Pederson EA, Morrall RAA (1994) Effect of non-host and fungicide treated barriers on horizontal spread of *Ascochyta* blight of lentil. Can J Plant Pathol 16:317–325

Prasada R, Singh SP (1975) Sexual behaviour of *Uromyces fabae*. Indian J Mycol Plant Pathol 5:139–144

Prasada R, Verma UN (1948) Studies on lentil rust, *Uromyces fabae*. Indian Phytopathol 1:142–146

Roundhill SJ, Fineran BA, Cole ALJ, Ingerfeld M (1995) Structural aspects of *Ascochyta* blight. Can J Bot 73:485–497

Russell AC, Hagerty G (1992) Fungicide seed treatments for the control of *Ascochyta fabae* f.sp. *lentis* on lentil. In: Popay AJJ (ed) Proceeding of 44th New Zealand plant protection conference, Wellington, New Zealand, 11–12 August 1992, Rotorua, New Zealand, pp 72–74

Russell AC, Cromey MG, Jermyn WA (1987) Effect of seed treatment on seedborne *Ascochyta* of lentil. Proc Ann Can Agron Soc N Z 17:15–18

Sattar A (1934) A comparative study of the fungi associated with blight disease of certain cultivated leguminous plants. Trans Br Mycol Soc 18:278–301

Saxena HC (1987) Biological control of wilt complex of lentil by *Trichoderma harzianum*. M.Sc. thesis, G.B. Pant University of Agriculture and Technology, Pantnagar, 100 p

Saxena DP, Khare MN (1988) Factors influencing vascular wilt of lentil. Indian Phytopathol 41:69–74

Saxena DR, Saxena M, Sexena RR, Khare MN (1992) Root morphology anatomy in relation to wilt incidence in lentil. LENS Newsl 19:46–49

Sharma KB, Agnihotri JP (1972) Pathogenicity and variability of lentil wilt. Indian J Mycol Plant Pathol 2:170

Sharma HC, Deshkar MV, Khare MN (1971) Factors affecting *Fusarium* wilt of lentil (*Lens esculenta*

Medik.). In: Proceedings of the pathological wilting of plants, University of Madras, India, 42 p

Shukla P, Amin KS (1991) Integrated disease management of rust of lentil (*Lens culinaris* L.) caused by *Uromyces fabae* (Pers.) de Bary. Indian J Mycol Plant Pathol 21:107

Singh K (1985) Effect of seed treatment on lentil rust development. Lens Newsl 12:26–27

Singh D (2003) Epidemiology and management of rust of field pea. Ph.D. thesis submitted to G.B. Pant University of Agriculture and Technology, Pantnagar, 220 p

Singh G, Dhingra KK (1980) Effect of sowing dates and varietal reaction on rust of lentil. J Res PAU 17:233–235

Singh G, Mukhopadhyay AN (2000) Biological potential of mutants of *Gliocladium virens* for wilt complex of lentil. Legume Res 23:133–135

Singh SN, Srivastava SK, Agarwal SC (1985) Relative efficacy of fungicides in the control of seedling blight and rust of lentil. Indian J Agric Sci 55:284–286

Sugha SK, Chauhan RS, Singh BM (1994) Sensitivity of aeciospores and uredospores of the pea rust pathogen to selected systemic fungicides. Trop Agric 71:27–30

Ujevic I, Kovacikovo E, Urosevic B (1965) *Botrytis parasitica* and *Fusarium oxysporum* as dangerous parasites of lentil. Polnopospodastvo 11:584–594

Vasudeva RS, Srinivasan KV (1952) Studies on the wilt disease of lentil (Lens esculenta Moench.). Indian Phytopathol 5:23–32

Vishunavat K, Shukla P (1982) Effect of seed treatment on lentil mycoflora. Indian Phytopathol 35:132–133

Yuen GY, Steadman JR, Lindgren DT, Schaff DA, Jochum C (2001) Bean rust biological control using bacterial agents. Crop Prot 20:395–402

Botrytis Gray Mold of Chickpea (*Cicer arietinum* L.)

7

H.S. Tripathi and Usha Suyal

7.1 Introduction

Chickpea (*Cicer arietinum* L.) commonly known as gram is a versatile crop among the grain legumes and ranks first among the pulses both in acreage and production. It is an important pulse crop in over 45 countries of Asia, Africa, the Americas, and Oceania, with an annual production of 8.62 million tonnes from 11.12 million hectares (FAO 2005). Chickpea is used as an important source of protein in human nutrition and cattle feed and is also used to improve soil fertility by biological nitrogen fixation. Chickpea usually receives few inputs other than labor, insecticides, and seed. The major constraints to production include disease susceptibility of local varieties, environmental stresses, drought, diseases, pests, and poor crop management. Worldwide losses from this fungus account for 20 % of the harvest of the affected crops, and their cost is estimated at 10–100 billion euros per year (Genoscope 2008). The fertile land and frequent rain during crop growth are responsible for dense canopy creating a microenvironment highly variable to growth and development of the pathogen. Several epidemics of BGM causing complete crop loss in the major chickpea-producing countries have been reported. Botrytis gray mold (BGM) caused by *B. cinerea* Pers. Ex. Fr. is the second most potentially important disease of chickpea after Ascochyta blight caused by *Ascochyta rabiei* [Pass] Lab. BGM can devastate chickpea, resulting in complete yield loss in years of extensive winter rains and high humidity (Reddy et al. 1993; Pandey et al. 1982).

7.2 Geographical Distribution, Economic Importance and Losses

The occurrence of Botrytis gray mold on chickpea was first reported by Shaw and Ajrekar in 1915. Joshi and Singh (1969) observed BGM of chickpea in epiphytotic form in Nainital *Tarai* area of Uttar Pradesh, now part of Uttarakhand state. Since 1967–1968 *Botrytis cinerea* has caused vast devastation in chickpea crop grown in parts of West Bengal, Bihar, Uttar Pradesh, Rajasthan, Haryana, Punjab, and Himachal Pradesh (Singh 1997). The disease was responsible for heavy losses in the Indo-Gangetic plains of India during 1979–1982 (Grewal and Laha 1982) and caused 70–100 % losses in yield at Central State Farm in Hissar and several parts of Punjab. During 1978–1979 an epidemic of BGM destroyed chickpea crop completely over an area of 20,000 ha in the states of Punjab, Haryana,

H.S. Tripathi (✉) • U. Suyal
Centre of Advance Studies in Plant Pathology, G.B. Pant University of Agriculture and Technology, Pantnagar 263 145, Uttarakhand, India
e-mail: tripathihs@gmail.com

L.P. Awasthi (ed.), *Recent Advances in the Diagnosis and Management of Plant Diseases*,
DOI 10.1007/978-81-322-2571-3_7

Uttar Pradesh, and Bihar (Singh et al. 1982; Grewal and Laha 1982). In Bangladesh, the damage caused by the disease was estimated to be 70–80 % in 1989 (Bakr et al. 1997). Mahmood and Sinha (1990) reported more than 60 % losses in grain yield. Reddy et al. (1993) observed an epiphytotic of gray mold in *Tarai* region of Nepal, where the yield losses were recorded up to 100 %. In Nepal the disease occurs almost every year, and an estimated loss of 66 % in the experimental field and about 15 % in the farmers' field was reported (Joshi 1992). The effects of BGM on pod yield depend on the onset of the disease in relation to crop growth and disease severity, both of which depend largely on weather conditions and inoculum level of the pathogen.

7.3 Causal Organism

BGM of chickpea is caused by *Botrytis cinerea* Pers. Ex. Fr. The asexual stage of the necrotrophic fungus *B. cinerea* (Moniliaceae, Hyphales) is dominant on chickpea crops. *B. cinerea* grown on potato dextrose agar (PDA) has a white, cottony appearance, which turns light gray with age. The mycelium is septate, brown, and 8–10 μm wide. Young hyphae are thin and hyaline. Conidia and conidiophores are not in pycnidia or acervuli. Conidiophores are lighter brown than hyphae, with hyaline tip, septate, and 8–24 μm wide. Tips of conidiophores or their branches are slightly enlarged and bear small pointed sterigmata. Conidia are hyaline, one-celled oval or globose or short cylindrical, and born in clusters at the tips of conidiophore branches.

Conidia from chickpea host measure 4–24 × 4–18 μm (average 14.9 × 8.4 μm) and from potato dextrose agar 4.16 × 4–10 μm (average 7.4 × 6.1 μm). Conidia readily germinate in water or even on the host surface under conditions of high humidity by producing 1–3 thin, hyaline germ tubes. The sporodochia on the host surface are fairly large, measure 0.5–5 μm in diameter consisting of densely interwoven brown septate hyphae, and produce round to oval unicellular conidia measuring 4–8 μm (av. 5.26 μm) diameter. These conidia do not germinate; the sporodochia soon become non-soporiferous and change into hard sclerotia measuring 2–11 × 2–7 (5–6 × 3–8) μm (Joshi and Singh 1969). However, no white hyphal matrix is produced and these sclerotial structures resemble those of *Sclerotinia* spp. The perfect stage of *Botrytis cinerea* is *Botryotinia fuckeliana* (Groves and Loveland 1985). However, there is no report of the occurrence of perfect stage of the fungus on chickpea. Joshi and Singh (1969) found that sometimes only sclerotia are found in old cultures of *B. cinerea* and sporodochia were not seen. These may or may not be surrounded by white mycelial growth characteristics of *Sclerotinia* sp. In 2-week-old cultures, they form an abundance of sporodochia and sclerotia (Joshi and Singh 1969). The sexual stage germinates from fertilized sclerotia by the emergence of apothecia that release sexually produced ascospores (Faretra and Grindle 1992). Apothecia formation requires either two sexually compatible isolates (MAT 1–1 and Mat 1–2) or a pseudo-homothallic isolate (MAT-1/2) (Faretra and Grindle 1992). There are no reports of the sexual state of *B. cinerea* occurring naturally on chickpea stubbles. However, it has been produced under laboratory conditions in India (Singh 1997).

7.4 Disease Diagnosis

7.4.1 Characteristic Symptoms

All the aerial parts of chickpea are susceptible to the disease with growing tips and flowers being the most vulnerable (Bakr et al. 1992; Grewal and Laha 1982; Haware and McDonald 1992; Haware 1998; Bakr et al. 1997). Drooping of the infected tender terminal branches is a common field symptom (Haware and McDonald 1992; Pande et al. 2005a, b). According to Joshi and Singh (1969), initial symptoms appear on stem, leaves, inflorescence, and pods as gray or dark brown lesions covered with erect hairy sporophores. Stem lesions are 10–30 mm long which later girdles the stem completely. Tender branches break off at the points where gray mold causes rotting. Affected leaves and flowers turn into a rotting mass. In the field, the disease first appears in isolated patches when the crop has achieved maximum canopy, and in the morning relative humidity is very high with low temperature. As the disease advances, patches of disease plant become more prominent, spreading slowly in the entire field. According to Laha and Grewal (1983), symptoms appeared on leaflets, petioles, and growing tips as water-soaked lesions. The lesions are brown and limited in size. However, under conditions of high humidity, leaflets got blighted and bear abundant fungal fructifications.

On thick, hard stems, the gray mold growth is gradually transformed into a dirty, gray mass containing dark green to black sporodochia. The sclerotia are small, dark bodies and should not be confused with larger, black or dark brown sclerotia embedded in white mycelium of *Sclerotinia sclerotiorum* (Lib.) de Bary (Joshi and Singh 1969).

7.4.2 Seedling Rot

The pathogen *Botrytis cinerea* is one of the many fungi associated with seedling disorders of chickpea (Cother 1977; Bretag and Mebalds 1987), creating a soft rot (Burgees et al. 1997). In most chickpea-growing regions of the world, foliar infection is considered most important, whereas in Australia, soft rot of young seedlings resulting from seed-borne infection is also important and can result in total crop failure (Burgees et al. 1997). Symptoms include poor emergence, yellowing, wilting, and death of seedlings and pale yellow to light tan discoloration of the taproot. Most plants that develop soft rot become flaccid and then die within a few days. Plants seldom recover from the disease.

7.5 Epidemiology

Reports on epiphytotics of botrytis gray mold from different parts of the world indicated the existence of definite and efficient mechanisms of survival of the pathogen from one season to another. The information regarding the survival and epidemiology is scanty as far as the botrytis gray mold of chickpea is concerned.

7.5.1 Survival

B. cinerea may survive in nature from one season to the next in several ways in the absence of chickpea crop. The survival may be through continued activity and growth, either parasitically on other hosts or saprophytically on available dead material or through entering upon an inactive phase of the life cycle or a dormant resting structure.

7.5.2 Crop Debris

Several workers have indicated the importance of infected debris as a source of survival of the pathogen and infection of the crop in the next season (Mahmood and Sinha 1990; Nene and Reddy 1987; Grewal 1988; Singh and Kaur 1989; Singh and Tripathi 1992). Coley-Smith (1980) found that conidia buried beneath the soil surface showed a rapid drop in viability and disappeared within 10–12 weeks, but those placed on soil surface survived for about 50 weeks. Being basically a saprophyte, it has been observed to survive saprophytically in soil and/or on infected crop debris (Mahmood and Sinha 1990; Grewal 1988; Singh and Kaur 1989).

According to Meeta et al. (1986), *B. cinerea* survives in infected plant debris for 180 days at a soil depth of 6 cm but not at 2 or 4 cm. Grewal (1988) indicated that the fungus remained viable on infected plant debris present as admixture in seed lots. The fungus was found to be viable in infected seed and plant debris stored at 18 °C for 5 years (Grewal 1988). But, Singh and Kaur (1990) found that *B. cinerea* did not survive beyond 8 months in chickpea debris at a depth of 10 cm in the soil and that the survival and recovery of the fungus was more at 5–10 °C. Singh and Tripathi (1992, 1993) observed the survival of *B. cinerea* in infected chickpea debris at a depth of 25 cm for 8 months, i.e., until the following season.

7.5.3 Survival in Seed

Cother (1977) was the first to demonstrate the seed-borne nature of *B. cinerea*. He showed that the majority of seeds inoculated with *B. cinerea* failed to germinate in soil as they were colonized by the fungus and covered with sclerotia. It has been observed by several workers that in the event of severe infections, pods are also attacked resulting in no seeds or only small and shriveled infected seeds. In nature, thus, infected seed is thought to be the primary source of infection (Cother 1977; Grewal and Laha 1983; Haware et al. 1997; Meeta et al. 1988; Guraha et al. 2003; Singh and Tripathi 1993).

Fungus was found to be externally and internally seed-borne to the extent of 8.2 and 2.5 %, respectively, in naturally infected seeds stored at 18 °C for 5 years (Grewal 1988). In another study seeds collected from naturally infected plants from three cultivars G-543, H-208, and H-355 of chickpea, 8.0–18.5 % seeds were found infected externally and internally with *B. cinerea* (Grewal and Laha 1983).

Haware et al. (1999) reported that *B. cinerea* was recorded up to 56 % in seed samples collected from Faisalabad (Pakistan). They found that naturally infected seed showed 7–17 % external and 1–1.5 % internal seed infection. Despite the fact that infected seeds carry on the inoculums, for a considerable period, the role of seed-borne inoculums, after getting exposed to soil environment, in epidemiology, is not yet been established (Tripathi and Rathi 1992).

7.5.4 Sclerotia and Chlamydospores

The fungus *B. cinerea* is known to produce sclerotia on crop stubbles of many host species. The sclerotia are thought to be the main means of the fungal long-term survival (Coley-Smith 1980). In Europe, apothecia emerge from the fertilized sclerotia, and wind-dispersed ascospores are released mainly in the spring after chilling and periods of high rainfall on *Vicia* beans (Harrison 1988). Sclerotia develop on the previous season's chickpea stubble in Australia after exposure to cold (> 10 °C) winter temperatures. As daytime temperatures increase in spring, sclerotia germinate asexually, forming conidia on conidiophores. The sclerotia remain viable for the rest of the growing season but do not survive the following hot, dry summer conditions; hence, sclerotia are not considered to be a means of long-term survival in Australia.

7.5.5 Alternative Hosts

Due to the wide host range of this pathogen, the role of alternative hosts is likely to play an important part in survival from one chickpea crop to another (Coley-Smith 1980; Haware 1998; Knights and Siddique 2002; Pande et al. 2005a, b). However, further studies are required to understand the host-specific pathogenicity of *Botrytis* isolates of chickpea.

7.5.6 Disease Development

There is a wealth of literature available on the temperature and relative humidity requirements of *B. cinerea* on many crops of importance. It should, however, be noted that the temperature and relative humidity requirements for *B. cinerea* appear to be influenced by the host plant and even by the plant part being infected (Elad et al. 1992). On chickpeas, the optimum temperature for sporulation and conidial germination is 25 °C (Mahmood and Sinha 1990; Singh 1997) and 20 °C (Rewal and Grewal 1989a), respectively, with 5 and 30 °C being the minimum and maximum extremes for conidial germination. However, different isolates were found to require differential light intensities and relative humidity for conidial germination (Rewal and Grewal 1989a).

BGM may develop rapidly over time and space, depending on the environmental conditions. Relative humidity, leaf wetness, and temperature are the most important factors (Tripathi and Rathi 1992; Butler 1993; Pande et al. 2002). Bakr and Ahmed (1992) found that disease increased at temperatures of 17–28 °C and

70–97 % relative humidity. In Bangladesh, maximum disease severity was recorded at a temperature range of 20–28 °C (Bakr et al. 1997) and 25–30 °C in India (Reddy et al. 1990; Tripathi and Rathi 1992). In the Indian subcontinent, BGM epidemics have occurred in years with high rainfall and a high number of rainy days (Bakr and Ahmed 1992; Joshi 1992; Tripathi and Rathi 1992; Davidson et al. 2004). The duration of leaf wetness appears to have some influence on the development of BGM on chickpea. Disease severity increased with leaf wetness periods greater than 12 h/day (Singh and Kapor 1984). The epidemics can spread rapidly at 95 % or above relative humidity and up to a maximum temperature of approximately 25 °C in a dense crop canopy. Under such conditions, the disease cycle can be completed in 7 days (Haware 1998).

7.5.7 Host Range

B. cinerea is a facultative parasite with a very wide host range in the temperate and subtropical regions (Singh 1970). The inoculums are always more or less present in the environment waiting for the congenial weather to become active (Nene 1984). It causes gray mold diseases in a number of crop plants, such as strawberry, grapevine, apple, cabbage, carrot, cucumber, eggplant, lettuce, pepper, squash, tomato, and several ornamentals such as chrysanthemum, dahlia, lily, roses, gladiolus, tulips, etc. (Agrios 1978). Rathi and Tripathi (1991) reported several hosts (cultivated and weeds) of *B. cinerea* in Nainital *Tarai*.

7.5.8 Pathogenic Variability

The pathogen *B. cinerea* is reported to have extreme variability and adaptability to a wide range of environmental conditions. Joshi and Singh (1969) and Singh (1970) observed the formation of sclerotial and/or sporodochial bodies on *B. cinerea*-infected chickpea plants in the *Tarai* region of Nainital, India, which were not found later from the same area (Pandey 1988). Singh and Bhan (1986) and Rewal and Grewal (1989b) identified 4 and 5 pathotypes, respectively, among the *B. cinerea* isolates collected from Northern India. Kishore (2005) differentiated eight chickpea isolates of *B. cinerea* collected from India and Nepal into distinct pathotypes based on their morpho-cultural characters and reaction on 39 differential lines and RAPD markers.

Molecular markers such as microsatellites are powerful tools for accurate detection of genetic diversity because they are highly polymorphic across numerous loci and are reproducible. In chickpea, microsatellites have revealed genetic variation among isolates of *Ascochyta rabiei*. A recent study that used microsatellite DNA markers developed specifically for the *B. cinerea* genome (Fournier et al. 2002) revealed genetic variation in *B. cinerea* isolates of chickpea from 4 regions of Bangladesh, India, and Nepal (Isenegger et al. 2005).

A UPGMA tree revealed that isolates from Bangladesh are quite diverse and several were closely related to isolates from India and Nepal (Isenegger et al. 2005). Isolates from subpopulations from Bangladesh showed potential for a highly adapted pathogenic group to chickpea, which can threaten (or break down) long-term control with fungicides.

Previously, molecular evidence revealed the role of genetic recombination in *B. cinerea* from grapevine in France (Giraud et al. 1997). This is important, as genetic recombination can generate new genotypes; hence, genetic diversity can spread quickly via asexual conidia. In other studies in *B. cinerea*, molecular markers have revealed high genetic diversity and high gene flow among populations from vegetable crops in Europe (Alfonso et al. 2000; Moyano et al. 2003).

7.5.9 Biochemical Basis of Host Plant Resistance

Unlike other major plant pathogen system of crop plants, detailed investigations have not been undertaken on the infection process of *B. cinerea* on chickpea plant, and the biochemical basis of BGM resistance in chickpea has not been determined. Preliminary investigations on the infection process recorded that inoculated spores of the fungus germinate within 6–8 h and the germ tube proliferates saprophytically and forms a mycelial mat on the leaf surface. During the period of proliferation and formation of a mycelial mat, the hyphal tips, in direct contact with the host surface, swell to form appressoria and form the infection hyphae, which penetrate directly through the cuticle and form the subcuticular and subepidermal mycelium. In some cases, the hyphae penetrate directly through the host surface, although penetration through stomata has also been observed (Pandey 1988). After penetration the infection hyphae grow and ramify in the leaf tissue subcuticularly and subepidermally. Mycelium grows within the mesophyll cells, which thickens and branches after penetration. The pathogen causes extensive damage to the leaf tissue by destroying epidermal and mesophyll cells, most probably by degrading the cell walls even in advance of invading hyphae.

It was observed that palisade and spongy parenchyma cells in the resistant genotype (ICC 10302) were more compact than in the moderately resistant (GG 588) and susceptible genotype (H 355). No significant anatomical alterations were observed in any of the genotypes up to 48 h of germination. Degradation of mesophyll cells was quite evident in most parts of the susceptible cultivar 72 h after inoculation, which became more pronounced after 96 h resulting in complete necrosis of the leaf after 120 h. In moderately resistant and resistant parents, the breakdown of mesophyll cells was first recorded 96 h after inoculation. Consequently, yellowing was observed after 120 h and complete degradation of mesophyll cells was quite pronounced after 120 or 144 h (Pandey 1988). It was observed that under high humidity, even the field resistant genotypes were infected by *B. cinerea*. However, the infection and colonization in resistant cultivars were delayed by 24–48 h as compared with the susceptible cultivars. Mohhamadi (1987) reported that leaf surface inhibitors, probably phenolic in nature, are important in resistance in chickpea under field conditions, and a high correlation was observed between total phenolic content of the leaf washings and degree of resistance of the genotypes. In the presence of inhibitors, spore germination and germ tube growth were delayed for 6–8 h, and this time is sufficient for desiccation of spores under tropical conditions. However, under humid conditions, there was no desiccation of the spores and germ tubes on the leaf surface; hence, even the field tolerant varieties became susceptible.

Total phenolic content, sugars, antifungal peptides, and phytoalexins are observed to be associated with BGM resistance. *Botrytis* spp. is known to be "high-sugar" pathogen that usually attacks plant tissues with more sugar content (Horsfall and Diamond 1957). Mitter et al. (1977) observed that healthy chickpea plants of a BGM-resistant genotype ICC 1069 had significantly lower total soluble sugars and free amino acids and higher total phenol level than the susceptible var. BGM 408. The amount of sulfur-containing amino acids, methionine and cysteine, was almost double in genotype ICC 1069 compared with BGM 408. Further, shoot tips of both the cultivars had higher quantities of sugar and free amino acids and low content of phenols compared with the middle and lower leaves.

Two antifungal peptides with novel N-terminal sequences, designated as cicerin and arietin with molecular weights of 8.2 and 5.6 kDa, respectively, were found in chickpea seeds. Arietin exhibited a higher translation-inhibiting activity in a rabbit reticulocyte lysate system and was found to be highly antifungal to *B. cinerea*, *Mycosphaerella arachidicola*, and *F. oxysporum* (Ye et al. 2002).

A pterocarbon phytoalexin, maackiain, was found associated with BGM resistance in *C. bijugum* Rech. f., a wild relative of chickpea. The concentration of maackiain in *C. bijugum* foliage was 200–300 μg/g, compared with <70 μg/g in susceptible species. After inoculation with *B. cinerea*, maackiain concentration increases to >400 μg/g in *C. bijugum*, whereas no significant increase was recorded in the susceptible species.

7.6 Disease Management

7.6.1 Cultural Practices

Cultural management of BGM in recent years has been very well demonstrated in India, Bangladesh, and Nepal. Early sowing, high seed rates, closer row spacing, and traditional use of bushy genotypes result in dense canopies, which favors development of BGM in chickpea. Thus, techniques which minimize canopy density may alleviate disease intensity. Haware and McDonald (1992) reported that delayed sowings reduced BGM incidence even in susceptible cultivars, but significantly reduced the grain yields. Singh (1997) also observed that the late sown crop (around November 20) in Punjab, India, showed significantly low incidence of BGM. Bakr et al. (1993) reported that wider spacings alone reduced BGM disease significantly in Bangladesh. Haware et al. (1997) and Bakr et al. (1997) also reported that wider row spacing reduced BGM disease incidence and resulted in higher grain yields. Combination of wider row spacing, intercropping with linseed, and two spray applications of carbendazim at 0.2 % significantly reduced BGM severity and increased grain yield of chickpea and linseed.

7.6.2 Fungicides

Bakr et al. (1993) reported that seed treatment with bavistin + thiram (1:1), indofil M-45, thiabendazole, ronilan, rovral, and bavistin at 0.3 % controls seed-borne inoculum of *B. cinerea*.

Foliar spray with ronilan, bavistin + thiram combination at 0.1 %, or bavistin alone at 0.2 % provided complete protection to chickpea plants against aerial infection by *B. cinerea* (Grewal and Laha 1983). Leaf spot of French bean caused by *B. cinerea* and *B. fabae* was successfully controlled by the use of Dithane M-45+ Triton in Egypt (Mansour 1980). Singh and Kaur (1990) reported that seed treatment with triadimefon (0.1 %), bavistin + thiram (0.3 %), Dithane M-45 (0.3 %), or Baytan (0.1 %), together with one foliar spray of Dithane M-45, hexacap, thiram, thiabendazole, baytan, or triadimefon at 50 days after sowing or at the appearance of first symptoms, completely controlled both primary infection and secondary infection of BGM. Haware and McDonald (1993) suggested the judicious of vinclozolin (0.2 %) in the integrated management of BGM. Singh et al. (1986) reported that foliar spray with indofil M-45, thiabendazole, baytan, bayleton, or thiram during crop growth stages in February–March controlled foliar infection.

Haware et al. (1997) reported that one spray with vinclozolin (0.2 %) at the time of flowering in the integrated management system reduced BGM incidence. However, Bakr et al. (1997) observed that two foliar sprays with vinclozolin (0.2 %) were required to control BGM in Bangladesh.

7.6.3 Biological Control

Although repeated fungicide application can alone achieve effective management of BGM in chickpea, biological control of *B. cinerea* using species of *Trichoderma* has been reported in some fruit and vegetable crops (Tronsmo 1986; Nelson and Powelson 1988; Elad 1994). Mukherjee and Haware (1993) isolated species of *Trichoderma* from the rhizosphere of chickpea, tested them against *B. cinerea* in the laboratory, and identified the most effective isolate of *Trichoderma viride* in controlling BGM of chickpea. Mukherjee et al. (1995) in further tests isolated fungicide (vinclozolin)-tolerant isolates for use along with vinclozolin in integrated BGM management system. Haware et al. (1997) reported that there were no difference in BGM incidence and grain yield of chickpea between three sprays of *T. viride* (10^7–10^8 spores/ml) and three sprays of vinclozolin (0.2 %). Burgees et al. (1997) reported that seed treatment with *Gliocladium roseum* suppressed the sporulation of *B. cinerea* on chickpea seed. They also found that sporulation of *B. cinerea* on chickpea seed naturally infected or inoculated with *B. cinerea* was suppressed by seed treatment with conidial suspension of *Gliocladium roseum* at 10^7 and 10^8 conidia/ ml, respectively. There was no significant effect of *Rhizobium* on disease suppression by *G. roseum*, and treatment with *G. roseum* at 10^8 did not reduce nodulation. Haware et al. (1999) reported biocontrol potential of *T. viride* isolate T-15 (isolated from chickpea rhizosphere) on *B. cinerea* in chickpea under controlled environmental conditions.

7.6.4 Host Plant Resistance

Extensive screening of the available chickpea material has been done at ICRISAT and some other agricultural universities to obtain resistance against Botrytis gray mold (Haware and Nene 1982; Pandey et al. 1982; Chaubey et al. 1983; Singh et al. 1982); more than 600 genotypes were screened. Six lines, GL-635, GL-699, GL-907, GL-926, GL-929, and GL-930, were rated as resistant. At Pantnagar nine other lines, viz., ICC-466, ICC-478, ICC-662, ICC-755, ICC-756, ICC-799, ICC-800, ICC-1591, and NEC-138-2, were also found resistant in (rating 3), and eight lines, viz., GL-776, GL-777, GL-784, ICC-4000, ICC-4950, ICC-5033, P-1528-1-1, and P-0.9, were moderately resistant (rating 5). Pandey et al. (1982) evaluated chickpea germplasms and varieties and classified GNG-3, C-235, and BG-249 as resistant and GNG-16, H-77-4, BG-243, H-78-84, BG-406, DG-77-27, BC-250, BG-413, BG-407, BG-408, Type-3, DG-77-29, RSG-47, DG-77-40, and Radhey as tolerant for gray mold. Chaubey et al. (1983) and Rathi et al. (1993) screened chickpea for *Botrytis* gray mold resistance and reported two lines, viz., ICC-1069 and ICC-7574, as resistant and one line, ICC-7574, as tolerant. Lines ICC-1069, ICC-7574, and NEC-138–2 also have been found tolerant to disease in an isolation plant propagator at ICRISAT (Haware and Nene 1982).

Tripathi and Rathi (1992) screened 10,000 chickpea (*Cicer arietinum* L.) accessions/germplasm lines against *Botrytis cinerea*. The following ICC lines were resistant: 466, 478, 662, 755, 756, 799, 800, 1069, 1591, 7574, and 10302. ICCL-87322 was also resistant. Lines GL-776, GL-777, GL-784, ICC-4000, ICC-4950, ICC-5033, P-1528-1-1, and P-0-9 were moderately resistant. Most of these resistant/moderately resistant accessions were *kabuli* type with erect habit.

Pande et al. (2006) screened a chickpea minicore collection composed of 211 germplasm accessions representing the diversity of global chickpea germplasm collection of 16,991, maintained at the International Crops Research Institute for the semiarid tropics to identify sources of multiple-disease resistance. They observed high level of resistance to *Fusarium* wilt, where 21 accessions were asymptomatic and 25 resistant. In all, 355 and 6 accessions were moderately resistant to *Ascochyta* blight, BGM, and dry root rot, respectively.

7.6.5 Integrated Disease Management

An adequate level of genetic resistance to BGM is not available in the cultivated genotypes, and fungicides become ineffective during conditions of high disease pressure. Hence, integrated disease management (IDM) using the available management options is essential to successfully manage the disease and mitigate yield losses. Chemical control of BGM combined with wider row spacing (Reddy et al. 1993) or the use of *T. viride* (Agarwal et al. 1999; Agarwal and Tripathi 1999; Haware et al. 1999) as a biocontrol agent has been attempted, and use of tolerant genotype ICCL 87322 in combination with wider row spacing and spraying with bavistin was the best combination followed by the use of the tolerant genotype ICCL 87322 in combination with wider row spacing and intercropping with linseed. Judicious use of fungicides as a seed treatment and/or foliar spray in an IDM system could be economical and affordable to the resource-poor farmer.

An IDM program involving cultivation of a BGM-tolerant Avarodhi, soil application of diammonium phosphate, wider row spacing (60 cm), seed treatment with carbendazim + thiram (3 g/kg seed), and need-based foliar application of carbendazim have been devised. This IDM program was evaluated in farmers' participatory research in two districts of Nepal during 1998–1999 crop seasons as a collaborative research activity between ICRISAT, Nepal Agricultural Research Council (NARC), and Natural Resources Institute (NRI), UK, and has resulted in a 400 % increase in grain yields and 300 % increase in net income.

7.7 Future Outlook

In chickpea, BGM is a devastating disease and extensive studies on the biology of the pathogen and screening programs to identify host-plant resistance have failed. Despite the extensive investigations in other hosts, the infection process of *B. cinerea* on chickpea has not been studied. Also, very little is known about the resistance mechanisms of chickpea against *B. cinerea*. Knowledge of the infection process and host defense mechanisms will help in devising management strategies for BGM. Hence, IDM programs suitable for adoption by resource-poor farmers should be emphasized. It is advised that BGM management in chickpea should be based on the location-specific disease predictive models. Transgenic plant technology using PGIPs and other antifungal proteins could be the possible approach for imparting disease resistance to commonly adapted cultivars in the future.

References

Agarwal A, Tripathi HS (1999) Biological and chemical control of botrytis grey mould of chickpea. J Mycol Plant Pathol 29:52–56

Agarwal A, Tripathi HS, Rathi YPS (1999) Integrated management of gray mould of chickpea. J Mycol Plant Pathol 29:116–117

Agrios GN (1978) Plant pathology. Academic, London, p 576

Alfonso C, Raposo R, Melgarejo P (2000) Genetic diversity in *Botrytis cinerea* populations on vegetable crops in greenhouses in south-eastern Spain. Plant Pathol 49:243–251

Bakr MA, Ahmed F (1992) Botrytis gray mold of chickpea in Bangladesh. In: Haware MP, Faris DG, Gowda CLL (eds) Botrytis gray mold of chickpea. Summary proceeding of the BARI/ICRISAT working group meeting. ICRISAT, Patancheru, pp 10–12

Bakr MA, Rahman MM, Ahmed F, Kumar J (1993) Progress in the management of Botrytis gray mold of chickpea in Bangladesh. In: MP Haware, CLL Gowda, D McDonald (eds) International Crops Research Institute for the Semi-Arid Tropics (ICRISAT). Recent advances in research on Botrytis gray mold of chickpea: summary proceedings of the second Working Group Meeting to discuss collaborative research on Botrytis Gray Mold of Chickpea, ICRISAT, Patancheru, pp 17–18

Bakr MA, Hossain MS, Ahmed AU (1997) Research on Botrytis grey mould of chickpea in Bangladesh. In: Haware MP, Lenne JM, Gowda CLL (eds) Recent advances in research on Botrytis grey mould of chickpea. ICRISAT, Patancheru, pp 15–18

Bretag TW, Mebalds MI (1987) Pathogenicity of fungi isolated from *Cicer arietinum* L. grown in North-Western Victoria. Aust J Exp Agri 27:141–148

Burgees DR, Bretag T, Keane PJ (1997) Biocontrol of seed borne *Botrytis cinerea* in chickpea with *Gliocladium roseum*. Plant Pathol 46:298–305

Butler DR (1993) How important is crop microclimate in chickpea botrytis gray mold? In: Haware MP, Gowda CLL, McDonald D (eds) Recent advances in research on botrytis gray mold of chickpea. ICRISAT, Patancheru, pp 7–9

Chaubey HS, Beniwal SPS, Tripathi HS, Nene YL (1983) Field screening of chickpea for resistance to Botrytis gray mold. Int Chickpea Newsl 8:20–21

Coley- Smith JR (1980) Sclerotia and other structures in survival. In: Coley- Smith JR, Verhoeff K, Jarvie WR (eds) The biology of Botrytis. Academic, London, pp 85–114

Cother EJ (1977) Isolation of important pathogenic fungi from seeds of chickpea. Seed Sci Technol 5:593–597

Davidson JA, Pande S, Bretag TW, Lindbeck KD, Kishore GK (2004) Biology and management of *Botrytis* spp. In legume crops. In: Eland Y, Williamson B, Tudzynski P, Delen N (eds) Botrytis: biology, pathology and control. Kluwer Academic Publishers, Dordrecht, pp 295–318

Elad Y (1994) Biological control of grape grey mould by *Trichoderma harzianum*. Crop Prot 13:35–38

Elad Y, Shtienberg D, Yunis H, Mahrer Y (1992) Epidemiology of grey mould, caused by Botrytis. Research. In: Verhoeff K, Malthrakis NE, Williamson B (eds) Proceedings of the 10th international Botrytis symposium. Heraklion, Crete, Greece. Pudoc Scientrific Publishers, Wageningen, pp 147–158

FAO (2005) FAO bulletin of statistics. Food and Agricultural Organizations of the United Nations, Rome

Faretra F, Grindle M (1992) Genetic studies of *Botryotinia fuckellana* (*Botrytis cinerea*). In: Verboeff K, Malathrakis NE, Williamson B (eds) Recent advances in Botrytis symposium. Heraklion, Crete, Greece. Pudoc Scientific Publishers, Wageningen, pp 7–16

Fournier E, Giraud T, Loiseau A, Vautrin D, Estoup A, Solignac M, Cornuet JM, Brygoo Y (2002)

Characterization of nine polymorphic microsatellite loci in the fungus *Botrytis cinerea* (Ascomycota). Mol Ecol Note 2:253–255

Genoscope (2008) *Botrytis cinerea*, estimated losses for vineyards in France amount to 15–40 % of the harvest, depending on climatic conditions. In: Sequencing projects of *Botrytis cinerea* [online]

Giraud T, Fortini D, Levis C, Leroux P, Brygoo Y (1997) RFLP markers show genetic recombination in *Botryotinia fuckeliana (Botrytis cinerea)* and transposable elements reveal tow sympatric species. Mol Biol Evol 14:1177–1185

Grewal JS (1988) Diseases of pulse crops an overview. Indian Phytopathol 41:1–14

Grewal JS, Laha SK (1982) Chemical control of botrytis grey mould of chickpea. Indian Phytopathol 36:516–520

Groves JW, Loveland CA (1985) The connection between *Botrytinia fukeliana* and *Botrytis cinerea*. Mycologia 45:415–425

Guraha SN, Singh G, Sharma YR (2003) Diseases of chickpea and their management. In: Massod A, Shiv K, Singh NB (eds) Chickpea research in India. Indian Institute of Pulses Research, Kanpur, pp 195–227

Harrison JG (1988) The biology of *Botrytis* spp. on *Vicia* beans and chocolate leaf spot disease-a review. Plant Pathol 37:168–201

Haware MP (1998) Diseases of chickpea. In: Allen DJ, Lenne JM (eds) The pathology of food and pasture legumes. ICARDA/CAB International, Wallingford, pp 473–516

Haware MP, McDonald D (1992) Integrated management of botrytis gray mould of chickpea. In: Haware MP, Faris DG, Gowda CLL (eds) Botrytis gray mould of chickpea. ICRISAT, Patancheru, pp 3–6

Haware MP, Nene YL (1982) Screening chickpea for resistance to botrytis grey mould. Int Chickpea Newsl 6:17–18

Haware MP, Tripathi HS, Rathi YPS, Lenne JM, Jayanthi S (1997) Integrated management of botrytis gray mould of chickpea: cultural, chemical, biological, and resistance options. In: Haware MP, Lenne JM, Gowda CLL (eds) Recent advances in research on Botrytis grey mould of chickpea. ICRISAT, Patancheru, pp 9–12

Haware MP, Mukherjee PK, Lenne JM, Jayanthi S, Tripathi HS, Rathi YPS (1999) Integrated biological-chemical control of Botrytis gray mould of chickpea. India Phytopath 52:174–176

Horsfall JG, Diamond AE (1957) Interaction of tissues sugar, growth substances and disease susceptibility. A Pflanzenkel Pflanzenpathol Pflanzenschutz 64:415–421

Isenegger DA, MacLeod WJ, Ford R, Pande S, Abu Bakr M Taylor PWJ (2005) Genetic structure of *Botrytis cinerea* that causes Botrytis grey mold disease of chickpea in Bangladesh. In: Innovations for sustainable plant health-15th Australian plant pathological society handbook, pp 224

Joshi S (1992) Botrytis grey mould of chickpea in Nepal. In: Haware MP, Faris DG, Gowada CLL (eds) Botrytis gray mould of chickpea. ICRISAT, Patancheru, pp 12–13

Joshi MM, Singh RS (1969) A Botrytis grey mould of gram. Indian Phytopathol 22:125–126

Kishore GK (2005) Cultural, morphological, pathogenic and genetic variation in *Botrytis cinerea*, causal agent of gray mold in chickpea. PhD thesis, Jawaharla Nehru Technological University, Hyderabad, Andhra Pradesh, India

Knights EJ, Siddique KHM (2002) Manifestation of *Botrytis cinerea* on chickpeas in Australia. In: Workshop proceedings integrated management of Botrytis Grey Mould of Chickpea in Bangladesh and Australia. Bangladesh Agricultural Research Institute, Joydebpur, pp 70–77

Laha SK, Grewal JS (1983) Botrytis blight of chickpea and its perpetuation through seed. Indian Phytopathol 36:630–634

Mahmood M, Sinha BK (1990) Gray mould disease of Bengal gram in Bihar, Final technical bulletin. Tirhut College of Agriculture, Dholi, p 35

Mansour K (1980) Chemical control of rust and leaf spots of field beans (*Vicia faba* L.). Agri Res Rev 58:49–56

Meeta M, Bedi PS, Kumar K (1986) Chemical control of gray mold of gram caused by *Botrytis cinerea* in Punjab. J Res 23:435–438

Meeta M, Badi PS, Jindal KK (1988) Host range of *Botrytis cinerea* the incitant of grey mould of gram. Plant Dis Res 3:77–78

Mitter N, Grewal JS, Pal M (1977) Biochemical changes in chickpea genotypes resistant and susceptible to grey mould. Indian Phytopahtol 50:490–498

Mohhamadi AG (1987) Phenols in relation to resistance against ascochyta blight and botrytis grey mould of Chickpeas. M.Sc.(Ag.) thesis, Govind Ballabh Pant University of Agriculture and Technology, Pantnagar, India

Moyano C, Alfonso C, Gallego J, Raposo R, Melgarejo P (2003) Comparison of RAPD and AFLP marker analysis as a means to study the genetic structure of *Botrytis cinerea* populations. Eur J Plant Pathol 109(5):515–522

Mukherjee PK, Haware MP (1993) Biological control of botrytis gray mold of chickpea. Int Chickpea Newsl 28:14–15

Mukherjee PK, Haware MP, Jayanthi S (1995) Preliminary investigations in integrated biocontrol of Botrytis gray mold of chickpea. Indian Phytopathol 48:141–149

Nelson ME, Powelson ML (1988) Biological control of gray mould of snap bean by *Trichoderma hamatum*. Plant Dis 72:727–729

Nene YL (1984) A proposed list of common names for diseases of chickpea. Int Chickpea Newsl 4:31–32

Nene YL, Reddy MV (1987) Chickpea diseases and their control. In: Saxena MC, Singh KB (eds) The chickpea. CAB International, Wallingford, pp 233–270

Pande S, Singh G, Rao JN, Bakr MA, Chaurasia PCP, Joshi S, Johansen C, Singh SD, Kumar J, Rahman MM, Gowda CLL (2002) Integrated management of botrytis gray mold of chickpea, Information bulletin No. 61. ICRISAT, Andhra Pradesh

Pande S, Neupane RK, Bakr MA, Stevenson PC, Rao JN, MacLeod WJ, Siddique KHM, Kishore GK, Chaudhary RN, Hoshi S, Johansen C (2005) Rehabilitation of chickpea in Nepal and Bangladesh through integrated management of Botrytis gray mold. In: Paper presented in the 15th Australasian plant pathology conference on innovations for sustainable plant health. 26–29 September 2005, Deakin University Waterfront Campus, Geelong, Vic. pp. 224, Australian Plant Pathology Society, Canberra

Pande S, Stevenson PC, Rao JN, Neupane RK, Chaudhary RN, Grzywacz D, Baurai VA, Kishore GK (2005b) Reviving chickpea production in Nepal through integrated crop management, with emphasis on Botrytis gray mold. Plant Dis 89:1252–1262

Pande S, Sharma M, Pathak M, Rao JN (2006) Comparison of greenhouse and field screening techniques for Botrytis gray mold resistance. Int Chickpea Pigeonpea Newsl 13:27–29

Pandey BK (1988) Studies on botrytis grey mould of chickpea. Ph.D. thesis, G. B. P. U. A & T. Pantnagar, Uttar Pradesh, India

Pandey MP, Beniwal SPS, Arora PP (1982) Field reaction of chickpea varieties to chickpea grey mould. Int Chickpea Newsl 7:13

Rathi YPS, Tripathi HS (1991) Host range of *Botrytis cinerea,* the causal agent of grey mould of chickpea. Int Chickpea Newsl 24:37–38

Rathi YPS, Tripathi HS, Chaube HS, Beniwal SPS, Nene YL (1993) Screening chickpea for resistance to botrytis grey mould. Int Chickpea Newsl 11:31–33

Reddy MV, Nene YL, Singh G, Bashir M (1990) Strategies for management of foliar diseases of chickpea. In: Van Rheenen HA, Saxena MC, Walley BJ, Hall SD (eds) Chickpea in the nineties: proceedings of the second international workshop on chickpea improvement. ICRISAT, Patancheru, pp 117–127, 4–8 Dec 1989

Reddy MV, Ghanekar AM, Nene YL, Haware MP, Tripathi HS, Rathi YPS (1993) Effect of vinclozolin spray, plant growth habit and inter-row spacing on botrytis gray mold and yield of chickpea. Indian J Plant Prot 21:112–113

Rewal N, Grewal JS (1989a) Effect of temperature, light and relative humidity on conidial germination of three strains of *Botrytis cinerea* infecting chickpea. Indian Phytopathol 42:265–268

Rewal N, Grewal JS (1989b) Differential response of chickpea to grey mould. Indian Phytopathol 42:265–268

Singh RB (1970) Studies on the control of blight and wilt of gram. M. Sc (Ag.) thesis, G. B. P. U. A & T, Pantnagar, UP, India

Singh G (1997) Epidemiology of Botrytis grey mould of chickpea. In: Haware MP, Lenne JM, Gowda CLL (eds) Recent advances in research on Botrytis grey mould of chickpea. International Crop Research Institute for Semi Arid Tropics, Patancheru, pp 47–50

Singh G, Bhan LK (1986) Chemical control of grey mould in chickpea. Int Chickpea Newsl 15:18–20

Singh G, Kapor S (1984) Role of incubation and photoperiod on the intensity of botrytis grey mould of chickpea. Int Chickpea Newsl 12:23–24

Singh G, Kaur L (1989) Genetic variability studies and scope for improvement in chickpea, Punjab, India. Int Chickpea Newsl 20:7–8

Singh G, Kaur L (1990) Chemical control of grey mould of chickpea. Plant Dis Res 5:132–137

Singh MP, Tripathi HS (1992) Effect of temperature and depth of burial on survivability of *Botrytis cinerea* Pers. ex. Fr. causal agent of grey mould of chickpea (*Cicer arietinum* L.). Indian J Mycol Plant Pathol 22:39–43

Singh MP, Tripathi HS (1993) Effect of storage temperatures on survivability of *Botrytis cinerea* in chickpea seeds. Indian J Mycol Plant Pathol 23:177–179

Singh G, Kapoor S, Singh K (1982) Screening of chickpea for grey mould resistance. Int Chickpea Newsl 7:13–14

Tripathi HS, Rathi YPS (1992) Epidemiology of botrytis grey mould of chickpea. In: Haware MP, Faris DG, Gowda CLL (eds) Botrytis grey mould of chickpea. ICRISAT, Patancheru, pp 8–9

Tronsmo A (1986) *Trichoderma* used as a biocontrol agent against *Botrytis cinerea* rots on strawberry and apple. Sci Rep Agric Univ Norwa 61:1–22

Ye XY, Ng TB, Rao PF (2002) Cicerin and arietin, novel chickpea peptides with different antifungal potencies. Peptides 23:817–822

8 Fungal Diseases of Okra (*Abelmoschus esculentus* L.) and Their Integrated Management (IDM)

C.P. Khare, Sushma Nema, J.N. Srivastava, V.K. Yadav, and N.D. Sharma

Okra (*Abelmoschus esculentus* (L.) Moench) is an important vegetable crop grown mainly for its tender green fruits in India. The green fruits are rich in vitamins A and C and minerals like Ca, Mg, and Fe. In home consumption, India tops the world (Dhankhar and Mishra 2004). It is a multipurpose crop due to its various uses. Okra seeds are also good sources of protein and vegetable oil (Yadav and Dhankhar 2001). Okra crop is grown throughout the year and is susceptible to many fungal pathogens. Fungal diseases are a major constraint next to the yellow vein mosaic virus (YVMV) in all areas of the country okra producing. It is suffered by fungal diseases which are belonging to 23 genera and 31 species of fungal pathogens (Table 8.1).

Important fungal diseases of okra (*Abelmoschus esculentus* L.) discussed as under.

C.P. Khare • N.D. Sharma
Division of Plant pathology, Indira Gandhi Agricultural University, Raipur, Chhattisgarh, India

S. Nema
Division of Plant pathology, Jawaharlal Krishi Vishwa Vidyalaya, Jabalpur, Madhya Pradesh, India

J.N. Srivastava (✉)
Department of Plant Pathology, Bihar Agricultural University, Sabour, Bhagalpur, Bihar, India
e-mail: j.n.srivastava1971@gmail.com

V.K. Yadav
JNKVV, College of Agriculture, Kundeshwar, Tikamgarh, Madhya Pradesh, India

8.1 Damping-Off of Seedling and Root Rots

Damping-off kills seedlings before or soon after they emerge. It is observed in severe form in Karnataka, Assam, and on early sown crops in Northern parts of India. It is coupled with others and complete failures occur in such conditions: cool, cloudy weather, high humidity, wet soils, compacted soil, and overcrowding especially favors the development of damping-off. Infection before seedling emergence results in poor germination. If the decay is after seedling emergence, they fall over or die which is referred to as "damp-off." Seedlings that emerge develop a lesion near where the tender stem contacts the soil surface.

Damping-off of seedlings caused by *Pythium* spp. may affect okra in the late-sown summer crop. *Fusarium solani*, *Rhizoctonia solani*, and *Macrophomina phaseolina* proved to be the causal organisms of okra damping-off and root rot diseases. *F. solani* proved to be the most aggressive fungi in okra crops.

8.1.1 Management

Over-irrigation should be avoided to reduce humidity around the crop. The efficacy of clean fallow and rotational crop reduces population densities of *Pythium aphanidermatum*, *P. myriotylum*, and *Rhizoctonia solani* in soil (Johnson

L.P. Awasthi (ed.), *Recent Advances in the Diagnosis and Management of Plant Diseases*,
DOI 10.1007/978-81-322-2571-3_8

Table 8.1 Fungal diseases of okra and their pathogens

SN.	Name of diseases	Causal organisms
1	Damping-off	*Pythium* spp., *Rhizoctonia* spp.
2	Fusarium wilt of okra	*Fusarium oxysporum* f.sp. *vasinfectum*
3	Verticillium wilt	*Verticillium albo-atrum/V. dahliae*
4	Powdery mildew	*Erysiphe cichoracearum*, *Sphaerotheca fuliginea*, and *Oidium abelmoschi*
5	Cercospora leaf spots	*Cercospora abelmoschi*, C. malayensis, *C. hibisci*, *C. hibiscina*
6	Phyllosticta leaf spots	*Phyllosticta hibiscini*
7	Alternaria leaf spots	*Alternaria hibiscinum*
8	Seedling blight/ collar rot	*Macrophomina phaseolina* and *Colletotrichum dematium*
9	Dieback	*M. phaseolina*, *C. dematium*
10	Southern blight	*Sclerotium rolfsii*
11	Anthracnose	*Colletotrichum capsici*, *C. hibisci*
12	Root and stem rot	*Phytophthora palmivora*
13	Wet rot	*Choanephora cucurbitarum*
14	Stem canker	*Fusarium chlamydosporum*
15	Rust of okra	*Uromyces heterogeneous*
16		*Pleospora infectoria*
17	Fruit rots	*Pythium* spp. and *Phytophthora* spp.
18	Pod spots	*Ascochyta* spp.
19	Seed rots	*Colletotrichum dematium*, *Fusarium* spp. (*F. oxysporum*, *F. moniliforme*)
20		*Alternaria alternata*, *Cladosporium cladosporioides*, *Aspergillus* spp., and *Penicillium* spp.

et al. 1997). The field should be regularly inspected for the disease-affected seedlings. Such seedlings should be removed and destroyed. Anitha and Tripathi (2001b) screened fungicides against *R. solani* and *P. aphanidermatum*, which cause seedling mortality in okra; carbendazim, thiophanate-methyl, carboxin, thifluzamide, and captan were effective against *R. solani*, while metalaxyl, captan, carboxin, and iprodione were effective against *P. aphanidermatum*. Seed treatment with antagonist fungal culture of *Trichoderma viride* (3–4 g/kg of seed) or thiram (2–3 g /kg of seed) will be used against the disease. Soil drenches with benomyl, captan, and vitavax (carboxin) were effective against *R. solani*. *Fusarium* infection was reduced 38 % by benomyl. Soil drenching with Dithane M 45 (0.2 %) or Bavistin (0.1 %) affords protection against the disease. A combination of benomyl + captan was effective against all three root rot-causing organisms. Seed treatment with carbendazim followed by soil application of *Trichoderma viride* was proved effective controlling the seedling diseases of okra caused by *Rhizoctonia solani* and *Pythium aphanidermatum* (Johnson et al. 1997). Organic amendments increased seed germination and reduced *R. solani* infection. *Trichoderma harzianum* showed complete reduction in growth of *R. solani*, and plant guard (containing of *T. harzianum*) reduced the growth of all pathogens *F. solani*, *M. phaseolina*, and *Fusarium oxysporum*. Rhizo-N (containing *Bacillus subtilis*) and *B. subtilis* reduced the growth of all pathogens. The fungicide Rizolex-T (tolclofos-methyl) caused reduction on the growth of *R. solani*, *M. phaseolina*, and *F. solani* (Johnson et al. 1997).

Among the plant growth-promoting rhizobacteria, *Bacillus pumilus* (SE34), *B. pasteurii* (T4), *B. subtilis* (IN937b), and *B. subtilis* (GBO3) strains significantly improved the crop and reduced the incidence of seed mycoflora (Mashooda et al. 2003). Strains of *Rhizobium* and *Bradyrhizobium* species were effective in controlling the soilborne fungi for their biological control potential against soilborne, root-infecting fungi (*Fusarium* spp., *Macrophomina phaseolina*, and *Rhizoctonia solani*) on okra (Shahnaz et al. 2005).

Rhizobium meliloti used as a seed dressing or as a soil drench inhibited growth of the soilborne root-infecting fungi *Rhizoctonia solani* and *Fusarium solani* (Ehteshamul and Ghaffar 1993). *Rhizobium meliloti* was antagonistic to *Rhizoctonia solani* and *Fusarium* spp. in okra, when applied as seed or soil treatments (Ghaffar 1993). *Bradyrhizobium* sp. and *R. meliloti* either used as seed dressing or as soil drench significantly suppressed root rot infection caused by *M. phaseolina*, *F. solani*, and *R. solani* in okra (Siddiqui et al. 2000).

Integrated disease management strategy for controlling the seedling diseases of okra caused by *Rhizoctonia solani* and *Pythium aphanidermatum* by using seed treatment with carbendazim and carboxin followed by soil application of *Trichoderma viride* was proved effective (Anitha and Tripathi 2000). The efficiency of bacterial antagonists was more toward *P. aphanidermatum* than toward *R. solani*. *T. viride* was effective against *R. solani* (Anitha and Tripathi 2001a).

8.2 Cercospora Leaf Spots

This is a serious disease in the month of August when there is high humidity (Jhooty et al. 1977; Sohi and Sokhi 1974). Several species of *Cercospora viz, Cercospora abelmoschi*, *C. hibisci*, *C. hibiscina*, and *C. malayensis* are reported on leaves of okra causing leaf spots and sometimes blight. Leaf spots by *Cercospora* spp. have no definite shape, size, or margin. The causal fungus appears as an olivaceous to sooty-colored growth on the lower leaf surface. Injured leaves will often roll, wilt, and abscise.

C. abelmoschi causes no definite spots but grows as sooty to dark olivaceous mold on the lower surface of the leaf. Badly affected leaves roll, wilt, and fall down (Sohi and Sokhi 1972).

The spots caused by *C. hibiscina* produce dark olivaceous patches of moldy growth on the lower surface of the leaf. The spots caused by *C. malayensis* are brown, irregular with gray center and darker colored margins.

These fungi survive through conidia and stromata on crop debris in soil and cause maximum infection at 15–29 ° C.

8.2.1 Management

The control of leaf spots requires regular spraying with fungicides such as copper oxychloride, zineb, maneb, ziram, or captan. Bavistin (0.1 %) applied at 15 days interval had the lowest disease incidence followed by thiovin + Bavistin + diazinon at 15 or 30 days interval (Rahman et al. 2000). Prior to appearances of the disease spraying of Kavach (0.8 %) and Bavistin (0.5 %) were most efficient in controlling the disease (Dharam et al. 2001).

8.3 Other Leaf Spots

1. *Phyllosticta leaf spots caused by Phyllosticta hibiscini* are sparingly observed on leaves along with *Cercospora* leaf spots. The spots are large with gray center and later produce shot holes. The pycnidia appear as minute black dots on both leaf surfaces. Spores are hyaline and cylindrical.
2. *Alternaria leaf spots* caused by *Alternaria hibiscinum* appear as brown, subcircular spots of varying size and sometimes with concentric rings. Such spots are formed only on senescent leaves or when the plant is weakened.

8.4 Powdery Mildew

The disease is known to occur in severe southern parts of India. It is not reported to be severe in northern plain as it normally occurs very late in northern parts where major cultivation of this crop is done.

Disease is characterized by the obvious white coating of fungal mycelium on lower and upper leaf surface. Severe infection will cause the leaf to roll upward and result in leaf scorching. A large part of the talc-like powder on the leaf surface is composed of spores. These spores are easily blown by winds to nearby susceptible plants. Heavily infected leaves become yellow and then become dry and brown. The disease is found mainly on the older leaves and stems of plants. Yields of many of the infected crops are reduced due to premature foliage loss (Sohi and Sokhi 1974).

Powdery mildew is caused by *Erysiphe cichoracearum* and *Sphaerotheca fuliginea* and also

Oidium abelmoschi and *Leveillula taurica* (Souza and CafeFilho 2003).

An outbreak of *Erysiphe cichoracearum* on okra was associated with unusually dry weather. However, the disease was associated with high rainfall (Diaz 1999).

8.4.1 Management

Plants under nutritional stress in most cases will develop powdery mildew much sooner than plants of the same age grown under a good nutritional program. Hence, the plant should be well manured, and application of fertilizers should be done on the basis of standard recommendations. Seed dressing with thiophanate and benomyl conferred resistance to the disease for a few days. Application of wettable sulfur (0.2 %) or Bavistin (0.1 %) on 1-week interval showed effective control the disease. The incidence of disease was reduced by application of Karathane (dinocap), Solbar (barium polysulfide) 1 %, Orthophaltan (folpet) 0.2 %, and also sulfur dust (0.2 %). Spray of wettable sulfur (0.5 %) at 50 and 65 days after sowing reduces the disease (Prabhu et al. 1971). Though a dimethirimol-resistant *Oidium* sp. appeared on okra during 1970s (Omer 1972), the use of systemic fungicides for disease control is still reliable. Spraying with 0.01 % Topas recorded no incidence of powdery mildew. However, the efficacy was comparable to that of carbendazim (0.1 %). Topas (0.01 %) can be recommended as an alternative of carbendazim (Naik and Nagaraja 2000). Three sprays of 0.05 % tridemorph or 0.2 % sulfur after the appearance of disease symptoms can be recommended for disease control in okra crops (Singh et al. 1998). Four sprays of penconazole (0.05 %) and cyproconazole (0.03 %) at 15 days interval were the most effective in reducing the disease (Ragupathy et al. 1998).

As an alternative method to control powdery mildew, use of cattle urine at 30 % was efficient to control disease (Broek et al 2002). The ladybird (*Psyllobora bisoctonotata*) is a mycopredator of powdery mildew. Third and fourth larval stages were found to be the most efficient feeding stages where the insect eats vegetative (mycelium) structures and develops their reproductive structures, such as conidia and conidiophores, produced by the fungus on the surface of leaves (Soylu and Yigit 2002).

Disease-tolerant hybrids Vijaya, JNDOH-1, NOH-15, AROH-47, and HYOH-1 were found capability to reducing the disease (Neeraja et al. 2004).

8.5 Wilt of Okra

The disease is found wherever okra is grown intensively. In the recent past occurrence of wilt disease, some severity has appeared. In field surveys (Turki) about 60 % of okra plants were found to be infected with okra wilt pathogen, *Verticillium dahliae* (Esentepe et al. 1972). The *V. dahliae* infects cacao crop also (Emechebe et al. 1972), while *R. solani* isolates from winged beans (*Psophocarpus tetragonolobus*) were found to infect okra (Singh and Malhotra 1994).

The disease can appear at any stage of plant growth. Crops sowing in May to June suffer more than the crops sowing in February to March. Often, 20–30 % plants die due to its attack. Younger plants are more susceptible than the maturing plants.

The conspicuous symptom appears as yellowing and stunting of the plant followed by wilting and rolling of the leaves. Finally, the plant dies. However, before appearance of typical wilting, the leaves showed veinclearing. They lose turgor. Often, the leaves hang down in daytime to recover again in the night, but ultimately, they wilt and the plant dies. Sometimes, the plant may look healthy but the apical buds and fruits dry. If a diseased stem or root is cut longitudinally, the vascular bundles appear as dark streaks. In severe attacks, the whole stem is blackened.

The causal organism *Fusarium oxysporum* f. sp. *Vasinfectum* (Atkinson) Snyder and Hansen is known to cause the cotton wilt. The race may be a different one. The *F. solani* f. sp. *hibisci* is seed transmissible and infects okra only (Ribeiro et al. 1971).

It is mainly a soilborne fungus. But there is one time reported that fungus is found on seeds also. It survives for sometime as a saprophyte on colonized roots and then as chlamydospores in the soil. In contact with host roots, the chlamydospores or conidia germinate and penetrate to roots. After some growth in the cortex, the fungus reaches the xylem where it multiplies very rapidly. Dissemination of the fungus can occur by any method that can transfer the soil from one place to another (Jadhav et al. 2000).

8.5.1 Management

Use of certified seed and choice of date of sowing in affected fields are very important. February-March sowing in affected fields enables the plants to escape infection. Applying crop rotation for removal of roots of diseased plants and deep summer plowing reduce disease incidence. Mulching with transparent polyethylene sheets reduced populations of *R. solani*, and *Fusarium* spp. to zero in soil to a depth of 15–20 cm, and soil fumigation with 2,3-dibromopropionitrile and trichloronitroethylene is recommended. Seed treatment with Bavistin at 2 g/kg seeds was the most effective seed treatment followed by Agrozim, Derosal, and Pausin-M at the same rate for chemical control of okra wilt (*Fusarium solani*) (Patel et al. 2004).

The effectiveness of plant growth-promoting rhizobacteria isolates was tested among them. *Bacillus pumilus* (SE34), *B. pasteurii* (T4), *B. subtilis* (IN937b), and *B. subtilis* (GBO3) strains are significant against some seed-borne fungal diseases of okra (Mashooda et al. 2003).

The efficacy of plant extracts in inhibiting the growth of *F. solani* (causing wilt in okra) was investigated. Garlic extract (unsterilized) produced the maximum inhibition while extracts of *Allium cepa* (bulb), *Celsia coromandeliana*, *Ipomoea fistula* [I. carnea], *Jatropha curcas*, and *Ocimum sanctum* [*O. tenuiflorum*] showed slight inhibition (Patel and Vala 2004). The growth of the fungal species from rhizospheric soil and rhizoplane okra was also remarkably reduced by the garlic extract (Muhsin et al. 2001).

Maximum germination was recorded when seeds were treated with *Trichoderma viride* (25 g/kg seeds) high antagonistic potential against *Rhizoctonia solani*. Treating seeds with high doses of biofungicide (50 or 100 g/kg seeds) did not inhibit germination (Mathivanan et al. 2000). In vitro, *Trichoderma hamatum* and *T. harzianum* were antagonistic to *R. solani*. The highest germination of treated seeds with *T. viride* was recorded against *Fusarium pallidoroseum*, followed by *F. oxysporum* and *F. moniliforme* (Gurjar et al. 2004).

Seed dressings with antagonists, using gum arabic as a sticker, reduced infection by the root rot fungi. Combined use of antagonists and organic fertilizers was better than their separate use. *Paecilomyces lilacinus* was more effective than several chemical treatments against root rot and root-knot disease complex (Ghaffar 1988).

Varieties/lines okra I.S. 6653, 7194, 9273, 9857, C.S. 3232, 8899, Pusa Sawani, and Pusa Makhamali are resistant to wilt.

8.6 Collar Rot Disease

Naturally infected seeds of okra (*Abelmoschus esculentus*) with *M. phaseolina* appear brown to black and show dieback, root, and collar rot diseases. The incidence of the disease ranged from 12.7 to 58.3 % (Jha and Dubey 2000). Infected seeds were symptomatic with or without micro-slerotia. In asymptomatic seeds, the mycelium was confined to the seed coat and endosperm only, whereas mycelium and micro-sclerotia occurred in the seed coat, endosperm, and embryo of symptomatic seeds. Extraembryonal infection resulted in disease transmission to seedlings whereas intra-embryonal infection mostly caused pre- and postemergence mortality (Agrawal and Singh 2000). *M. phaseolina* was found to be present in the seed coat and embryo; seed infection due to *M. phaseolina* led to both pre- and postemergence mortality of okra; transmission of the pathogen from seed to seedling occurs (Shahid et al. 2001).

8.6.1 Management

The later planting was associated with the pre- and post-germination mortality and development lowest incidence of collar rot disease (*Macrophomina phaseolina*) (Dubey and Jha 1999).

Jha et al. (2000) were evaluated in vitro against *Macrophomina phaseolina*. Leaf extract of *Eclipta alba* showed maximum inhibition of mycelial growth at 5 % concentration, whereas, at 10 % concentration, leaf extract of *Argemone mexicana* showed maximum inhibition of mycelial growth, followed by *Eclipta alba* leaf extract. Among oil cakes tested, *Brassica juncea* cake exhibited maximum inhibition of mycelial growth at 5 % concentration. At 10 % concentration, cake of *Pongamia glabra* showed maximum mycelial growth inhibition, followed by *Azadirachta indica* cake. Fungicides were tested against fungal and bacterial antagonists in the laboratory, carboxin and metalaxyl did not inhibit the fungal antagonists *Trichoderma viride*, while little inhibition of *Gliocladium virens* was noticed at 0.1 % concentration. However, carbendazim and thiophanate-methyl inhibited both the fungal antagonists. The fungicides affected none of the bacterial antagonists (Anitha and Tripathi 2001a). The antagonism of *Trichoderma viride*, *Pseudomonas fluorescens*, and *Bacillus subtilis* was effective to *R. bataticola* (*Macrophomina phaseolina*) (Kaswate et al. 2003).

T. viride combined with neem cake was better in controlling the disease complex of root-knot nematode (*M. incognita*) and root rot fungus (*R. bataticola* [*Macrophomina phaseolina*]) than *T. viride* combined with groundnut cake (Chaitali et al. 2003).

8.7 Stem Canker

During the rainy season, a severe outbreak of stem canker was observed in okra in Maharashtra, India. This is thought to be the first report of *Fusarium chlamydosporum* causing canker in okra (Fugro 1999).

Fugro and Jadhav (2003) recorded stem canker of okra in Konkan, Maharashtra, India. The pathogen was isolated from the infected stem, branches, calyx, immature pods, and leaf petioles of okra showing the typical disease symptoms. A total of 23 plant species belonging to 8 different families were studied to determine the host range of the pathogen.

Dark brown to black, circular to elongated lesions developed on the stem of young okra seedlings. Many infected seedlings girdled at the point of infection which ultimately led to death of the seedlings. Under high rainfall and humid conditions, many infected plants showed splitting of the bark exposing inner cortex tissues. The leaf petioles, flower buds, calyx, and immature pods were also infected showing small, circular to elongated dark brown lesions (Fugro 1999).

The fungus was highly pathogenic to plant species belonging to the family Malvaceae. Some of the plant species belonging to Amaranthaceae, Solanaceae, and Caricaceae were also susceptible, but disease incidence was lower than that observed in Malvaceae. The identified hosts of the pathogen were *Abelmoschus tetraphyllus*, *Abelmoschus tuberculatus*, *Abelmoschus ficulneus [Abelmoschus ficulneus]*, *Abelmoschus moschatus*, cotton, ambadi, hollyhock (Malvaceae), Amaranthus, tomato, and pawpaws.

8.7.1 Management

Among the cultivars, only KS-404, KS-410, and JNDO-5 showed field resistance to stem canker.

8.8 Fruit Rots

Okra fruit rot disease is caused by *Pleospora infectoria*. Fruit rots caused by *Pythium* and *Phytophthora* occur when fruits are mishandled, bruised, packed tightly, and transported or stored in humid and ward conditions. *Ascochyta* also attacks on okra causing pod spots (Kumar and Rao 1976). The fruits develop lesions with ash gray centers bearing minute fructifications of the fungus. They shrivel and dry. Abou (1985) recorded *Alternaria radicina*, *Botrytis.* sp., and *Fusarium* sp. from okra fruits. In Kuwait, okra were collected from several markets; *Alternaria alternata* and *Sclerotina* spp. were the most common fungi isolated (Abdel 1988). Premature fruit abortion disease of okra in Nigeria was reported to cause by

Choanephora cucurbitarum. Presence of seed-borne pathogens like *Botryodiplodia theobromae* showed the suppressive effect on seed germinability and okra seedlings (Ndzoumba et al. 1990).

8.9 Seed-Borne Fungal Infection of Okra

Eight seed-borne fungal diseases were recorded in okra from Nigeria, none of which has previously been reported as seed-borne in okra. The most common fungi that were isolated from seeds of okra are *Alternaria alternata*, *Curvularia lunata [Cochliobolus lunatus]*, *Cladosporium cladosporioides*, *Rhizopus nigricans [R. stolonifer* var. *stolonifer]*, *Aspergillus niger*, *Aspergillus flavus*, *Penicillium citrinum*, *Trichoderma harzianum*, and a dark sterile mycelium (Esuruoso et al. 1975). The prominent fungi *Alternaria alternata*, *Cladosporium cladosporioides*, *Fusarium* spp., *Aspergillus* spp., and *Penicillium* spp were isolated from field of okra (Asha et al. 2001), while Jamadar et al. (2001) recorded 27 fungi associated with the different colored seeds, among which *Aspergillus flavus*, *A. niger*, *Colletotrichum gloeosporioides [Glomerella cingulata]*, *Fusarium moniliforme [Gibberella fujikuroi]*, *Rhizoctonia solani*, *Rhizopus nigricans [Rhizopus stolonifer], and Phomopsis* sp. were the predominant fungi. Black-colored seeds had the highest percentage of seed mycoflora (15.20 %), while white-colored seeds had the lowest (8.7 %). Vigor and seed weight was also low in black seeds while high in white seeds with low mycoflora association.

8.9.1 Management

Benomyl was the most efficient seed treatment, followed by copper oxychloride + zinc and mancozeb to elucidate the fungi associated with seed (Al-Kassim 1996). Common fungicides, i.e., Dithane M-45 (mancozeb), Bavistin (carbendazim), Agrosan GN (phenylmercury acetate and ethylmercury chloride), and thiram, at 0.1, 0.2, and 0.3 % were investigated to reduce the seed mycoflora (Asha et al. 2001). The incidence of five fungal seed-borne diseases of okra, viz., foot and root rot, anthracnose and dieback, *Cercospora* leaf spot, *Corynespora* leaf spot, and leaf blight, respectively, caused by *Fusarium oxysporum*, *Colletotrichum dematium*, *Cercospora abelmoschi*, *Corynespora cassiicola*, and *Macrophomina phaseolina* diseases, has been found to be reduced by the use of clean apparently healthy seeds and seeds treated with Vitavax 200 (Anam et al. 2002). The combination of fungicides, like Anucop + Bavistin, Anucop + Dithane, Bavistin + Dithane, Anucop + Captan + Vitavax, Bavistin + Captan + Vitavax, were most effective against the seed-borne fungal diseases of the crop both in greenhouse and field conditions. Plant latices as biopesticide against seed-borne fungi of okra were tested (Agarwal and Singh 2002). The maximum control of seed-borne fungal infection (75 %) was observed in diluted *M. champaca* latex against seed-borne fungi on okra.

8.10 Some Other Fungal Diseases of Okra

Colletotrichum capsici and *C. hibisci* cause anthracnose of okra stems, fruits, and leaves. Spraying with zineb, captan, or mancozeb can reduce their occurrence. Rust of okra is of rare occurrence. It is caused by *Uromyces heterogeneous*.

8.11 Conclusion

Excessive pesticides load on the crops can be minimized to adopt IDM for fungal disease management. Application of plant/herbal extracts, other alternative of fungicides, and use of trace elements have to be utilized for disease management. Biocontrol agents blended with fungicides are promising to seed and soilborne disease management. Crop rotation and many other cultural practices seem to have little effect on foliar disease management. However, mycoparasite like *Ampelomyces quisqualis* and mycopredator like ladybird (*Psyllobora bisoctonotata*) could be utilized for powdery mildew disease management. Nontarget effect of pesticides on foliar diseases and resistant material exploitations may be beneficial component of IDM.

Refrences

Abdel-Rahim AM (1988) Post-harvest fungal diseases of some vegetables in Kuwait. Arab J Plant Prot 6(2):83–87

Abou-Heilah AN (1985) Postharvest fungal diseases of some vegetables in the two main markets of Riyadh (Saudi Arabia). J Univ Kuwait Sci 12(1):103–111

Agarwal S, Singh T (2002) Plant latices as biopesticide against seed-borne fungi of okra. J Mycol Plant Pathol 32(1):135

Agrawal S, Singh T (2000) Effect of extra- and intra-embryonal infection of *Macrophomina phaseolina* on disease transmission in okra seeds. J Mycol Plant Pathol 30(3):355–358

Al-Kassim MY (1996) Seed-borne fungi of some vegetables in Saudi Arabia and their chemical control. Arab Gulf J Sci Res 14(3):705–715

Anam MK, Fakir GA, Khalequzzaman KM, Hoque MM, Rahim A (2002) Effect of seed treatment on the incidence of seed-borne diseases of okra. Pak J Plant Pathol 1(1):1–3

Anitha K, Tripathi NN (2000) Integrated management of seedling diseases of okra caused by Rhizoctonia solani Khun and Pythium aphanidermatum (Edson) Fitzp. Indian J Plant Prot 28(2):127–131

Anitha K, Tripathi NN (2001a) Laboratory screening of fungal and bacterial antagonists against *Rhizoctonia solani* (Khun) and *Pythium aphanidermatum* (Edson) inciting seedling diseases of okra. Indian J Plant Prot 29(1/2):146–148

Anitha K, Tripathi NN (2001b) Screening of fungicides against seedling mortality of okra caused by *Rhizoctonia solani* and *Pythium aphanidermatum*. Plant Dis Res 16(1):52–56

Broek RVD, Iacovino GD, Paradela AL, Galli MA (2002) Alternative control of *Erisiphe cichocearum* on Okra crop. Ecossistema 27(1/2):23–26

Chaitali, Singh L, Singh S, Goswami BK (2003) Effect of cakes with *Trichoderma viride* for the management of disease-complex caused by *Rhizoctonia bataticola* and *Meloidogyne incognita* on okra. Ann Plant Prot Sci 11(1):178–180

Dhankhar BS, Mishra JP (2004) Objectives of okra breeding. J New Seeds 6(2/3):195–209

Dharam S, Maheshwari VK, Gupta A (2001) Fungicidal control of Cercospora leaf spot in seed crop of okra [*Abelmoschus esculentus* (L.) Moench.]. Seed Res 29(2):254–256

Diaz FA (1999) Okra (*Abelmoschus esculentus*) powdery mildew in Mexico. Revista Mexicana de Fitopatologia 17(1):44–45

Dubey SC, Jha AK (1999) Influence of dates of sowing and environmental factors on collar rot of okra. Indian Phytopathol 52(3):291–293

Ehteshamul H, Ghaffar S (1993) Use of rhizobia in the control of root rot diseases of sunflower, okra, soybean and mungbean. J Phytopathol 138(2):157–163

Emechebe AM, Leakey CLA, Banage WB (1972) Verticillium wilt of cacao in Uganda: the relationship between Verticillium dahliae and cacao roots. Ann Appl Biol 70(2):157–162

Esentepe M, Karcilioglu A, Sezgin E (1972) The first report of Verticillium wilt on sesame and okra. J Turkish Phytopathol 1(3):127–129

Esuruoso OF, Ogundiran SA, Chheda HR, Fatokun DO (1975) Seedborne fungi and some fungal diseases of okra in Nigeria. Plant Dis Rep 59(8):660–663

Fugro PA (1999) A new disease of okra (*Abelmoschus esculentus* L.) in India. J Mycol Plant Pathol 29(2):264

Fugro PA, Jadhav NV (2003) Stem canker of okra in Konkan region of Maharashtra. J Mycol Plant Pathol 33(2):288–289

Ghaffar A (1988) Soilborne diseases research centre of Bangladesh. Final research report, 1st January–30th June, pp 102–110

Ghaffar A (1993) Rhizobia as biocontrol organisms. BNF Agric Bull 12(2):6

Gurjar KL, Singh SD, Rawal P (2004) Management of seed borne pathogens of okra with bio-agents. Plant Dis Res 19(1):44–46

Jadhav NV, Fugro PA, Sawant GG (2000) Effect of media, pH, carbon and nitrogen sources on the growth and sporulation of Fusarium chlamydosporum causing stem canker of Okra. Indian J Environ Toxicol 10(2):81–83

Jamadar MM, Ashok S, Jahagirdar S (2001) Studies on seed mycoflora and nematodes and their effect on germination and vigour index of colour graded okra (*Abelmoschus esculentus* L.). Crop Res Hisar 22(3):479–484

Jha AK, Dubey SC (2000) Occurrence of collar rot of okra (*Abelmoschus esculentus*) in the plateau region of Bihar. J Res Birsa Agric Univ 12(1):67–72

Jha AK, Dubey SC, Jha DK (2000) Evaluation of different leaf extracts and oil cakes against *Macrophomina phaseolina* causing collar rot of okra. J Res Birsa Agric Univ 12(2):225–228

Jhooty JS, Sokhi SS, Bains HS, Rewal HS (1977) Evaluation of germplasm of Okra against powdery mildew and Cercospora blight. News Letter No.2 Vegetable and Hort. for Humid Tropics pp 30–32

Johnson AW, Burton GW, Sumner DR, Handoo Z (1997) Coastal Bermudagrass rotation and fallow for management of nematodes and soilborne fungi on vegetable crops. J Nematol USA 29(4):710–716

Kaswate NS, Shinde SS, Rathor RR (2003) Effect of biological agents against different isolates of *Rhizoctonia bataticola* (Taub.) Butler in vitro. Ann Plant Physiol 17(2):167–168

Kumar CSKV, Rao AS (1976) A report of leaf spot diseases on some vegetable, fodder and ornamental plants. Curr Sci 45(8):309–310

Mashooda B, Rai VR, Lokesh S (2003) Effect of plant growth promoting rhizobacteria on seed-borne fungal pathogens in okra. Indian Phytopathol 56(2):156–158

Mathivanan N, Srinivasan K, Chelliah S (2000) Biological control of soil-borne diseases of cotton, eggplant, okra and sunflower by *Trichoderma viride*. Zeitschrift fur Pflanzenkrankheiten und Pflanzenschutz 107(3):235–244

Muhsin TMA, Zubaidy SR, Ali ET (2001) Effect of garlic bulb extract on the growth and enzymatic activities of rhizosphere and rhizoplane fungi. Mycopathologia 152(3):143–146

Naik KS, Nagaraja A (2000) Chemical control of powdery mildew of okra. Indian J Plant Prot 28(1):41–42

Ndzoumba B, Conca G, Portapuglia A (1990) Observations on the mycoflora of seeds produced in Gabon. FAO Plant Prot Bull 38(4):203–212

Neeraja G, Vijaya M, Chiranjeevi C, Gautham B (2004) Screening of okra hybrids against pest and diseases. Indian J Plant Prot 32(1):129–131

Omer MEH (1972) Chemical control of powdery mildews of cucumber and okra in the Sudan. Wad Medani Sudan Exp Agric 8(3):265–270

Patel NN, Vala DG (2004) Evaluation of phyto extracts against the growth of Fusarium solani. Plant Dis Res 19(2):204

Patel NN, Gohel VP, Vala DG (2004) Control of okra wilt with chemical seed treatment. Plant Dis Res 19(1):47–48

Prabhu AS, Phatak KD, Singh RP (1971) Powdery mildew of 'bhindi' (*Abelmoschus esculentus* L.) in Delhi State. Indian J Hortic 28(4):310–312

Ragupathy N, Thiruvudainambi S, Thamburaj S (1998) Chemical control of powdery mildew disease of bhendi. South Indian Hortic 46(1/2):102–103

Rahman MA, Ali M, Mian IH, Begum MM, Kalimuddin M (2000) Pesticidal control of Pseudocercospora leaf spot and shoot and fruit borer of okra seed crop. Bangladesh J Plant Pathol 16(1/2):31–34

Ribeiro RLD, Robbs CF, Akiba F, Kimura O, Sudo S (1971) Arquivos da Universidade Federal Rural do Rio de Janeiro. Inst Biol 1(1):9–13, UFRRJ, Brazil

Shahid A, Khan A, Chauhan SS (2001) Studies on seed-borne nature of *Macrophomina phaseolina* in okra. Ann Plant Prot Sci 9(1):152–154

Shahnaz D, Perveen F, Atif D (2005) Effect of different strains of Rhizobium spp. in the control of root infecting fungi and growth of crop plants. Int J Biol Biotechnol 2(2):415–418

Siddiqui IA, Ehteshamul-Haque S, Zaki MJ, Ghaffar A (2000) Greenhouse evaluation of rhizobia as biocontrol agent of root-infecting fungi in okra. Acta Agrobot 53(1):13–22

Singh A, Malhotra SK (1994) Host range studies of *Rhizoctonia solani* causing web blight in winged bean. Bhartiya Krishi Anusandhan Patrika 9(2):113–116

Singh NB, Sharma HK, Srivastava KJ (1998) Chemical control of powdery mildew in okra seed crop. News Letter National Horticultural Research and Development Foundation. 18(4): 1–4

Asha S, Rai JP, Singh HK (2001) Seed mycoflora of Okra and its control by fungicides. Progress Hortic 33(1):84–89

Sohi HS, Sokhi SS (1972) Disease of bhendi (Okra) and their control. The Lal Baugh 14(3):1–6

Sohi HS, Sokhi SS (1974) Behavior of okra varieties to damping of powdery mildew and cercospora blight. Indian Phytopath 27:90–91

Souza VLD, CafeFilho AC (2003) Resistance to Leveillula taurica in the genus Capsicum. Plant Pathol 52(5):613–619

Soylu S, Yigit A (2002) Feeding of mycophagous ladybird, *Psyllobora bisoctonotata* (Muls.), on powdery mildew infested plants. OILB/SROP Bull 25(10):183–186

Yadav SK, Dhankhar BS (2001) Correlation studies between various field parameters and seed quality traits in okra cv. Varsha uphar. Seed Res 29:84–88

Sheath Blight Disease of Paddy and Their Management

9

Ashok Kumar Singh and J.N. Srivastava

Sheath blight of paddy is one of the most widely spreading diseases of paddy. This disease was first time reported from Japan in 1910 by Miyaki. But in India, Paracer and Chahal reported this disease from Gurdaspur (Punjab) only in 1963. Intensive and changed cultivar practices have intensified the severity of the disease. It is a potentially devastating disease of rice in all temperate and tropical rice production regions, especially in irrigated production systems (Dath 1990; Rush and Lee 1992b). Several estimates of yield reduction due to sheath blight have been reported, ranging from 5.2 to 69 % (Hori 1969; Kannaiyan and Prasad 1978; Naidu 1992); crop losses usually ranged from negligible to 50 % depending on the extent of severity and crop stages at which the disease appears and the environmental condition (Laha and Venkataraman 2001). Cultivar "Mahsuri" suffered as much as 69 % reduction in grain yield due to sheath blight. Mostly high yielding semi-dwarf varieties are susceptible to this disease, epidemic in West Godavari district of Andhra Pradesh during 1985 (Sinha and Prasad 2008). Recently, banded leaf blight symptoms of this disease have been reported from Uttar Pradesh. This disease also occurs in Kerala and parts of Tamil Nadu. It is becoming one of the important diseases of paddy.

9.1 Symptoms

Symptoms of the disease usually appear when plants are in the late tillering or early internode elongation growth stage. Small, water-soaked spots (about 1 cm long) first appear on the leaf sheath at or above water level as water soaked, circular to oblong, ellipsoid to ovoid shapes, and somewhat irregular greenish gray in color. Older lesions are elliptical or ovoid with a grayish white center and light brown to dark brown margin. They enlarge to approximately 1 cm in width and 2–3 cm in length. Lesions may coalesce forming bigger lesions with irregular outline, and this lesion interrupts the flow of water and nutrients to the leaf tip and the tip may die. As the plant grows and the canopy closes, the humidity inside the canopy increases. In this humid environment, the fungus grows upward inside the plant and on the plant's surface, causing new lesions. The fungus also can spread to nearby plants. Severely damaged plants may lodge. Eventually, the hole sheath rots and the affected leaf can easily be pulled off from the plant. In severe cases, all the

A.K. Singh
Division of Plant Pathology, Sher-e-Kashmir University of Agricultural Sciences & Technology, Jammu, J & K, India
e-mail: ashoks286@gmail.com

J.N. Srivastava (✉)
Department of Plant Pathology, Bihar Agricultural University, Sabour, Bhagalpur, Bihar, India
e-mail: j.n.srivastava1971@gmail.com

L.P. Awasthi (ed.), *Recent Advances in the Diagnosis and Management of Plant Diseases*,
DOI 10.1007/978-81-322-2571-3_9

leaves of a plant are blighted resulting in death of the plant. Severely infected plants produced poorly filled or empty grains, especially those on the lower portion of the panicles. Poorly developed grain usually breaks up during milling, thus reducing quality of rice.

Sheath blight symptoms at seedling stage of crops

Sheath blight symptoms at boot stage of crops in sever condition

Sheath blight symptoms before maturity of crops

Causal Organism – *Rhizoctonia solani* Kuhn.

Perfect Stages – *Thanatephorus cucumeris* (Frank) Donk/*Corticium sasakii* (Shirai) Matsumoto. *Hypochnus sasakii* Sharai (Rangaswami 1993).

The perfect stage forms under high humidity and appears as a thin, mildew-like growth on soil, leaves, and infected stems just above the ground line. Basidia are on a membranous layer of mycelium and have four sterigmata, each bearing one basidiospore (Agrios 2005).

The fungus belongs to division – Eumycota
Subdivision – Deuteromycotina
Class – Hyphomycetes
Subclass – Agonomycetidae
Order – Agonomycetales
Family – Agonomycetaceae

The fungi included in this order are often referred to as Mycelia Sterilia as they lack even the imperfect state, i.e., in which no spores of any kind are produced and reproduce only by fragmentation of the mycelium (Mehrotra and Aneja 1990). The hyphae, which are initially hyaline but later brown, are characteristically branched. The branches arise at right angles (90°) from below the septa and show distinct constrictions at the point of origin. The branching characteristics are usually the only ones available for identification of the fungus as *Rhizoctonia*. The spongy *sclerotia* are formed in great numbers on the infected leaf sheath and are clearly visible.

9.2 Disease Cycle

The sheath blight organism survives between crops in plant debris and as sclerotia, which are initially small white but turning brown to black at maturity, rocklike reproductive structures. Sclerotia consist of compact masses of mycelia. They are irregular, hemispherical, and flattened at the bottom; individual sclerotia are 1–6 mm in diameter. They may unite to form a larger mass. Large sclerotia are significantly more virulent than smaller ones.

The sclerotia can survive in the soil for several years. Sclerotia are loosely attached and easily dislodge from the plant. When a succeeding crop of rice is flooded, the sclerotia float to the surface of the water during soil puddling, leveling, weeding, and other operations and infect the plants with which they come into contact. The fungus penetrates through the cuticle or the stomatal slit. Infection pegs are formed from each lobe of the lobate appressorium of infection cushion. The mycelium grows from the outer surface of the sheath going through the sheath edge and finally through the inner surface. Primary lesions are formed while the mycelium grows rapidly on the surface of the plant tissue and inside its tissue. It proceeds upward and laterally to initiate the formation of the secondary lesions on the sheaths of lower leaves near the waterline.

9.3 Host Range

The fungus *Rhizoctonia solani* has vide host range and cause infection, e.g., *Echinochloa colonum*, *E. crus-galli*, *Cyperus iria*, *Arachis hypogaea* L. (groundnut), *Capsicum annuum* L. (chili), *Daucus carota* L. (carrot), *Glycine max* (soya bean), *Gossypium* sp. (cotton), *Hordeum vulgare* (barley), *Lactuca sativa* L. (lettuce), *Lycopersicon esculentum* Mill. (tomato), *Sorghum bicolor* (sorghum), *Triticum* sp. (wheat), *Tulipa* sp. (tulips), and *Zea mays* L. (maize). It can also attack other weeds/crops such as *Zea mays*, *Pennisetum americanum*, *Vigna radiata*, *V. mungo*, *Solanum tuberosum*, *Cynodon dactylon*, *Digitaria adscendens*, *Panicum crus-galli*, *and Cyperus rotundus*. However, the color and shape of lesions appeared on leaves/sheaths varied. Sclerotial inoculum of the fungus in the soil or the presence of infected weeds, viz., *Cynodon dactylon*, *Cyperus rotundus*, *Echinochloa crus-galli*, *Digitaria adscendens*, etc.in and around the fields and infected seedlings in the nursery may help in perpetuating the disease (Roy 1973; Kannaiyan and Prasad 1979). This is the region that symptoms normally are first observed near the bunds of the rice fields where infected weeds were growing.

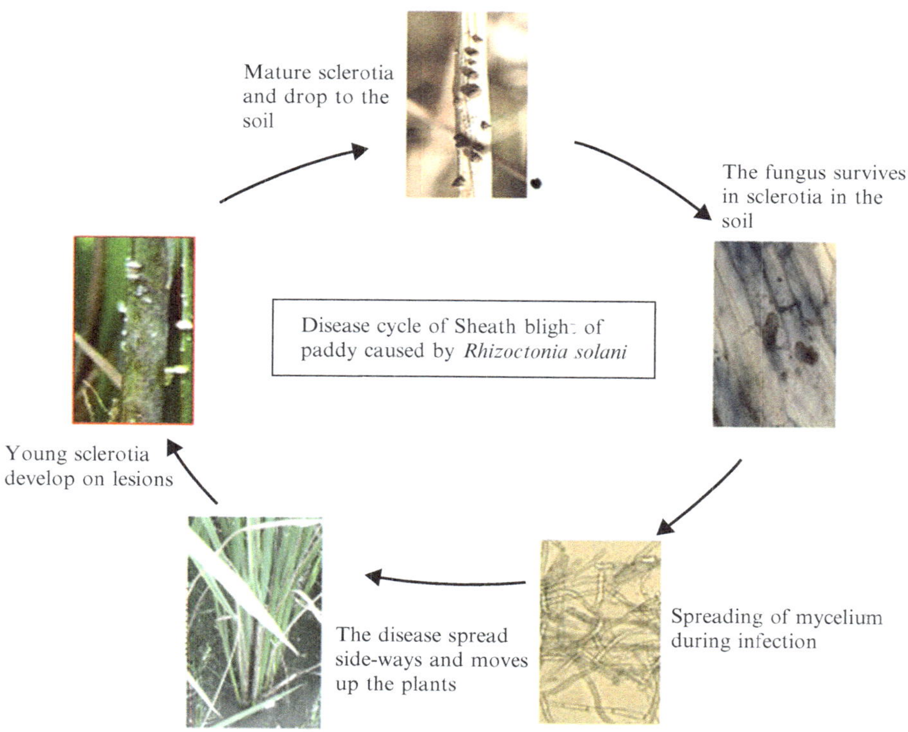

9.4 Disease Management

9.4.1 Through Cultural Practices

1. *Resistant varieties*. There are no highly resistant varieties, but research indicates moderately susceptible and moderately resistant varieties are measurably better than a susceptible variety. In fields with a history of losses due to severe sheath blight, varieties less susceptible to sheath blight may be planted. Sheath blight may be severe on semidwarf varieties because of the short distance between waterline (site of infection) and panicles (Marchetti 1983). Morphological features imparting resistance (Bharti, CR 1014, Nalini, Pankaj, Ratna, Tetap) were found to be less susceptible to sheath blight. On artificial inoculations of rice cultivars, maximum disease severity was recorded in PR 103 (97 %) with a yield loss of 13.4 %. Disease severity in other varieties was 91% in PR 109 and Pusa Basmati 1, 87 % in PR 108, 79 % in Jaya and PR 106, 78.8 % in PR 110, and 65 % in IR 8. Maximum yield reduction (58 %) was recorded in Pusa Basmati (Singh et al. 2003). Under field and laboratory conditions, no variety of rice was found resistant to *R. solani* as reported (Kang et al. 1965; Amante et al. 1990; Roy 1977).
2. *Crop rotation*. Rotation is another sound strategy, but rotating of rice with soybeans doesn't break the disease cycle. And rotation to other crops is not always practical for many growers. The probability of sheath blight is less in fields planted to rice for the first time, unless water drains from other fields that have been planted to rice or if rice is flooded from a surface water source. If sheath blight has been a problem in a field in the past and is again planted to rice, you can expect sheath blight to recur even if the field has been planted to soybeans for 2 years. If sheath blight occurred in a field the previous year, expect an even greater problem if a susceptible variety is planted the following year. It is safest to plant susceptible varieties, such as Mahsuri, in fields where sheath blight has not previously occurred. A rotation of 2 years in soybeans and 1 year in rice helps control sheath blight. However, since the fungus can survive in the soil or in plant debris for years and can reproduce on soybeans, rotation will not assure you that sheath blight will not be a problem.
3. *Spacing*. Avoiding high seeding rates in nursery and dense planting of seedlings in field of rice cause less air movement among the plants. This results in more moisture on the plants from dew and rain and promotes conditions that favor sheath blight development. Transplanting rice seedlings at spacing of 25×25 cm against 15×15 cm, the favorable effect of high N application in disease development could be reduced (Roy 1978).
4. *Nitrogen rates and applications*. The new varieties respond to heavy nitrogen applications in order to achieve their high-yielding potential. Split nitrogen applications and avoid excessive nitrogen rates. Excessive nitrogen promotes succulent, dense growth of rice that encourages sheath blight. Split nitrogen application at pre-flood and at midseason, or split the midseason application into two one-fourth rate applications to discourage sheath blight development. High levels of nitrogenous fertilizers, double-cropping, high plant populations per unit area, and utilization of early maturing, short stature, high tillering compact, susceptible cultivars have intensified the severity of the disease in most rice-growing regions in the world (Ou 1985; Rush and Lee 1992a, b). Heavy dose of nitrogen with phosphorus and potash or normal rate of nitrogenous fertilizers without P and K application enhances the disease; it is clear that application of heavy dose of nitrogenous fertilizers distinctly increasing the severity of sheath blight (Basu and Sengupta 1996).
5. *Other fertility practices*. Proper balance of nutrients is important in maintaining plant health. Test soils for fertility and apply fertilizer based on recommendations from those tests. Reduction in sheath blight with potash application has also been observed. Disease incidence was maximum in zinc deficient and minimum with soil treated with zinc sulfate

without soil incorporated of the bioagent (Khan and Sinha 2005a).

6. *Other cultural practices*. Control weeds by destroying grasses and other collateral hosts and burning infected straw and stubbles reduce infection. Papavizas and Lewu (1979) suggested burial of residue of the previous crop to a depth of 20–25 cm by a moldboard plow and then application of carboxin. Introduction of minimum tillage which merely stirs the soil rather than inverting it improves condition for survival of sclerotia and drain fields as soon as possible to reduce conditions that favor high humidity and more severe disease. Soil type might be one of the important factors in determining the spread and incidence of the disease; sheath blight incidence had been positively correlated with the sandiness of the soil (Sarkar and Sen Gupta 2002). The beneficial effect of soil amendment in plant disease management lies in their ability to suppress pathogen populations through the stimulation of soil saprophytes and also by toxic substances produced during their decomposition (Khan and Sinha 2005b).

9.4.2 Through Biological Approach

Irrational use of chemicals for plant disease control has often caused hazardous effects such as pollution of the environment, toxicity in food materials, development of resistant races of pathogens, etc. Methyl bromide or carbon disulfide is one of the conventional effective fumigants against several soilborne pathogens. Methyl bromide residue in soil after fumigation may cause phytotoxicity, accumulation of bromide in leaves of vegetable crops, contamination of underground water, etc. (Hoffman and Malkomes 1974). Such perception caused major changes in pesticide use in plant disease management (Gullino and Kuijpers 1994).

Biological has emerged as an alternative and most promising means of the management of plant pathogens. The possible uses of fungal antagonists of rice pathogen have been viewed as an alternative disease management strategy. Among the several antagonist tested by various scientists, species of *Trichoderma*, *Gliocladium*, *Aspergillus*, etc. have been found effective in reducing the sheath blight and extensively explored for the control of soilborne plant pathogens (Khan and Sinha 2005a).

Recent year's considerable success has been achieved by introducing antagonists to soil or to specific court of infection and also by seed treatments (Singh et al. 2003). Antagonistic effects of *Bacillus subtilis*, *B. cereus*, *Enterobacter* sp., *Pseudomonas fluorescens*, *P. putida*, and *P. aureofaciens* on the growth of *R. Solani* in vitro have been demonstrated (Gnanamanickam and mew 1990; Lee et al. 1990; Singh and Sinha 2004). Seed bacterization with fluorescent and non-fluorescent bacteria suppressed the sheath blight disease and protected the plant from infection.

Various fungi such as *Aspergillus niger*, *A. terreus*, *Gliocladium virens*, and *Trichoderma* spp. inhibited mycelial growth of *Rhizoctonia solani* in vitro (Khan and Sinha 2007a; Vaish and Sinha 2004, 2006). Rice leaf sheath isolate (*T. harzianum*) was found most effective, in reducing disease severity and incidence and increasing grain yield, against sheath blight pathogen *R. solani* in vitro screening (Khan and Sinha 2007b). In pot culture, soil amendment with *T. atroviride* reduced the incidence of sheath blight in rice (Mamian and Paulsamy 1987). Application of *T. harzianum* 7 days before inoculation of *R. solani* resulted in maximum reduction of the disease and gave maximum increase in grain yield as compared to other sequences. Higher rates of *T. harzianum* (4 or 8 g/l) were found highly effective in reducing disease severity and incidence and increasing grain yield over control (Khan and Sinha 2007b). Two fungal antagonists (*G. virens* and *T. longibrachiatum*) applied to the soil as wheat bran dust preparation survived well in soil and reduced the pathogen population. When the antagonists were supplemented with organic substrates, an increase in the colony farming units of the antagonists and a marked reduction in the pathogen survival were noticed (Baby and Manibhushan Rao 1993). Four antibiotics – two developed in Japan, viz., Validamycin and polyoxin, and two developed in China, viz., Jingganycin and Chingfengmeisu – have been found effective against sheath blight (Gangopadhyay and Chakraborti 1982; Singh et al. 2003).

Pathogen is reported to survive in the soil during winter as sclerotia or mycelium. However, only little information is available on the role of seed-borne inoculum in transmission of *R. solani* from seed in disease development in the field (Roy 1989; Acharya and Gupta 1996). Seed treatment with *Trichoderma virens* or carbendazim reduced disease incidence on seedlings (Shivalingam et al. 2006). Seed treatment with *T. harzianum* and foliar spray with bavistin was most effective in lowering down the disease severity of sheath blight (14.7 %) (Biswas et al. 2008).

9.4.3 Through Chemical Approach

Fungicides have been in use since ancient times for the control of plant pathogens; the role of chemicals as fungicides came into recognition only after the fungi were clearly regarded as a cause of plant diseases. A lot of information has accumulated on fungicides during the last two centuries, mainly because of the advancements in the knowledge of fungicidal chemistry. Some alternative methods of disease control, besides the chemical control, have also received due recognition during this period. However, in the present day intensive agricultural system, fungicides play a key role in the disease management strategy and are instrumental in boosting production level of crops by minimizing the attack of dangerous plant pathogens (Shokhi et al. 1996). However, with the introduction of pencycuron, a phenylurea-based fungicide, the control of seed-borne *Rhizoctonia* infections would become easy in many crops (Thind et al. 2002). Validamycin A and aureofungin have given good result in India. Sclerotia viability was reduced by applying herbicides particularly paraquently at and thiram and thiobencarb under field conditions (Pathak 1990). Fungicides, viz., hexaconazole, propiconazole, folicur, and thifluzamide, were found highly effective against *R. solani* (Vaish and Sinha 2003). Kannaiyan and Prasad (1979) concluded that soil application of edifenphos, kitazin, and carboxin can completely inhibit sclerotial germination and inactivate the mycelium of *R. solani*. Quintozene, thiabendazole, edifenphos, chlorothalonil, and chloroneb completely arrested growth and sclerotial production even at 0.025 % (Kannaiyan and Prasad 1979). Dev (1980) applied soil fungicides thiram and quintozene before planting (20 kg/ha) to kill soilborne sclerotia. A single application of edifenphos (0.5 1) may be made at maximum tillering stage or benomyl; carboxin or carbendazim at 0.5 kg/ha may be sprayed three times at 14 days intervals. Thiram and edifenphos were most effective followed by quintozene and edifenphos for soil application. Benomyl killed sclerotia when dipped for 48 h but spraying or soil application of benomyl and polyoxin could not kill the sclerotia on rice plants or the soil surface in Taiwan (Leu and Yang 1979). The proper control of sheath blight of rice by two sprays of monceren at boot stage and panicle emergence (Lore et al. 2005). It is concluded from the studies that Tilt 25 EC at 0.1 % can be used for effective control of major fungal diseases of rice such as sheath blight, sheath rot, and brown spot along with grain discoloration and help in producing good quality of rice grain (Lore et al. 2007). This completely contradictory result suggests that chemotherapy control may vary between localities or that development of resistant strains of the fungus or adverse effects of soil and environmental on the chemical may be related.

Acibenzolar-S-methyl (Bion, Syngenta) has the unique ability to protect plants from fungal, bacterial, and viral infections (Ruess et al. 1996). Also known as benzothiadiazole, it activates systemic acquired resistance against fungal pathogens belonging to oomycetes, ascomycetes, and deuteromycetes as well as bacteria and certain viruses. It shows potential for practical application against several bacteria and fungi in vegetables and range of other crops and act at low rates (less than 50 g a.i./ha). Probenazole (Oryzamate, Meiji Seika Kaisha), an other plant activator, is widely used in Japan against rice blast and bacterial leaf blight (Iwata 2001; Mahmood et al. 2007).

References

Acharya S, Gupta PKS (1996) Seed borne nature and seed transmission of rice sheath blight. Int Rice Res Note 21:76–77

Agrios GN (2005) Plant pathology, 5th edn. Elsevier Academic Press, New York, pp 596–598

Amante AD, De la Pena R, Sitch LA, Leung H, Mew TW (1990) Sheath blight resistance in wild rice. In: Kuldip Singh Chahal, Shokhi SS, Rattan GS (eds) Investigations on sheath blight of rice in Punjab. Indian Phytopath.56: 22–26

Baby VI, Manibhushan Rao K (1993) Control of rice sheath blight through the integration of fungal antagonists and organic amendments. Trop Agric 70:240–244

Basu A, Sengupta Prasanta K (1996). Effect of forms of nitrogen fertilizers on sheath blight of rice. In: Ashraf Ali Khan, Sinha AP (eds) Influence of soil and nutritional factors on the effectivity of *Trichoderma harzianum* against sheath blight of rice. Indian Phytopath. 58: 276–281

Biswas SK, Ratan V, Srivastava SSL, Singh R (2008) Influence of seed treatment with biocides and foliar spray with fungicides for management of brown leaf spot and sheath blight of paddy. Indian Phytopath 61(1):55–59

Dath AP (1990) Sheath blight disease of rice and its management. Associated Publishing Company, New Delhi, pp 80–87

Dev VPS (1980) Sheath blight control with soil fungicides. Int Rice Res Newsl 5:14–15

Gangopadhyay S, Chakraborti NR (1982) Sheath blight of rice. Rev Plant Path 61:451–460

Gnanamanickam SS, Mew TW (1990) Biological control of rice diseases (blast and sheath blight) with bacterial antagonists. An alternate strategy for disease management. In: Grayson BT, Green MB, Copping LG (eds) Pest management in rice. Elsevier Applied Science Pub. Ltd, New York, pp 87–110

Gullino ML, Kuijpers LAM (1994) Social and political implication of managing plant diseases with restricted fungicides in Europe. In: Sokhi SS, Thind TS (eds) Rational use of fungicides in plant disease control. National Agri. Information Center, Ludhiana, pp 183–194

Hoffman GM, Malkomes HP (1974) Bromide residues in vegetable crops after soil fumigation with methyl bromide. In: Sokhi SS, Thind TS (eds) Rational use of fungicides in plant disease control. National Agri. Information Center, Ludhiana, pp 183–194

Hori M (1969) On forecasting the damage due to sheath blight of rice plants and the critical point for judging the necessity of chemical control of the disease. Rev Plant Prot Res 2:70–73

Iwata M (2001) Probenazole- a plant defence activator. In: Thind TS (ed) Changing over of fungicide umbrella in crop protection. Indian Phytopath. 60(4): 421–433

Kang IM, Lee EK, Lee SC, Lim YK (1965) Studies on the resistance of rice variety to sheath blight (*Corticium sasakii*) in field. In: Kuldip Singh Chahal, Shokhi SS, Rattan GS (eds) Investigations on sheath blight of rice in Punjab, Indian Phytopath.56: 22–26

Kannaiyan S, Prasad NN (1978) Studies on the variability of sclerotia of *R. solani* Kuhn in soil and water. Madras Agric J 65:741–742

Kannaiyan S, Prasad NN (1979) Sheath blight incidence on weed hosts. In: Kuldip Singh Chahal, Shokhi SS, Rattan GS (eds) Investigations on sheath blight of rice in Punjab, Indian Phytopath.56(6): 22–26

Khan AA, Sinha AP (2005a) Influence of soil and nutritional factors on the effectivity of *Trichoderma harzianum* against sheath blight of rice. Indian Phytopath 58(3):276–281

Khan AA, Sinha AP (2005b) Influence of different factors on the effectivity of fungal bioagents to manage rice sheath blight, in nursery. Indian Phytopath 58(3):289–293

Khan AA, Sinha AP (2007a) Factors effecting of *Trichoderma spp.* as biocontrol agents against sheath blight of rice. Indian Phytopath (Accepted). In: Proceeding of the 19th training on "Plant Disease Management on Small Farms" organized by Center of Advanced Studies in Plant Pathology, G.B. Pant University and Technology, Pantnagar, pp. 177–182

Khan AA, Sinha AP (2007b) Biocontrol potential of *Trichoderma* species against sheath blight of rice. Indian Phytopath 60(2):208–213

Laha GS, Venkataraman S (2001) Sheath blight management in rice with biocontrol agents. Indian Phytopath 54:461–464

Lee YH, Shim GY, Lee EJ, Mew TW (1990) Evaluation of biological activity of fluorescent pseudo monads against some rice fungal diseases in vitro and glasshouse. Kor J Plant Paihol 6:73–78

Leu LS, Yang HC (1979) Comparative study on the effectiveness of the eight recommended fungicides to control rice sheath blight. Plant Proto Bull Taiwan 21:323–329

Lore JS, Thind TS, Mohan C (2005) Performance of some new fungicide formulations against sheath blight of rice. Pest Res J 17:49–51

Lore JS, Thind TS, Hunjan MS, Goel RK (2007) Performance of different fungicides against multiple diseases of rice. Indian Phytopath 60(3):296–301

Mahmood T, Kakishima M, Komatsu S (2007) Role of probenazole in the defense against bacterial blight of rice. In Thind TS (ed) Changing over of fungicide umbrella in crop protection. Indian Phytopatho. 60(4): 421–433

Mamian S, Paulsamy S (1987) Biological control of sheath blight disease of rice. J Bio Control 1:57–59

Marchetti C (1983) Screening of rice cultivars against sheath blight of paddy. Int Rice Rec Newl 7:19

Mehrotra RS, Aneja KR (1990) An Introduction to Mycology. Wiley India Pvt. Ltd. New Delhi. India, pp 610–611

Miyaki I (1910) Studies uber die reisptlance in Japan. Jl Cell Agric Imp Univ 2:237–276

Naidu VD (1992) Influence of sheath blight of rice on grain and straw yield in some popular local varieties. J Res Publ 10:78–80

Ou SH (1985) Rice disease. In: Ashraf Ali Khan, Sinha AP (eds) Screening of *Trichoderma* spp. against *Rhizoctonia solani* the causal agent of rice sheath blight. Indian Phytopatho. 60: 450–456

Papavizas GC, Lewu JA (1979) Integrated control of *Rhizoctonia solani*. In: Scippers B:, Gams W (eds) Soil-borne plant pathogens. Academic, London, pp 415–424

Paracer CS, Chahal DS (1963) Sheath blight of rice caused by *R. solani* Kuhn. A new record of India. Curr Sci 32:328–329

Pathak RA (1990) Chemical management of sheath blight disease of rice. Int Rice Res Newsl 11:24

Rangaswami G (1993) Diseases of crop plants in India. Prentice Hall of India Private Limited, New Delhi, pp 169–170

Roy AK (1973) Natural occurrence of *Corticium sasakii* on some weeds. In: Kuldip Singh Chahal, Shokhi SS, Rattan GS (eds) Investigations on sheath blight of rice in Punjab, Indian Phytopath. 56: 22–26

Roy AK (1977) Note on screening of rice cultures against sheath blight. Indian J Agric Sci 47:259–260

Roy AK (1978) Percent of rice plants infected by sheath blight in different treatment of spacing and nitrogen application. Curr Sci 47:307–308

Roy AK (1989) Source of seed-borne infection of sheath blight of rice. Oryza 26:111–112

Roy AK (1997) Note on screening of rice cultivars against sheath blight. In: Kuldip Singh Chahal, Shokhi SS, Rattan GS (eds) Investigations on sheath blight of rice in Punjab, Indian Phytopath. 56: 22–26

Ruess W, Mueller K, Knauf-Beiter G, Kunz W, Staub T (1996) Plant activator CGA 245704, an innovative activator for disease control in cereals and tobacco. In: Thind TS (ed) Changing over of fungicide umbrella in crop protection. Indian Phytopath. 60(4): 421–433

Rush MC, Lee FN (1992a) Sheath blight. In: Ashraf Ali Khan, Sinha AP (eds) Screening of *Trichoderma* spp. Against *Rhizoctonia solani* the causal agent of rice sheath blight, pp. 22–23. Indian Phytopath. 60: 450–456

Rush MC, Lee FN (1992b) Sheath blight. In: Webster RK, Gunnell PS (eds) Compendium of rice diseases. APS Press, St. Paul, pp 22–23

Sarkar SC, Sen Gupta PK (2002) Some soil factors influencing incidence of sheath blight of rice caused by *Rhizoctonia solani* Kuhn. Ann Plant Prot Sci 10:299–301

Shivalingam PN, Vishwakarma SN, Singh US (2006) Role of seed-borne inoculum of *Rhizoctonia solani* in sheath blight of rice. Indian Phytopath 59(4):444–452

Shokhi SS, Singh PP, Singh PJ (1996) Progressive development of fungicides- past and present. In: Sokhi SS, Thind TS (eds) Rational use of fungicides in plant disease control. National Agri. Information Center, Ludhiana, pp 1–18

Singh RV, Sinha AP (2004) Management of rice sheath blight by *Pseudomonas fluorescens and grain yield.* Ann Plant Prot Sci 13:410–414

Singh SK, Shukla V, Singh HP, Sinha AP (2003) Current status and impact of sheath blight in rice (*Oryza sativa* L). Agri Rev 25:289–297

Sinha AP, Prasad Gokil (2008) Integrated management of sheath blight disease of rice. In: Proceeding of the 19th training on "Plant Disease Management on Small Farms" organized by Center of Advanced Studies in Plant Pathology, G.B. Pant University and Technology, Pantnagar, pp. 177–182

Thind TS, Mohan C, Raj P (2002) Competitive fitness of metalaxyl resistant population of Phytophthora infestans and its cross resistance to strobilurins and other fungicides. Indian J Mycol Plant Pathol 32:122–124

Vaish DK, Sinha AP (2003) Determination of tolerance in *Rhizoctonia solani*, *Trichoderma virens* and *Trichoderma* sp. (Isolate 20) to systemic fungicides. Indian J Plant Pathol 21:48–50

Vaish DK, Sinha AP (2004) Characterization of isolates of *Trichoderma spp.* for their biocontrol ability against *Rhizoctonia solani. Indian J.* Plant Pathol 22:123–127

Vaish DK, Sinha AP (2006) Evaluation of fungal antagonists against *Rhizoctonia solani* causing sheath blight of rice. Indian J Agri Res 40:79–85

10 Diagnosis and Management of Bacterial Plant Diseases

B.S. Thind

10.1 Diagnosis of Bacterial Plant Diseases

For an effective disease management, the correct diagnosis of a disease based on the identification of its causal agent is very essential. The diagnosis of some diseases on the basis of symptoms can be made to a satisfactory level, but the identification of the causal agent is necessary for confirmation. For the diagnosis of a plant disease, a comprehensive host list that covers a known disease, its typical symptoms and its known potential pathogens for a specific host is required. The compendia of different crop diseases published by American Phytopathological Society are very valuable publications for this purpose. A list of plant diseases found on the website (http://www.apsnet.org/online/common/top.asp) of American Phytopathological Society is also a source of valuable information. *Westcott's Plant Disease Handbook* is also useful because specific symptoms associated with each disease are given (Horst 2001). The photographs of symptoms, especially the coloured ones, also aid in the identification of plant diseases. A list of known pathogens for a given crop greatly reduces the choices to one or two suspected genera for the given symptoms.

The symptoms of plant diseases caused by different groups of pathogens, namely, fungi, bacteria and viruses, have many similarities, and overlapping of the symptoms usually occurs. However, two signs, i.e., *water-soaking* and *bacterial exudation*, are associated only with most but not all the bacterial plant diseases. The amount of bacterial ooze varies with the nature of disease and the environmental conditions. The ooze is generally more forceful in vascular diseases and in freshly caused lesions. The intensity of ooze is also greater under humid conditions. In a dry weather, the bacterial ooze is converted into beads and scales/flakes, often becoming detectable only with the help of a magnifying glass, especially when it is sparse.[1]

Riley et al. (2006) have also emphasised the importance of signs by stating that signs are much more specific to disease-causing agents than are symptoms and are extremely useful in the diagnosis of a disease and identification of the agent causing the disease.

In bacterial wilt of solanaceous plants, the surface of a transverse section is generally moister in appearance in comparison to a fungal wilt. White

B.S. Thind (✉)
Ex-Professor-cum-Head, Department of Plant Pathology, Punjab Agricultural University, Ludhiana 141012, India
e-mail: thind_bs@yahoo.com

1 Abbreviations of bacterial genera given here in their respective bracts, i.e., *Agrobacterium (Ag.) Acidovorax (Ac.)*, *Burkholderia (B.)*, *Clavibacter (C.)*, *Erwinia (E.)*, *Pantoea (Pa.)*, *Pectobacterium (Pe.)*, *Pseudomonas (Ps.)*, *Ralstonia (R.)*, *Streptomyces (S.)* and *Xanthomonas (X.)* are used in this publication.

L.P. Awasthi (ed.), *Recent Advances in the Diagnosis and Management of Plant Diseases*, DOI 10.1007/978-81-322-2571-3_10

droplets of bacterial exudates generally appear on the cut surface and the ooze intensifies on squeezing the infected plant tissue. A diagnostic feature of bacterial wilt of cucurbits is coming out of the bacterial ooze from the cut ends of the infected stems. This sticky bacterial mass can be drawn into fine threads by touching the cut end of the stem with a finger/knife blade and gently pulling them away from the cut end. The fine threads of bacterial ooze can also be seen by pulling away the cut ends of the infected stem (Fig. 10.1).

Microscopic examination of diseased plants is extremely helpful in establishing bacterial nature of the diseases. The bacterial streaming is found in almost all the bacterial diseases except the crown gall. A portion of the lesion cut from the border adjacent to the healthy tissue is mounted in a drop of water and placed under the microscope, without covering it with a coverslip. Clouds of bacteria streaming out of the cut end of the mounted tissue can be seen. The intensity of ooze is greater in vascular diseases like bacterial blight of rice and wilt of solanaceous plants. Phase contrast microscopy is more suitable for this purpose because starch or latex particles can be distinguished from the bacterial cells. Bacterial ooze test can also be performed in the field by cutting an infected leaf (bacterial blight of rice) or stem (bacterial wilt of solanaceous plants) and placing it in a beaker/glass of clean water. After a few minutes, threads of bacterial ooze coming out of the cut end can be seen. In diseases caused by fastidious bacteria such as spiroplasmas, phytoplasmas and *Xylella fastidiosa*, the vascular bundles of the sectioned tissues are examined with an electron microscope to detect the presence of these pathogens.

It is very important to confirm the diagnosis of a disease by proving Koch's postulates. To fulfil the Koch's postulates, the causal bacterium should be isolated and purified. After inoculation, it should produce the typical symptoms of the disease on the host, identical to those found in the nature. The bacterium should be reisolated from the inoculated plants, and the reisolated bacterium should be identical to the inoculated bacterium in all respects. Generally, the isolation of the bacterium should be made on a general, non-selective medium instead of a semi-selective medium. A general purpose, non-selective but

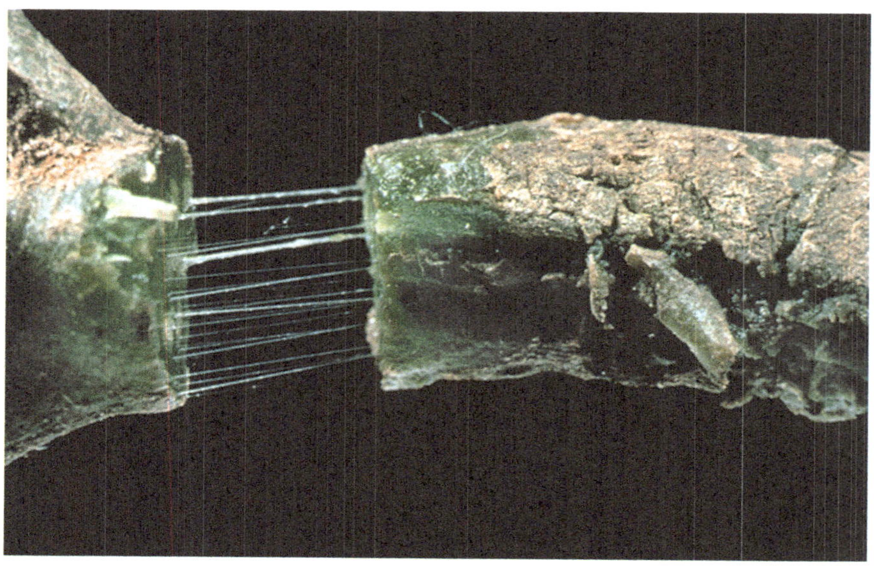

Fig. 10.1 Bacterial wilt of cucurbits: Strands of viscous bacterial ooze drawn from cut stem ends (Courtesy: M.P. Hoffman; www.apsnet.org)

differential medium containing tetrazolium chloride is useful for distinguishing a potential pathogen from saprophytes during the initial isolation, and for this purpose, the concentration of the tetrazolium chloride should be reduced to 0.001 % to avoid inhibitory effects on members of *Xanthomonas* genus. However, the semi-selective media are useful for confirming the presence of suspected pathogens. Semi-selective media are more helpful for isolations from soil but generally not from plant tissues because the latter are generally surface sterilised before isolation. A few of the bacterial plant pathogens such as phytoplasmas and papaya bunchy top pathogen have not been cultured on artificial media. The presence of such pathogens in the host plants should be confirmed by other methods such as electron microscopy.

The final confirmation of a plant pathogen is done by proving its pathogenicity. However, this requires time, availability of host plants and optimum environmental conditions for disease development. Moreover, in some bacterial pathogens, the time required to produce the disease symptoms is too long. Some strains of *Clavibacter michiganensis* subsp. *michiganensis* require 21–28 days to produce the symptoms. On the other hand, the confirmation by pathogenicity tests is often quick for leaf-spotting pathogens as they take only 3–4 days to produce the symptoms under optimum conditions. For some bacterial pathogens, hypersensitivity test on tobacco, as an alternative to pathogenicity test, is quite useful, while it does not work with others. Lelliot and Stead (1987) have given a detailed account for the diagnosis of bacterial diseases of plants by giving their detailed symptoms, diagnostic procedures, host inoculation tests, composition of culture media and procedures for staining techniques and biochemical tests.

The diagnosis of an undescribed or a new disease is certainly more complicated and requires more skill and expertise than confirming the presence of an already known disease. Therefore, the entire set of diagnostic principles should be used to narrow down the choices, as no simple tests are suitable. More emphasis should be put on microscopic examination and the initial association of the pathogen with the disease syndrome. Before spending a lot of time on numerous phenotypic and genotypic tests for pathogen identification, some basic information about the pathogen should be obtained to decide which tests are to be used for further identification. The tests for basic information include Gram-staining reaction, oxidative versus fermentative metabolism and presence or absence of spores and flagella. To further characterise the bacterium, the relatively inexpensive tests include metabolic tests (API strip tests, bioMerieux, Inc.), metabolic substrate analysis (Micrology TM, Biolog, Inc., Hayward, CA), analysis of fatty methyl esters (MIDI, Newark, DE) or 16S rDNA sequence analysis. An extreme caution must be exercised in interpreting results of any one test, especially for identification of an unknown disease caused by an unidentified bacterial pathogen. The past experience emphasises the importance of performing basic bacteriological tests before accepting a name that comes from a database, regardless of the refined analytically methods used to generate a similarity index.

A polyphasic analysis still appears to be the most reliable approach for identification of new pathogens. After the pathogenicity has been proved and the genus determined with relatively simple basic bacteriological tests, the final identification of the bacterium is interpreted from the results of various genotypic methods. Hu et al. (2001) reported that comparison of 16S rDNA sequence data with phenotypic data for type strains assisted in selection of determinative tests that may discriminate distinct taxa for simplified laboratory analyses.

10.2 Monoclonal Antibodies

With the development of hybridoma technology, considerable improvement has been made in immunodiagnostic techniques. Monoclonal antibodies (MAbs) produced by this method have a defined specificity to a single epitope. Hybridomas produce consistent antibodies, which give consistent result in different laboratories. Several MAbs produced by plant pathogenic bacteria and used in their epidemiological studies are listed in Table

10.1. Many of these antibodies have been characterised by testing specificities with large number of strains of the target pathogens. Some bacterial taxa are relatively homozygous in the sense that they possess common antigenic determinants and one antibody generally reacts with all or nearly all strains of the taxon. *C. michiganensis* subsp. *michiganensis* and *X. axonopodis* pv. *pelargonii* fall in this category. On the other hand, many plant pathogenic bacteria like *Pectobacterium atrosepticum* and *Pe. carotovorum* subsp. *carotovorum* are serologically heterogeneous; therefore, not all members of the population react with a polyclonal antiserum or a taxon-specific MAb. Pathogen-specific MAbs that react with heat-stable lipopolysaccharides enable development of robust detection kits for use in rapid laboratory and field diagnosis. MAbs directed towards capsule and/or extracellular polysaccharides in pathogens, such as *C. michiganensis* subsp. *michiganensis* and *Ralstonia solanacearum*, are useful in a number of immunodiagnostic formats.

Most of the MAbs produced for different bacterial plant pathogens have been tested using ELISA. The ability of ELISA assays to detect 10^{5-6} cfu ml^{-1} is adequate for the identification of bacterial pathogens from symptomatic plants

Table 10.1 Taxon-specific monoclonal antibodies produced for plant pathogenic bacteria[a]

Genus, species/subspecies or pathovar	MAb designation	No. of target strains tested	Reference[c]
Acidovorax avenae subsp. *citrulli*	3D1F3	26	M. Bandla (personal communication)
Agrobacterium tumefaciens biovar 3	A6F21-1D3G7C8		Bishop et al. (1989). *Phytopathology* 79: 995–998
Clavibacter michiganensis subsp. *michiganensis*	Cmm1	88	Alvarez et al. (1993). *Phytopathology* 83: 1405
C. michiganensis subsp. *sepedonicus*	McAb1 to McAb-5	19	De Boer and Wieczorek (1984). *Phytopathology* 74: 1431–1434
Erwinia carotovora subsp. *atroseptica*	14/18.6, 14/2, 14/8.6	3	Vernon-Shirley and Burns (1992). *J. Appl. Bacteriol.* 72: 97–102
E. chrysanthemi	6A6	36	Singh et al. (1999). *Plant Dis.* 84: 443–448
E. stewartii	C/G7/B2	43	Lamka et al. (1991). *Phytopathology* 81: 839–846
Pseudomonas avenae	Pa1 to Pa-5	20	Alvarez et al. (1993). *Proc. Symp. Seed Health Test. 1st*, Ottawa, Ont., Canada
Ps. syringae	Six MAbs[b]	223	Ovod et al. (1995). *Phytopathology* 85: 226–232
phaseolicola Ps. syringae pv.	AG-1, AG-2	9	Wong, W.C. (1990). *Lett. Appl. Microbiol.* 10: 241–244
Ralstonia solanacearum species-specific	Rs1, Rs1a	75	Alvarez et al. (1992). Presented at Bact. Wilt Int. Conf. Kaohsiung, Taiwan
Strain-specific	MAb		Griep et al. (1998). *Phytopathology* 88: 795–803
Xanthomonas albilineans	12 MAbs	1	Tsai et al. (1990). *Plant Prot. Bull.* 32: 125–135
	7 MAbs	38	Alvarez et al. (1996). *Plant Pathol.*45: 358–366
X. axonopodis pv. *begoniae*	Xcb-1	26	Benedict et al. (1990). *Appl. Environ. Microbiol.* 56: 572–574
X. axonopodis pv. *citri*	X-4600	30	Permar and Gottwald (1989). *Phytopathology* 79: 780–783

(continued)

Table 10.1 (continued)

Genus, species/subspecies or pathovar	MAb designation	No. of target strains tested	Reference[c]
X. axonopodis pv. *dieffenbachiae*	Xcd 1, Xcd 3, Xcd7	329	Lipp et al. (1992). *Phytopathology* 82: 677–682
	Xcd 108	302	Norman and Alvarez (1994). *Plant Dis.* 78: 954–958
X. campestris pv. *campestris*	X9, X13, X17	200	Alvarez et al. (1985). *Phytopathology* 75: 722–728
	10C5, 20H6, 16B5, 17C12, 10H12, 11B6	37	Franken A.A.J.M. (1990). *Proc. Symp. Perspect. Monoclon. Antib. Agric.* PUDOC Wageningen, Neth.
X. hortorum pv. *pelargonii*	Xcp-1	76	Benedict et al. (1990). *Appl. Environ. Microbiol.* 56: 572–574
X. hortorum pv. *phaseoli*	XP2	18	Wong, W.C. (1990). *Lett. Appl. Microbiol.* 10: 241–244
X. oryzae pv. *oryzae*	Xco1, Xco2, Xco5	178	Benedict et al. (1989). *Phytopathology* 79: 322–328
X. oryzae pv. *oryzicola*	Xco1a	8	Benedict et al. (1989). *Phytopathology* 79: 322–328
X. translucens pv. *undulosa*	AB3-B6	44	Bragard et al. (1993). *Fitopathol. Bras.* 18: 42–50
X. campestris pv. *mangiferaeindicae*	XCM-1-XCM-6	4	Sanders et al. (1994). *J. Appl. Bacteriol.* 77: 509–518
Xylophilus ampelinus		63	Gorris et al. (1989). *Proc. 7th Int. Conf. Plant Path. Bact.* Budapest, pp. 913–921

[a]Data summarised from Alvarez (2004)
[b]Two MAbs (Ps core-1 and Ps core-2), specific to core lipopolysaccharide and four O-chain-specific MAbs (Ps-O:2–1, Ps-O:2–2, Ps-O:2–3 and Ps-O:3–1) were used to classify 223 strains belonging to 19 pathovars of *Ps. syringae*
[c]References included in the table are not given under the references.

and colonies on the selective media. The sensitivity of ELISA can be enhanced tenfold by using an extraction buffer containing ethylenediaminetetra acetic acid and lysozyme, which releases lipopolysaccharides into solution, thereby enhancing the antibody-antigen reaction without increasing background readings. ELISA techniques using both polyclonal and monoclonal antibodies are available for several taxa of phytopathogenic bacteria, and rapid detection kits are also commercially available. Multitarget ELISA, which can detect more than one species in the same ELISA plate well using different enzyme labels, has been developed by Agdia, Inc. A multiple ELISA for detection of *C. michiganensis* subsp. *michiganensis* and *X. axonopodis* pv. *vesicatoria* is currently available (Alvarez 2004).

10.3 Flow Cytometry

The development of flow cytometry has enhanced the immunodiagnostic detection of bacteria. In this technique, as the bacterial cells or other particles pass individually through a sensor in a liquid stream, they are rapidly identified and quantified. Cells are identified by conjugation of fluorescent dyes to specific antibodies, and multiple cellular parameters are determined simultaneously based on cell's fluorescence and its ability to scatter light. The cells may be sorted electronically, permitting purification and/or culture of subpopulations of selected cells for further confirmatory tests (Alvarez 2001; Alvarez and Adams 1999). Flow cytometry has been used for the detection of *C. michiganensis* subsp. *michiganensis* in tomato seed extracts, of *X. axonopodis* pv. *dieffenbachiae* in anthurium (Alvarez and Adams

1999) and of *X. campestris* pv. *campestris* in seed extracts of *Brassica* sp. (Chitarra et al. 2002) and to determine the viability of *R. solanacearum* in seed potatoes (Van der Wolf et al. 2004).

Other methods such as immunofluorescence, immunomagnetic separation and lateral flow devices are also used for the detection of plant pathogenic bacteria. Immunofluorescence is widely used in Europe to detect bacterial pathogens in seed and propagating materials. In the Netherlands, it is used to screen 60,000 potato seed pieces annually to determine the presence of *R. solanacearum* (Van der Wolf and Schoen 2004). It is also used to detect *C. michiganensis* subsp. *sepedonicus* in potato seed pieces, and in France, it is used to screen tomato seed lots for the presence of *C. michiganensis* subsp. *michiganensis*. Lateral flow devices use the principles, primarily those of ELISA, but various types of filters used act as solid support for the initial binding reaction (Danks and Barker 2000). A lateral flow device test kit, developed by Central Science Laboratory, UK, detects *R. solanacearum* in a 3-min single-step process. Rapid ImmunoStrip® kits for detection of *R. solanacearum*, *C. michiganensis* subsp. *michiganensis* and *X. hortorum* pv. *pelargonii* are sold by Agdia, Inc. In immunomagnetic separations, target cells are isolated from a mixed solution using paramagnetic polystyrene beads coated with specific antibodies. After washing, bound cells can be used for polymerase chain reaction or they can be grown on semi-selective media, as they are viable.

10.4 Genomic Techniques

During the last two decades, many genomic techniques have been developed for the detection of known pathogens in plant samples. Several workers have reported the development and/or application of pathogen-specific primers for the detection of bacterial plant pathogens in heterogeneous mixtures. Most of these involve amplification by PCR coupled with one or more other techniques. Multiplex PCR is used to identify several pathogens simultaneously (Bertolini et al. 2003). A new cooperational PCR (Co-PCR) has also been developed to detect *R. solanacearum* in water (Caruso et al. 2003). Another PCR technique, real-time PCR, is rapid and more accurate for the detection of bacterial pathogens and has several other advantages. Other genomic methods include DNA-hybridisation, dot blots and nucleic acid sequence-based amplifications.

The main emphasis has been on the development of specific probes for the detection of target pathogens. In 1994, Maulis and associates developed pathovar-specific probes and PCR tests for *X. campestris* pv. *pelargonii*, which could detect 10–50 cfu of the bacterium per sample. Therefore, these tests should be able to detect the pathogens in asymptomatic plant tissue also. The primers and a probe were also developed for the detection of *Ps. syringae* pv. *actinidiae* in asymptomatic kiwi fruits (Koh and Nou 2002). Cubero and Graham (2002) developed primer sets for *X. axonopodis* pv. *citri* that can distinguish pathotype A from *X. aurantifolii* pathotypes B and C. Primers were based on sequence differences in the ITS region and *pthA* gene. Primer sets based on ribosomal sequences had high level of specificity for *X. axonopodis* pv. *citri* A (of citrus canker type A), whereas those based on *pthA* were universal for pathogens of all types of citrus bacterial cankers.

Different methods developed for the detection and identification of plant pathogenic bacteria have merits and demerits. Genomic techniques are quick and have more specificity and sensitivity, but they have not been adopted in all the laboratories due to certain limitations. For some laboratories, the cost of chemicals and equipment and the requirement of trained personnel for DNA extraction are the limitations for using PCR in the routine. Therefore, keeping in view the merits and demerits of all the available methods, Alvarez (2004) has rightly advocated the adoption of integrated approaches for the detection of plant pathogenic bacteria and diagnosis of bacterial diseases.

10.5 Enrichment Techniques

The enrichment technique has frequently been used to increase the sensitivity of IF, ELISA and/or PCR. An enrichment-ELISA assay using pathogen-specific MAbs was used to detect *X. axonopodis* pv. *dieffenbachiae* in leaf samples and *Pe. atrosepticum* or *R. solanacearum* in potato seed pieces. Alvarez et al. (1997) reported that even a limited culturing step, which is still insufficient for colony identification, increases the sensitivity of ELISA by about 10,000-fold. Enrichment assays have the added advantage that only the viable cells multiply to give a positive signal. Immunofluorescence colony-staining (IFC) technique combines an enrichment step in pour plates for colony detection with fluorochrome-labelled antibodies and also increases the sensitivity of IF approximately 10^4 times (Veena and Van Vuurde 2002). The micro-colonies formed due to the multiplication of viable cells can be seen with an epifluorescence stereomicroscope at 40–60X, whereas individual (dead) cells are not visible at this low magnification. IFC is used mainly for epidemiological studies and for confirmation of results achieved with other detection methods, which do not distinguish between living and dead cells. Optimisation of buffers and conjugates also improves test results. Antibodies of subclass IgG are highly suitable for IFC formats because their molecules are smaller than those of IgAs (dimers) and IgMs (pentamers) and they diffuse readily through agarose gel.

The enrichment of target bacteria on semi-selective media also improves the sensitivity of PCR reaction. Such tests are often termed 'BIO-PCR' because living cells of colonies harvested from culture plates are used in the PCR reaction. The major bands result from amplifications of DNA from living cells, although small amplicons may also be formed from dead bacteria. Sakthivel et al. (2001) were able to detect 55 fg of DNA per reaction tube which was equivalent to 70 cfu of *X. oryzae* pv. *oryzae* ml^{-1} of original sample and about 7 cells per reaction tube. To detect *Burkholderia cepacia* by a 'BIO-PCR' assay, contaminated samples were incubated only for 24 h in broth; DNA was extracted and added to PCR beads with specific primers. The entire process was completed in 27 h in comparison to 5–6 days required for isolation and identification of the target bacterium.

Certain compounds that interfere with DNA amplification have limited the use of direct PCR for the detection of target bacteria from natural samples. In a number of such cases, immunocapture or immunomagnetic separation has been employed to solve the problem. The optimum time for immunocapture of *Ps. syringae* pv. *syringae* with Advanced Magnetics™ beads coated with a polyclonal antiserum was only one-hour incubation. The detection limits were significantly improved when immunocapture was followed by a specific PCR test. Walcott and Gitaitis (2000) reported that immunomagnetic separation plus PCR resulted in a 100-fold increase in detection of *Ac. avenae* subsp. *citrulli* over direct PCR. Van Overbeek et al. (2002) also supported the polyphasic approach for studying the interaction between *R. solanacearum* and its biocontrol agent, *Ps. corrugata*.

Schaad et al. (2003) have reviewed the advances made in molecular-based diagnostic techniques used for the detection of plant pathogens including bacteria. They have compared the efficiency and sensitivity of conventional PCR, real-time PCR, BIO-PCR and immunocapture PCR. Real-time BIO-PCR is much more sensitive than conventional real-time PCR, and it can detect 2 cfu ml^{-1} of *Ps. syringae* pv. *phaseolicola* in seed extracts and of *C. michiganensis* pv. *sepedonicus* in potato tuber extracts. Multiplex PCR can be used to detect more than one species of bacteria in the same reaction tube using probes labelled with different fluorescent dyes. To avoid cross absorption, the wavelength of each dye must be well separated.

Recently, Miller et al. (2009) have stressed the importance of early and accurate diagnoses and pathogen surveillance on local, regional and global scales to predict outbreaks and allow time for development and application of mitigation strategies. Plant disease diagnostic networks have developed worldwide to address the problems of efficient and effective disease diagnosis and pathogen detection, engendering cooperation of institutes and experts within countries and across national borders. Internet has opened global avenues for access to databases, communication and cooperation not thought possible decades ago. Without this innovation, the organisation of effective networks to meet the challenge of invasive pathogens was unlikely. Case studies of regional, national and international diagnostic networks are presented. The authors have also emphasised the importance of morphological identification and systematics for the detection and identification of the pathogens in spite of frequent use of molecular methods in developed countries.

10.6 Management of Bacterial Plant Diseases

Plant pathogenic bacteria cause devastating losses to crop plants and ultimately huge sufferings to the mankind. Bacterial plant pathogens are very difficult to control due to their rapid spread in the crops. The problem is further compounded due to the limited number of chemicals available for their control and development of strains of plant pathogenic bacteria resistant to the chemicals. The reports of plant pathogenic bacteria developing resistance to copper fungicides and streptomycin have appeared in the literature. Hence, there is an urgent need to develop new management strategies free from these drawbacks. Nowadays, more emphasis is laid on the biological control and the use of bacteriophages and bacteriocins for the management of bacterial plant pathogens. Bacteriophages, bacteriocins and siderophores are also components of the biological control. A brief account of management of bacterial plant pathogens achieved through the use of biological control (including bacteriophages, bacteriocins and siderophores), chemicals and heat therapy is given below.

10.6.1 Biological Control

Bacterial diseases of plants are managed using various management strategies including use of chemicals. There are many drawbacks associated with chemical use. The major problem is the development of pathogen strains resistant to the chemicals. The occurrence of resistant strains of *E. amylovora*, the cause of fireblight of pome fruits, and of *X. vesicatoria*, the cause of bacterial spot of tomato and pepper, to streptomycin is well known. The strains of plant pathogenic bacteria, resistant to copper fungicides, are also found where these fungicides have been repeatedly used for a long time. Behlau et al. (2013) reported that multiple and independent introductions of copper resistance genes took place for strains of *X. citri* subsp. *citri* from Argentina and strains of *X. alfalfae* subsp. *citrumelonis* from Florida. The use of chemicals also causes environmental pollution and residue problems. Moreover, nowadays, a demand for pesticide-free products is also increasing rapidly. In certain cases, the use of chemicals is not permitted (on peach fruits before harvest) due to residue problem or if the product is to be certified as organic. Keeping in view the above-mentioned limitations associated with chemical use, there is an urgent need to develop an eco-friendly alternative to chemical use. Biological control offers an attractive alternative to chemical use.

Biocontrol is fundamentally an applied ecology. More specifically, the goal is to manage a microbial community to favour the biocontrol agent(s) and disfavour the pathogens. Exclusion of latecomers by a pioneering species is a general principle of community ecology and a fundamental tenet of biological control (Andrews 1992).

Biological control of plant diseases is gaining importance. It has been tried successfully in many diseases and some of the diseases are being controlled on commercial scale. For a successful biological control, biocontrol agent must colonise the host surface. The establishment of a biocontrol agent is necessary before the arrival of the pathogen when the antagonist is not an aggressive coloniser. Colonisation by an antagonist is a general requirement for biological control, and thus, microorganisms that grow well on the host surface are usually better candidates than those that do not. It is generally believed that resident microorganisms generally work better as biocontrol agents. However, in certain cases, the introduced microorganisms do equally well. Soilborne bacteria like *Ps. fluorescens* can colonise pear flowers and give control of fireblight almost as good as commercial bactericides. The deciding factor for residency vs. transiency is whether the invaders can survive the unfavourable weather conditions and not whether they can grow under favourable conditions.

The search for biocontrol agents should be made in the suppressive soils, from the rhizosphere of healthy plants in an infected crop or from a healthy crop growing in a pathogen-infected area. For the control of postharvest diseases of fruits, microorganisms, which quickly and thoroughly colonise fruit surface, should be selected. Biocontrol agents are generally screened in vitro before testing them in the greenhouse or in the field. In vitro tests are generally poor predictors of field performance. Andrews (1992) has given a scheme for screening of biocontrol agents.

The understanding of the mechanism(s) of a biocontrol agent can facilitate optimisation of control, as well as help to screen for more efficient strains of the agent. Biocontrol agents employ different mechanisms to check the growth of the pathogens. These mechanisms include competition for nutrients and host space; production of antibiotics, bacteriocins and siderophores; induction of systemic resistance; etc. Most of the biocontrol agents employ more than one mechanism to antagonise the target pathogen.

10.6.2 Plant Growth-Promoting Rhizobacteria

Many plant-associated bacteria improve plant health by inhibiting the growth of major and minor (does not attack plants directly but produce metabolites harmful to the plants) pathogens. These bacteria are called *plant growth-promoting bacteria* (PGPB). The most widely studied group of PGPB are the plant growth-promoting rhizobacteria (PGPR) that colonise the root surfaces and the rhizosphere. Kloepper et al. (1999) and Gray and Smith (2005) have reported that some of these PGPR can also enter inside the roots and establish as endophytes. Many of them are able to cross the endodermis barrier, moving from the root cortex to the vascular system, and ultimately live as endophytes in the stem, leaves, tubers and other organs. The major mechanisms of biocontrol employed by PGPB include competition for an ecological niche or a substrate, production of inhibitory metabolites and induction of systemic resistance in host plants to a broad spectrum of pathogens. Compant et al. (2005) have published an excellent review on biocontrol of plant diseases through the use of PGPB, highlighting the principles, mechanisms of action and future prospects.

PGPB control the plant pathogens by the synthesis of allelochemicals, including iron-chelating siderophores, antibiotics, biocidal volatiles, lytic enzymes and detoxification compounds. A variety of antibiotics including amphisin, 2,4-diacetylphloroglucinol (DAPG), oomycin A, phenazine, HCN, tropolone, pyoluteorin, pyrrolnitrin, tensin and cyclic lipopeptides are produced by pseudomonads, and oligomycin A, zwittermicin A, xanthobaccin and kanosamine are produced by *Bacillus*, *Streptomyces* and *Stenotrophomonas* spp. (Compant et al. 2005). An esterase produced by *Pantoea dispersa* causes an irreversible detoxification of albicidin produced by *X. albilineans*, the cause of leaf scald of sugarcane. The detoxification coupled with biocontrol efficacy leads to an excellent control of the disease (Amusa and Odunbaku 2007). Velusamy et al. (2006) reported the production of DAPG inhibitory to *X. oryzae* pv. *oryzae*, by plant-associated *Ps. fluorescens* strains, and the production was detected on the basis of PCR screening and confirmed by HPLC, HNMR and IR analyses. They further reported that DAPG suppressed bacterial blight of rice up to 59–64 % in net house and field experiments and Tn5 mutants defective in DAPG production (Phl^-) were much less effective in suppressing the disease. Biocontrol activity through the synthesis of allelochemicals, commonly found in free-living rhizobacteria, has also been found in endophytic bacteria. Sessitsch et al. (2004) reported that certain endophytic bacteria isolated from field-grown potato plants reduced in vitro the growth of *S. scabiei* and *X. campestris* through production of siderophores and antibiotics. An endophytic *Ps. fluorescens* strain FPT 9601 has been reported to synthesise DAPG and deposit DAPG crystals in and around roots of tomato.

10.6.3 Induced Systemic Resistance

Certain PGPB trigger a phenomenon known as induced systemic resistance (ISR) to protect the plants from different types of pathogens, i.e., fungal, bacterial and viral. Manifestation of ISR is dependent on the combination of host plant and bacterial strain. Most reports of PGPB-mediated ISR pertain to free-living rhizobacteria, but endophytic bacteria have also been reported to trigger ISR. *Systemic acquired resistance* (SAR), a phenomenon similar to ISR, is triggered in plants by pathogens or weakly virulent strains of pathogens. In this case, the defence mechanisms of the host react to the invasion of the pathogen leading to hypersensitive reaction due to which pathogen remains confined to local necrotic lesions of brown necrotic desiccated tissue. A major distinction, often drawn between ISR and SAR, is that the latter is dependent on accumulation of salicylic acid (SA). Majority of PGPB that activate ISR appear to do it via a SA-independent pathway involving jasmonate and ethylene signals. However, some PGPB do trigger a SA-dependent signalling pathway by producing very small amounts (nanograms) of SA in the rhizosphere.

Fravel (2005), in his review article, has emphasised the need to intensify the research for the production, formulation and delivery of biocontrol agents to boost their commercialisation. He has also suggested combining biocontrol agents with each other and with other management approaches, integrating biocontrol into an overall system. *Ag. radiobacter* strain K 84, the first biocontrol agent used commercially, was registered with the US Environmental Protection Agency (EPA) in 1979 for the control of crown gall. Ten years later, the first fungus, *Trichoderma harzianum* ATCC 20476, was registered with EPA for the control of a plant disease. Till 2001, a total of 14 bacteria and 12 fungi were registered with EPA for the control of plant diseases and one bacterium has been removed from the EPA list. Commercial products used for biocontrol of plant diseases account about 1 % of agricultural chemical sales (Lidert 2001), while the fungicides represent approximately 15 % of pesticide sales (http://www.epa.gov.). However, in spite of their extremely low percentage in pesticide sales, the biopesticides have contributed immensely to the control of plant diseases.

The formulation of a biocontrol agent can affect its performance. The populations of a *Streptomyces* sp. remain stable in talcum powder and starch granules for 10–14 weeks, and the stability is greater at 4 °C than at 24 °C. The methods and strategies for introducing and maintaining of biocontrol agents also affect the outcome of biocontrol. The ideal biocontrol introduces or promotes the antagonists only when and where they are needed or are most effective and minimises wasteful application of biocontrol agents to nontargets.

Biological control of plant diseases has been tried successfully in many diseases, and crown gall of stone fruits and fireblight of pome fruits are being controlled on commercial scale. *Ag. radiobacter* strain K 84 is very effective in controlling crown gall. The seeds, cuttings and roots of young seedlings are dipped in cell suspension of strain K 84 before sowing. Since its first release in 1973 in Australia, it is being used on commercial scale for the control of crown gall on stone fruits and roses in Europe, Africa, North America and South America. It was the first microorganism to be used commercially for the control of a plant disease. However, in 1973, some reports from Greece indicated its ineffectiveness for controlling crown gall disease (Kerr 1980).

Ag. radiobacter strain K 84 produces a bacteriocin called agrocin 84, which is inhibitory to many pathogenic strains of *Agrobacterium.* The biosynthesis of agrocin 84 is encoded by a 48 kb plasmid, called pAgK84. It is a conjugative plasmid and can be transferred to the agrobacteria including pathogens. The transfer of this plasmid makes the recipient bacterium resistant to agrocin 84, thereby resulting in non-control of the recipient bacterium with strain K 84. This might has been one of the reasons for the failure of control of crown gall in Greece. Recently, a failure of biological control in one peach nursery in Italy, due to transfer of pAgK84 from *Ag. rhizogenes* K 84 to natural pathogenic agrobacteria, has also occurred (Raio et al. 2009).

The transfer of pAgK84 to plant pathogenic agrobacteria posed a potential threat to the continued success of the biological control of crown gall. Prof. Allen Kerr's team from Waite Agricultural Research Institute, Adelaide, Australia, deleted this region and constructed a deletion mutant (transfer deficient, Tra^-) and named it as strain K 1026. In June 1987, strain K 1026 was released for experimental field trials in Waite Agricultural Research Institute, Adelaide, Australia. It was only the third recombinant DNA organism to be field-tested anywhere in the world. In 1988, the Department of Agriculture, New South Wales, released strain K 1026 for commercial use as a pesticide, under the trade name *No Gall*. Strain K 1026 was the first recombinant DNA organism to be released for commercial use anywhere in the world. Strain K 1026 has not been approved for release in any other country, except Australia. Strain K 1026 is not recommended for use on fruit-bearing crops. Out of two biovars of *Ag. tumefaciens*, biovar 2 is more sensitive to biocontrol by strain K 84 or K 1026 of *Ag. radiobacter*.

Ag. vitis causes crown gall of grapes. F2/5, a non-tumorigenic strain of *Ag. vitis*, is effective for controlling crown gall of grapes. F2/5 produces an antibiotic that inhibits growth of many tumorigenic *Ag. vitis* strains in vitro, but antibiotic minus mutants of F2/5 were found to be as effective as the wild-type strain for controlling crown gall (Burr et al. 1997). F2/5 gives better disease control when applied 24 h prior to pathogen inoculation. Most recently, Kaewnum et al. (2013) have confirmed the biocontrol potential of F2/5 strain against *Ag. vitis*.

Johnson and Stockwell (1998) have highlighted the use of biocontrol agents for the management of fireblight. BlightBan A506 (a product of *Ps. fluorescens* strain A506) is now available commercially for its control. Two applications of BlightBan A506 caused 40–60 % reduction in blossom blight. In another trial, its single application caused 50 % reduction in fireblight, but when this application was followed by weakly applications of streptomycin, the reduction in disease was 70 %. The combination treatment also caused more reduction in frost injury. Johnson et al. (2004) optimised timings of sprays of *Ps. fluorescens* (BlightBan A506) and *Pa. agglomerans* to control fireblight on apple and pear blossoms by adopting fireblight forecasting concepts.

Biocontrol agents can be applied with bactericides, SAR inducers or other methods of plant disease management. Obradovic et al. (2004) obtained better control of bacterial spot of tomato with the combination of phage and acibenzolar-S-methyl (ASM) than phage, ASM or copper-mancozeb alone. Obradovic et al. (2005) also suggested the integrated use of acibenzolar-S-methyl and phages as an alternative management strategy for the disease. Ji et al. (2006) also obtained significant control of bacterial spot and speck of tomato with combined application of *Ps. syringae* strain Cit7 and *Ps. fluorescens* strain 89B-61.

Acibenzolar-S-methyl (ASM, trade names, Actigard 50 WG and CGA 245704) is a plant activator that induces systemic acquired resistance in many crops against a number of pathogens. Louws et al. (2001) showed that it (35 g a.i. ha^{-1}) could be used as a viable alternative to copper-based bactericides for field management of bacterial spot and speck of tomato, where copper-resistant populations of these pathogens predominate. Moss et al. (2007) found 75-3S hrpG mutant of *X. campestris* pv. *vesicatoria* as effective as ASM in controlling bacterial spot of tomato under field conditions. They further reported that it caused a mean reduction in foliar disease severity of ~76 % compared to the mean of ~29 % for *Ps. syringae* Cit7, a most effective biocontrol agent found previously.

10.6.4 Bacteriophages

Bacteriophages (shortly called phages) are viruses, which kill bacteria. Their ability to kill bacteria has prompted the scientists to use them for the control of bacteria pathogenic to humans and plants, since their discovery in the second decade of the twentieth century. Among the earliest reports on the use of phages for plant disease control are those of Coons and Kotila (1925), Kotila and Coons (1925) and Moore (1926). Coons and Kotila (1925) controlled rotting of carrot discs by combined inoculation of *E. carotovora* subsp. *carotovora* and its phage. In the same year, Kotila and Coons checked the rotting of potato tubers by co-inoculation of *E. carotovora* subsp. *atroseptica* and its phage. In 1934, a report by Massey reported the low incidence of bacterial blight of cotton in fields flooded by Nile river water. He suggested that the low incidence of the disease was due to the presence of a phage in river water. There is also a report on control of Stewart's wilt of corn with phages. Phage treatment of infected seed reduced the disease incidence from 18 to 14 %.

In spite of these and other successful reports indicating the success of phages for plant disease control, their use on field scale did not become a reality for many years. Several workers, including Okabe and Goto (1963), Vidaver (1976) and Goto (1992), questioned their feasibility and effectiveness for controlling bacterial diseases on field scale. The control of phage-resistant bacterial strains with host range (h) mutant phages, development of protective phage formulations and integration of biocontrol agents and systemic acquired resistance inducers with phage use have enhanced the efficacy of phages for plant disease management. Moreover, the remarkable success achieved in controlling many important bacterial human pathogens has also given impetus to successful plant disease management by phages. Jones et al. (2007) have given a comprehensive account of work done on the use of phages for disease management and have clearly showed the practicability of phage use for controlling bacterial plant diseases citing some important success stories. Besides achieving highly effective disease control, there are several below given potential advantages in using phages in disease control:

1. Phages are self-replicating; hence, there is no need of repeated applications. In many cases, one application will do the job.
2. Phages are natural components of biosphere; they can be easily isolated from wherever their hosts, i.e., bacteria, are present.
3. Phage receptors on bacteria are generally pathogenicity determinants; the development of phage-resistant mutants usually leads to attenuation in bacterial virulence.
4. As phages are non-toxic to eukaryotic cells, they can be used in cases where use of chemical is prohibited such as treatment of peach fruits before harvest.
5. Phages are specific and they eliminate only target bacteria in contrast to broad-spectrum antibiotics. Therefore, they do not harm other members, including beneficial indigenous flora.
6. Phages are not affected by most agrochemicals; hence, they can be used in combination with these chemicals and also with biocontrol agents.
7. It is much easier and cheaper to develop a mutant phage against a phage-resistant bacterium than to develop an antibiotic against an antibiotic-resistant bacterium.
8. It is fairly easy and inexpensive to produce phages, and these can be stored at 4 °C for months without significant reduction in titre.

The development of bacterial strains resistant to phages was a major limiting factor in using phages for plant disease control. Jackson (1989) developed a technique to deal with occurrence of phage-resistant mutants by preparing a mixture of host range mutant (h-mutants) phages for disease control. H-mutants (h = host range or extended host range) possess the ability to lyse bacterial strains that are resistant to the parent phage while still being capable of lysing the wild-type bacterium. Therefore, they have an extended host range compared to the parent phage.

The application of a mixture of four phages, including wild-type and h-mutant phages, twice a week in the early morning prior to sunrise, gave significantly better control of bacterial spot of tomato. The phage mixture caused 17 % reduction in disease severity, while copper-mancozeb application caused 11 % reduction. Moreover, the phage application also gave significantly higher yield of extra large fruits than the copper-mancozeb treatment (Flaherty et al. 2000). Flaherty et al. (2001), used a mixture of h-mutants, developed from five phages, exhibiting the broadest host range, which reduced the incidence of bacterial blight of geranium caused by *X. campestris* pv. *pelargonii* by 50 % or more compared to control, and the disease incidence was also significantly less than those of recommended bactericides.

Another strategy to enhance plant disease control with phages is to apply them through bacteria that are able to persist in the plant environment and that are also host to these phages. A strain of *Pa. agglomerans* was used to deliver and sustain a mixture of four phages to control fireblight of pear caused by *E. amylovora*. *Pa. agglomerans* is also a biological control agent of *E. amylovora*. The strains of both the bacteria were susceptible to these phages (Svircev et al. 2006). Tanaka et al. (1990) also used the similar strategy to control tobacco bacterial wilt by applying an avirulent strain of *R. solanacearum* M4S and its bacteriophage. The phages that are better adapted to targeted environment and multiply rapidly are more suitable for disease control.

Timings of phage application in relation to arrival of the pathogen on the host also influence the disease control in several instances. The application of phage simultaneously with the pathogen inoculation or shortly before usually gives better results. Phage application 1 h or 1 day before inoculation of peach bacterial spot pathogen gave best control, while the fireblight of apple was best controlled when the phage mixture and *E. amylovora* were applied at the same time. Black rot of cabbage and bacterial spot of pepper are best controlled when the phages are applied on the same day of pathogen inoculation. The field application of phage during the daytime should be made when the sunlight irradiation is minimum to achieve better results. The control of bacterial spot of tomato was better with phage applications in the evening than morning (Balogh et al. 2003). While in another study, control was better when phage was applied at dawn in comparison to late morning application. For an effective control of tomato bacterial leaf spot, the phage mixture should contain 10^{6-8} plaque-forming units per ml.

Tomato has been used as a model system for developing an integrated management strategy to control foliar bacterial diseases. Various combinations involving bacterial antagonists, phages, plant growth-promoting rhizobacteria and systemic acquired resistance (SAR) inducers have been compared in greenhouse experiments. Acibenzolar-S-methyl (ASM), a SAR inducer, significantly reduced bacterial spot of tomato in field, but the combination of ASM and phage provided an additional reduction in disease pressure and resulted in more effective foliar disease control than ASM, phage or copper-mancozeb alone (Obradovic et al. 2004). Phage treatment, integrated with other practices, is currently widely used in greenhouse and production fields in Florida as a part of standard integrated management programme for the disease.

The successful use of phages for controlling bacterial plant diseases is evident from the fact that the phages are commercially produced and used on the large scale in the greenhouse and production fields. Agri Phi, Inc., established by L. E. Jackson, was the first company to commercially produce phages, specifically for control of bacterial plant diseases (Jackson 1989). Based on greenhouse and field trials, OmniLytics, Inc., in Salt Lake City, UT (formerly Agri Phi, Inc.), was the first to receive US Environmental Protection Agency's registration (registration No. 67986–1) to use phages in agriculture. The registration is for using host-specific phages on tomatoes in greenhouse and production fields as a part of standard integrated management programme to control tomato bacterial spot (Jones et al. 2006). Lang et al. (2007) have recommended the integration of phage mixtures with ASM for the management of onion leaf blight caused by *X. axonopodis* pv. *allii* to reduce the growers' reliance on conventional copper bactericides applied with mancozeb.

10.6.5 Bacteriocins

As defined by Nomura (1967), bacteriocins are non-replicating, bactericidal protein-containing substances, which are produced by certain strains of bacteria and are active against some other strains of the same or closely related species. This definition includes heterogeneous substances ranging from low-molecular-weight compounds to high-molecular-weight particles resembling bacteriophage protein components. A. Gratia was the first person to discover bacteriocins in 1925. The term *bacteriocine* was coined in 1953; the terminal 'e' is omitted in the current terminology of bacteriocins. They differ from traditional antibiotics that they have a narrow killing range and are toxic to bacteria closely related to the producing strain. Bacteriocins kill bacteria in several ways, i.e., by affecting protein synthesis, DNA stability and energy flux or membrane integrity. Agrocin 84 specifically inhibits DNA synthesis because of its being a structural analogue of adenine.

The control of phytopathogenic bacteria with bacteriocins has received the attention of many research workers. The control of crown gall of stone fruits with *Ag. radiobacter* strain K 84/agrocin 84 has already been covered above. Purified syringacin 4-A has been tried to control *Ps. syringae* pv. *phaseolicola* on bean. Its pre-inoculation spray @ 3 ng per leaf reduced the lesion count from 250 to zero. Its application to soybean seed protected the seed against *Ps. savastanoi* pv. *glycinea* infection and also increased the seed germination by 20 %. The cost of treatment was 0.49 US dollar per hectare (Vidaver 1976). Lavermicocca et al. (2002) reported that application of crude bacteriocin produced by *Ps. syringae* pv. *ciccaronei* @ 6000 arbitrary units ml^{-1} reduced the severity of olive knots caused by *Ps. savastanoi* pv. *savastanoi* by reducing the knot weight by 81 and 51 % on leaves and leaf scar sites, respectively. Bacteriocin application also reduced the epiphytic population of the pathogen, which was at least 350 and 20 times lower than the control populations on twigs and on leaves, respectively.

In spite of limited use of bacteriocins in plant disease control, they have a great potential to be used as prophylactic treatment for seed- or tuber-borne pathogens, prevention of secondary spread of bacteria from infected plants and protection of high-value crops like apple and pear against fireblight and stone fruits against crown gall. An important recommendation in the use of bacteriocins is that at least two and preferably three serologically unrelated bacteriocins must be used simultaneously. It will reduce the chances of development of bacterial strains resistant to the bacteriocins.

10.6.6 Siderophores

It is a Greek word meaning iron carrier. Siderophores are low-molecular-weight, high-affinity iron (III) chelators secreted by bacteria, fungi and many plants. Iron Fe^{3+} ions have a very low solubility at neutral pH and, therefore, cannot be utilised by organisms. Siderophores dissolve these ions by chelation as soluble Fe^{3+} complexes that can be taken up by active transport mechanisms. One major group of siderophores, i.e., derivatives of hydroxamic acid, chelates ferric iron very strongly. Kloepper et al. (1980) were the first to demonstrate the importance of siderophore production as a mechanism of biological control. As siderophores sequester the limited supply of iron (III) in the rhizosphere, they limit its availability to the pathogens and ultimately suppress their growth. Fluorescent pseudomonads have been shown to act as biocontrol agents against certain soil-borne plant pathogens. They produce yellow-green pigments (pyoverdines), which fluoresce under UV light and function as siderophores. Pyoverdines chelate iron in the rhizosphere and deprive pathogens of iron, which is required for their growth and pathogenesis.

Jagadeesh et al. (2001) found that siderophore-deficient mutants of a fluorescent *Pseudomonas* sp. did not control bacterial wilt of tomato caused by *R. solanacearum* as much as the wild-type strain, thus confirming the role of siderophores in the biocontrol mechanism. Siderophore-deficient mutants gave 18.75 % disease control in comparison to 75.00 % given by wild-type strain.

10.6.7 Heat Therapy

Heat therapy has been successfully used for controlling bacterial diseases of plants. Hot water treatment, hot air treatment and moist-hot-air treatment are used for managing ratoon stunting disease of sugarcane, but among them, the last method is more effective and, hence, widely used. In India, the recommended temperature is 54 °C for 4 h, while in the USA, the treatment is done at 51 °C for 4 h. The treatment is also effective for grassy shoot and leaf scald of sugarcane.

Seed treatment of cowpea with hot water at 50 °C for 30 min and of green gram at 52 °C for 30 min has been recommended for the control of bacterial blight of cowpea and bacterial leaf spot of green gram, respectively. Solar heat treatment of cowpea and green gram seeds (soaking of seed in tap water from 8.00 A.M. to 12.00 noon and then spreading the seed on the ground in a thin layer exposed directly to sunrays for 5 h on a bright sunny day in summer) is also effective for controlling both the above-mentioned diseases (Jindal et al. 1989; Thind 2012). Most recently, Hoffman et al. (2013) have reported that continuous thermal exposure of 40–42 °C for a minimum of 48 h was sufficient to significantly reduce titre or entirely eliminate *Candidatus* Liberibacter asiaticus from huanglongbing-affected citrus seedlings.

References

Alvarez AM (2001) Differentiation of bacterial populations in seed extracts by flow cytometry. In: De Boer SH (ed) Plant pathogenic bacteria. Kluwer, Dordrecht, pp 393–396

Alvarez AM (2004) Integrated approaches for detection of plant pathogenic bacteria and diagnosis of bacterial diseases. Annu Rev Phytopathol 42:339–366

Alvarez AM, Adams PD (1999) Flow cytometry: a promising tool for seed health testing. In: Proceedings of the Third International Seed Health Symposium, International Seed Test Association, Ames, pp 110–114

Alvarez AM, Rehman FU, Leach JE (1997) Comparison of serological and molecular methods for detection of *Xanthomonas oryzae* pv. *oryzae* in rice seed. Presented at Seed Health Test: Progr. Towards the 21st Century, Cambridge

Amusa NA, Odunbaku OA (2007) Biological control of bacterial diseases of plants in Nigeria: problems and prospects. Res J Agric Biol Sci 3:979–982

Andrews JH (1992) Biological control in the phyllosphere. Annu Rev Phytopathol 30:603–635

Balogh B, Jones JB, Momal MT, Olson SM, Obradovic A (2003) Improved efficacy of newly formulated bacteriophages for management of bacterial spot on tomato. Plant Dis 87:949–954

Behlau F, Hong JC, Jones JB, Graham JH (2013) Evidence for acquisition of copper resistance genes from different sources in citrus-associated xanthomonads. Phytopathology 103:409–418

Bertolini E, Olmas A, Lopez MM, Cambra M (2003) Multiplex nested RT-PCR in a single closed tube for sensitive and simultaneous detection of four RNA viruses and *Pseudomonas savastanoi* pv. *savastanoi* in olive trees. Phytopathology 93:286–292

Burr TJ, Reid CL, Tagliati E, Bazzi C, Sule S (1997) Biological control of grape crown gall by strain F2/5 is not associated with agrocin production or competition for attachment sites on grape cells. Phytopathology 87:706–711

Caruso P, Bertolini E, Cambra M, Lopez MM (2003) A new and sensitive Co-operational polymerase chain reaction (Co-PCR) for rapid detection of *Ralstonia solanacearum* in water. J Microbiol Methods 55:257–272

Chitarra LG, Langerak CJ, Bergervoet JH, van den Bulk RW (2002) Detection of the plant pathogenic bacterium, *Xanthomonas campestris* pv. *campestris* in seed extracts of Brassica sp. applying fluorescent antibodies and flow cytometry. Cytometry 47:118–126

Compant S, Duffy B, Nowak J, Clement C, Bark EA (2005) Use of plant growth-promoting bacteria for biocontrol of plant diseases: principles, mechanisms of action and future prospects. Appl Environ Microbiol 71:4951–4959

Coons GH, Kotila JE (1925) The transmissible lytic principle (bacteriophage) in relation to plant pathogens. Phytopathology 15:357–370

Cubero J, Graham JH (2002) Genetic relationship among worldwide strains of *Xanthomonas* causing canker in citrus species and design of new primers for their identification by PCR. Appl Environ Microbiol 68:157–164

Danks C, Barker I (2000) On-site detection of plant pathogens using lateral flow devices. Bull OEPP/EPPO 30:421–426

Flaherty JE, Jones JB, Harbaugh BK, Somodi GC, Jackson LE (2000) Control of bacterial spot on tomato in the greenhouse and field with H-mutant bacteriophages. Hortic Sci 35:882–884

Flaherty JE, Harbaugh BK, Jones JB, Somodi GC, Jackson LE (2001) H-mutant bacteriophages as a potential biocontrol of bacterial blight of geranium. Hortic Sci 36:98–100

Fravel DR (2005) Commercialization and implementation of biocontrol. Annu Rev Phytopathol 43:337–359

Goto M (1992) Fundamentals of bacterial plant pathology. Academic Press, Inc., San Diego, 342 pp

Gray EJ, Smith DL (2005) Intracellular and extracellular PGPR: commonalities and distinctions in the plant-bacterium signalling processes. Soil Biol Biochem 37:395–412

Hoffman MT, Doud MS, Williams L, Zhang M-Q, Ding F *et al* (2013) Heat treatment eliminates '*Candidatus*

Liberibacter asiaticus' from infected citrus trees under controlled conditions. Phytopathology 103:15–22
Horst RK (2001) Westcott's plant disease handbook, 6th edn. Kluwer Academic Publishers, Boston
Hu FP, Young JM, Triggs CM, Park DC, Saul DJ (2001) Relationships within the Proteobacteria of plant pathogenic *Acidovorax* species and subspecies, *Burkholderia* species and *Herbaspirillum rubrisubalbicans* by sequence analysis of 16S rDNA, numerical analysis and determinative tests. Anton Leeuw J Microbiol Serol 80:201–214
Jackson LE (1989) U. S. Patent No. 4828999
Jagadeesh KS, Kulkarni JH, Krishnaraj PU (2001) Evaluation of the role of fluorescent siderophore in the biological control of bacterial wilt in tomato using Tn[5] mutants of fluorescent *Pseudomonas* sp. Curr Sci 81:882–883
Ji P, Campbell HL, Kloepper JW, Jones JB, Suslow TV *et al* (2006) Integrated biological control of bacterial speck and spot of tomato under field conditions using foliar biological control agents and plant growth-promoting rhizobacteria. Biol Control 36:358–367
Jindal KK, Thind BS, Soni PS (1989) Physical and chemical agents for the control of *Xanthomonas campestris* pv. *vignicola* from cowpea seeds. Seed Sci Technol 17:371–382
Johnson KB, Stockwell VO (1998) Management of fire blight: a case study in microbial ecology. Annu Rev Phytopathol 36:227–248
Johnson KB, Stockwell VO, Sawyer TL (2004) Adaptation of fire blight forecasting to optimize the use of biological controls. Plant Dis 88:41–48
Jones JB, Iriarte FB, Obradovic A, Balogh B, Momol MT *et al* (2006) Management of bacterial spot on tomatoes with bacteriophages. In: Proceedings of the international symposium on biological control of bacterial plant diseases, 1st, Darmstadt, Germany, 408:154, Land-Forstwirtsch: Mitt. Biol. Bundesanst
Jones JB, Jackson LE, Balogh B, Obradovic A, Iriarte FB *et al* (2007) Bacteriophages for plant disease control. Annu Rev Phytopathol 45:245–262
Kaewnum S, Zheng D, Reid CL, Johnson KL, Gee JC *et al* (2013) A host-specific biological control of grape crown gall by *Agrobacterium vitis* strain F2/5: its regulation and population dynamics. Phytopathology 103:427–435
Kerr A (1980) Biological control of crown gall through production of agrocin 84. Plant Dis 64:25–30
Kloepper JW, Leong J, Teintze M, Schroth MN (1980) Enhanced plant growth by siderophores produced by plant growth-promoting rhizobacteria. Nature 286:885–886
Kloepper JW, Rodriguez-Kabana R, Zehnder GW, Murphy JF, Sikora E *et al* (1999) Plant root-bacterial interactions in biological control of soil borne diseases and potential extension to systemic and foliar diseases. Australas Plant Pathol 28:21–26
Koh YJ, Nou IS (2002) DNA markers for identification of *Pseudomonas syringae* pv. *actinidiae*. Mol Cell 13:309–314
Kotila JE, Coons GH (1925) Investigations on the blackleg disease of potato. Mich Agric Exp Sta Tech Bull 67:3–29
Lang JM, Gent DH, Schwartz HF (2007) Management of *Xanthomonas* leaf blight of onion with bacteriophages and a plant activator. Plant Dis 91:871–878
Lavermicocca P, Lonigro SL, Valerio F, Evidente A, Visconti A (2002) Reduction of olive knot disease by a bacteriocin from *Pseudomonas syringae* pv. *ciccaronei*. Appl Environ Microbiol 68:1403–1407
Lelliot RA, Stead DE (1987) Methods for the diagnosis of bacterial diseases of plants. Blackwell Scientific Publications, Oxford, 215 pp
Lidert Z (2001) Biopesticides: is there a path to commercial success? In: Vurro M, Gressel J, Butt T, Harman GE, Pilegram A *et al* (eds) Enhancing biocontrol agents and handling risks. Ios Press, Amsterdam, p 283
Louws FJ, Wilson M, Campbell HL, Cuppels DA, Jones JB *et al* (2001) Field control of bacterial spot and bacterial speck of tomato using a plant activator. Plant Dis 85:481–488
Miller SA, Beed FD, Harmon CL (2009) Plant disease diagnostic capabilities and networks. Annu Rev Phytopathol 47:15–38
Moore ES (1926) d' Herelle's bacteriophage in relation to plant parasites. S Afr J Sci 23:306
Moss WP, Byrne JM, Campbell HL, Ji P, Bonas U *et al* (2007) Biological control of bacterial spot of tomato using *hrp* mutants of *Xanthomonas campestris* pv. *vesicatoria*. Biol Control 41:199–206
Nomura M (1967) Colicins and related bacteriocins. Annu Rev Microbiol 21:257–284
Obradovic A, Jones JB, Momol MT, Balogh B, Olson SM (2004) Management of tomato bacterial spot in the field by foliar applications of bacteriophages and SAR inducers. Plant Dis 88:736–740
Obradovic A, Jones JB, Momol MT, Olson SM, Jackson LE *et al* (2005) Integration of biological control agents and systemic acquired resistance inducers against bacterial spot on tomato. Plant Dis 89:712–716
Okabe N, Goto M (1963) Bacteriophages of plant pathogens. Annu Rev Phytopathol 1:397–418
Raio A, Peluso R, Puopolo G, Zoina A (2009) Evidence of pAgK84 transfer from *Agrobacterium rhizogenes* K84 to natural pathogenic *Agrobacterium* spp. in an Italian peach nursery. Plant Pathol 58:745–753
Riley MB, Williamson MR, Maloy O (2006) Plant disease diagnosis. APSnet Feature Story. http://www.apsnet.org/online/feature/plantdisease/
Sakthivel N, Mortensen CN, Mathur SB (2001) Detection of *Xanthomonas oryzae* pv. *oryzae* in artificially inoculated and naturally infected rice seeds and plants by molecular techniques. Appl Microbiol Biotechnol 56:435–441
Schaad NW, Frederick RD, Shaw J, Schneider WL, Hickson R *et al* (2003) Advances in molecular-based diagnostics in meeting crop biosecurity and phytosanitary issues. Annu Rev Phytopathol 41:305–324

Sessitsch A, Reiter B, Berg G (2004) Endophytic bacterial communities of field-grown potato plants and their plant growth-promoting and antagonistic abilities. Can J Microbiol 50:239–249

Svircev AM, Lehman SM, Kim W, Barszcz E, Schneider KE *et al* (2006) Control of the fire blight pathogen with bacteriophages. In: Proceedings of international symposium on biological control of bacterial plant diseases, 1st, Seeheim/Darmstadt, Germany, 259, Berlin: Dtsch. Bibl., CIP-Einh. aufn

Tanaka H, Negishi H, Maeda H (1990) Control of tobacco bacterial wilt by an avirulent strain of *Pseudomonas solanacearum* M4S and its bacteriophage. Ann Phytopathol Soc Jpn 56:243–246

Thind BS (2012) Phytopathogenic procaryotes and plant diseases (refer Bacterial blight of cowpea, pages 405–410 and Bacterial leaf spot of green gram, pages 410–415). Scientific Publishers, Jodhpur, 545 pp

Van der Wolf JM, Schoen CD (2004) Bacterial pathogens: detection and identification methods. Marcel Dekker, New York, pp 1–5

Van der Wolf JM, Sledz V, Van Elsas JD, Van Overbeek L, Van Bergervoet JHW (2004) Flow cytometry to detect *Ralstonia solanacearum* and to assess viability. In: Allen C, Prior P, Hayward AC (eds) Bacterial wilt: the disease and the Ralstonia solanacearum species complex. APS Press, St. Paul, p 6

Van Overbeek LS, Cassidy M, Kozdroz J, Trevors JT, Van Elsas JD (2002) A polyphasic approach for studying the interaction between *Ralstonia solanacearum* and potential control agents in the tomato phytosphere. J Microbiol Methods 48:69–86

Veena MS, van Vuurde JW (2002) Indirect immunofluorescence colony staining method for detecting bacterial pathogens of tomato. J Microbiol Methods 49:11–17

Velusamy P, Immanuel JE, Gnanamanickam SS, Thomashow L (2006) Biocontrol of rice bacterial blight by plant-associated bacteria producing 2,4-diacetylphloroglucinol. Can J Microbiol 52:56–65

Vidaver AK (1976) Prospects for control of phytopathogenic bacteria by bacteriophages and bacteriocins. Annu Rev Phytopathol 14:451–465

Walcott RR, Gitaitis RD (2000) Detection of *Acidovorax avenae* subsp. *citrulli* in watermelon seed using immunomagnetic separation and the polymerase chain reaction. Plant Dis 84:470–474

11 Recent Developments in Bacterial Blight of Pomegranate and Its Management

K.K. Sharma, Jyotsana Sharma, and V.T. Jadhav

11.1 Introduction

Pomegranate (*Punica granatum* L.) is an important ancient fruit known for its high nutritive and medicinal values. It is a native of Iran and adjoining areas and is mainly cultivated in subtropical and tropical regions of the world comprising countries like India, Iran, the USA, Afghanistan, Turkey, Spain, Tunisia, and Israel. In India, which is one of the largest pomegranate-producing countries of the world, the fruit is commercially cultivated mainly in the states of Maharashtra, Karnataka, Andhra Pradesh, Gujarat, Tamil Nadu, and Rajasthan. Besides, small-scale cultivation of pomegranate is also observed in the states of Himachal Pradesh and Uttarakhand.

Bacterial blight, in recent years, has become the most important disease of pomegranate due to its epidemic occurrence in all the major pomegranate-growing states of Maharashtra, Karnataka, and Andhra Pradesh inflicting enormous losses on growers. Although bacterial blight has been known to be present in India since 1952 (Hingorani and Mehta 1952), the disease was of minor importance until 1998. Apart from its first report from Delhi in 1952, bacterial blight has been reported by various workers from different parts of the country, namely, Bangalore in Karnataka (Hingorani and Singh 1959; Chand and Kishun 1991), Tamil Nadu (Rangaswami 1962), Himachal Pradesh (Sohi et al. 1964), Haryana (Kanwar 1976), Maharashtra (Kamble 1990), and Punjab (Rani and Verma 2001). Recent epidemics of bacterial blight have been reported from Maharashtra, Karnataka, and Andhra Pradesh (Dhandar et al. 2004; Sharma et al. 2008; Benagi and RaviKumar 2009). Pomegranate bacterial blight is not reported from any country outside India.

K.K. Sharma (✉) • J. Sharma • V.T. Jadhav
ICAR-National Research Centre on Pomegranate, Solapur, (Maharashtra), India
e-mail: kksharma231@gmail.com

11.2 Distribution and Losses

Recent surveys revealed blight prevalence in all the major pomegranate states of Maharashtra, Karnataka, and Andhra Pradesh in mild to severe form. In Maharashtra the disease was prevalent in 41.5 % orchards in the districts of Solapur, Sangli, Pune, Ahmednagar, Nashik, Osmanabad, Aurangabad, Jalna, and Latur. In Karnataka disease prevalence was 58.33 % with disease occurring in the districts of Bagalkot, Gadag, Koppal, and Bijapur. Besides, blight has also been reported from Bellary, Belgaum, Chitradurga, Tumkur, and Davanagere districts of Karnataka. In Andhra Pradesh, the disease was prevalent in Anantapur and Mahaboobnagar districts. In Anantapur district which has more than 75 % area under pomegranate cultivation in the state, blight prevalence was 43.47 % (NRCP *Ann.*

L.P. Awasthi (ed.), *Recent Advances in the Diagnosis and Management of Plant Diseases*,
DOI 10.1007/978-81-322-2571-3_11

Report 2007–2008, 2008–2009; Benagi and RaviKumar 2009). Besides, bacterial blight has also been recorded in Himachal Pradesh on prominent cultivars Bhagwa, Ganesh, Arakta, and Mridula causing huge losses to the growers (Khosala et al. 2009).

In Maharashtra, blight resulted in losses up to 80.0 % in some orchards of Solapur District in 2006–2007 (NRCP *Ann. Report* 2006–2007). According to Srivastava (2008) fifty percent of the total area under pomegranate cultivation has been affected by the disease, and farmers have reported yield reduction by 60–80 %. As per Maharashtra state statistics, about 31,000 ha area has been affected by blight out of a total area of 93,000 ha under pomegranate cultivation in 2007–2008 (Jadhav and Sharma 2009). In Karnataka, production of pomegranate declined from 1.8 lakh tonnes per annum to less than 10,000 tonnes in 2007–2008 per annum in a span of 4 years, thereby causing a revenue loss of about Rs 200 crores (Benagi and RaviKumar 2009).

11.3 Diagnostics and Detection

Bacterial blight can be diagnosed by symptoms which are conspicuous on leaves, twigs, and fruits.

11.3.1 Symptoms

Leaves Initially, minute, circular water-soaked lesions are observed on undersurface of the foliage which later on are seen on both surfaces and become regular to irregular in shape and necrotic with brownish black centers and translucent yellow halo and vary in size between 1 mm and 5 mm. Normally, there are 4–15 spots per leaf; however, at times leaves having 28 lesions have also been observed. Lesion may or may not coalesce, and their development may extend to veins and midrib. Blighted leaves often turn yellow and fall off prematurely (Fig. 11.1).

Stems Brownish black lesions are observed on twigs and stems which extend along the bark of the stem. Though most of the time infections remain restricted to bark and cortex portion only, in rare instances infections extending to vascular tissues have also been observed. Blight lesions often girdle the stems/twigs resulting in their breaking off at the girdled portion with a slight mechanical or wind pressure. As infections are quite frequent at nodes of the stem, disease is also known as nodal blight. Advanced blight infections on main trunk and other branches often reveal canker formation due to the activity of cork cambium, thereby restricting further growth of the pathogen (Fig. 11.2).

Fruits Fruits are the most susceptible part of the plant to bacterial infections. Like leaves, initially water-soaked lesions are observed on young to mature fruits which turn necrotic with brownish black discoloration and generally coalesce resulting in blighted appearance. Blight lesions on fruits normally reveal minute fissures (cracks) which are L or Y shaped and discern bacterial blight from other fungal spots which do not show such fissures. Blight infections on fruits are restricted to rind (pericarp) portion only and are not observed in locules or arils of the fruit.

Fig. 11.1 Bacterial blight symptoms on leaves

Fig. 11.2 Blight symptoms on stem. (**a**) Initial blight lesion. (**b**) Blight lesions at advanced stage. (**c**) Blighted stem revealing girdling. (**d**) Canker formation in blighted stem

Infected fruits with blight lesions often result in splitting, rendering them unfit for consumption due to secondary infections (Fig. 11.3). However, such cracked fruits may be procured by processing units at lower price for postharvest products like fruit juice, ready to serve (RTS) drink, juice concentrate, seed for seed oil etc.

Blight-infected leaves in the field can be diagnosed by the presence of translucent margins which are not visible in spots caused by fungal pathogens. However, old bacterial blight spots may not reveal such translucent margins. Infected fruits in field can be diagnosed by placing a drop of water on blight spot which after a few minutes becomes slimy due to exudation of bacterial ooze. In laboratory, blight infections on leaves, fruits, and stems can be easily diagnosed and pathogen detected by mounting section of the diseased tissue on a glass slide in a drop of water and examining it under the microscope. Exudation of bacterial ooze from such diseased tissues confirms the presence of bacterial blight pathogen. The blight bacterium in culture can be detected and identified by the application of advanced and sensitive molecular techniques such as PCR-based methods.

11.3.2 Etiology

On the basis of disease symptoms, cultural aspects, and biochemical and pathogenicity tests, causal organism of pomegranate blight has been identified as *Xanthomonas axonopodis* pv. *punicae* by various workers (Hingorani and Singh 1959; Rangaswami 1962; Chand and Kishun 1991; Rani and Verma 2002; Dhandar et al. 2004; NRCP *Ann. Report* 2006–2007). The identity of the pathogen has been further confirmed and established using latest PCR-based techniques as

Fig. 11.3 Infected fruits revealing blight symptoms. (**a**) Initial infections. (**b**) Spots revealing fissures. (**c**) Blighted fruit with initial splitting. (**d**) Severely blighted fruits with conspicuous splitting

Xanthomonas axonopodis pv. *punicae* (NRCP. *Ann. Report* 2008; Mondal and Singh 2009). Until 1995, the blight pathogen was classified as *Xanthomonas campestris* pv. *punicae* (*Hingorani & Singh*) Dye. However, in 1995, Vauterine, Hoste, Kersters, and Swings reclassified the bacterium on the basis of DNA hybridization and named it *Xanthomonas axonopodis* pv. *punicae* (Hingorani & Singh) (Vauterine et al. 1995).

Pathogen The bacterium *Xanthomonas axonopodis* pv. *punicae* is a gram-negative rod, motile with single polar flagellum, and nonspore forming, produces nondiffusible yellow pigment, and measures 0.4 – 0.75 × 1.0 – 3.0 μm. The colonies are smooth, circular light yellow, glistening mucoid, butyrous, and convex with entire margins. The isolates are positive in milk proteolysis, H_2S production, KOH test, gelatin liquefaction, and tyrosinase and catalase activity. Young growth on nutrient glucose agar medium is translucent and does not impart any foul odor. Bacterial growth was better at temperature range of 23.0–30.0 °C, whereas it could not grow below 10.0 °C and above 40.0 °C (Mogle et al. 2009). On the other hand, Chand and Kishun (1991) observed bacterial growth at temperature range of 4.0–35.0 °C.

Pomegranate bacterial blight pathogen has been cloned and sequenced using enterobacterial repetitive intergenic consensus (ERIC) primers (Mondal and Singh 2009), and the study could be useful in detecting and identifying virulent strains of the bacterium.

11.4 Epidemiology

Primary Source of Inoculum and Its Survivability Blight bacterium survives in infected plant stems, buds, and plant debris in the soil. Studies on pathogen's survivability have revealed that blight bacterium could survive in infected plant parts for a period of 7 months to 1 year depending on survivability conditions.

Studies carried out at NRCP, Solapur, revealed that diseased leaves collected from severely blight-affected orchard and kept at room temperature (25.0–38.0 °C) continued to produce bacterial ooze as observed under the microscope, and the bacterium could be isolated on nutrient agar glucose medium after 1 year of incubation, thereby revealing the survivability of the bacterium for1 year. Although, even after 2 years, the bacterium continued to produce ooze from such lesions of infected leaves, the bacterium neither could be isolated on medium nor could it infect the plants on inoculation, thereby indicating the survivability of the pathogen in infected leaves up to 1 year only (NRCP, *Ann. Report*, 2008–2009). According to Rani and Verma (2002) infected leaves lying on the ground revealed bacterial growth in culture till 7 months, though recovery of the bacterium was cent percent up to 30 days followed by gradual decline up to 7 months. The same authors, however, were able to recover the pathogen up to 8 months when blight-infected leaves, fruits, and intact cankers on branches were incubated under laboratory conditions or blighted leaves kept protected under field conditions.

It is apparent that under natural conditions infected plant debris in the orchard is exposed to various microbial activities leading to rapid decomposition of leaves and also some microorganisms might be antagonistic to blight pathogen, consequently reducing the survivability chances of the bacterium in nature. Also, *Xanthomonas* is known to be a poor saprophyte as all species of the pathogen are found only in association with host or plant debris. It is also possible that because of the perennial nature of the pomegranate, there is continuous availability of the host for the bacterium to survive for long periods and as such the bacterium shows less or no requirement for survival in soil.

Dissemination The dissemination of bacteria from primary source of inoculum to different plant parts mainly takes place through rain splashes, irrigation water, pruning tools, insect vectors, and man. Insects like butterfly, aphids, and larvae of fruit borer have been reported to disseminate bacterial blight (Benagi and RaviKumar 2009). Insects not only carry the bacteria to plants but also inoculate them into requisite sites. Man helps in dissemination by his cultural practices or by transporting infected plants or plant parts over long distances to new areas.

The bacterium enters and infects different plant parts through natural openings like stomata, lenticels, hydathodes or wounds.

Incubation Period After the bacterium enters the host and starts multiplying in different tissues, symptoms start developing in 3–4-day period and continue to develop up to 30 days depending on incubation conditions. Successful and early symptoms in artificially inoculated plant parts developed at incubation temperatures of and around 30.0 °C under moist conditions. Kanwar (1976) reported that infections occur rapidly on injured leaves, flowers, and fruits than on uninjured parts and symptoms developed within 4–7 days on the injured portion and in 8–12 days on the uninjured portion. Chand and Kishun (1991) tried different inoculation techniques (pinprick, rubber block press, and leaf-cut method) and observed infections within 21 days through leaf-cut method. Rani and Verma (2001) recorded disease symptoms on injured surfaces of flowers, fruits, and leaves within 7–10 days, whereas in uninjured plant parts, symptoms developed in 12–15 days of inoculations both under artificial conditions. Studies carried out on incubation period at NRCP, Solapur, revealed symptom initiation after 3–4 days of inoculation which continued to develop even up to 30 days depending upon incubation factors (NRCP, *Ann. Report.*, 2006–2007, 2007–2008).

Transmission of Bacterial Blight Through Planting Material It has been observed that planting material (stem cuttings and air-layered cuttings) obtained from diseased plants (made apparently healthy by pruning of diseased parts) may carry the blight pathogen in latent form particularly in buds, resulting in infection of new plants produced from the planting material. Studies conducted at NRCP, Solapur, revealed that inoculated plants showing severe blight symptoms when pruned by removing diseased and old twigs revealed blight symptoms in all (100 %) the pruned plants and in 40 % of the plants obtained from cuttings of apparently healthy plants after about 7 months of incubation, whereas no disease was observed in healthy plants kept as check during the same period of time. The symptoms were mainly observed on the nodes and leaves adjacent to the nodes (NRCP*Ann. Report*, 2009). Chand and Kishun (1993) reported that necrosis of the leaves was very late and was far behind the forward movement of the bacterium which indicated that during its movement the bacterium shows biotrophic nature for some time than necrotrophic nature.

Blight Development in Relation to Meteorological Factors Bacterial blight was prevalent throughout the year at a temperature range of 9.0–43.0 °C and varying humidities (30.0–>80.0 %) under Solapur conditions (Sharma et al. 2009). However, blight development was brisk during the rainy season from July to September as compared to December to March due to availability of free water and humid conditions required for infections and further multiplication and dissemination of the pathogen. Surveys of Solapur District revealed 48.9 % orchards revealing severe blight infections during rainy season from July to September as compared to December to February when only 10.5 % orchards had severe blight infections (NRCP, *Ann. Report*, 2006–2007). Benagi and RaviKumar (2009) also reported more disease severity in rainy season than in autumn or summer season in Karnataka.

Correlation between blight severity and different meteorological parameters revealed that blight progress was positively and significantly correlated with relative humidity and rainfall, whereas temperature had nonsignificant correlation with disease under Solapur conditions (NRCP, *Ann. Report.*, 2009). Rani and Verma (2002) reported gradual decline in maximum and minimum atmospheric temperature, sharp increase in maximum and minimum relative humidity, and moderate rainfall favored disease severity.

11.5 Bacterial Blight Management

Effective management of bacterial blight can be achieved by employing cultural and chemical methods in an integrated manner.

11.5.1 Cultural Methods

(a) *Healthy and disease-free planting material*: It is of utmost importance that healthy and disease-free planting material be procured from a disease-free certified nursery for establishing an orchard.
(b) *Avoidance of rainy season crop*: As pomegranate tree flowers throughout the year in Central and southern India, normally three crops can be regulated in a year, viz., spring crop also known as Ambe bahar (January–July), rainy season crop known as Mrig bahar (June–December), and autumn season crop known as Hasta bahar (September–March). The rainy season crop (Mrig bahar), due to high disease pressure during the season, should be avoided as far as possible and instead autumn crop or spring crop be preferred.
(c) *Rest period*: After harvest, plants should be provided rest period for 3–4 months by restricting the irrigation to minimum and supplying organic manure and nutrients so as to encourage vigorous growth during the ensuing season. Defoliation at the end of the rest period prior to crop regulation further ensures reduction of pathogen inoculum.
(d) *Sanitation measures*: Sanitation measures should be strictly adhered to, and all infected plant parts fallen on the ground be collected and burnt. Diseased and dead twigs/branches

should be properly pruned and Bordeaux paste be applied to cut ends of the stems or plants be sprayed with Bordeaux mixture (1.0 %)/copper oxychloride (0.2 %)/copper hydroxide (0.2 %). Pruning tools should be disinfected with a solution of sodium hypochlorite (2.5 %) or calcium hypochlorite (2.5 %) solution before proceeding to the next tree. Dusting the orchard, particularly soil below tree canopy, with copper oxide dust (4 %) at 20 kg/ha or bleaching powder at 20 kg/ha (or drenching with bleaching powder at 2.5 %) once or twice in a year would ensure reduction of bacterial inoculum.

11.5.2 Chemical Control

Different workers have reported the efficacy of the antibiotic streptocycline in bacterial blight mitigation. Bacterial blight has been managed effectively with sprays of streptocycline (500 ppm) alone or in combination with fungicides like copper oxychloride (0.2 %)/carbendazim (0.1 %) followed by spray schedule with another antibiotic, Bactrinol (2-bromo-2-nitropropane-1,3-diol) at 500 ppm in combination with copper oxychloride (0.2 %) at 15 days interval (NRCP, *Ann. Report*. 2006–2007, 2007–2008). Benagi and RaviKumar et al. (2009) also reported the efficacy of streptocycline (500 ppm)+copper oxychloride (0.2 %) followed by Bronopol (2-bromo-2-nitropropane-1,3-diol) (500 ppm)+copper oxychloride (0.2 %) in controlling disease and increasing yield. According to Yenjerappa et al. (2009), least disease was observed in plots sprayed with streptocycline (0.05 %)+copper oxychloride (0.2 %) followed by Bactrinol (0.05 %)+copper oxychloride (0.2 %).

11.5.3 Biological Control

Among the various bioagents studied, sprays of *Pseudomonas fluorescens* (0.1 %) have been found effective in blight control under field conditions; however, the bioagent was found not superior to antibiotic streptocycline (NRCP, *Ann. Report*,2008–2009). Similarly, Yenjerappa et al. (2009) in Karnataka also reported the efficacy of *Pseudomonas fluorescens* (0.5 %) in managing bacterial blight in field but found the bioagent inferior to antibiotics streptocycline and Bactrinol.

11.5.4 Integrated Nutrient Management

Application of organic manures like vermicompost and neem cake to plants during the rest period or prior to flowering and nutrients, namely, calcium, magnesium, zinc, boron, iron, and manganese, during the flower and fruit development stages either through soil application/drip irrigation or foliar sprays improves plant vigor and yield and minimizes disease.

11.5.5 Resistant Varieties

At present no pomegranate variety is available which imparts resistance to bacterial blight, and all present-day varieties, namely, Ganesh, Bhagwa, Arakta, Mridula, and Ruby, are either highly susceptible or susceptible to blight pathogen. As such there is a need to develop disease-resistant variety for economical and effective management of bacterial blight. Breeding work on disease resistance including screening of germplasm against blight pathogen, hybridization, and molecular breeding (transgenics) is already in progress at IIHR, Bangalore; UAS, Dharwad; MPKV, Rahuri; and NRCP, Solapur. Some germplasm accessions like Nana and Daru have revealed some degree of resistance against the blight pathogen, both under controlled and field conditions, and resistance is being exploited in further breeding programs (NRCP, *Ann. Report*, 2009)

11.5.6 Integrated Disease Management

Bacterial blight can be effectively managed by adopting integrated management practices including selection of healthy and disease-free planting material, avoidance of rainy season crop, following stringent sanitation measures, providing rest period to the crop, applying organic manures and nutrients, and spraying the crop with antibiotics like streptocycline (500 ppm) along with copper-based fungicides like copper oxychloride (0.2 %)/copper hydroxide (0.2 %) at 15-day interval as mentioned above.

References

Benagi VI, RaviKumar MR (2009) Present status of pomegranate bacterial blight and its management. In: Souvenir & abstracts, 2nd international symposium on pomegranate and minor including Mediterranean fruits, org. ISHS, Belgium, June 23–27, 2009 at UAS, Dharwad. Karnataka, India, pp 53–58

Chand R, Kishun R (1991) Studies on bacterial blight (*Xanthomonas campestris* pv. *punicae*) on pomegranate. Indian Phytopathol 44(3):370–372

Chand R, Kishun R (1993) Systemic movement of *Xanthomonas campestris* pv.*punicae* (Hingorani and Singh) dye from leaf to node in pomegranate. Int J Trop Plant Dis 11:85–90

Dhandar DG, Nallathambi P, Rawal RD, Sawant DM (2004) Bacterial leaf and fruit spot. A new threat to Pomegranate orchards in Maharashtra. J Mycol Plant Pathol 34(3):971

Hingorani MK, Mehta PP (1952) Bacterial leaf spot of pomegranate. Indian Phytopathol 5:55–56

Hingorani MK, Singh JN (1959) *Xanthomonas punicae* sp nov. on *Punica granatum* L. Indian J Agric Sci 29:45–48

Jadhav VT, Sharma KK (2009) Integrated management of diseases in pomegranate. In: Souvenir and abstracts 2nd international symposium on pomegranate and minor including Mediterranean fruits, UAS Dharwad, 23–27 June 2009, pp 48–52

Kamble KG (1990) Investigations on bacterial disease of pomegranate (*Punica granatum*). M.Sc thesis, MAU Parbhani. M.S.

Kanwar ZS (1976) A note on bacterial disease of pomegranate (*Punica granatum* L.) in Haryana. Haryana J Hortic Sci 5:177–180

Khosala K, Kumar J, Bhardwaj SS (2009) Management of leaf and fruit diseases of pomegranate through chemicals. In: Souvenir and abstracts 2nd international symposium on pomegranate and minor including Mediterranean fruits, UAS Dharwad, 23–27 June 2009, p 141

Mogle TR, MagarSJ, Suryawanshi AP, Somwanshi SD (2009) Pathogenicity, morphology and cultural characteristics of *Xanthomonas axonopodis* pv.*punicae*, causing oily spot of pomegranate. In: Souvenir and abstracts 2nd international symposium on pomegranate and minor including Mediterranean fruits, UAS Dharwad, 23–27 June 2009, p 137

Mondal KK, Singh D (2009) Bacterial blight-an emerging disease of pomegranate. In: Souvenir and abstracts 5th international conference on plant pathology in the globalized era, IARI, New Delhi, 10–13 November 2009, p 40

NRCP (2006) Annual report, 2006–07. National Research Centre on Pomegranate, Solapur, pp 13–24

NRCP (2007) Annual report, 2007–2008. National Research Centre on Pomegranate, Solapur, pp 24–36

NRCP (2008) Annual report, 2008–2009. National Research Centre on Pomegranate, Solapur, pp 36-47

NRCP (2009) Annual report, 2009–2010. National Research Centre on Pomegranate, Solapur, p 69

Rangaswami G (1962) Pomegranate. In: Bacterial plant diseases in India. Asian publication House, Bombay, pp 129–130, 830 pp

Rani U, Verma KS (2001) Pathogenic potential of *Xanthomonas axonopodis* pv. *punicae* and field response of different pomegranate cultivars. Plant Dis Res 15(2):198–202

Rani U, Verma KS (2002) Occurrence of black spot of pomegranate in Punjab. Plant Dis Res 17(1):93–96

Sharma KK, Sharma J, Jadhav VT, Chandra R (2008) Bacterial blight of pomegranate and its management. Indian Phytopathol 61(3):380–381

Sharma KK, Sharma J, Jadhav VT, Chandra R (2009) Present status of pomegranate bacterial blight caused by *Xanthomonas axonopodis* pv. *punicae* and its management. In: Souvenir and abstracts 2nd international symposium on pomegranate and minor including Mediterranean fruits, UAS Dharwad, 23–27 June 2009, pp 135–136

Sohi HS, Jain SS, Sharma SL, Verma BR (1964) New records of plant diseases from H.P. Indian Phytopathol 17(1):42–45

Srivastava UC (2008) Horticulture in India: a success story? Crop Care 34:23–32

Vauterine L, Hoste B, Kersters K, Swings J (1995) Reclassification of *Xanthomonas*. Int J Syst Bacteriol 45:472–489

Yenjerappa ST, Nargund VB, Rafee CM (2009) Field performance of bactericides, bio-agents and botanicals against the bacterial blight of pomegranate. In: Souvenir and abstracts 2nd international symposium on pomegranate and minor including Mediterranean Fruits, UAS Dharwad, 23–27 June 2009, p 138

Role of Defensive Antiviral Proteins from Higher Plants in the Management of Viral Diseases

12

L.P. Awasthi, S.P. Singh, and H.N. Verma

12.1 Introduction

Plants, animals, and other microorganisms are provided, in their genetic makeup, with a certain range of antimicrobial compounds. With respect to viruses, a few plants show resistance to their infection. This resistance, in many cases, has been associated with the protective chemicals within the plant cells which are known for their antifungal or antimicrobial property and reported to be proteinaceous in nature. Many higher plants have developed a variety of defense systems to combat pathogen attack which is essential for their survival. Some of these plants possess endogenous proteins that act as virus inhibitors. They are generally basic proteins with molecular weight ranging from 24 to 32 kDa and effective against a wide range of plant viruses. The viral inhibitors are well studied in *Phytolacca americana*, *Dianthus caryophyllus*, *Boerhaavia diffusa*, *Cuscuta reflexa* *Mirabilis jalapa*, *Bougainvillea spectabilis*, and *Celosia cristata*. These viral inhibitors are most effective when mixed with the virus inoculum or when they are applied one day before or shortly after mechanical inoculation.

12.2 History

Duggar and Armstrong (1925) reported for the first time that the crude sap extract of Pokeweed (*Phytolacca decandra* L.) markedly inhibited the infectivity of tobacco mosaic virus (TMV). Kuntz and Walker (1947) made first attempt to investigate the nature and property of the spinach extract. A variety of plants belonging to different taxonomic families were subsequently used for viral disease management. Loebenstein and Ross (1963) demonstrated formation of virus interfering substances in sap extracted from resistant apical uninoculated halves of *Datura* leaves, whose basal halves had been inoculated ten days earlier with TMV. The sap from resistant halves of leaves when mixed with virus, reduced the infectivity of TMV, as compared to control sap. Verma et al. (1979a, b, c) and Verma and Awasthi (1979a, b, c) conducted experiments with antiviral substance of plant origin and found considerable reduction in infection of viruses. Awasthi and Mukerjee (1980) found protection of potato virus infection by extract from some medicinal plants. The control of viral diseases of some cucurbitaceous crops was also reported by the same group (Verma et al. 1980). Awasthi et al. (1984) observed that pre-inoculation sprays of

L.P. Awasthi (✉) • S.P. Singh
Department of Plant Pathology, N.D. University of Agriculture and Technology, Kumarganj, Faizabad 224229, Uttar Pradesh, India
e-mail: lpawasthi14@gmail.com

H.N. Verma
Jaipur National University, Jaipur, India

L.P. Awasthi (ed.), *Recent Advances in the Diagnosis and Management of Plant Diseases*,
DOI 10.1007/978-81-322-2571-3_12

Boerhaavia diffusa root extract were effective against *tobacco mosaic virus* in tobacco and tomato, *cucumber mosaic virus* in cucumber, *Cucumber green mottle mosaic virus* in melon, sunn hemp rosette virus in *Crotalaria juncea*, and *Gomphrena globosa*. Verma et al. (1985) suggested possible control of natural infection of *Mung bean yellow mosaic virus* (MYMV) in mung bean and urdbean by plant extracts.

Zaidi et al. (1988) reported inhibitory effect of neem extract (*A. indica*) against *Spinach mosaic virus* in *Chenopodium amaranticolor*. Verma et al. (1994) observed the efficacy of leaf extracts of different species of *Clerodendrum*, when applied on to the leaves of several hypersensitive hosts. The aqueous leaf extract prevented the infection of viruses by increasing the resistance of the host plants towards subsequent virus infection. Verma and Varsha (1995) used *Clerodendrum aculeatum* alone and with certain proteinaceous modifiers (CA-M) against sunn hemp rosette virus (SHRV) in *Crotalaria juncea* and observed that in CA-M (with papain) sprayed plants, disease incidence was much lower when treated plants were challenged with SHRV 6 days after the treatment. Verma et al. (1996) purified a non-phytotoxic systemic resistance inducer from *C. aculeatum* leaves. A water-soluble basic protein of mol. wt. 34 kDA present in *Clerodendrum aculeatum* (Ca-SRI), when applied prior to virus inoculation, reduced more than 90 % of local lesions in *N. glutinosa* by TMV.

Bharathi (1999) reported that extract of *Mirabilis jalapa* completely inhibited *Cucumber mosaic virus* in brinjal (*Solanum melongena* L.), while the inhibition of CMV by the plant extract of *Prosopis chilensis*, *Bougainvillea spectabilis*, and *Eucalyptus citriodora* was 83 %, 75 %, and 58 %, respectively. In pre-inoculation treatments with *M. jalapa*, the percent infection of CMV on brinjal ranged from 0 to 56 % (Awasthi and Rizvi 1999). They also found that infection of *Tomato yellow leaf curl virus*, a vector-borne virus, was checked significantly by the application of *B. diffusa* root extract. Jayashree et al. (1999) studied the efficacy of 10 plant extracts against *Pumpkin yellow vein mosaic virus* in pumpkin and observed maximum inhibition of virus transmission by insect vector *Bemisia tabaci* by *Bougainvillea spectabilis* extract followed by *B. diffusa*. Surendran et al. (1999) observed the antiviral activity of plant extracts (*Azadirachta indica*, *Clerodendrum infortunatum*, *Ocimum sanctum*, and *Vitex negundo*) against *Brinjal mosaic virus* on local lesion host *Datura stramonium*. The pre-inoculation sprays of 10 % leaf extract or oil formulations of *A. indica* were found effective in reducing the virus infection under field conditions. Singh (2002) and Singh and Awasthi (2002) reported that aqueous root extract of *B. diffusa* effectively reduced mung bean yellow mosaic and bean common mosaic virus disease in mung bean and urdbean along with increased grain yield in field conditions. Later, Awasthi and Kumar (2003a, b), Kumar and Awasthi (2003a, b) revealed that weekly sprays of aqueous root extract of *B. diffusa* significantly prevented infection, multiplication, and spread of *Cucumber mosaic virus*, *Bottle gourd mosaic virus*, *Cucumber green mottle mosaic virus*, and *Pumpkin mosaic virus* in cucurbitaceous crops. Kumar and Awasthi (2008) were able to prevent infection and spread of cucumber mosaic disease in cucumber through plant proteins. Singh and Awasthi (2009) tested various medicinal plants for the management of yellow mosaic disease of mung bean (*Vigna radiata*) Yadav et al. (2009). Awasthi and Yadav (2009) worked on the management of viral diseases of tomato by seed treatment and foliar sprays of *Boerhaavia diffusa* root extract and *Clerodendrum aculeatum* leaf extract.

12.3 Virus Inhibitors and Their Characteristics

Antiviral resistance-inducing proteins act on any step of virus synthesis, i.e., from the uncoating of viral proteins to the appearance of symptoms. Proteins inhibit virus infection or multiplication when applied before or after virus multiplication. Virus inhibitory property of virus inhibitors depends on their concentration and time of application. Functions of proteins are also affected by temperature and pH. For example, virus inhibitor in *Boerhaavia diffusa* roots was inactivated at 95 °C and pH 4 but not at pH 10 (Verma and Awasthi 1979c).

Types of virus inhibitors: On the basis of mode of action, the virus inhibitors may be grouped into two types:

(A) Inhibitors of virus infection
(B) Inhibitors of virus multiplication

12.4 Inhibitors of Virus Infection

Occurrence of highly potent inhibitors of virus infection has been reported from different plants. A number of reviews have adequately listed various plants showing virus inhibitory activity (Bawden 1954; Ragetli 1975; Verma 1982; Awasthi and Singh 2009). Duggar and Armstrong (1925) observed that when the extract from pokeweed was mixed with infective sap of TMV, there was complete inhibition of the virus. They also found that extracts of *Datura stramonium* and *Pelargonium* sp. also inhibited viral infectivity when mixed with the virus. Kuntz and Walker (1947) made first attempt to investigate the nature and property of the spinach extract. A variety of plants belonging to different taxonomic families were used for viral disease management. Inhibitors present in extracts of a particular plant species are effective only when host species were inoculated with virus along with inhibitors. Loebenstein and Ross (1963) demonstrated the formation of virus interfering substance(s) in sap extracted from resistant apical uninoculated halves of *Datura* leaves, whose basal halves had been inoculated ten days earlier with TMV. The sap from resistant halves of leaves when mixed with virus reduced the infectivity of TMV, as compared to control sap. Subsequently, Loebenstein and Ross (1963) studied the characteristics of the induced interfering substances or agents from resistant leaves of *Datura* that interfered with infection by TMV. The agent was a protein, with a molecular weight considerably less than those characteristic of viruses. The inhibitory activity was lost on heating at 78 °C for 50 min and by aging for 5 days at 3 °C. The agent was non-dialyzable and partially sedimented at 93,000 g. It did not inactivate virus in vitro, since from a mixture of interfering agent, an infectious preparation of TMV can be recovered after centrifugation at 59, 000 g for 60 min. Crude AVF preparation retained the activity for several months when stored at 4–10 °C, and for several days at room temperature. It was further suggested that AVF acts as an antimetabolite to the biosynthesis of virus nucleic acid, or it blocks some sites essential for virus multiplication.

Verma et al. (1979a, b, c) and Verma and Awasthi (1979a, b, c) conducted experiments with antiviral substances of plant origin and found considerable reduction in infection by the viruses. Later on, Awasthi and Mukerjee (1980) found protection of potato virus infection by extract from some medicinal plants. The control of viral diseases of some cucurbitaceous crops was also reported by the same group (Verma et al. 1980).

Awasthi et al. (1984) observed that pre-inoculation sprays of *Boerhaavia diffusa* root extract were effective against wide range of viruses in different susceptible hosts. Verma et al. (1985) also suggested possible control of natural infection of *Mung bean yellow mosaic virus* (MYMV) in mung bean and urdbean by plant extracts. The infection, on these crops, by MYMV, was suppressed by aqueous partially clarified leaf extract of *Clerodendrum fragrans*, *Aerva sanguinolenta*, and root extract of *B. diffusa*. The treatments were administered as foliar sprays after 3–4 days from the seedling stage. The extract from *C. fragrans* reduced the virus infection, delayed the appearance of disease symptoms, and promoted flowering and consequent fruiting. The treatment also increased the nodulation and yield.

Prevention of *Oat sterile dwarf virus* infection and suppression of disease symptoms were observed by some phytochemicals (Awasthi et al. 1989). Verma and Verma (1993) revealed that leaf extract of *C. aculeatum* along with soil amendment with dry leaf powder showed two-fold increase in nodulation and grain yield with 50 % reduction in disease incidence caused by *Mung bean yellow mosaic virus*. Verma and Singh (1994) reported that *C. aculeatum* may be a possible prophylactic agent against natural viral infection in mung bean plants. The plants grown in pots and kept in the field were protected against natural viral infection by spraying with leaf extract of *C. aculeatum*, together with soil amendment with dry leaf powder or fresh extract. Unsprayed plants showed severe disease symptoms, while treated plants showed only mild symptoms. Soil treatment with dry leaf powder + sprays with fresh leaf extract were effective in increasing the yield as well as in reducing disease incidence and severity.

Verma et al. (1994) observed the efficacy of leaf extracts from different species of *Clerodendrum*, when applied on to leaves of several hypersensitive hosts. The extracts prevented the infection of viruses by increasing the resistance of the host plants. The numbers of local lesions produced on treated leaves were much lower as compared to untreated leaves. The decrease in lesion number by different species of *Clerodendrum* was variable.

Verma and Varsha (1995) used the leaf extract of *Clerodendrum aculeatum* alone and also with certain proteinaceous modifiers (CA-M) against sunn hemp rosette virus (SHRV) in *Crotalaria juncea* and observed that in CA-M (modifies – papain) sprayed plants, disease incidence was very low when treated plants were challenged with SHRV 6 days after the treatment. Verma et al. (1996) purified a non-phytotoxic systemic resistance inducer from *C. aculeatum* leaves. The purified basic protein (CA-SRIP) having a molecular weight of 34 kDa completely prevented virus infection in *N. glutinosa*, when sprayed prior to virus inoculation.

The prevention of *Tomato yellow leaf curl, a* vector-borne virus, was checked significantly by the application of *B. diffusa* root extract (Awasthi and Rizvi 1999). Jayashree et al. (1999) studied the efficacy of 10 plant extracts against *Pumpkin yellow vein mosaic virus* in pumpkin. They observed maximum inhibition of virus transmission by *Bemisia tabaci*, through *Bougainvillea spectabilis* extract followed by *B. diffusa*.

Surendran et al. (1999) observed antiviral activity of plant extracts (*Azadirachta indica*, *Clerodendrum infortunatum*, *Ocimum sanctum*, and *Vitex negundo*) against *Brinjal mosaic virus* on local lesion host *Datura stramonium*. The pre-inoculation sprays of 10 % leaf extract or oil formulations of *A. indica* were found effective in reducing the number of local lesions and also in preventing virus infection, under field conditions.

Singh (2002) and Singh and Awasthi (2002) reported that aqueous root extract of *B. diffusa* effectively reduced mung bean yellow mosaic and bean common mosaic virus disease in mung bean and urdbean along with increased grain yield in field conditions. Later, Awasthi and Kumar (2003a, b), Kumar and Awasthi (2003a, b) revealed that weekly sprays of aqueous root extract of *B. diffusa* significantly prevented infection, multiplication, and spread of *Cucumber mosaic virus*, *Bottle gourd mosaic virus*, *Cucumber green mottle mosaic virus*, and *Pumpkin mosaic virus* in cucurbitaceous crops.

Singh et al. (2004), Singh and Awasthi (2004) and Singh et al. (2005) reported the prevention of yellow mosaic disease of mung bean and urdbean by clarified aqueous root extract of *B. diffusa*. Six sprays of *B. diffusa* root extract (10 %) reduced 80–90 % disease incidence and increased nodulation, plant height, primary and secondary branches, pod formation, and grain yield. Awasthi and Singh (2006) reported that the most effective treatment was seed treatment with *B. diffusa* root extract + three foliar sprays, which exhibited 70 % reduction in disease incidence.

Inhibitory effect of the extract of *A. indica* was reported against *Spinach mosaic virus* in *Chenopodium amaranticolor* (Zaidi et al. 1988). Spraying with neem leaf extract on upper surface of the test plant was effective up to 4 h and the efficacy decreased gradually with increase in time interval between treatment and inoculation (Sangar and Dhingra 1982). Aqueous neem extract was more active when mixed with virus inoculum of *Spinach mosaic virus*. An aqueous extract of neem leaf also inhibited ring mosaic of pea caused by tomato spotted wilt virus (TSWV) under laboratory conditions (Ganapathy and Narayanaswamy 1990). Singh et al. (1988) reported that leaf and bark extract of neem inhibited the infection of *C. amaranticolor* by *Cowpea mosaic comovirus*. Louis Vimi and Balakrishan (1996) reported that five medicinal plants, viz., *Basella alba*, *Glycyrrhiza glabra*, *Phyllanthus fraternus*, *Plumbago rosea*, and *Thespesia populnea*, decreased *Pumpkin mosaic virus* infection in systemic hosts.

Bharathi (1999) reported that extract of *Mirabilis jalapa* completely inhibited *Cucumber mosaic virus* in brinjal (*Solanum melongena* L.) while the inhibition of CMV by the extract of *Prosopis chilensis*, *Bougainvillea spectabilis*, and *Eucalyptus citriodora* was 83 %, 75 %, and 58 %, respectively. In pre-inoculation treatments with *M. jalapa* the percent infection of CMV on brinjal ranged from 0 to 56 % as compared to control.

Awasthi and Singh (2004) reported that the leaf extract of *Clerodendrum aculeatum* significantly reduced infection of a mosaic disease in *Amorphophallus campanulatus*. Further the molecular studies on the antiviral protein from *C. aculeatum* were conducted by Kumar et al. (1997). Efforts were made by Singh and Awasthi (2008) to manage ring spot disease of papaya through antiviral agents of plant origin along with milk protein. Similarly, Kumar and Awasthi (2008) were able to prevent infection and spread of cucumber mosaic disease in cucumber through plant proteins. Recently, Singh and Awasthi (2009) tested various medicinal plants for the management of yellow mosaic disease of mung bean (*Vigna radiata*). Yadav et al. (2009) and Awasthi and Yadav (2009) worked on the management of viral diseases of tomato by seed treatment and foliar sprays of *Boerhaavia diffusa* root extract and *Clerodendrum aculeatum* leaf extract. Awasthi and Singh (2008) reported a possible mechanism of action for the inhibition of the plant viruses by an antiviral glycoprotein isolated from *B. diffusa* roots. Baranwal et al. (2002) purified antiviral protein from *Celosia cristata*. The protein inhibited the lesion formation by TMV, sunn hemp rosette virus, and *Potato virus X* (PVX) in a few hypersensitive hosts.

A large number of plants have been tested for their antiviral activity using different host-virus combinations. The investigations have revealed that they were not identical in chemical composition and behavior. Marked fluctuations of the inhibitor contained in many plants occurred during different seasons and various stages of plant growth. Although substances that interfere with the virus infection have been reported to occur in several plants, even so, only a few of the inhibitory substances have been isolated and characterized.

12.5 Purification of Virus Inhibitors/Resistance Inducers from Plants

Purification of antiviral agents from different plants involved different purification techniques, depending on the nature of the compound present in the crude extract of plants. For purification of antiviral substances (mostly polysaccharides, proteins, or glycoproteins), the protocol adopted by various workers varied, however. In general, fresh or dried leaves/roots from outdoor plants have been used. The leaves/roots were ground in suitable buffer; the juice was expressed through cheese cloth and then centrifuged at low speed (5000–7000 rpm for 15–30 min). The supernatant fluid was clarified either by high-speed centrifugation, heat treatment, or organic solvents. Afterwards, polysaccharide inhibitors were precipitated with ethanol and proteinaceous or glycoproteinaceous inhibitors with different saturations of ammonium sulfate (40–100 %). The precipitate was dissolved in low ionic strength buffer and dialyzed. Subsequently, the solution was passed through a DEAE-cellulose column or Sephadex G-25 column to remove pigmented material.

First inhibitor purified and characterized was from carnation plant by Ragetli and Weintraub (1962). The scheme for purification of the inhibitor involved steps like low-speed centrifugation, dialysis through semipermeable membrane, DEAE treatment, and exclusion chromatography over Sephadex G-75 column. The DEAE treatment completely eliminated all RNase activity. The characteristic features of the inhibitor from carnation were:

- The inhibitor was proteinaceous and contained 16 amino acids.
- Its molecular weight was 14, 000 Da.
- The protein showed positive charges up to pH 7.8.
- At a concentration of 0.6 μg inhibitor/ml, 100 % inhibition of TMV was observed.
- The inhibitor was inactivated at a temperature of 80 °C.

Verma and Awasthi (1979a, b, c) isolated a strong and highly potent inhibitor of virus from roots of *B. diffusa*. The partial purification by organic solvent, protein precipitants and Sephadex gel filtration, revealed that the inhibitor was a glycoprotein and had molecular weight of 16–18 kDa. The purified preparation contained 70–80 % protein and carbohydrate.

Leaf extract from *Capsicum* applied to the under surface of bean (*Phaseolus vulgaris*) leaves inhibited alfalfa mosaic virus (AMV) infection on the upper surface. Inhibitors from plant extract do not irreversibly inactivate viruses, because the original virus regains its infectivity when the mixture is diluted or ultracentrifuged (Fischer and Nienhaus 1973). Baranwal et al. (2002) used ammonium sulfate to sediment *Celosia cristata* antiviral protein (25 kD) (CCP25). The sediment was dissolved in buffer, dialyzed and subjected to DEAE-cellulose column chromatography. Infectivity of the viruses was completely lost by leaf extract of *Pelargonium hortorum*, *Chenopodium album*, and *C. amaranticolor*. The *Pelargonium* juice was resistant to heating at 100 °C for 10 min.

12.6 Characteristics of Virus Inhibitors/Resistance Inducers

Kassanis and Kleczkowski (1948) for the first time purified virus inhibitors from pokeweed, *Phytolacca americana* (*esculenta*). It was found to contain 8–12 % carbohydrate and 14–15 % nitrogen. They suggested that the inhibitor was probably a glycoprotein. It was basic in nature and combined reversibly with purified TMV, like pancreatic ribonuclease – another inhibitor of plant viruses. Since carbohydrate was constantly associated with the protein, the inhibitor was presumed to be a glycoprotein. However, Benda (1956) later on found two types of substances in the sap of New Zealand spinach (*Tetragonia expansa*). A relatively stable protein was an inhibitor and the other a soluble oxalate salt was an augmenter and increased the number of local lesions. Subsequently, it was found that inhibitors from a several other plants also had no effect on viruses but their action was on the host plants. This was shown by applying the inhibitory sap before and after virus inoculation. It was also believed that the inhibition took place as a result of competition between virus and inhibitor. Presumably, infection in such cases was prevented either by blocking entrance of the virus to a susceptible region of the host plant or tying up some constituents within cell required for virus multiplication.

Francki (1964) stated that loss of infectivity of *Cucumber mosaic virus* (CMV) on exposure to cucumber leaf extracts could be due to the aggregation of virus particles and the formation of a complex between some host materials and virus particles, thus preventing infection. Infection of *Gomphrena globosa* with *Potato virus X* (PVX) was inhibited by leaf extract from all potato varieties that are tolerant, hypersensitive, or immune to the virus. There was no indication that the inhibitors from different resistant types of potato differed in their effectiveness. Therefore, degree of host resistance has no direct relationship to the inhibitory capacity of the extract. Inhibitors introduced within host tissues probably produce some stimulatory effect, which translocates through cells to the upper epidermis. The chemical nature of some of the virus inhibitors present in healthy plants has been elucidated. The well-known inhibitor from pokeweed (*Phytolacca americana*) is a basic protein consisting of about 116 amino acid residues and possessing a molecular weight of 13, 000 Da (Wyatt and Shepherd 1969). McKeen (1956) and Rao and Raychaudhuri (1965) attempted to investigate the nature of inhibitors present in the extracts of cucumber, tobacco, and *Datura*. They suggested that the inhibitory substances in the extracts were proteinaceous in nature.

Ragetli (1957) working on a potent inhibitor from carnation (*Dianthus caryophyllus*) made a detailed attempt to study the mode of action of the inhibitor. The inhibitor was effective when applied to leaf surface simultaneously with virus or prior to it. When it was administered after virus inoculation, marked interference with the infection process was not observed beyond 15–30 min. Virus and inhibitors could be separated easily by centrifugation in vitro. Thus, it was concluded that the inhibitor acted at the time, the virus came in contact with the plant, presumably by binding the receptor site and preventing the infection.

12.7 Physical Properties of the Virus Inhibitory Plant Proteins

1. *Dilution*: The inhibitory property of the plant extracts was reduced greatly by dilution. Tenfold dilutions of plant extracts with distilled water in most of the cases removed the inhibition or decreased it remarkably. The juice from *Dianthus barbatus*, *D. caryophyllus*, *Boerhaavia diffusa*, etc. appears to be very powerful, since their action is still apparent at dilution of 1:2000 or more (Verma and Awasthi 1979b).
2. *Effect of heat*: The activity of inhibitors present in different plants was found to be greatly influenced by heating the crude extract. Heating plant extracts for 10 min at 60–70 °C removed inhibitory activity partially or completely in many cases but was generally less efficient than dilution. On the basis of thermal inactivation of inhibitors present in saps of different plants, inhibitors have been classified into two types:
   (a) Inhibitory activity lost after 10 min at 60–80 °C. For example, *Amaranthus*, *Basella*, *Cuscuta*, *Datura stramonium*, *Hablitzia*, *Beta* sp., etc.
   (b) Inhibitory activity lost after 10 min at 80–100 °C. For example, *A. retroflexus*, *Chenopodium amaranticolor*, *C. quinoa*, *C. album*, *Atriplex*, *Pelargonium*, *Salsola*, etc.

   Heating had not much effect on the inhibitory activity of extracts or juices from *Amaranthus mangostanus*, *Boerhaavia diffusa*, *Clerodendrum aculeatum*, *C. indicum*, *C. phlomoides*, *C. inerme*, etc.
3. *Longevity in vitro*: The inhibitors containing preparations of crude sap could be stored at room temperature for different periods before losing activity. For most, inhibitory activity was lost after a week of storage; some extracts retained activity up to one month. The most stable inhibitors were generally those which possessed high thermal stability. Plant extracts of many species retained their full inhibitory action after storage for several weeks at or near 4 °C.
4. *Effect of chemicals*: The various inhibitors present in the crude extracts of plants were mostly insoluble in organic solvents such as petroleum ether, chloroform, benzene, and diethyl ether. In several plants, the inhibitors contained in crude extract could be precipitated with 90–95 % ethanol with only slight loss in activity, whereas after precipitation with 10 % TCA activity was generally lost.
5. *Effect of dialysis*: Mostly, the inhibitors contained in crude extracts of plants were non-dialyzable, indicating thereby that they had a molecular weight of more than 10,000 Da. This is in sharp contrast to most antifungal and antimicrobial substances occurring in plants whose molecular weights are typically lower.
6. *Effect of various enzymes*: The effect of enzymes such as trypsin, chymotrypsin, papain, pronase, and RNase has been tested on a few plant extracts. The results following treatment with enzymes varied with different plant extracts. In some cases, activity was abolished by incubation in the presence of those enzymes, whereas in other cases it was not destroyed. Normally, the proteinaceous substances were influenced by proteolytic enzymes, whereas glycoproteinaceous substances remained unaffected.
7. *Effect of pH*: The stability of inhibitors in crude extract was greatly influenced by pH. Mostly, the inhibitors remained active between pH 5 and 7. However, in some cases, such as *Acacia arabica*, *Basella alba*, *Clerodendrum aculeatum*, *Datura metel*, and *Syzygium cumunis* inhibitors were stable between pH 4 and 10. The presence of mercaptoethanol in the solution helped to increase the activity of the inhibitory extracts. The activity was considerably decreased, however, after treatment with SDS or 6 M urea.
8. *Effect of high-speed centrifugation*: Inhibitors in plant extracts generally did not sediment on ultracentrifugation up to 40, 000 rpm or 120, 000 g for 2 hours. Activity following ultracentrifugation always remained in the supernatant and was unaffected biologically by ultracentrifugation.

Various plants and the nature and characteristics of the inhibitors present in them

Name	Characteristics of the purified inhibitors	Action	References
Abutilon striatum	Polysaccharide	Inhibits *Abutilon mosaic virus* and tobacco virus infection, forms an unstable complex, and aggregation of virus particles was observed	Flores et al. (1967)
Boerhaavia diffusa	Glycoprotein MW 20,000; carbohydrate 8–13 %; protein 70–80 %	Inhibits the infectivity of many plant viruses	Verma and Awasthi (1979a, b, c)
		Induces systemic resistance reversible by actinomycin D	
		Provokes formation of antiviral agent which inactivates virus in vitro	
Brassica oleracea	Polysaccharide MW 23,000	Alters the susceptibility of host by changing the cell wall permeability	Varma (1973)
Chenopodium amaranticolor	Basic protein MW 29,000	Inhibits infectivity of many plant viruses	Singh et al. (1988)
Clerodendrum aculeatum	Basic protein MW 32,000; resistant to proteases	Induces systemic resistance, reversible by actinomycin D. Inhibits infectivity of many plant viruses	Verma et al. (1991)
Dianthus caryophyllus	Glycoprotein; dianthin 30 MW 29,500; dianthin 32 MW 31,700	Induces systemic resistance and inhibits infectivity of 17 plant viruses, including TMV RNA	Ragetli and Weintraub (1962)
Mirabilis jalapa	Basic protein MW 24,200	Inhibits infectivity and mechanical transmission of many plant viruses	Habuka et al. (1990)
	MAP has compressed structure which confers resistance to proteases		
	Inhibitory activity of MAP is substantially increased (22 times) by elimination of disulfide bonds with genetic engineering		
Phytolacca americana	Basic protein MW 29,000; pl 8.1	Inhibits infectivity of many plant viruses	Wyatt and Shepherd (1969)
PAP	Basic protein MW 30,000; pl 8.3. Contains greater proportion of basic amino acid residues as compared to PAP	Like TMV, CMV, WMV, and *Sugarcane mosaic virus*	
PAP-11	Contains higher concentration of tyrosine	Ribosome-inactivating protein (RIP)	
PAP-11	Does not cross react with anti-Pap antibodies		
PAP-s			
Spinacia oleracea	Basic protein MW 29, 000; pl 10.3. Serologically related to inhibitory proteins occurring in *Phytolacca dianthus* and *Chenopodium amaranticolor*	Inhibits infectivity of many plant viruses	Kuntz and Walker (1947)
Yucca recurvifolia	Basic protein MW 23, 000; pl 9.4. Exhibits amino acid composition similar to PAP	Inhibits infectivity of many plant viruses	Okuyama et al. (1978)

12.8 Detailed Studies on Virus Inhibitors from

12.8.1 *Boerhaavia diffusa*

The plant was named in honor of Herman Boerhaave, a famous Dutch physician of the eighteenth century (Chopra 1969). *Boerhaavia*, a herbaceous plant, belongs to the Nyctaginaceae (four o'clock) family, order Thymilae, group Dicotyledons, and phylum Angiosperms (Rendle 1925). Six species are found in India: *B. diffusa*, *B. chinensis*, *B. erecta*, *B. repens*, *B. rependa*, and *B. rubicunda* (Chopra 1969; CSIR 1988). The whole plant or its specific parts (leaves, stem, and roots) are known to have medicinal properties and have a long history of use by indigenous and tribal people in India. It has many ethnobotanical uses (the leaves are used as vegetable; the root juice is used to cure asthma, urinary disorders, leucorrhea, rheumatism, and encephalitis) and is medicinally used in the traditional Ayurvedic system. Besides, *B. diffusa* shows potent antiviral efficacy of this plant against phytopathogenic viruses. Antiviral agent isolated from this plant was found to be a glycoprotein with a molecular weight of 16–20 kDa (Verma and Awasthi 1979a, b, c).

12.8.1.1 Chemical Composition of *Boerhaavia diffusa*

The *Boerhaavia diffusa* plant contains a large number of compounds such as flavonoids, alkaloids, steroids, triterpenoids, lipids, lignins, carbohydrates, proteins, glycoproteins, punarnavin, and punaravoside (Agrawal and Dutt 1936; Basu et al. 1947; Surange and Pendse 1972; Ahmad and Hossain 1968; Jain and Khanna 1989). A glycoprotein having a molecular weight of 16 kDa was isolated and studied in detail for its biological activity (Mishra and Tiwari 1971; Verma et al. 1979a, b, c).

12.8.1.2 Biological Activity

As Medicine in the Traditional System *B. diffusa* plants have been widely used by indigenous tribes in the traditional system of medicine. The roots have been widely used for the treatment of dyspepsia, jaundice, enlargement of spleen, abdominal pain, abdominal tumors, and cancers (Kirtikar and Basu 1956).

As Medicine in the Ayurvedic System The roots and leaves with flowers have been found to be highly potent in Ayurvedic medicine; different parts of this plant were reported to have various medicinal properties (CSIR 1988).

Pharmacological and Clinical Properties Pharmacological studies have demonstrated that *Punarnava* possesses punarnavoside, which exhibits a wide range of properties – diuretic, anti-inflammatory, antifibrinolytic, anticonvulsant, antibacterial, anti-stress agent, antihepatotoxic, antiasthmatic, antiscabies, and anti-urethritis.

Antiviral Activity of ***Boerhaavia diffusa*** The roots of *B. diffusa* are a rich source of a basic protein, which is used for inducing systemic resistance in many susceptible crops against commonly occurring viruses (Verma and Awasthi 1979a, b,c, 1980; Verma et al. 1979a, b, c; Awasthi et al. 1984, 1985, 1989). Maximum antiviral activity, in each case, was recorded with the aqueous extract of dried root powder applied before virus inoculation. The active principle was purified and isolated (Verma et al. 1979a, b, c). This protein or antiviral agent was active against tobacco mosaic virus in *Nicotiana glutinosa*, *Datura metel*, *Chenopodium amaranticolor*, and *Nicotiana tabacum* (Ky58 White Burley and NP31); sunn hemp rosette virus in *Cyamopsis tetragonoloba*, *Vigna unguiculata*, and *Crotalaria juncea*; Gomphrena mosaic virus in *Chenopodium amaranticolor*, *Vigna unguiculata*, and *Gomphrena globosa* when applied a few hours (2–24 h) before inoculation by the respective inoculum of viruses (Verma and Awasthi 1979a, b, c; Awasthi et al. 1984). The antiviral agent was a basic glycoprotein (70–80 % protein and 8–13 % carbohydrates) with a molecular weight of 16–20 kDa as determined by gel filtration chromatography (Verma et al. 1979a, b, c). After application of systemic resistance-inducing protein, the susceptible healthy hosts produced a virus inhibitory agent

(VIA). The VIA showed the characteristics of protein, and upon incubation with the viruses, reduced their infectivity of both in vitro and in vivo. The biophysical characteristics of induced VIA were also studied and it was found to be a basic protein. The glycoprotein occurring in *B. diffusa* roots functions as a signal molecule and is of great interest as it has a role in stimulating the defense systems of plants against viruses (Verma and Awasthi 1980; Awasthi et al. 1987).

Prevention and management of viral diseases of crops in fields by *Boerhaavia diffusa* inhibitor/resistance inducer

Virus	Crop	Disease protection (%)	Increase in yield (%)	Reference
Potato virus X	Potato	68	22	Awasthi and Mukerjee (1980)
Tomato leaf curl virus	Tomato	71	24	Awasthi et al. (1984), Awasthi and Rizvi (1999)
Complex infection of tomato	Tomato	75	16	Awasthi et al. (1985)
Mosaic and *Cucumber green mottle mosaic virus*	Cucumber	62	9	Awasthi et al. (1985)
Tomato mosaic virus	Tomato	78	12	Awasthi et al. (1985)
Brinjal mosaic virus	Brinjal	64	9	Awasthi et al. (1985)
Oat sterile dwarf virus	Oats	42	NA	Kempiak et al.(1991)
Tomato yellow mosaic virus	Tomato	68	29	Awasthi and Rizvi (1998)
Bean common mosaic virus	Black gram	42	28	Singh and Awasthi (2002)
Bottle gourd mosaic virus	Bottle gourd	68	42	Kumar and Awasthi (2003a, b)

12.8.2 *Clerodendrum aculeatum*

The genus *Clerodendrum* L. [Family Lamiaceae (Verbenaceae)] is very widely distributed in tropical and subtropical regions of the world and comprises of small trees, shrubs, and herbs. The first description of the genus was given by Linnaeus in 1753, with identification of *C. infortunatum*. After a decade later, in 1763, Adanson changed the Latin name "*Clerodendrum*" to its Greek form "*Clerodendron*." *Clerodendrum* is a very large and diverse genus and till now 580 species of the genus have been identified and are widely distributed in Asia, Australia, Africa, and America.

Some of the major chemical constituents of Clerodendrum genus: Hispudilin, –*O*-ethylclerodendricin, Iridiod diglucoside, Colebrin, Clerodermic acid, Jionoside D, Uncinatone, Apigenin, Clerostero, Serratagenic acid, and Scutellarin.

12.8.2.1 Phytochemistry

Clerodendrum is reported in various indigenous systems of medicine throughout the world for the treatment of various diseases. Efforts have been made by various researchers to isolate and identify biologically active principle and other major chemical constituents from various species of the genus. It has been reported that the major class of chemical constituents present are steroids in various *Clerodendron* species such as *C. inerme*, *C. phlomidis*, *C. infortunatum*, *C. paniculatum*, *C. cyrtophyllum*, *C. fragrans*, *C. splendens*, and *C. campbellii* (Bolger et al. 1970; Abdul-Alim 1971; Joshi et al. 1979; Sinha et al. 1980, 1982; Singh and Singhi 1981; Singh and Prakash 1983; Singh and Singh 1983; Pinto and Nes 1985; Akihisa et al. 1989; Atta-Ur-Rehman et al. 1997; Goswami et al. 1996; Yang et al. 2000, 2002; Kanchanapoom et al. 2001, 2005; Gao et al. 2003; Pandey et al. 2003; Lee et al. 2006).

A few species of genus *Clerodendrum* and their distribution in the world

Scientific name	Synonym	Distribution
C. inerme Gaertn. *Clerodendrum aculeatum*		India, Sri Lanka, Southeast Asian Countries
C. phlomoidis Linn. f.	*C. multiforum* Burm	India
C. serratum Spreng		India
C. siphonanthus R. Br.	*C. indicum* (Linn) Kuntze	India
C. colebrookianum		Tropical regions of Asia
C. myricoide		India
C. commersonii Spreng		China
C. bungei Steud		Japan
C. glabrum E. Mey		Southern Africa

12.8.2.2 Biological Activity

1. *Anti-inflammatory activity* – Inflammation is a very complex pathophysiological process involving a variety of biomolecules responsible for causing it, such as leucocytes, macrophages, mast cells, platelets, and lymphocytes by releasing eicosanoids and nitric oxide. Pro-inflammatory cytokines such as TNF-α and IL-1β are also responsible for various inflammatory conditions.
2. *Antimicrobial activity*. Anti-infective compounds from natural resources are of great interest as the existing drugs are getting less effective due to increased tolerance of microorganisms. Essential oil obtained from leaves of the plant showed antifungal activity against variety of fungal species such as *Alternaria* species, *Aspergillus* species, *Cladosporium herbarum*, *Cunnimghamella echinulata*, *Helminthosporium sacchari*, *Microsporum gypseum*, *Mucor mucedo*, *Penicillium digitatum*, and *Rhizopus nigricans* (Sharma and Singh 1979).
3. *Other biological activities of Clerodendrum genus*. Other major biological activities reported for this genus are antihypertensive, antitumor, antidiabetic, antihyperlipidemic, larvicidal, and antidiarrheal activities.
4. *Antiphytoviral activity*. An endogenous agent that occurs in *Clerodendrum aculeatum* leaves induced a very high degree of systemic resistance (CA-SRI) against virus infection in plants when lower leaves were treated with *Clerodendrum aculeatum* leaf extract. The induction of systemic resistance by CA leaf extract was very fast, was reversed by actinomycin D, and was associated with the development of a virus inhibitory agent (VIA) in the extract-treated healthy susceptible plants. The VIA was present both in treated and non-treated leaves of plants treated with *C. aculeatum* leaf extract. Such endogenously occurring substances from plants, which can function as signal molecules, are of particular interest and deserve greater emphasis, because they are not antiviral themselves but they act by inducing the hosts to produce VIA(s) (Verma et al. 1995).

12.8.3 *Phytolacca americana*

American Pokeweed (*Phytolacca americana*) is a large herbaceous perennial plant. It is also known as American nightshade, cancer jalap, coakum, garget, inkberry, pigeonberry, pocan bush, pokeroot, pokeweed, redweed, scoke, red ink plant, and chui xu shang lu (in Chinese medicine). Broadly distributed in fields and waste places. *Phytolacca americana* was the first plant species shown to contain an inhibitor (Duggar and Armstrong 1925). The inhibitor in *Phytolacca* sap is probably the most potent.

Chemical Composition The plant has been reported to contain triterpenes, saponins, Phytolaccoside A,B,C,D,E,F,G (esculentoside E), phytolaccagenin, jaligonic acid, esculentic acid, 3-oxo-30-carbomethoxy-23-norolean-12-en-28-oic acid, phytolaccagenic acid, oleanolic acid, phytolaccatoxin, canthomicrol, astragalin, protein PAP-R, mitogen (a series of glycoproteins), caryophyllene.

Antiviral Property *Phytolacca* species are of economic and medicinal interest. The medicinal value is attributed to the antibacterial, antifungal, and antiviral activities (Batista et al. 1995). These activities mainly depend on the phytochemical constituent characteristic of this genus (Abdel-Mogib et al. 2002).

Anticancer The anticancer effects appear to work primarily based upon antitumor and anti-inflammatory properties, along with immune stimulant functions. Anticancer, antileukemic, or antitumor constituents include: ascorbic acid, astragalin, beta-carotene, caryophyllene, isoquercitrin, oleanolic acid, riboflavin, tannin, and thiamine.

Anti-Inflammatory Constituents include saponins, alpha-spinasterol, ascorbic acid, calcium oxalate, caryophyllene, isoquercitrin, jialigonic acid, and oleanolic acid in the roots and berries.

Action Crude extract acts as an antiphytoviral agent, against different plant viruses like tobacco necrosis virus (TNV), tobacco mosaic virus (TMV), and tomato spotted wilt virus (TSWV). When it was applied onto *Phaseolus vulgaris*, *Datura stramonium*, and *Chenopodium amaranticolor* as pre-inoculation spray (in vivo), it reduced the infectivity of above viruses up to 90 %. However, when the extract was mixed with the virus inoculum (in vitro), it inhibited the local lesion development by 100 % after one hour of mixing with TNV, and three hours for both TMV and TSWV (Allam et al. 1979; Verma and Baranwal 1983; Barakat 1988; Hansen 1989; Takanami et al. 1990; Othman et al. 1991; Meyer et al. 1995; Yordanova et al. 1996; El–Dougdoug 1997; Shoman 2002).

12.8.4 *Mirabilis jalapa*

Mirabilis jalapa belongs to family Nyctaginaceae. Ithails from tropical South America, but has become natural throughout tropical and warm temperate regions and in cooler temperate regions. The whole plant or its specific parts (leaves and roots) are known to have medicinal properties. *Mirabilis jalapa* contains a ribosome-inactivating protein (RIP), called *Mirabilis* antiviral protein (MAP); the protein was tested against infection by *Potato virus X*, *Potato virus Y*, *Potato leafroll virus*, and *Potato spindle tuber viroid*. Root extracts of *M. jalapa* sprayed on test plants 24 h before virus or viroid inoculation inhibited infection by almost 100 %, as corroborated by infectivity assays and the nucleic acid spot hybridization test. Antiviral activity of MAP extracts was observed against mechanically transmitted viruses but not against aphid transmitted viruses. Purified MAP showed the same antiviral effect as the crude extract. MAP was purified to homogeneity and was found to be lysine rich and basic (pI 9.8), with a molecular weight close to 24.2 kDa. Purified MAP has been shown to inhibit the mechanical transmission of tomato mosaic virus (TMV) in tobacco, tomato, and pepper plants and cucumber green mottle mosaic virus in cucumber plants. Moreover, MAP was also shown to inhibit protein synthesis in *Escherichia coli* as well as in eukaryotes and to possess repellent properties against aphids and white flies. Kataoka et al. showed that MAP was compartmentalized in *M. jalapa* vacuoles, sequestering its ribosome-inactivating activity away from its own ribosomes.

Inhibitory Activity of *M. jalapa* Extracts against PVX and PVY *M. jalapa* root extracts were applied to the leaves of *G. globosa*, an indicator plant which reacts hypersensitively to PVXCP infection. Results show that the root and leaf extracts diluted 1:5 (vol/vol) in sterile water were strongly inhibitory to PVX infection, because almost 100 % inhibition was observed. The inhibitory activity of MAP was not affected by dilution even extracts diluted with tap water gave an inhibitory effect. Similar effects were found by using leaf or root tissues. Purified MAP showed high antiviral activity.

12.8.5 *Tagetes minuta* L.

Tagetes minuta also known as Mexican marigold, mint marigold, wild marigold, or stinking roger. The volatile oils of plants have been recognized since antiquity to possess biological activity and a number of plant fractions and pure isolates have been mentioned as containing substances which interfere with or inhibit infection of viruses. *Tagetes minuta* oil and its components act as potent antiviral agent.

Compounds Present in *Tagetes minuta* Z-β-ocimene and dihydrotagetone present in *Tagetes minuta* oil have been found to inhibit carnation ring spot (CaRSV) and carnation vein mottle viruses (CaVMV) (Matthews 1991). The freshly distilled *Tagetes minuta* oil contains ocimene 55 % and dihydrotagetone 33 % (Singh et al. 1992). The whole oil of *Tagetes minuta* and its pure components, i.e., ocimene and dihydrotagetone were tested individually with virus cultures of CaVMV and CaRSV on *Chenopodium amaranticolor*.

12.8.5.1 Screening of Antiviral Activity

Activity of volatile oils was tested against carnation ringspot and carnation vein mottle virus in different dilutions. Most of the tests were performed by using 0.5 % and 2.5 % concentration of essential oils as beyond this concentration phytotoxic effect appeared on *Chenopodium amaranticolor* leaves, at higher concentrations. The 0.5 and 2.5 % concentration of essential oils was mixed with crude sap containing each virus and incubated at room temperature for 24 h. After incubation, sap containing virus was inoculated individually on bioassay host *Chenopodium amaranticolor* after adding Celite (as abrasive) to monitor the inhibitory effect.

Tagetes minuta plant grows wild in the hilly areas like Himachal Pradesh, Jammu and Kashmir, Uttar Pradesh, and North Eastern States of India and cultivated as commercial *Tagetes* oil crop, hence easily available in bulk quality. The oil and pure isolates are natural products and hence no threat to environment. Application of oil and pure isolates ensure quick and efficient recovery from viral infections. It also helps in the plant virus management. Since *Tagetes* crop grows wild and can be distilled in rich pockets/places with prototype distillation unit, hence the oil will be a cheap, eco-friendly, and easily available antiviral natural product.

12.8.6 *Bougainvillea spectabilis*

Bougainvillea spectabilis belongs to family Nyctaginaceae; it also contains an endogenous virus inhibitor which confers resistance to *Tospovirus*, tobacco mosaic virus (TMV), cucumber mosaic virus (CMV), and cowpea aphid-borne mosaic virus (CAMV) in their respective susceptible hosts. The viral inhibitor in *Bougainvillea spectabilis* is very potent, stable, and is 28 kDa basic protein (BAP). The partial cDNA encoding the *Bougainvillea* antiviral protein was synthesized from the leaf of *Bougainvillea spectabilis*, cloned, and sequenced. Homology with other antiviral proteins was studied.

Sequence Homology Homology search was performed using National Center for Biotechnology Information Basic Local Alignment Search Tool (NCBI BLAST) program. The putative region was chosen among the three open reading frames for further homology studies. Alignment of the peptide sequences with known antiviral/ribosome-inactivating protein sequences revealed weak homology of BAP-cDNA sequence with the reported AVP/RIP sequences, viz., *Mirabilis* antiviral protein, Pokeweed antiviral protein, and *Clerodendrum aculeatum* AVP. The results of the sequence homology analysis infer that the cDNA may be specific to *Bougainvillea spectabilis*.

12.8.7 *Dianthus caryophyllus* L. (carnation)

Dianthin 30 and dianthin 32, two proteins isolated from the leaves of *Dianthus caryophyllus* (carnation), were purified to homogeneity by chromatography on nitrocellulose. The molecular weight of dianthin 30 is 29,500 and that of dianthin 32 is 31,700. Both dianthins are glycoproteins containing mannose.

Antiviral Activity Tobacco mosaic virus was mixed with the substances to be tested or with an equal volume of water as a control. Inoculum, containing 600 grit Carborundum as an abrasive, was rubbed on to leaves of the local lesion host

Nicotiana glutinosa in a glasshouse at 20–40°C. Each treatment was replicated 10 times and randomized on whole leaves of the test plants. Lesions were counted after 3 days of infection. Dianthin 30 and dianthin 32, mixed with tobacco mosaic virus before infection, prevented local lesions in the leaves of *Nicotiana glutinosa* by more than 50 % at concentrations of 0.5 and 1 ug/ml, respectively (Stevens et al. 1981). Both dianthins markedly decreased the production of lesions by tobacco mosaic virus, and this presumably account for the antiviral properties of carnation leaf extract (Van Kammen et al. 1961; Ragetli and Weintraub 1962) and have been compared with interferon (Fantes and O'Neill 1964).

12.8.8 *Satureja montana* L. ssp. Variegate

It belongs to family Lamiaceae and is a very important source of essential oils and other biologically active molecules. Essential oils are variable mixtures, principally of terpenoids and specifically of monoterpenes and sesquiterpenes, although diterpenes may also be present. Monoterpenes are detected in every essential oil comprising from as little as 1 % to more than 95 % of the oil and are usually present as main constituents in oil fractions of *Satureja* plants. They play an important role in the resistance against diseases and insects. Essential oils and their components exhibit antiviral, antimycotic, antioxygenic, antiparasitic, and insecticidal properties. The phenol components with hydroxyl groups were found to posses the major antimicrobial activity. Carvacrols had anti-inflammatory activity and limonenes showed antiviral activity.

12.8.8.1 *S. montana* Essential Oil and Its Major Components

Thymol and carvacrol affected the development of local lesions caused by tobacco mosaic virus and cucumber mosaic virus. Both phenolic compounds are biologically active – thymol has antiseptic and carvacrol possesses antifungal properties. Thymol and carvacrol are structurally very similar, having the hydroxyl group at a different location on the phenolic ring. Although, among the essential oil constituents, phenolic compounds with hydroxyl groups were previously described as antimicrobial agents and antiphytoviral agents. When the oil was applied onto *N. glutinosa* plants as a pre-inoculation spray, the number of local lesions was significantly inhibited. Crude extract and the essential oil of *Plectranthus tenuiflorus* also showed inhibitory effect against tobacco necrosis virus, tobacco mosaic virus, and tomato spotted wilt virus. Monoterpenes were responsible for the antiviral activity of the oil and may show synergism in their antiviral effect. When the oil was applied on local hosts simultaneously with the infecting virus, the number of local lesions was reduced by TMV infection and CMV infection. When applied individually, thymol and carvacrol reduced the number of local lesions on both CMV- and TMV-infected plants of *Chenopodium amaranticolor*.

12.9 Conclusions

It has been demonstrated that interferon-like native inhibitors of plant virus infection occur in a few plants, growing wild in nature or grown for ornamental purposes, which prevented virus infection in healthy susceptible hosts prior to virus infection. The endogenous virus inhibitors having strong antiviral property lack virus specificity and had an association with DNA-dependent protein synthesis. They are pH and heat stable, like interferon, found in vertebrate system. The endogenous virus inhibitors themselves have no direct effect on the virus. Their treatment on plants results in the production of the actual virus inhibitory substances, like PR proteins, which later on circulate in the whole plant system to cause systemic resistance against viruses.

Possibilities of using biological proteins in the treatment of plant virus diseases under field diseases are undergoing serious evaluations. Although the present work may not be of great commercial importance just now, its achievement itself is vital. The knowledge gained will spawn more effective virus disease control methods. The intention has been to combine the features of inducer yielding plants as well as other biological agents with the virus protective agricultural plants. The use of natural resources from plant species in the treatment of plant viral diseases has not been extensively explored and may provide some new information about antiphytoviral activity of plants.

References

Abdel-Mogib M, Albar HA, Batterjee SM (2002) Chemistry of the Genus *Plectranthus*. Molecules 7:271–301

Abdul-Alim MA (1971) A chemical study of the leaves of *Clerodendron inerme*. Planta Med 19:318–321

Agrawal RR, Dutt SS (1936) Chemical examination of *Punarnava or Boerhaavia diffusa* Linn. II. Isolation of an alkaloid punarnavine. Chem Abstr 30:3585

Ahmad K, Hossain A (1968) Isolation, synthesis and biological action of hypoxanthine-9-L-arabinofuranoside. J Agric Biol Sci 11:41

Akihisa T, Matsubara Y, Ghosh P, Thakur S, Tamura T, Matsumoto T (1989) Sterols of some Clerodendrum species (Verbenaceae) occurrence of the 24-α and 24-β epimers of 24-ethylsterols lacking a Δ25-bond. Steroids 53:625–638

Allam EK, Morsy AA, Ali MDH, Abo El-Ghar AI (1979) Inhibitors from some higher plants inhibiting TMV CMV infection. Egypt J Phytopath 10:9–14

Atta-Ur-Rehman, Begum S, Saied S, Choudhary MI, Farzana A (1997) A steroidal glycoside from Clerodendron inerme. Phytochemistry 45:1721–1722

Awasthi LP, Yadav CP (2009) Induction of systemic resistance in tomato against viral diseases through botanicals. In: Indian Phytopathological Society, 5th international conference, plant pathology in the globalized era, Division of Plant Pathology, Indian Agricultural Research Institute, New Delhi, India. 10–13 November, 2009. Abstract No.401 (S-08)

Awasthi LP, Singh PK (2004) Management of mosaic disease of Amorphophallus campanulatus through leaf extract of Clerodendrum aculeatum. Indian Phytopathol 57(3):345

Awasthi LP, Singh S (2006) Management of papaya ringspot virus in nursery by phytochemicals. Ind J Virol 17(2):138

Awasthi LP, Singh S (2008) Eco-friendly management of viral diseases of papaya (*Carica papaya* L.). Ind Phytopathol 61(3):378

Awasthi LP, Singh S (2009) Management of ring spot disease of papaya through plant products. Ind Phytopathol 62(3):369–375

Awasthi LP, Kumar P (2003a) Prevention of infection and multiplication of cucumber green mottle mosaic virus in muskmelon treated with *Boerhaavia diffusa*. Ind Phytopathol 56:362

Awasthi LP, Kumar P (2003b) Protection of some cucurbitaceous crops against natural infection of virus through *Boerhaavia diffusa*. Ind Phytopathol 56:317

Awasthi LP, Mukerjee K (1980) Protection of potato virus X infection by plant extracts. Biol Plant 22:205–206

Awasthi LP, Rizvi SMA (1998) Prevention of infection by a vector borne virus of Tomato by *Boerhaavia diffusa* glycoprotein. In: National conference on integrated pest management. 23–35, February, 1998. Sultan Qaboos University. Sultanate of Oman (Abstract)

Awasthi LP, Rizvi SMA (1999) Effect of *Boerhaavia diffusa* glycoprotein on the transmission of Tomato yellow leaf curl virus by *Bemisia tabaci* Gen. In: National symposium on vectors of plant diseases. 11–13 November, 1999. N.D.U.A.&T Kumarganj, Faizabad, p. 56

Awasthi LP, Kluge S, Verma HN (1987) Characteristics of an antiviral agent induced by *Boerhaavia diffusa* glycoprotein in host plants. Ind J Virol 3:156–169

Awasthi LP, Chowdhury B, Verma HN (1984) Prevention of plant virus disease by *Boerhaavia diffusa* inhibitor. Ind J Trop Plant Dis 2:41–44

Awasthi LP, Kluge S, Verma HN (1989) Characteristics of antiviral agents induced by *B. diffusa* glycoprotein in host plants. Ind J Virol 3:156–169

Awasthi LP, Pathak SP, Gautam NC, Verma HN (1985) Control of virus diseases of vegetable crops by a glycoprotein isolated from *B. diffusa*. Ind J Plant Pathol 3:311–327

Barakat A (1988) Studies on plant virus inhibitors from certain species of the Sinai flora. Microbiol Lett 38:123–130

Baranwal VK, Tumer NE, Kapoor HC (2002) Depurination of ribosomal RNA and inhibition of viral RNA translation by an antiviral protein of Celosia cristata. Ind J Exp Biol 40:1195–1197

Basu NK, Lal SB, Sharma SN (1947) Investigations on Indian medicinal plants. Q J Pharm Pharmacol 20:38–42

Batista O, Simoes MF, Duarte A, Valderia ML, Delatorre MC, Rodriguez B (1995) An antimicrobial abietane from the root of *Plectranthus hereroensis*. Photochemistry 38:167–169

Bawden FC (1954) Inhibitors and plant viruses. Adv Virus Res 2:31

Benda GTA (1956) The effect of New Zealand Spinach juice on the infection of cowpeas by tobacco ringspot virus. Virology 2(4):438–454

Bharathi M (1999) Effect of plant extract and chemical inhibitors on cucumber mosaic virus of brinjal. J Mycol Plant Pathol 29:57–60, Contact: Bharathi, M.; Sarathi Cooperative Housing Society, Barabanda Road, Erragadda, 8-4-371/B/86, Plot No.66, Hyderabad, Andhra Pradesh, 500 018, India

Bolger LM, Rees HH, Ghisalberti EL, Goad LJ, Goodwin TW (1970) Isolation of two new sterols from Clerodendrum campbellii. Tetrahedron Lett 11:3043–3046

Chopra GL (1969) Angiosperms. Systematics and life cycle. S. Nagin & Co., Jalandhar, pp 361–365

CSIR (1988) The wealth of India: raw materials, vol VII B. CSIR, New Delhi, p 174

Duggar BM, Armstrong JK (1925) The effect of treating the virus of tobacco mosaic with juices of various plants. Ann Mo Bot Gard 12:359–366

El-Dougdoug KA (1997) Antiphytoviral activity of Khella and Black Commun on infectivity and chemical structure of To MV. In: Proceedings of the 9th conference of microbiology. Cairo, 25–27 March, 1997, pp 203–221

Fantes KH, O'Neill CF (1964) Some similarities between a viral inhibitor of plant origin and chick interferon. Nature London 203:1048–1050

Fischer H, Nienhaus F (1973) Virus inhibitors in pepper *Capsicum annuum*. Phys Z 78:25–41

Flores EW, Walkyria BC, Maneghini M (1967) An inhibitory activity of leaf extract of *Abutilon striatum* Dicks on the infectivity of Abutilon mosaic virus and tobacco mosaic virus. Arc Inst Biol SPaulo 34:83–89

Francki RIB (1964) Inhibition of cucumber mosaic virus infectivity by leaf extracts. Virology 24:193–199

Ganapathy T, Narayanaswamy P (1990) Effect of plant products on the incidence of major diseases of groundnut. Int Arachis Newslett 7:20–21

Gao LM, Wei XM, He YQ (2003) Studies on chemical constituents in leafs of *Clerodendron fragrans*. Zhongguo Zhong Yao Za Zhi 28:948–951

Goswami P, Kotoky J, Chen ZN, Lu Y (1996) A sterol glycoside from leaves of *Clerodendron colebrookianum*. Phytochemistry 41:279–281

Habuka N, Akiyama K, Tsuge H, Miayamo M, Matsumoto T, Noma M (1990) Expression and secretion of Mirabilis antiviral protein in Escherichia coli and its inhibition of in vitro eukaryotic and prokaryotic protein synthesis. J Biol Chem 265:10988–10992

Hansen AJ (1989) Antiviral chemicals for plant disease control. Plant Sci 8:45–88

Jain GK, Khanna NM (1989) Punarnavoside: A new antifibrinolytic agent from *Boerhaavia diffusa* Linn. Indian J Chem 28(B):163–166

Jayashree K, Pan KB, Sabitha D (1999) Effect of plant extracts and derivatives, butter milk and virus inhibitory chemicals on *Pumpkin yellow vein mosaic virus* transmission. Indian Phytopathol 52(4):357–361

Joshi KC, Singh P, Mehra A (1979) Chemical investigation of the roots of different *Clerodendron* species. Planta Med 37:64–66

Kanchanapoom T, Chumsri P, Kasai R, Otsuka H, Yamasaki K (2005) A new iridoid diglycoside from *Clerodendrum chinense*. J Asian Nat Prod Res 7:269–272

Kanchanapoom T, Kasaia R, Chumsric P, Hiragad Y, Yamasaki K (2001) Megastigmane and iridoid glucosides from *Clerodendrum inerme*. Phytochemistry 58:333–336

Kassanis B, Kleczkowski A (1948) The isolation and some properties of a virus-inhibiting protein from *Phytolacca esculenta; J.* Gen Microbiol 2:143–153

Kempiak G, Schuster G, Awasthi LP, Kluge S (1991) Attempts to reduce damage caused by oat sterile dwarf virusinoatsusingvirazole,2,4-dioxohexahydrotriazine, *Boerhaavia* inhibitor and alkane-monosulfonate. Acta Phytopathologica et Entomologica Hungarica 26:219

Kirtikar KR, Basu BD (1956) Indian medicinal plants, vol III, 2nd edn. Lalit Mohan Basu, Allahabad, pp 2045–2048

Kumar D, Verma HN, Narendra T, Tewari KK (1997) Cloning and characterisation of a gene encoding an antiviral protein from *Clerodendrum aculeatum* L. Plant Mol Biol 33:745–751

Kumar P, Awasthi LP (2008) Prevention of infection and spread of *Cucumber mosaic virus* disease in cucumber (*Cucumis sativus* L.) through plant products. Indian J Virol 19(1):107

Kumar P, Awasthi LP (2003a) Prevention of *Cucumber mosaic virus* infection and spread in cucumber plants, treated with *Boerhaavia diffusa* inhibitor. Indian Phytopathol 56(2):318

Kumar P, Awasthi LP (2003b) Management of infection and spread of bottle guard mosaic virus disease in bottle gourd through botanicals. Indian Phytopathol 56(2):361

Kuntz JE, Walker JC (1947) Virus inhibition by extracts of spinach. Phytopathology 37(8):561–579

Lee JH, Lee JY, Kang HS, Jeong CH, Moon H, Whang WK, Kim CJ, Sim SS (2006) The effect of acteoside on histamine release and arachidonic acid release in RBL-2H3 mast cells. Arch Pharm Res 29:508–513

Loebenstein G, Ross AF (1963) An extractable agent, induced in uninfected tissues, by localized virus infection, that interfere with infection by tobacco mosaic virus. Virology 22:507–517

Louis Vimi, Balakrishan S (1996) Effect of application of selected medicinal plant extracts on the incidence of pumpkin mosaic virus. Ind Phytopathol 49(4):373–377

Matthews REF (1991) Plant virology, 3rd edn. Academic Press INC, New York/London

McKeen CO (1956) The inhibitory activity of extract from *Capsicum frutescens* on plant virus infections. Can J Bot 34:891–903

Meyer G, De Dan, Allan Z, (1995) Antiviral proteins in higher plants. pp. 119–130. Library of congress cataloying in public data Boca Raton Ann. Arab. London. Tokyo

Mishra AN, Tiwari HP (1971) Constituents of the roots of *Boerhaavia diffusa*. Phytochemistry 10:3318

Okuyama T, Takemi K, Saka H, (1978) Sic Rep Fac Agr Ibaraki Univ, 26: 49 (In Japanees)

Othman BA, El-Dougdoug K, Abo El-Nasr M (1991) Effect of garlic bubbilies extraction on tomato mosaic virus. Ann Agric Sci 36:423–430

Pandey R, Verma RK, Singh SC, Gupta MM (2003) 4α-methyl-24β-ethyl- 5α-cholesta-14,25-dien-3β-ol and 24β-ethylcholesta-5, 9(11), 22e-trien-3β-ol, sterols from *Clerodendrum inerme*. Phytochemistry 63:415–420

Pinto WJ, Nes WR (1985) 24β-ethylsterols, n-alkanes and n-alkanols of *Clerodendrum splendens*. Phytochemistry 24:1095–1097

Ragetli HWJ (1957) Behavior and nature of a virus inhibitor occurring in *D. caryophyllus*. Tijdschr Planteziekten 63:245–344

Ragetli HWJ (1975) The mode of action of natural plant virus inhibitors. Curr Adv Plant Sci 19:321–334

Ragetli HWJ, Weintraub M (1962) Purification and characteristics of a virus inhibitor from *Dianthus caryophyllus* L. Virology 18:241–248

Rao DG, Raychaudhuri SP (1965) Further studies on the inhibition of ring spot strain of potato virus X by plant extracts, culture filtrates of *Trichothecium roseum* and chemicals. Indian J Microbiol 5:9–12

Rendle AB (1925) The classification of flowering plants, vol 2, Dicotyledons. Campridge University Press, London

Sangar RBS, Dhingra MK (1982) Potato virus inhibitor from Neem leaf extract. J Indian Potato Assoc 9:143–149

Sharma SK, Singh VP (1979) The antifungal activity of some essential oils. Indian Drug Pharm Ind 14:3–6

Shoman SA (2002) Role of salicylic acid in plant resistance to tobacco necrosis and tobacco mosaic viruses infection. Az J Microbiol 58:178–191

Singh AK, Singh M, Singh AK (1988) Antiviral activity and physical proportion of the extract of *Azadirachta indica*. Indian J Virol 4:76–81

Singh B, Sood RP, Singh V (1992) Chemical composition of *Tagetes minuta* L. oil from Himachal Pradesh (India). J Ess Oil Res 4:525–526

Singh P, Singhi CL (1981) Chemical investigation of *Clerodendron fragrans*. J Indian Chem Soc 58:626–627

Singh R, Prakash L (1983) Chemical examination of stems of *Clerodendron inerme* (L) Gaertn. (Verbenaceae). Pharmazie 38:565

Singh S (2002) Studies on management of yellow mosaic disease of mungbean (*Vigna mungo* (L.) Hepper) through botanicals. M.Sc. (Ag.) thesis, N.D. University of Agriculture & Technology, Kumarganj, Faizabad (U.P.), pp. 1–100

Singh S, Awasthi LP (2002) Prevention of infection and spread of *Bean common mosaic virus* disease of mungbean and urdbean through botanicals. Indian J Plant Pathol 11(1 & 2):63–65

Singh S, Awasthi LP (2004) Prevention of infection and spread of mungbean yellow mosaic virus (MYMV) on urdbean (*Vigna mungo*) through *Boerhaavia diffusa* root extract. Indian J Plant Pathol 22(1&2):50–55

Singh S, Awasthi LP (2008) Management of ring spot disease of papaya (*Carica papaya* L.) through antiviral agents of plant origin along with milk protein. Indian J Virol 19(1):106–107

Singh S, Awasthi LP, Khan MN (2005) Management of yellow mosaic disease of mungbean and urdbean through aqueous root extract of *Boerhaavia diffusa*. New Bot XXXII:55–62

Singh S, Awasthi LP, Verma HN (2004) Prevention and control of yellow mosaic disease of mungbean by application of aqueous root extract of *Boerhaavia diffusa*. Indian Phytopathol 57(3):303–304

Singh Sanjay, Awasthi LP (2009) Evaluation of medicinal plants against yellow mosaic disease of mungbean [*Vigna radiata* (L.)]. In: National conference on herbal and traditional medicine. Department of Botany, Dapoli Urban Bank Senior Science College, Dapoli, Dist. Ratnagiri (Maharastra), 6–7 February, 2009 (Abst. No. M-P.-18)

Sinha NK, Pandey VB, Dasgupta B, Higuchi R, Kawasaki T (1982) Acteoside from the flowers of *Clerodendron infortunatum*. Indian J Chem 22B:97–98

Sinha NK, Pandey VB, Shah AH, Dasgupta B (1980) Chemical constituents of the flowers of *Clerodendron infortunatum*. Indian J Pharm Sci 42:21

Stevens WA, Spurdon C, Onyon LJ, Stirpe F (1981) Effect of inhibitors of protein synthesis from plants on tobacco mosaic virus infection. Experientia 37:28–29

Surange SR, Pendse GS (1972) Pharmacognostic study of roots of *Boerhaavia diffusa* Willd. (*punarnava*). J Res India Med 7:1

Surendran M, Shanmugam V, Rajagopalan B, Ramanian N (1999) Efficacy of botanicals on *Brinjal mosaic virus*. Plant Dis Res 14(1):63–66

Takanami Y, Kuwata S, Ideda T, Kubo S (1990) Purification and characterization of the antiplant viral protein from *Mirabilis Jalapa* L. Ann Phytopath Soc Jpn 56:488–494

Van Kammen A, Noordam D, Thung T (1961) The mechanism of inhibition of infection with tobacco mosaic virus by an inhibitor from carnation sap. Virology 14:100–108

Varma JP (1973) Isolation and characterization of a virus inhibitor from cabbage (*Brassica oleracea* var.wirsing) leaves. Indian Phytopathol 26:713–722

Verma A, Singh RB (1994) *Clerodendrum aculeatum* a possible prophylactic agent against natural viral infection in mungbean. Ann Plant Prot Sci 2(2):60–63

Verma A, Verma HN (1993) Management of viral disease of mungbean by *Clerodendrum* leaf extracts. Indian J Plant Pathol 11(1 & 2):63–65

Verma HN (1982) Inhibitor of plant viruses from higher plants. In: Singh BP, Raychoudhury SP (eds) Current trends in plant virology. Today and Tomorrow's Printers and Publishers, New Delhi, pp 151–159

Verma HN, Awasthi LP (1979a) Prevention of virus infection and multiplication by leaf extract of *Euphorbia hirta* and the properties of the virus inhibitor. New Bot 6:49–59

Verma HN, Awasthi LP (1979b) Further studies on a mosaic virus of *Gomphrena globosa*. Phytopath Z 95:178–182

Verma HN, Awasthi LP (1979c) Antiviral activity of *Boerhaavia diffusa* root extract and the physical properties of the virus inhibitor. Can J Bot 57:926–932

Verma HN, Awasthi LP (1980) Occurrence of a highly antiviral agent in plants treated with Boerhaavia diffusa inhibitor. Can J Bot 58:2141–2144

Verma HN, Baranwal VK (1983) Antiviral activity and the physical properties of the leaf extract of *Chenopodium ambrosoides* L. Proc Indian Acad Sci (Plant Sci) 92:461–465

Verma HN, Varsha (1995) Prevention of natural occurrence of tobacco leaf curl disease by primed *Clerodendrum aculeatum* leaf extracts. In: Verma JP, Verma A, Kumar D (eds) Detection of plant pathogens and their management. Angkor Publishers (P) Ltd, New Delhi, pp 202–206

Verma HN, Awasthi LP, Mukerjee K (1979a) Prevention of virus infection and multiplication by extracts from medicinal plants. Phytopathol Z 96:71–76

Verma HN, Awasthi LP, Mukerjee K (1979b) Induction of systemic resistance by antiviral plant extracts in non-hypersensitive hosts. Zeitschrift Pflanzenk Pflanzenschutz 86:735–740

Verma HN, Awasthi LP, Saxena KC (1979c) Isolation of the virus inhibitor from the root extract of *Boerhaavia diffusa* inducing systemic resistance in plants. Can J Bot 57:1214–1217

Verma HN, Varsha, Srivastava S (1991) Antiviral agents from plants for control of viral diseases, Abstracts: international conference on virology in the tropics. Lucknow, India, p 250

Verma HN, Awasthi LP, Kumar V, Chaudhary B, Rastogi P, Dwivedi SD (1980) Control of plant virus diseases by extract from higher plants. J Indian Bot Soc 59:30

Verma HN, Chowdhury B, Rastogi P (1994) Antiviral activity in leaf extracts of different *Clerodendrum* species. Z Pflanzenk Pflanzenschuz 91(1):34–41

Verma HN, Rastogi P, Prasad V, Srivastava A (1985) Possible control of natural virus infection on *Vigna radiatus* and *Vigna mungo* by plant extracts. Ind J Plant Pathol 3:21–24

Verma HN, Srivastava S, Varsha, Kumar D (1996) Induction of systemic resistance in plants against viruses by a basic protein from *Clerodendrum aculeatum* leaves. Phytopathology 86:485–492

Verma HN, Varsha, Baranwal VK (1995) Agricultural role of endogenous antiviral substances of plant origin. In: Chessin M, De Borde D, Zipf A (eds) Antiviral proteins in higher plants. CRC Press, Boca Raton, pp 23–37

Wyatt SD, Shepherd RJ (1969) Isolation and characterization of a virus inhibitor from *Phytolacca americana*. Phytopathology 69:1787–1794

Yadav CP, Awasthi LP, Singh S (2009) Management of viral diseases of tomato through biopesticides: an eco-friendly approach. Indian J Virol 20(1):42

Yang H, Hou A-J, Mei S-X, Sun H-D, Che C-T (2002) Constituents of *Clerodendrum bungei*. J Asian Nat Prod Res 4:165–169

Yang H, Jiang B, Hou A-J, Lin Z-W, Sun H-D (2000) Colebroside A, a new diglucoside of fatty acid ester of glycerin from *Clerodendrum colebrookianum*. J Asian Nat Prod Res 2:177–185

Yordanova A, Korparov NE, Stomenova, Starcheva M (1996) Antiphytoviral activity of 1–morpholinomethyl tetrahydro 2–Pyrimidinone (DDB). Plant Pathol 45:547–551

Zaidi ZB, Gupta VP, Samad A, Naqvi QA (1988) Inhibition of spinach mosaic virus by extracts of some medicinal plants. Curr Sci 57(3):151–152

13 Yellow Vein Mosaic Disease: A New Threat to Mesta (*Hibiscus* sp.) Cultivation

Arpita Chatterjee and Subrata Kumar Ghosh

13.1 Introduction

Hibiscus cannabinus L. is popularly known as kenaf in the Western world and belongs to the family Malvaceae. *Hibiscus sabdariffa* L., another species of *Hibiscus*, is a close relative of kenaf and commonly known as roselle, Java jute, Thai jute, Pusa hemp, etc. Both of them are collectively regarded as mesta. *H. cannabinus* grows faster and attains maturity at around 150 days, while *H. sabdariffa* is late maturing and is ready for harvest at around 180 days of crop age. The plants are bisexual, are annual and produce large cream-coloured flowers. The stem of roselle is flexible while it is more or less rigid in kenaf (Singh 1997). Mesta is grown under a wide range of agro-climatic conditions. It is cultivated on different types of soil ranging from light clay loam to sandy loam. Both the plant species grow well in warm and humid climate with annual rainfall ranging from 100 to 160 cm. The temperature for optimum growth is in between 25 and 40 °C.

Mesta crops are cultivated primarily for obtaining bast fibres of commerce from the stems upon retting. These fibres are also used as substitute for jute fibres and are likewise used for making rope, cordage, canvas, sacking, carpet backing, fishing nets, etc., and thus possess high potentiality of diversification according to the need of end users. Recently, mesta plants are being used in paper-making. Mesta, in general, is grown in area of more than 26 lakh hectares producing fibre over 12 lakh bales annually (Singh 1997). In India the average productivity of mesta is highest in Tripura (1441 kg/ha), followed by Andhra Pradesh (1399 kg/ha), West Bengal (1334 kg/ha), Bihar (1220 kg/ha), Orissa (1083 kg/ha), Assam (880 kg/ha) and Meghalaya (844 kg/ha). India is the chief importer of the fibres produced out of these crops, because the average yield of mesta in India is quite low (900–1800 kg/ha). Average low yield of fibre from these crops is mainly due to attack by a number of diseases, of which viral diseases are notable (Brunt et al. 1996). A number of viruses, namely, Brazilian tobacco streak, Cotton leaf curl, groundnut rosette, Cotton anthocyanosis virus and yellow vein mosaic virus from mesta, have been isolated by different workers (Chatterjee 2007).

The occurrence of yellow vein mosaic disease of mesta (MeYVMD) is a new entrant to disease scenario in India (Chatterjee et al. 2005, 2006, 2009; Ghosh et al. 2007, 2009). It is found in endemic form in different parts of India during

A. Chatterjee
Department of Botany, Barasat College, Barasat, Kolkata 700 124, India
e-mail: arpita10c@gmail.com

S.K. Ghosh (✉)
Division of Crop Protection, Central Research Institute for Jute and Allied Fibres, Barrackpore, Kolkata 700 120, India
e-mail: subratak_ghosh@yahoo.co.uk

L.P. Awasthi (ed.), *Recent Advances in the Diagnosis and Management of Plant Diseases*, DOI 10.1007/978-81-322-2571-3_13

the last few years and the disease has spread at a faster rate causing great reduction in yield and thus assumed a major threat to production. The association of a novel *Begomovirus*, namely, *Mesta yellow vein mosaic virus* (MeYVMV) with the disease, has been confirmed by electron microscopy and molecular techniques using PCR, sequence information and southern hybridization (Chatterjee and Ghosh 2007a).

13.2 Symptomatology

The symptoms started developing in the emerging leaves in the form of innumerable light-yellow coloured pinhead spots either in one or both halves of the lamina including veins. The spots gradually increased in size by coalescing with each other leading to the formation of yellow chlorotic flecks on veins and yellow mosaic symptoms on interveinal regions. With the advancement of the disease, the infected lamina showed yellow chlorotic discolouration. Gradually erratic chlorotic flakes, due to coalescing of pinhead spots, developed in the lamina including veins and veinlets. In severe cases, these flakes gradually increased in size and formed yellow netting and the infected leaves became completely chlorotic (Fig. 13.1). Gradual yellowing of petiole and crown portion was also noticed in infected plants. In case of severe infection, stunting of plant with defoliation of completely chlorotic leaves was observed. Occasionally, the stems of the diseased plants became partially or totally yellow. The flowers and fruits were malformed. In the event of infection by the virus at their early stage of growth, the infected plants did not flower even. The infected plants, in general, showed stunted growth with reduced leaf size. In addition to the above, the survival rate and life span of infected plants in glasshouse were also found less than those observed under field condition (Chatterjee et al. 2006, 2007).

13.3 Disease Development and Its Scoring Technique

To assess the extent of disease severity on a single plant and even on a single leaf, disease intensity index formula was developed based on the symptomatology. Disease development was noted chronologically by considering different grades, viz. (A) development of foliar symptom, (B) nature of discolouration, (C) extent of yellowing of veins, (D) area of discolouration and (E) extent of stunting (Chatterjee 2007). Total gradation was based on a numerical value of 100. In the scale, each symptom syndrome was given equal importance, and according to disease severity the symptom syndrome was divided into different gradations. Each grade was given a numerical value and noted with respect to each

Fig. 13.1 Yellow vein mosaic disease infected leaves showing the characteristic symptoms of *Hibiscus cannabinus* (*left*) and *H. sabdariffa* (*right*)

infected leaf. When the chlorotic spots started developing irregularly, the size of the individual area of discolourations was noted and their average was used to grade the area of discolouration. In this way numerical values for each leaf of the individual plant were recorded separately and indexing was done according to the following formula (Chatterjee 2007):

L_1, L_2, L_3, L_n = number of affected leaves on a plant varying from 1 to n

s = score of a diseased leaf varying from A to D

Lis = total score on diseased leaf of a plant in a particular stage/ period (i = 1/2/3/…./n)

E = growth habit of infected plant

T_N = total number of leaves in a single infected plant

$$\text{MeYVMD intensity index} = \frac{\left[\sum_{s=A}^{D} L_{1s} + \sum_{s=A}^{D} L_{2s} + \sum_{s=A}^{D} L_{3s} + \ldots\ldots\ldots\ldots \sum_{s=A}^{D} L_{ns}\right]}{T_N} + E$$

Gradual development in severity of the disease leading to the index value of 49.44 and 50.83 at field level and glasshouse condition, respectively, was noticed with the age of the plant. Mortality of the infected leaves was also noticed at the maximum infection level. The useful application of empirical formula, developed for assessing disease intensity of MeYVMD (cv. HC-583), revealed variation in symptom expression in plants grown in glasshouse condition and under natural condition. The lower index value obtained under field condition might be due to interaction of different epidemiological factors on plant growth and presence of lower population of the vector. Highest intensity or 100 % infection (MeYVMD index value) was never noticed under natural and glasshouse condition because all the infected plants died before it under both conditions. Hence, severity of 49.44 and 50.83 index value was sufficient to consider the whole plant totally infected at field level and glasshouse condition, respectively. In a single plant, not all the leaves had yellow mosaic discolourations, and even if discolourations were present, they were not necessarily of the same type because cells vary in their susceptibility. Furthermore, susceptibility of one leaf differs from another and the infected areas also differ between leaves. Hence, gradual development of the disease and the cumulative effect of all factors are the most important criteria to designate a single plant about its severity of infection (Chatterjee 2007).

13.4 Disease Scenario

13.4.1 Epidemiology

A detailed field survey was conducted during the cropping seasons of 2005–2007 on the incidence and severity of MeYVMD so as to understand the distribution of the disease throughout the mesta-producing regions of eastern, northern and southern India (Fig. 13.2).

The incidence and severity of MeYVMD in various fields at different locations ranged from 46 % to 93 % and 34 % to 78 %, respectively (Table 13.1). Highest incidences were observed in southern West Bengal followed by eastern Uttar Pradesh and northern West Bengal. The lowest incidence was noted in north-eastern Andhra Pradesh. Disease incidence was lower at Barrackpore than at Haringhata and Bongaon in southern West Bengal. Occurrence of MeYVMD was found to be very high in eastern areas of India. This situation might be due to the combination of different factors such as the cultivation of the highly susceptible cultivars as sole crop, the association of a begomovirus (MeYVMV) and a betasatellite (CLCuMV, *Cotton leaf curl Multan*

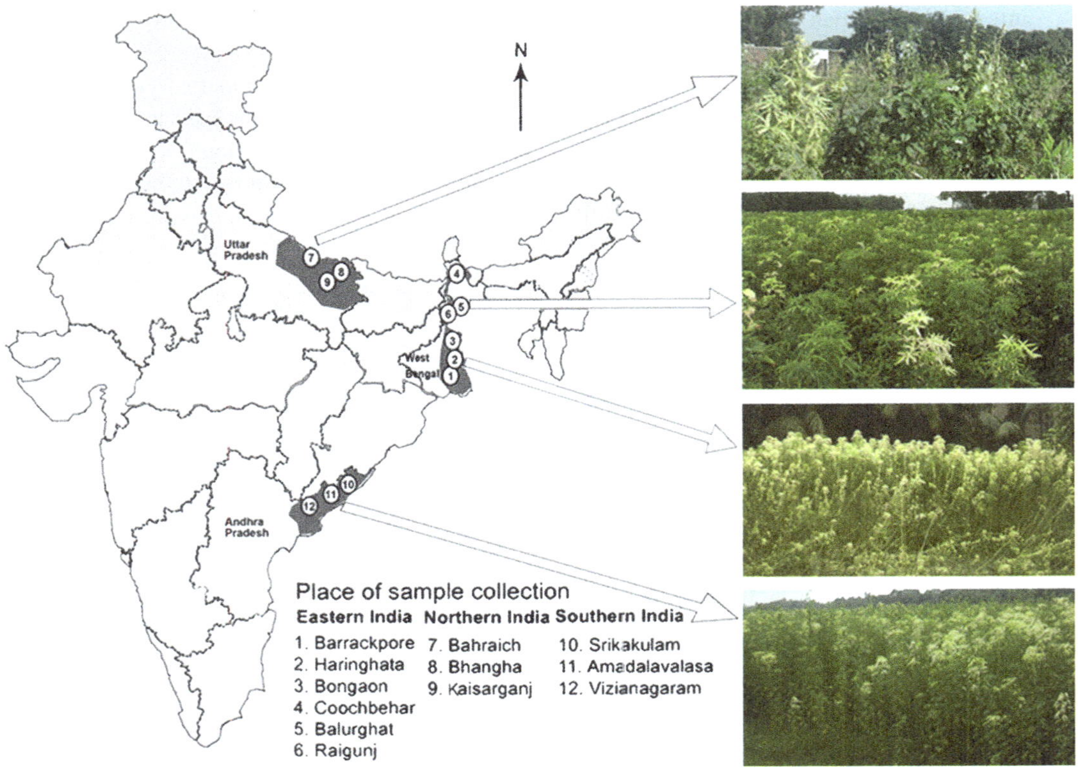

Fig. 13.2 Distribution of MeYVMD in India and locations sampled. Appearance of the disease in the farmers' fields in the corresponding regions is shown in the *right panels*

betasatellite) in most of the plants sampled, conducive environmental factors which facilitate the buildup of whitefly populations during the early crop growth period (Roy et al. 2009).

13.4.2 Observed Trend of Weather Factors Vis-à-vis Whitefly Populations

A model for prediction of the relationship of whitefly population (dependent variable) with different weather factors was developed through multiple regression analysis keeping maximum and minimum temperature, rainfall and relative humidity as predictors (constant), and the coefficients of multiple determination (R^2) were calculated from 2005 to 2007. Through ANOVA (SPSS Inc., USA), the goodness of fit of the model was determined, and the regression equation with respect to each sowing set was generated by estimating the beta-coefficients. The data sets for 2007 were chosen as representative as there was no significant difference with respect to either environmental parameters or whitefly populations studied for 3 years. Infestation of adult whiteflies on the mesta crops began in the second week of April and reached a peak during the second week of June (Roy et al. 2009). Heavy rainfall increased the humidity in the environment which again facilitated the multiplication of whiteflies. To express the whitefly population as the function of different weather factors, a predictive model was developed through multiple regression analysis. The estimated regression line was expressed as

Table 13.1 Average disease incidence and severity of MeYVMD after 100–120 days of crop growth in different geographical regions of India

Geographical region	Climate	Average disease incidence (%)[a]	Average disease severity (%)[b]
Eastern India: West Bengal			
Northern part	Humid subtropical		
Raigunj (11)		51.3 ± 1.0	35.6 ± 0.8
Balurghat (14)		59.3 ± 1.3	37.6 ± 1.0
Cooch Behar (9)		56.4 ± 1.3	33.6 ± 1.0
Southern part	Tropical wet and dry		
Barrackpore (8)		64.4 ± 1.5	43.0 ± 2.0
Bongaon (17)		90.1 ± 1.6	58.9 ± 0.2
Haringhata (22)		92.7 ± 0.4	59.7 ± 0.6
Northern India: Uttar Pradesh			
Eastern part	Humid subtropical		
Bahraich (8)		88.9 ± 1.3	76.0 ± 0.9
Kaisarganj (11)		91.5 ± 1.4	71.4 ± 0.5
Bhangha (14)		90.7 ± 0.9	77.6 ± 1.1
Southern India: Andhra Pradesh			
North-eastern part	Tropical wet and dry		
Amadalavalasa (7)		55.9 ± 1.9	45.2 ± 1.8
Srikakulam (4)		45.7 ± 2.5	47.9 ± 0.9
Vizianagaram (12)		46.8 ± 0.9	43.2 ± 1.3

Data within parentheses indicate the number of plots surveyed in each location
[a]Mean percentage of plants expressing symptoms ± S.E
[b]Mean percentage of leaves expressing symptoms ± S.E

$$Y = -58.639^{***} - 0.273^{**} X1 - 0.320 X2 + 0.050 X3 + 0.957^{***} X4 + e$$

where Y is adult whitefly population (nos.), $X1$ is rainfall (mm), $X2$ is maximum temperature (°C), $X3$ is minimum temperature (°C), $X4$ is relative humidity (%) and ε is the error factors not estimated in the model; *, **, *** indicates the F values were significant at 10 %, 5 % and 1 % level of significance, respectively.

The coefficient of multiple determinations (R^2) of the model was 0.809 implying that 80.9 % variation in whitefly populations is explained by the independent variables included in the model (rainfall, maximum temp, minimum temp and RH). The significant F value (at 1 % level of significance) indicated that the model is a good fit and the predicted adult whitefly populations showed a similar trend to that observed throughout the cropping season (Fig. 13.3).

In all 3 years symptoms of MeYVMD began to appear from the first week of May, fourth week of May and first week of June in plants sown in first week of April, first week of May and last week of May, respectively. The adult whitefly populations were highest in May and June and the maximum percent increments of disease incidence were recorded for each of these three different sowings during the latter part of June. The curve was almost linear for the April planting and strongly sigmoid for each of the later plantings, when disease saturation occurred during the final weeks of crop growth. The time to reach 50 % disease incidence was 15–16 weeks for the April planting and 7–8 and 3–4 weeks for those made in the first and last weeks of May, respectively. A regression approach was adopted to relate weekly disease increments and whitefly populations (Roy et al. 2009).

The values of the coefficients of determination (R^2) of these regression equations were estimated

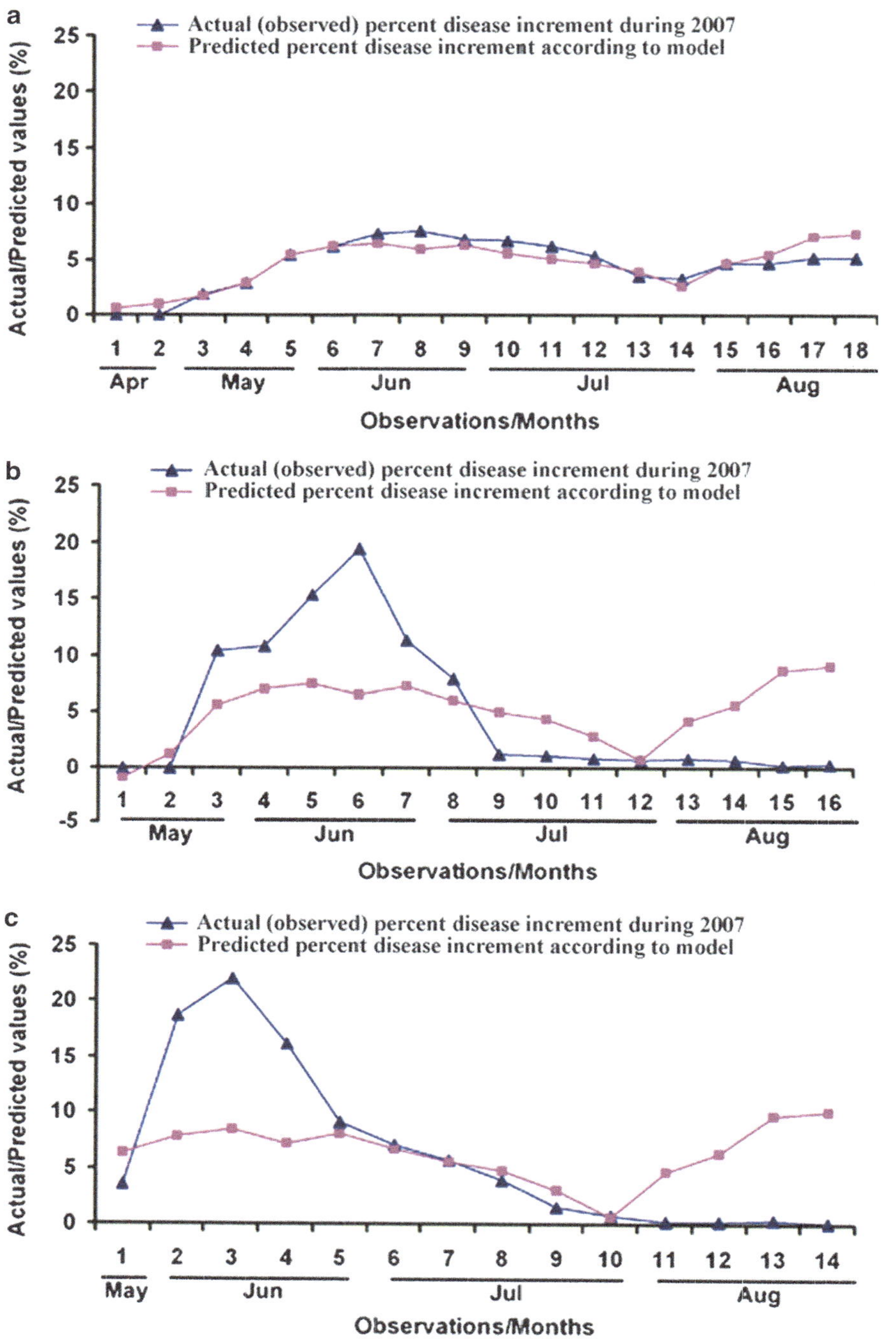

Fig. 13.3 Actual and predicted weekly increments in incidence of MeYVMD for plantings in (**a**) first week of April, (**b**) first week of May and (**c**) last week of May 2007

as 0.804, 0.198 and 0.117, respectively. *F* value of the estimated beta-coefficients was significant at the 1 % level for the first sowing set. The corresponding *F* values of other two sowing sets were either significant only at 10 % level (for second sowing set) or nonsignificant (for third sowing set) and thus indicated the model is a good fit only for the first sowing set. In the April planting the predicted early increments in disease incidence showed similar trend as observed, but in the later phase of the disease progress, the predicted values were somewhat higher than the observed values. The actual values of the disease increments for the other plantings were observed to be either higher or lower during the early and late stages of crop growth, respectively, than those predicted.

13.4.3 Economic Loss

The impact of MeYVMD (*H. cannabinus* cv. HC-583), on growth and yield, was greatest when the plants were inoculated at the earliest stage of growth. However, there were no significant differences in disease incidence amongst the groups of plants that were inoculated either at the early or later stages of growth. More than 90 % of the symptomatic plants that were inoculated 10–30 days after germination did not survive to maturity and produced very low fibre yield. However, when plants were inoculated 60 days after germination or later, MeYVMD did not significantly affect their height or yield.

Assessment of fibre quality in plants grown with varying levels of fertilizers in mesta under field condition revealed that the disease caused considerable loss in respect of both fibre fineness and strength. The reduction in fibre fineness was maximum at $N_0P_0K_2$ fertilizer level for HC mesta (16.40 %), whereas for HS mesta maximum reduction was observed at $N_{20}P_{10}K_{10}$ fertilizer level (23.68 %). Maximum reduction of fibre strength was noticed at $N_{40}P_{20}K_{20}$ fertilizer level for both the species and it was 50.88 % for HC mesta and 70.23 % for HS mesta (Chatterjee et al. 2008a).

Mesta is a fibre-producing crop and the amount of fibre production is a reflection of the rate of cambial activity, fibre layer initiation and the number of fibre bundles produced within the plant vascular system. Transverse sections through the internode regions of healthy diseased HC-583 stem revealed that the normal activity of cambium is altered due to viral infection and caused enormous changes in gross internal structures. The ultimate fibre cell number per bundle was lesser in diseased material and thus the fibre bundle was smaller in size. The fibre bundle layers were more or less parallel to each other in contrast to healthy mesta plants where the bundle layers converged to form pyramids. In diseased stem primary vascular tissues were higher in number as compared to smaller amount of secondary growth, which was prominent in healthy material. Histological evidence thus emphasized that malfunctioning of vascular system in host due to infection by this disease probably became a major factor for improper vascular tissue differentiation and that finally resulted into reduction in fibre yield in mesta plants (Chatterjee and Ghosh 2008b).

13.4.4 Varietal Response

Based on disease index scores (Chatterjee and Ghosh 2010) and areas under the disease severity curves (AUDSCs; Campbell and Madden 1990), there were significant differences in disease severity amongst the eight cultivars (*H. cannabinus* cv. HC-583, AMC-108; *H. sabdariffa* cv. HS-4288, HS-7910, AMV-1, AMV-2, AMV-3 and AMV-4). Cultivar HC-583 was the most sensitive to the disease with the highest AUDSC value of 1232, followed by AMC-108, HS-4288, AMV-1, AMV-2, AMV-3, AMV-4 and HS-7910. In all replications the AUDSC value for HS-7910 was lower than that of other cultivars indicating its reduced sensitivity to the disease (Chatterjee et al. 2008a; Roy et al. 2009).

13.5 Transmissibility

The disease was found to be transmitted by whitefly at a transmission efficiency of 85 % for *H. sabdariffa* and 79 % for *H. cannabinus*. A minimum of three whiteflies per plants was found to be effective for disease transmission (Chatterjee et al. 2008a). Typical symptoms appeared after a minimum incubation period of 9 days under glasshouse conditions. Minimum acquisition feeding period (AFP) for the vector was observed to be 12 h and minimum inoculation feeding period (IFP) was 4 h in both of the species. The disease transmissibility increased with the increase in IFP. In every case, 12 h IFP was good for symptom expression. At early stage of growth (when inoculation is done at 5–15 days after germination of plants belonging to both species), the transmission caused severity of symptom expression with severe stunting in inoculated plants. The disease was also observed to be transmitted by cleft grafting. In case of *H. sabdariffa* the transmissibility through cleft grafting was 80 %, while in *H. cannabinus* the transmissibility was 60 %. Symptoms appeared on *H. sabdariffa* within 7–9 days of grafting, while on *H. cannabinus* the symptoms appeared after 10–12 days of grafting.

Plants infected through whitefly and grafting gave positive Southern hybridization and nucleic acid spot hybridization (NASH) signal with α-^{32}P radiolabelled probes to *Cotton leaf curl Rajasthan virus* DNA-A and a positive PCR amplification with the begomovirus-specific primer sets confirming the transmissibility. No seed or soil transmission of the disease was noted (Chatterjee et al. 2008a).

13.6 Host Range

Besides five mesta varieties (HC-583, AMC-108, HS-4288, HS-7910 and AMV-4), only two species of *Vigna*, *V. umbellata* and *V. unguiculata* cv. V-240 of family Fabaceae, showed leaf crumpling symptom by experimental transmission with viruliferous whiteflies in glasshouse condition. The transmission efficacy was 13.3 % in case of *V. umbellata* and 43.7 % in case of *V. unguiculata*. These experimental host plants showed long incubation period of nearly 1 month (28–32 days for *V. umbellata*) or more than one month (32–42 days for *V. unguiculata*) to produce symptoms. Though both the species did not produce typical yellow vein symptoms, back-inoculation to HC-583 and HS-4288 developed typical yellow vein mosaic symptom. The efficacy of back-inoculation to HC-583 was higher (60 % for *V. umbellata* and 58.33 % for *V. unguiculata*) with a lesser incubation period than HS-4288 (28.57 % for *V. umbellata* and 21.43 % for *V. unguiculata*). Back-inoculation from symptomatic *Vigna* species needed more incubation periods than the normal transmission from diseased mesta plants. The disease transmission rate was observed to be higher (60–85 %) with less incubation period (6–12 days) for mesta varieties than the experimental host like *Vigna* (Chatterjee et al. 2008a).

13.7 Causal Agent

13.7.1 Identification

Electron Microscopy Transmission electron microscopy investigation with typical symptomatic leaves of *H. cannabinus* using 2 % uranyl acetate revealed the association of geminate particle, having size of 20 nm × 30 nm (Chatterjee et al. 2006).

Southern Hybridization Southern hybridization analysis and NASH test gave a positive hybridization signal with the infected mesta plants showing yellow vein mosaic disease symptom using α-^{32}P radiolabelled probes to *Cotton leaf curl Rajasthan virus* (India: Sriganganagar: 2001) DNA-A complete sequence (AF363011) and DNA-β associated with *Cotton leaf curl Rajasthan virus* (AY083590) and thus confirmed the involvement of a begomovirus with the disease (Chatterjee et al. 2005).

PCR-Based Detection MeYVMD sample exhibited a positive amplification of full-length DNA-A (~2.7 kb) and coat protein (CP) gene (~0.77 kb) with different primer sets (Table 13.2). The expected ~1.3 kb PCR amplicons was obtained from MeYVMD sample with primer specific to DNA-β (Briddon et al. 2002) and showed the

Table 13.2 Details of primers used for PCR amplifications

Target genomic fragment	Primer name	Sequence	Nt. ln.	Amplicon size (~kb)
Full-length DNA-A	P1 (virion-sense)	5′ – CATGAGTACGGAGATTGGGAC – 3′	21	2.7
(Chatterjee and Ghosh 2007a)	P2 (complementary-sense)	5′ – TCACACCAAAAGCATGAAGGGTCGAAGG – 3′	28	
Full-length DNA-A	FLD-F (F)	5′ – GARAGTACYCATGCYTCTAAYCC – 3′	23	2.7
(Chatterjee and Ghosh 2007a)	FLD-R (R)	5′ – AGTRTGRTTYTCRTACTTCCCAG – 3′ (where K = G/T, R = A/G, S = C/G, W = A/T, Y = C/T, B = C/G/T and V = A/C/G)	23	
Full-length DNA-A	Bcp1.f (F)	5′ – AATTAATAAAGTTTGAATTTTATATC – 3′	26	2.7
(Jose and Usha 2000)	Bcp2.r (R)	5′ – TCAATTCGTTACAGAGTC – 3′	18	
Full-length DNA-A	FL [H1] (F)	5′ – AAGCTTAAATAAATYTCCYGCYTAT – 3′	25	2.7
(Roy et al. 2009)	FL [H2] (R)	5′ – AAGCTTTGAGCGCGTCATATGATTG – 3′	25	
Full-length DNA-A	NIYVM (SP)-FL1 (F)	5′ – CAGAAGTCCGGATGTTCCAAG – 3′	21	2.7
(Roy et al. 2009)	NIYVM (SP)-FL2 (R)	5′ – TACATCCGATACATTCTGGGC – 3′	21	
Coat protein (CP) part of DNA-A	Bcp1H.f (F)	5′ – AAGCTTATGTCGAAGCGAGCTGCCG – 3′	25	0.77
(Jose and Usha 2000)	Bcp2.r (R)	5′ – TCAATTCGTTACAGAGTC – 3′	18	
Common region and	Primer A (F)	5′ – TAATATTACCKGWKGVCCSC – 3′	20	0.55
part of coat protein of DNA-A (Deng et al. 1994)	Primer B (R)	5′ – TGGACYTTRCAWGGBCCTTCACA – 3′ (where K = G/T, R = A/G, S = C/G, W = A/T, Y = C/T, B = C/G/T and V = A/C/G)	23	
DNA-β	β-1 (F)	5′ – GGTACCACTACGCTACGCAGC – 3′	21	1.3
(Briddon et al. 2002)	β-2 (R)	5′ – GGTACCTACCCTCCCAGGGGTACAC – 3′	25	

association of a satellite DNA-β molecule with the disease. Betasatellites were either absent or occurred only in very few samples collected from low-disease incidence locations, but almost all samples collected from high-disease incidence locations contained betasatellites (Chatterjee and Ghosh 2007a,b; Das et al. 2008a,b).

13.7.2 Characterization

The virus infecting mesta is a new recombinant monopartite begomovirus in nature. Cloning and sequence analysis data revealed association of different full-length DNA-A molecule and betasatellite of begomovirus with MeYVMD. No evidence for a second genomic component (DNA-B) was obtained, either by restriction mapping or by PCR using different sets of primers specific for DNA-B (Chatterjee and Ghosh 2007a,b).

Sequence Analysis of Full-Length DNA-A The potential full-length clones of DNA-A (EF373060) consisted of 2728 nucleotides. Nucleotide numbering proceeds from the 3′A in the conserved nonanucleotide sequence TAATATTAC. This forms the nick site of the *Rep* protein within the origin for virion-strand DNA replication. The genome was homologous to the DNA-A genomic component of bipartite begomoviruses and maintained the typical genome organization of begomovirus originating from Old World, with six conserved open reading frames (Chatterjee and Ghosh 2007a).

The DNA-A molecule (EF373060) showed highest nucleotide identity with *Cotton leaf curl Bangalore virus* [India: Bangalore: 2004] (AY705380; 83.5 %). It showed high degree of sequence identity with different isolates of *Cotton leaf curl virus*, namely, *Cotton leaf curl Rajasthan virus* [India: Hisar: 2003] (AY795607; 83.3 %), *Cotton leaf curl Rajasthan virus* [India: Sriganganagar: 2001] (AF363011; 83.1 %), *Cotton leaf curl Rajasthan virus* [India: Abohar: 2003] (AY795606; 83.1 %) and *Cotton leaf curl Rajasthan virus* [India: New Delhi: 2003] (AY795605; 83.0 %), out of the sequences reported from India. All the virus sequence reported and available had total nucleotide sequence identity less than 84 %. It was below the threshold value of 89 % which demarcated the present one as a distinct geminivirus species. Thus, the DNA-A of geminivirus infecting mesta is a novel one, and this begomovirus was considered as a new species and named *Mesta yellow vein mosaic virus* (MeYVMV) (Chatterjee and Ghosh 2007a).

The phylogenetic tree derived from complete nucleotide sequences of the DNA-A genomic components of begomovirus infecting mesta and other selected begomoviruses originated from same geographic region found more closely related to each other than viruses originated from other regions (Fig. 13.4). Sequence of this new DNA-A component was found to be grouped with the Old World monopartite begomoviruses and was distantly related with bipartite begomoviruses (*Tomato mottle virus*, L14460 and AY965900). This tree again showed the genetic difference between New World (*Squash leaf curl virus*, M38183; *Sida golden mosaic Costa Rica virus*, X99550) and Old World begomoviruses. The phylogenetic tree derived from the CP gene nucleotide sequences of begomovirus isolates infecting mesta with selected other begomoviruses showed a clear geographic relationship. The CP genes and its predicted amino acid sequences of begomovirus isolates infecting mesta originating from India are placed separately as a distinct group in the dendrogram (Chatterjee and Ghosh 2007a).

Sequence Analysis of DNA-β Molecule
The DNA-β molecule (DQ298137) was found to be approximately half the lengths (1354 nucleotides) of DNA-A molecules just like other reported DNA-β molecules and consisted of a functional ORF (βC1) in complementary-sense DNA. The coding region of this betasatellite (194–550 nt) is predicted to encode C1 protein with 118 amino acids having molecular weight of 13.678 kDa (Chatterjee and Ghosh 2007b).

The DNA-β molecule isolated from begomovirus infecting mesta showed highest nucleotide

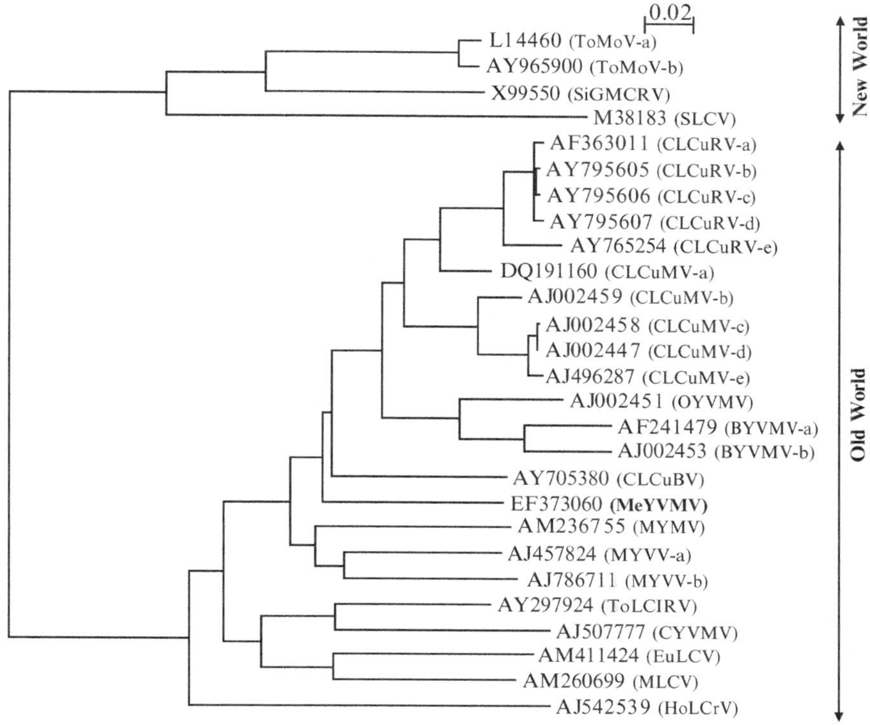

Fig. 13.4 Phylogenetic analysis of the complete MeYVMV DNA-A nucleotide sequences with other related DNA-A sequences of different begomoviruses. The viruses are *Cotton leaf curl Rajasthan virus* (CLCuRV), *Cotton leaf curl Multan virus* (CLCuMV), *Cotton leaf curl Bangalore virus* (CLCuBV), *Malvastrum yellow vein virus* (MYVV), *Malvastrum yellow mosaic virus* (MYMV), *Malvastrum leaf curl virus* (MLCV), *Croton yellow vein mosaic virus* (CYVMV), *Tomato leaf curl Iran virus* (ToLCIRV), *Bhendi yellow vein mosaic virus* (BYVMV), *Okra yellow vein mosaic virus* (OYVMV), *Euphorbia leaf curl virus* (EuLCV), *Hollyhock leaf crumple virus* (HoLCrV), *Tomato mottle virus* (ToMoV), *Squash leaf curl virus* (SLCV) and *Sida golden mosaic Costa Rica virus* (SiGMCRV). Vertical distances are arbitrary, horizontal distances are proportional to genetic distances

sequence similarity (86.1 %) with Indian isolates of begomovirus associated with cotton leaf curl disease (DQ191161 and AJ316038). This result supported the fact that DNA-β molecules showed relatedness based on geographic origin (Briddon et al. 2003; Zhou et al. 2003). Other begomovirus isolates infecting cotton also showed high percentage of sequence identity ranging from 77 to 86 % with this betasatellite. The highest percentage sequence identity of the C1 gene product of the DNA-β molecule was 85.5 % for begomovirus associated with tomato leaf curl disease (AJ316035), originating from Pakistan, and this isolate was the most prominent example of recombination, having a recombinant DNA-β molecule with an SCR (sequence conserved region) derived from a tomato DNA-β and the remainder of the molecule originating from a cotton-like DNA-β (Briddon et al. 2003; Chatterjee and Ghosh 2007b).

The putative phylogenetic relationships of DNA-β molecules based on the complete nucleotide sequences revealed two major groups (Fig. 13.5). The DNA-β molecule (DQ298137) showed close relation to DNA-β molecules originating from the same geographic region as well as host (Chatterjee and Ghosh 2007b). It grouped with DNA-β molecule isolated from *Hibiscus* originating from Pakistan (AJ297908) (i.e. same Indian subcontinent) within the Malvaceous

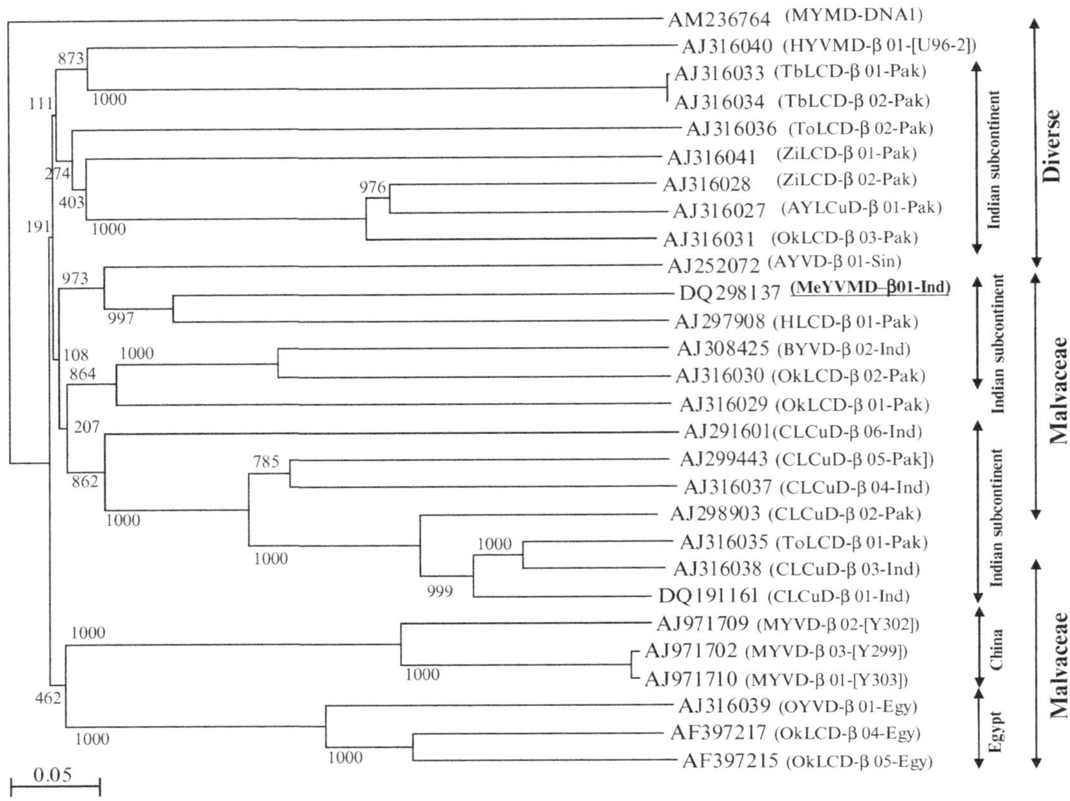

Fig. 13.5 Phylogenetic analysis of the DNA-β molecule of present study with other DNA-β molecules of different begomoviruses. The tree, drawn by neighbour-joining bootstrap method with the help of Clustal W 1.81 software package, was rooted on the sequence of DNA-1 molecule (MYMD-DNA1) of *Malvastrum yellow mosaic virus*. Vertical distances are arbitrary, horizontal distances are proportional to genetic distances. Numbers at nodes indicate percentage bootstrap scores (1000 replicates)

cluster (cluster 1) of second major group. This most likely represented the host adaptation by the associated helper begomovirus and coadaptation of its DNA-β rather than host adaptation of the satellite (Briddon et al. 2003). Here also the phylogenetic comparison of DNA-β satellites followed two paths leading to two major groups (Malvaceae and non-Malvaceae) as was observed by Briddon et al. (2003). The DNA-β molecule of MeYVMD is grouped with DNA-β associated with Old World begomoviruses and is found phylogenetically separated from the New World DNA-β molecule of *Honeysuckle yellow vein mosaic virus* (AJ316040) (Chatterjee and Ghosh 2007b).

Molecular Epidemiology

Sequence analysis of different begomoviruses and betasatellites, obtained from the symptomatic MeYVMD samples collected from different locations, along with the sequences of MeYVMV and MeYVMBV (*Mesta yellow vein mosaic Bahraich virus*) and their associated betasatellite sequences, revealed that different sequences of the associated begomovirus(es) and betasatellites were also present in infected samples (Table 13.3). All the begomovirus isolates obtained from eastern and southern regions of India showed highest average nucleotide sequence identity (96.0 %) with MeYVMV (EF373060) and appeared to be the isolates of this virus. In contrast, isolates collected from northern India

Table 13.3 Begomovirus and different betasatellite isolates associated with MeYVMD in India

Begomovirus	Geographical location	GenBank accessions
Mesta yellow vein mosaic virus (MeYVMV)		
MeYVMV-[Bar]	Barrackpore	EF373060
MeYVMV-[Bar2]	Barrackpore	EF428256
MeYVMV-[Bar3]	Barrackpore	EF432372
MeYVMV-[Bon]	Bongaon	FJ159262
MeYVMV-[Har]	Haringhata	FJ159263
MeYVMV-[Rai]	Raigunj	FJ159264
MeYVMV-[Coo]	Cooch Behar	FJ159265
MeYVMV-[Bal]	Balurghat	FJ159266
MeYVMV-[Ama]	Amadalavalasa	FJ159269
MeYVMV-[Viz]	Vizianagaram	FJ159270
MeYVMV-[Sri]	Srikakulam	FJ159271
Mesta yellow vein mosaic Bahraich virus (MeYVMBV)		
MeYVMBV-[Bah]	Bahraich	EU360303
MeYVMBV-[Kai]	Kaisarganj	FJ159267
MeYVMBV-[Bha]	Bhanga	FJ159268
Cotton leaf curl Multan betasatellite (CLCuMB)		
CLCuMB-[Har]	Haringhata	EF614159
CLCuMB-[Bon]	Bongaon	EF614158
CLCuMB-[Bas]	Basirhat	DQ298137
CLCuMB-[Rai]	Raigunj	FJ159274
CLCuMB-[Bal]	Balurghat	FJ159275
Ludwigia leaf distortion betasatellite (LuLDB)		
LuLDB-[Bah]	Bahraich	EF614160
LuLDB-[Kai]	Kaisarganj	EF614162
LuLDB-[Bha]	Bhanga	EF614161
LuLDB-[Ama]	Amadalavalasa	EU557374
LuLDB-[Viz]	Vizianagaram	FJ159272
LuLDB-[Sri]	Srikakulam	FJ159273

showed highest average nucleotide sequence identity (99.5 %) with MeYVMBV. These begomovirus isolates obtained from different locations of eastern, southern and northern India shared an average sequence identity of 98.8 %, 99.5 % and 99.6 % amongst their corresponding geographical groups, respectively (Roy et al. 2009). The isolates obtained from eastern and southern India shared an average sequence identity of 92.6 % between them, while with their north Indian counterparts they showed relatively lower sequence identity (83.8 % and 83.7 %, respectively).

All the betasatellite isolates obtained from eastern, northern and southern regions of India showed an average sequence identity of 94.1, 85.2 and 98.1 % amongst their respective geographical group. The betasatellite isolates obtained from northern and southern parts of India shared an average sequence identity of 85.2 % between them and collectively showed highest sequence identity (84.5 %) with different isolates of LuLDB (*Ludwigia leaf distortion betasatellite*) and are thus considered to be the isolates of LuLDB. Betasatellite isolates obtained from eastern India showed the highest average nucleotide sequence identity (85.4 %) with different isolates of CLCuMB and are considered to be the isolates of CLCuMB. East Indian betasatellite isolates shared an average sequence identity of 77.8 and 77.4 % with their counterparts obtained from northern and southern regions of India, respectively.

In the phylogenetic analysis all the begomovirus isolates formed two major clusters. The east and south Indian isolates formed the first major cluster with MeYVMV and north Indian isolates formed the second major cluster with MeYVMBV (Fig. 13.6). Within these clusters the isolates formed different subclusters according to their geographic locations. All the betasatellite isolates taken for this study also formed two major clusters. The first major cluster included south and north Indian betasatellite isolates, while the second cluster contained those from eastern India (Roy et al. 2009).

In eastern India the disease was associated with different variants of MeYVMV and CLCuMB, whereas in northern India it was associated with different variants of MeYVMBV and LuLDB. Interestingly, in southern India the begomovirus complex consisted of MeYVMV and LuLDB. Characterization of the begomoviruses and betasatellites associated with the MeYVMD from eastern, southern and northern India indicated the existence of two recently described begomovirus species, MeYVMV and MeYVMBV, and two isolates of CLCuMB and LuLDB in the symptomatic samples. Both MeYVMV and MeYVMBV probably evolved by natural recombination amongst the different

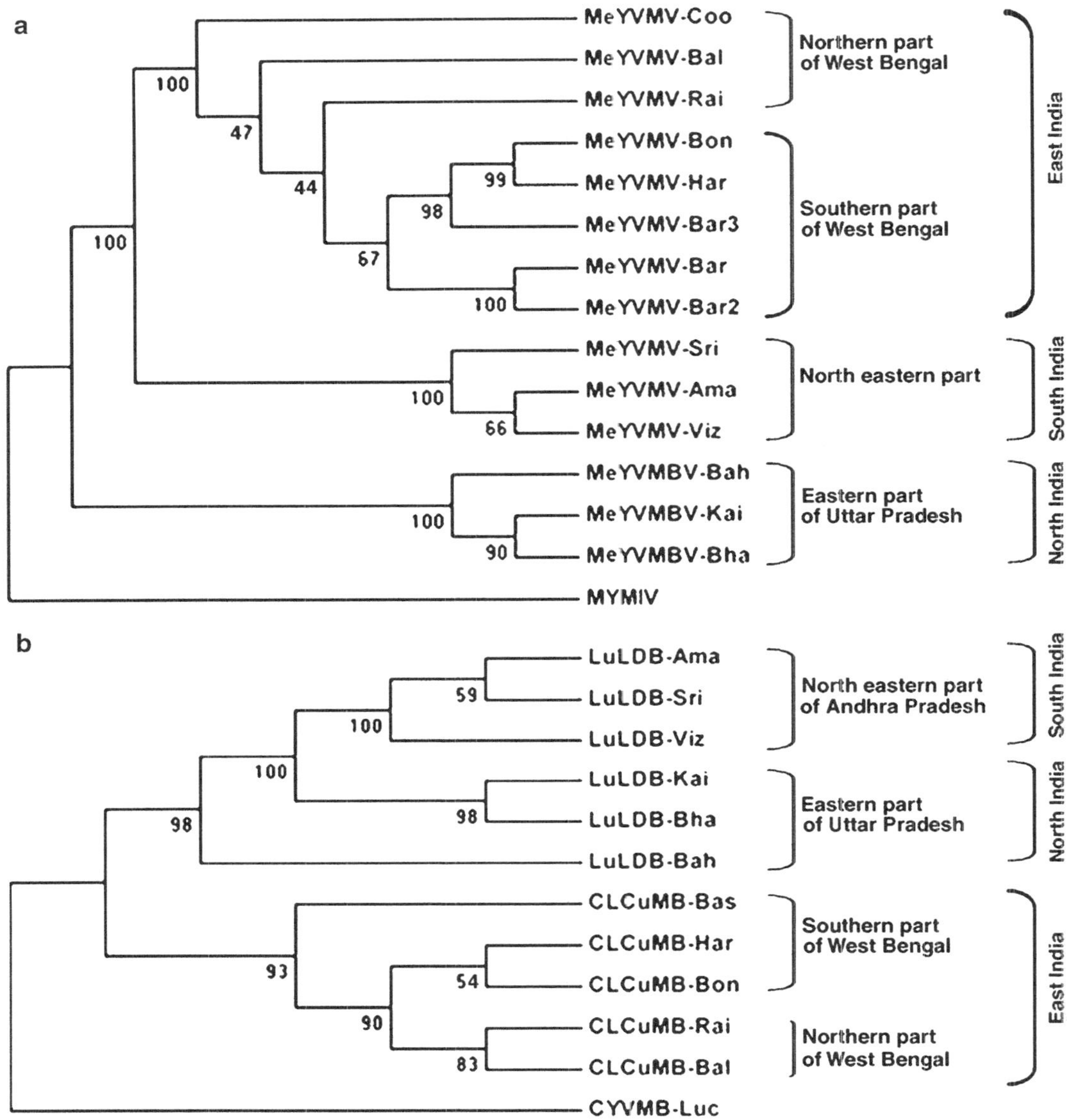

Fig. 13.6 Phylogenetic dendrograms based on alignments of the complete nucleotide sequences of (**a**) begomoviruses (rooted on the DNA-A sequence of *Mungbean yellow mosaic India virus*) and (**b**) betasatellites (rooted on the sequence of *Croton yellow vein betasatellite*) associated with MeYVMD from three distinct geographical regions in India. Dendrograms were constructed using the maximum parsimony method. Branches corresponding to partitions reproduced in less than 50 % bootstrap replicates are collapsed. The percentage of replicate trees in which the associated taxa clustered together in the bootstrap test (1000 replicates) is shown below the branches

begomovirus species occurring in the Indian subcontinent (Chatterjee and Ghosh 2007a,b; Das et al. 2008a; Roy et al. 2009).

The lesser affinity of these betasatellites with either of the two begomoviruses (MeYVMV or MeYVMBV) could also be explained by the fact that betasatellite amplification could not be obtained from many of the samples. It was observed that the samples which were collected from the fields with high-disease incidences gave maximum betasatellite amplification. Such observation gave rise to the possibility that only the begomovirus component is essential to cause the disease and the presence of betasatellite might contribute to the rapid spread of the disease. In general, the betasatellite contributed to pathogenicity, suppression of posttranscriptional gene silencing, up-regulation of viral DNA content in plants, binding of DNA and involvement in virus movement. Thus, it is involved in the spread of the helper begomovirus (Zhou et al. 2003).

13.8 Development of Diagnostics

13.8.1 Southern Blot Analysis

MeYVMD samples obtained from plants grown in glasshouse and different fields exhibited a positive hybridization signal in Southern blot analysis using specific probes prepared from clone of DNA-A (EF373060), CP gene (DQ298138) and DNA-β (DQ298137) and showed the association of DNA-A and betasatellite with the disease. The healthy sample failed to show hybridization signal. The viruliferous whiteflies showing positive hybridization signal using probe of DNA-A indicated the presence of the virus in them. The experimental *Vigna* hosts (*V. umbellata* and *V. unguiculata*) showing leaf crumpling symptom also showed the presence of DNA-A in them in Southern hybridization test (Chatterjee et al. 2007).

13.8.2 NASH Test

Infected mesta samples from glasshouse and different fields exhibited a positive hybridization signal in NASH test using specific probes of DNA-A (EF373060), CP gene (DQ298138) and DNA-β (DQ298137), indicating the association of DNA-A and betasatellite with the disease. The viruliferous whiteflies also showed positive hybridization signal with CP gene and DNA-β probes and thus indicated that whiteflies were confirmed vector of this disease. Other infected crop plants, ornamentals and weeds showed the presence or absence of begomoviruses having DNA-A and/or betasatellite in them by NASH test. In all the experiments DNA isolated from healthy leaf tissues did not gave hybridization signal (Chatterjee et al. 2007, 2008b).

13.9 Pathogen-Induced Biochemical Alterations In Situ

Enormous changes in biochemical components occur in mesta plants due to the infection with MeYVMV (Chatterjee and Ghosh 2008a). The gradual reduction in green pigments like chlorophyll (a, b and total) at different stages of pathogenesis in both species was determined at every 30-day interval. The disease development in mesta has also altered the ratio between chlorophyll a and b and thus hampered photosynthetic efficiency. Lower amount of phenolics (total phenols, ortho-dihydric phenols and bound phenols) in diseased plants was also noticed as compared with their respective control. This reduction of phenolics indicated a possible factor for disease development in these two species (Chatterjee and Ghosh 2008a).

The protein content was low in diseased plants over control, whereas in diseased leaves the free amino acid content was greater than the control ones. Higher amount of proline in diseased material was also observed. This indicated that the disease might have caused denaturation or breakdown of proteins as well as polypeptide chains and bound amino acids, and as a result free amino acid content in the host tissues might have been enhanced. When plants are exposed to microbial pathogens, they produce reactive oxygen species (ROS) that induce programmed cell death in the plant cells surrounding the infection site to effectively wall off the pathogen and terminate the disease process (Apel and Hirt 2004). The amino acid proline is a potent scavenger of ROS and this property of proline prevents the induction of programmed cell death by ROS (Chatterjee and Ghosh 2008a).

The SDS-PAGE protein profile of total soluble proteins from diseased leaves of both *H. cannabinus* and *H. sabdariffa* indicated differences in band patterns when compared with their respective healthy plants. Analysis of disease-related proteins revealed that those protein contents were greater in diseased *H. cannabinus* and *H. sabdariffa* as compared with those of respective controls. TLC separation and UV-spectrum analysis revealed the presence of higher amount of polyacetylenes in healthy plants than in the diseased ones, whereas concentration of isoflavones was found decreased in healthy ones as compared with diseased ones (Chatterjee and Ghosh 2008a).

Analyses of isozyme patterns and assays indicated alteration in activities of different enzymes due to the infection. Lower activity of catalase (CAT), acid phosphatase (ACP) and peroxidase (POD) enzymes was noticed in diseased plants as compared with healthy ones; in contrast, marked increase in levels of esterase (EST), polyphenol oxidase (PPO) and superoxide dismutase (SOD) activity was found in diseased plants. Altered zymogram pattern of isocatalases suggested inactivation of existing isocatalases, activation of inactive form and/or synthesis of new isocatalases. The higher EST and SOD activity in diseased leaves indicated a probable mechanism of overcoming the stress situation developed due to virus infection. The lower activity of POD enzyme, a key enzyme for lignin biosynthetic pathway, in diseased plants probably focused on the lowering down of metabolic pathway for lignocellulosic bast fibre formation, indicating a possible clue for reduction in fibre yield due to virus infection (Chatterjee and Ghosh 2008a).

13.10 Disease Management

The effect of insecticidal treatments on whitefly population and incidence of MeYVMD at farmers' field were observed using different biopesticides and chemical insecticides (Chatterjee and Ghosh 2007c). Amongst chemical insecticides tested, imidacloprid and thiamethoxam were found most effective against whitefly attack in plants at every stage of growth. Other insecticides like acetamiprid, carbosulfan and monocrotophos were also found effective to a lesser degree, but endosulfan showed lowest effect against whitefly vector. Amongst different biopesticides, Neem seed kernel extract and RNMV-energized mesta plants showed lowest disease incidence followed by Neem oil, *Prosopis* leaf extract and *Bougainvillea* leaf extract. In general, the effects of chemical insecticides on disease were higher than natural biopesticides used. But Neem seed kernel extract and RNMV-energized mesta plants showed significant effect in managing disease at par with the most effective insecticides used with maintenance of high yield and fibre quality. The effectiveness of Neem derivatives in reducing the disease incidence can be explained in terms of their direct interference with vector behaviour rather than their interaction with the virus. The F_1 plants raised from previously RNMV-inoculated mesta plants, designated as RNMV-energized mesta plants, not only showed improved plant health through reduced disease incidence on them but also showed finest fibre production amongst all the treatments used. This observation confirmed the view expressed with some fibre crop like jute with RNMV inoculation (Ghosh 2002). The reason behind reduction in disease incidence might be due to either development of inbuilt resistance in host or development of some situation in situ by RNMV inoculation and that became not conducive for multiplication of the present virus in mesta host (Chatterjee and Ghosh 2007c).

The healthy mesta plants kept within insect-proof net showed no disease symptom and thus the agronomic parameters like plant height, fibre yield and fibre quality were found best as compared to plants grown without net. The highest whitefly population was found in control plants (plants without insecticides or biopesticides) at every stage of growth and these plants showed highest disease incidence. The control plants were lowest in height as well as fibre yield. Since the control plants were shorter and thinner than those of plants grown with different treatments, it was clearly apparent that the infection substantially reduced the plant height as well as yield. The application of Neem derivatives and leaf extracts of *Bougainvillea spectabilis* and *Prosopis chilensis* proved beneficial in managing MeYVMD to a great extent and thus showed an avenue for effective utilization of biopesticides to check the spread of the disease under natural condition (Chatterjee and Ghosh 2007c).

References

Apel K, Hirt H (2004) Reactive oxygen species: metabolism, oxidative stress and signal transduction. Ann Rev Plant Biol 55:373–399

Briddon RW, Bull SE, Mansoor S, Amin I, Markham PG (2002) Universal primers for the PCR-mediated amplification of DNA β. Mol Biotech 20:315–318

Briddon RW, Bull SE, Amin I, Idris AM, Mansoor S, Bedford ID, Dhawan P, Rishi N, Siwatch SS, Abdel-Salam AM, Brown JK, Zafar Y, Markham PG (2003) Diversity of DNA β, a satellite molecule associated with some monopartite begomoviruses. Virology 312:106–121

Brunt AA, Crabtree K, Dallwitz MJ, Gibbs AJ, Watson L, Zurcher EJ (eds) (1996) Plant viruses online: description and lists from the VIDE Database (eds) URL: http://www.biology.Anu.edu.au/Groups/MES/vide

Campbell CL, Madden LV (1990) Temporal analysis of epidemics: introduction to plant disease epidemiology. Wiley, New York, pp 192–195

Chatterjee A (2007) Characterization of yellow vein mosaic disease of *Hibiscus cannabinus* and *H. sabdariffa* and its etiological agent. PhD thesis

Chatterjee A, Ghosh SK (2007a) A new monopartite begomovirus isolated from *Hibiscus cannabinus* L. in India. Arch Virol 152:2113–2118

Chatterjee A, Ghosh SK (2007b) Association of a satellite DNA β molecule with mesta yellow vein mosaic disease. Virus Genes 35:835–844

Chatterjee A, Ghosh SK (2007c) Seasonal variation of whitefly population and its management to prevent the spread of yellow vein mosaic disease in mesta. J Mycopath Res 45(2):189–194

Chatterjee A, Ghosh SK (2008a) Alterations in biochemical components in mesta plants infected with yellow vein mosaic disease. Braz J Plant Physiol 20(4):267–275

Chatterjee A, Ghosh SK (2008b) Histological changes in mesta plants infected with yellow vein mosaic disease. J Mycology Plant Pathol 38:132–134

Chatterjee A, Roy A, Padmalatha KV, Malathi VG, Ghosh SK (2005) Occurrence of a begomovirus with yellow vein mosaic disease of mesta (*Hibiscus cannabinus* and *H. sabdariffa*). Aus Plant Pathol 34(4):609–610

Chatterjee A, Roy A, Ghosh SK (2006) Yellow vein mosaic disease of kenaf. In: Rao GP, Paul Khurana SM, Lenardon SL (eds) Characterization, diagnosis & management of plant viruses, vol 1, Industrial crops. VI 22. Studium Press, Texas, pp 497–505

Chatterjee A, Sinha SK, Roy A, Sengupta DN, Ghosh SK (2007) Development of diagnostics for DNA A and DNA β of a begomovirus associated with mesta yellow vein mosaic disease and detection of geminiviruses in mesta (*Hibiscus cannabinus* L. and *H. sabdariffa* L.) and some other plants. J Phytopathol 155:683–689

Chatterjee A, Roy A, Ghosh SK (2008a) Acquisition, transmission and host range of a begomovirus associated with yellow vein mosaic disease of mesta (*Hibiscus cannabinus* and *H. sabdariffa*). Aus Plant Pathol 37:511–519

Chatterjee A, Sinha SK, Ghosh SK (2008b) Development and utilization of diagnostics for detection of geminiviral diseases in Eastern India. J Bot Soc Bengal 62(1):4–7

Chatterjee A, Roy A, Ghosh SK (2009) CAPS analysis – A possible tool to detect and group geminiviruses infecting some fibre crops and weeds. Plant Health Progress. doi:10.1094/PHP-2009-0202-01-RS

Das S, Ghosh R, Paul S, Roy A, Ghosh SK (2008a) Complete nucleotide sequence of a monopartite begomovirus associated with yellow vein mosaic disease of mesta from north India. Arch Virol 153(9): 1791–1796

Das S, Roy A, Ghosh R, Paul S, Acharyya S, Ghosh SK (2008b) Sequence variability and phylogenetic relationship of betasatellite isolates associated with yellow vein mosaic disease of mesta in India. Virus Genes 37(3):414–424

Deng D, Mc.Grath PF, Robinson DJ, Harrison BD (1994) Detection and differentiation of whitefly-transmitted geminiviruses in plants and vector insects by the polymerase chain reaction with degenerate primers. Ann Appl Biol 125:327–336

Ghosh SK (2002) Rice necrosis mosaic virus technology – a new low cost production technology for jute fibre production. J Mycopath Res 40:149–151

Ghosh R, Paul S, Roy A, Mir JI, Ghosh SK, Srivastava RK, Yadav US (2007) Occurrence of begomovirus associated with yellow vein mosaic disease of kenaf (*Hibiscus cannabinus*) in northern India. Online. Plant Health Progress. doi:10.1094/PHP-2007-0508-01-RS

Ghosh R, Paul S, Ghosh SK, Roy A (2009) An improved method of DNA isolation suitable for PCR-based detection of begomoviruses from jute and other mucilaginous plants. J Virol Methods 159:34–39

Jose J, Usha R (2000) Extraction of geminiviral DNA from a highly mucilaginous plant (*Abelmoschus esculentus*). Plant Mol Biol Rep 18:349–355

Roy A, Acharyya S, Das S, Ghosh R, Paul S, Srivastava RK, Ghosh SK (2009) Distribution, epidemiology and molecular variability of the begomovirus complexes associated with yellow vein mosaic disease of mesta in India. Virus Res 141:237–246

Singh DP (1997) Cultivation of Mesta for fibre production. In: Mesta (*Hibiscus cannabinus & Hibiscus sabdariffa*). Central Research Institute for Jute and Allied Fibres, Barrackpore, pp 12–28.

Zhou X, Xie Y, Tao X, Zhang Z, Li Z, Fauquet CM (2003) Characterization of DNA β associated with begomoviruses in China and evidence for co-evolution with their cognate viral DNA-A. J Gen Virol 84:237–247

Biology, Epidemiology, Resistance Genes, and Management of Soybean Mosaic Virus in Soybean (*Glycine max* (L.) Merrill)

14

Rajkumar Ramteke, G.K. Gupta, and S.K. Sharma

14.1 Introduction

Soybean is occupying the number one position among oilseeds not only in the whole world but also in India. Currently, area, production, and yield in the world are 108.16 mha, 268.02 mt, and 2.48 t/ha, and in India, it is 10.8 mha, 11.5 mt, and 1.06 t/ha, respectively (USDA 2013). Many pathogens inflict a huge yield loss in soybean, and the contribution of soybean mosaic virus (SMV) to this loss is substantial. According to the latest report (Wrather et al. 2010), a loss of 59.9 million metric tons was estimated in the 2006 harvest of soybeans in the top eight soybean-producing countries. Virus diseases caused a loss of 2.12 metric tons, out of which soybean mosaic virus was one of the major contributors. Soybean mosaic virus (SMV) is distributed worldwide and is one of the most important diseases of soybeans in many areas of the world (Sinclair and Shurtleff 1975) and has been reported wherever soybean is grown (Thottapilly and Rossel 1987; Giesler et al. 2002; Hill et al. 2007). SMV is recognized as a long-standing threat to the soybean industry. In India the disease is known to occur throughout the country (Gupta and Chauhan 2005).

It is a member of the genus *Potyvirus* in the family *Potyviridae* (Jayaram et al. 1992). This seed-borne and aphid-transmitted virus can induce a wide range of symptoms such as mosaic, mottling, chlorosis, and rugosity in leaves (Babu et al. 2008) as well as reductions in plant growth. It is one of the most widespread viruses and one of the 16 seed-transmitted viruses of soybean and devastating viral pathogens of soybean crops. SMV infection can result in severe yield loss, in some cases, up to 100 % (Arif and Hassan 2002; Liao et al. 2002; Gunduz et al. 2004). The primary inoculum source of SMV is thought to originate from an SMV-infected seed, though weeds may also serve as a reservoir of SMV. Further spread within and among soybean fields is through the activity of more than 30 different aphid species (Steinlage et al. 2002).

The use of genetic resistance to SMV has been considered the most effective means to control the disease (Wang et al. 2001). Extensive screening for soybean germplasm resistant to SMV resulted in identification of three independent resistance loci, i.e., *Rsv1*, *Rsv3*, and *Rsv4*, with different SMV strain specificities (Hayes et al. 2000; Gunduz et al. 2002; Liao et al. 2002; Zheng et al. 2005). The seven SMV strains, G1 through G7, are distinguished on the basis of differential

R. Ramteke (✉)
Scientist Genetics and Breeding, Crop Improvement Division, ICAR-DSR, Indore 452001, India
e-mail: ramtekeraj@gmail.com; ramtekeraj@rediffmail.com

G.K. Gupta
Plant Pathologist and Ex. Director, ICAR-DSR, Indore 452001, India

S.K. Sharma
ICAR, Indore 452001, India

L.P. Awasthi (ed.), *Recent Advances in the Diagnosis and Management of Plant Diseases*,
DOI 10.1007/978-81-322-2571-3_14

interactions with SMV-resistant cultivars (Cho and Goodman 1979). *Rsv4* is the only resistance gene that shows resistance to all the seven strains (Gunduz et al. 2004). The mechanism of *Rsv4* resistance seems to differ from that of *Rsv1* and *Rsv3* as it is not associated with extreme resistance or hypersensitive responses (Gunduz et al. 2004). Choi et al. (2005) reported the emergence of resistance-breaking isolates capable of overcoming resistance at all the three loci. Therefore, it is utmost important to search new durable genetic resistance either through screening existing germplasm or using biotechnology.

The virus can be identified by various techniques like symptomatology with differential hosts, tests of transmission through insects, and serological techniques. Immunoassay based on polyclonal antibodies such as ELISA (enzyme-linked immunosorbent assay) and monoclonal antibodies makes it possible to identify a specific strain.

14.2 Symptoms

In general, susceptible cultivars show transitory symptoms within 6–14 days after mechanical inoculation.

In the case of seed-borne infection, the seed either fails to germinate or produces diseased seedlings. Such seedlings have unifoliate chlorotic leaves which are curled longitudinally downward and are rugose, crinkled, and spindly. Subsequently, trifoliate leaves are more severely reduced in size, are mottled, and show rugosity. Plants infected early in the season have shortened petioles and internodes and are stunted. Leaves are reduced in size and generally puckered, occasionally having dark green enations on either side of the veins, and areas between enations also become chlorotic. Affected leaflets are generally asymmetric and curl downward at the margin. The youngest and most rapidly growing leaves show most severe symptoms. At maturity, the leaves of infected plants remain greener for a longer duration. The leaflets of infected plants are puckered, asymmetric, twisted, and curled downward at the margins.

Mottling of the seed coat sometimes referred to as hilum bleeding or hilum extension is a symptom of SMV on the seed (Kennedy and Cooper 1967; Koshimizu and Iizuka 1963). The severity and proportion of seed mottling vary according to the isolate and the soybean genotype. The disease also reduces the number, size, and weight of nodules on infected plants (Tu et al. 1970) and seed quality. Certain strains of SMV give rise to necrotic reactions of hypersensitivity in some cultivars and under particular environmental conditions.

14.3 Causal Virus

Like all potyviral genomes, the SMV genome is a single-stranded, positive-sense RNA molecule that is approximately 10 kb in length and contains a single open reading frame (Babu et al. 2008; Urcuqui et al. 2001; Chung et al. 2008). It encodes a large polyprotein that is co- and post-translationally cleaved into 11 final protein products (Babu et al. 2008; Urcuqui et al. 2001; Chung et al. 2008). An additional 25-kDa protein, termed P3-PIPO, has been discovered. This protein is derived from a frameshift on the P3 cistron (Chung et al. 2008). Virions contain 5.3 % nucleic acid, 94.7 % protein, and 0 % lipid.

The coat protein (CP) of the resistance-breaking (RB) isolates is diverse at the amino and carboxy termini and highly conserved in the core region. Resistance-breaking (RB) isolates reduce the yield of susceptible cultivars and cause mottling of the seed coat.

Soybean mosaic virus has a narrow host range and high rate of seed transmission in soybean, and it spreads in the field from infected seedlings by aphid vectors in a nonpersistent manner (CAB International 2003).

14.4 Epidemiology

The virus is normally transmitted in a nonpersistent manner by 16 species of insect vector including *Acyrthosiphon pisum*, *Aphis fabae*, and *Myzus persicae* (Edwardson and Christie 1986). The virus does not require a helper virus for vector transmission (Ross 1975). It is also transmitted by mechanical inoculation, by seed (not in

Glycine max cvs. Kawanggyo, Hill, or Bienville (Ross 1968)), by pollen to the seed, or by pollen to the pollinated plant.

As *soybean mosaic virus* (SMV) is an aphid- and seed-transmitted virus, it causes significant yield losses. Transmission of the virus by seeds was found to reach a value of 46 % (Laguna et al. 1999). However, in India, seed transmission to a rate of 7–52 % has been reported (Celia et al. 2004). Yield reduction as high as 86 % due to the virus has been reported (Kendrick and Gardner 1924; Ross 1968, 1977; Goodman and Oard 1980). Seed-borne infections are the primary sources of inoculum for SMV infections.

The strain specificity of SMV transmission through seed and SMV-induced seed-coat mottling is known, and isolate-by-soybean line interactions are reported to occur in both transmission rates and percentages of mottled seeds. SMVs that were transmitted poorly by the Asian soybean aphid, *Aphis glycines*, also were transmitted poorly through seed. No predicted amino acid sequences within the helper-component protease or coat protein coding regions differentiated the two groups of SMV strains. The loss of aphid and seed transmissibility by repeated mechanical transmission suggests that constant selection pressure is needed to maintain the regions of the SMV genome controlling the two phenotypes from genetic drift and loss of function (Domier et al. 2007).

14.5 Strainal Variations and Selective Strainal Resistance

Isolates of SMV differ in their pathogenicity on soybean cultivars (Ross 1969, 1975; Cho and Goodman 1979). A variety of symptoms ranging from mild mosaic to severe necrosis incited by various strains have been observed in various soybean cultivars (Cho and Goodman 1979; Ross 1969, 1975). It is probable that additional strains exist, particularly in the People's Republic of China and Japan. In USA, SMV was classified into seven strains (G1–G7) based on their virulence on eight soybean cultivars (Cho and Goodman 1979, 1982). Two additional distinct isolates, namely, G7A (Buzzell and Tu 1984) and C14 (Lim 1985), were later reported. In Brazil, a new isolate named SMV 95-1 was identified as a member of the SMV-G5 group (Almeida et al. 1995). This classification has been updated (Chen and Choi 2007). In China, SMV has now been grouped into 21 strains (SC1–SC21) (Wang et al. 2003; Guo et al. 2005; Li et al. 2010).

Kiihl and Hartwig (1979) recognized two types of resistance. The highest level of resistance gave complete protection against SMV-1 and a variant of the strain, SMV-1-B. A lesser level of resistance gave protection against SMV-1 in the homozygous condition, but all homozygous plants became necrotic after inoculation with SMV-1-B. They proposed the gene symbols Rsv Rsv for the highest level of resistance from PI 96983, rsv^t rsv^t for the lesser level of resistance from "Tokyo," and rsv rsv for the susceptible reaction. The three alleles formed an allelomorphic series with Rsv dominant to rsv^t and rsv^t dominant to rsv.

Rsv1 (Kiihl and Hartwig 1979) is the resistance gene most commonly found in commercially available cultivars (Chen et al. 1994). Of the reported resistance allele at this locus, none are effective against all SMV strains. Buzzell and Tu (1984) reported that the cultivar Raiden contained a resistance gene at a separate locus, Rsv2. Subsequent research (Wang et al. 1998; Buss et al. 1995) has shown that Raiden actually contains an Rsv1 allele. Another gene, Rsv3, from cultivar Columbia produces a necrotic reaction to SMV (Buzzell and Tu 1989). An SMV resistance locus was reported by Ma et al. (1995) that was considered as Rsv4 by Hayes et al. (2000) and confers resistance to all known strains of SMV. The Rsv4 allele reported by Ma et al. (1995) is derived from the line PI 486355, which was shown to contain two resistance loci, one which is allelic to Rsv1 and another (Rsv4) which is not allelic at either the Rsv1 or Rsv3 locus. The Rsv4 gene is completely dominant, in contrast to Rsv1 alleles, which show systemic necrosis in the heterozygous state (Chen et al. 1994). The Rsv4 locus from PI 486355 shows resistance without necrosis in both the heterozygous and homozygous states.

The first dominant gene identified in the soybean line PI 96983 was designated Rsv1. Single resistance genes in other cultivars, which confer differential reactions to strains G1 to G7, are alleles at the Rsv1 locus and have been designated as Rsv1y, Rsv1m, Rsv1t, Rsv1k, Rsv1s, and Rsv1n. Subsequently, Rsv3 and probably Rsv4 have been assigned to resistance loci independent of the Rsv1 locus. At 10 ° C, resistance conferred by Rsv1 is overcome and systemic spread of SMV strains occurs in lines that are resistant. In addition, apparent complementation among some pairs of virus strains, when used to inoculate the same plants, allows infection of lines containing the Rsv1 gene by strains to which they are normally resistant. Most sources of resistance are not effective against all SMV strains. Some virus strains interact with alleles of Rsv1 to develop a severe systemic necrosis that has the potential to be most damaging. The soybean line 8101 was identified as carrying all three SMV resistance genes: *Rsv1*, *Rsv3*, and *Rsv4* (Liao et al. 2011). Jindou 1 is a soybean cultivar from China, which is resistant to all seven soybean mosaic virus (SMV) strains identified in the USA (Shi et al. 2013). Wang et al. (2003, 2011) reported that the resistance to SC8 in Kefeng No. 1 was controlled by a dominant gene which was designated as Rsc8, which is located on chromosome 2 (MLG D1b). The details of comparison of differential reactions of soybean genotypes with SMV strains are shown in Table 14.1 and resistant cultivars in different countries in Table 14.2.

14.6 Molecular Marker Mapping Studies

A series of molecular markers were identified to be linked to Rsv1 by restriction fragment length polymorphism (RFLP), random amplification of polymorphic DNA (RAPD), simple sequence repeat (SSR), sequence characterized amplified region (SCAR), single nucleotide polymorphisms (SNPs), insertion/deletion (indel) polymorphisms, and allele-specific polymerase chain reaction (AS-PCR) (Gore et al. 2002; Hayes and Saghai Maroof 2000; Li et al. 1998, 2004; Jeong and Saghai Maroof 2004; Shi et al. 2008; Yu et al. 1994). The candidate gene, 3gG2, at Rsv1 locus was cloned from PI 96983, and its sequence was stored in GenBank with the accession AY518517 (Hayes et al. 2004). A PCR-based marker was also developed from the 3gG2 gene sequence and can be used to identify 3gG2 gene in soybean (Shi et al. 2008). Yu et al. (1994, 1996) mapped the Rsv1 locus on the soybean molecular linkage group F chromosome 13 to a cluster of resistance genes flanked by the RFLP markers K644H and B212V. Rsv3 has been reported as a gene from "Columbia" that conditions stem-tip necrosis in response to infection with SMV. L29 is an isoline of "Williams" that contains a gene from "Hardee" which provides resistance to SMV strain G7, but is susceptible to strain G1. The resistance gene in L29 is allelic to Rsv3 from Columbia and was not allelic to Rsv1 (Buss et al. 1999). F_2-derived lines from the L29 × Lee 68 cross were also used to map the Rsv3 gene using molecular markers. DNA of both parents was compared using 146 RFLP markers, and the parents differed at 36 of them. However, none exhibited linkage to Rsv3, based on a bulk segregant analysis. An alternative strategy using amplified fragment length polymorphism (AFLP) markers was successful in locating Rsv3 on linkage group "B." Since Rsv1 has been mapped to linkage group "F," this is additional evidence that the resistance gene in L29 is at a locus independent of Rsv1.

Rsv3 was mapped between the marker A519 and M3Satt located at a distance of 0.9 and 0.8 cM from the gene, respectively, on chromosome 14 (LGB2) (Jeong et al. 2002). Rsv3 in J05 was mapped on chromosome 14 with a distance of 1.5 cM from Sat_424 and 2.0 cM from Satt726 (Shi et al. 2008). Three SNP markers were developed from the sequences of A519 allele and confirmed by AS-PCR between Lee 68 and L29 (Jeong and Saghai Maroof 2004). Recently, based on genetic and physical maps in the vicinity of Rsv3, on the soybean chromosome 14 (MLG B2), five nucleotide-binding leucine-rich repeat (NB-LRR) genes, Glyma14 g38500, Glyma14 g38510, Glyma14 g38540, Glyma14 g38560, and Glyma14 g38590, were reported as the Rsv3 candidates (Suh et al. 2011).

Table 14.1 Comparison of differential reactions of soybean genotypes after inoculating with *soybean mosaic virus* strains

Genotypes	Gene/allele	Reaction to SMV strain[a]								MLG[b]	Reference
		G1	G2	G3	G4	G5	G6	G7	G7A		
Lee68/Essex	*rsv*	S	S	S	S	S	S	S	S	F,B2,D1b	Chen et al (1991)
York	*Rsv1-y*	R	R	R	N	S	S	S	S	F	Chen et al. (1991)
L88-8431/Raiden	*Rsv1-r*	R	R	R	R	N	N	R	N	F	Chen et al. (2001)
PI 96983	*Rsv1*	R	R	R	R	R	R	N	S	F (Chr.13)	Kiihl and Hartwig (1979); Chen et al. (1991)
Suweon 97	*Rsv1-h*	R	R	R	R	R	R	R		F	Chen et al. (2002)
Kwanggyo	*Rsv1-k*	R	R	R	R	N	N	N	N	F	Chen et al. (1991)
Ogden	*Rsv1-t*	R	R	N	R	R	R	N	S	F	Chen et al. (1991)
Marshall	*Rsv1-m*	R	N	N	R	R	N	N	S	F	Chen et al. (1991)
PI 507389	*Rsv1-n*	N	N	S	S	N	N	S	–	F	Ma et al. (2003)
Corsica (PI 559931)	*Rsv1-c*	S	R	S	–	R	S	R/N	–	F	Shakiba et al. (2013)
OX 686	*Rsv3*	N	N	N	N	R	R	R	–	B2(Chr.14)	Buzzell and Tu (1989)
OX 670	*Rsv3*	S	S	S	S	PR	PR	PR	–	B2	Gunduz et al. (2001)
L29	*Rsv3*	S	S	S	S	R	R	R	–	B2	Buss et al. (1999)
V94-5152	*Rsv4*	R	R	R	R	R	R	R	–	D1b(Chr.2)	Gunduz et al. (2001)
PI 88788	*Rsv4*	R	R	R	R	R	R	R	–		Gunduz et al. (2004)
Beeson (PI 548510)	*Rsv4-b*	ER	ER	S	–	R	ER	R	–	D1b	Shakiba et al. (2013)
J05	*Rsv1Rsv3*	R	R	R	R	R	R	R	–	F,B2	Zheng et al. (2006)
Jinpumkong 2	*Rsv1Rsv3*	R	R	R	R	R	R	R	–		Moon et al. (2009)
Hourei	*Rsv1Rsv3*	R	R	R	R	R	R	R	–		Gunduz et al. (2002)
Zao 18	*Rsv1Rsv3*	R	R	R	R	R	R	R	–		Liao et al. (2002)
PI 468355	*Rsv1Rsv4*	R	R	R	R	R	R	R	–	F, D1b	Ma et al. (1995)
Columbia	*Rsv3Rsv4*	R	R	N	R	R	R	R	–	B2, D1b	Ma et al (2002)
8101	*Rsv1Rsv3 Rsv4*	R	R	R	R	R	R	R	–	F,B2, D1b	Liao et al. (2011)

Modified after Shakiba et al. (2013)

[a]R = resistant (symptomless); ER = early resistance (delayed mosaic symptoms); N = necrosis; S = susceptible (mosaic); PR = partial resistant; – = data not available

[b]MLG = molecular linkage group

To map Rsv4, Hayes et al. (2000) used AFLP (Vos et al. 1995) and bulk segregant analysis (BSA) (Michelmore et al. 1991). AFLP is a PCR-based molecular marker that allows researchers to screen a large number of loci in a short period of time. By combining AFLP with BSA, one can quickly identify markers closely linked to a gene of interest. In an F_2 population segregating for a known major gene of interest, bulk segregant analysis is the most effective way to identify a closely linked marker. Rsv4 was mapped on chromosome 2(LG D1b) and flanked by two SSR markers, Satt542 at 4.7 cM and Satt558 at 7.8 cM genetic distance (Hayes et al. 2000). Three expressed sequence tag (EST) markers AI856415-g and AI856415-S and BF070293-S were mapped at 2.8 cM on one side of Rsv4, and additional two EST markers AW307114A (3.3 cM) and AW471852A (2.4 cM) were mapped on the other side of Rsv4 (Hwang et al. 2006). A fine map for Rsv4 has been constructed, and the gene was localized with 1.3-cM region with a physical

Table 14.2 List of soybean genotypes reported as resistant against *soybean mosaic virus*

Country	Cultivar	Worker(s)
USA	Marshall, "Ogden," Davis, York, Kwanggyo, Buffalo	Cho and Goodman (1979)
	PI 486355, Suweon 97 (PI 483084), PI 96983, Raiden (PI 360844)	Lim (1985)
	PI 406710	Chen et al. (1994)
	Columbia Holladay, Peking, Virginia, FFR-471, PI 507403, PI 556949	Yu et al. (1996)
Brazil	Campos Gerais, Bienville, Davis, PI 96983	Lima and Costa (1975)
Canada	OX615, L 78-379, Raiden, PI 96983	Buzzell and Tu (1984)
	OX686	Buzzell and Tu (1989)
Republic of Korea	"Pokwangkong," "Paldalkong"	Hong et al. (1987)
	"Eunhakong"	Shin et al. (1988)
	"Mallikong"	Hong et al. (1991)
	"Sinpaldalkong 2"	Kim et al. (1994)
	Sowonkong, Keunolkong, Sinpaldalkong, Jinpumkong 2	Moon et al. (2009)
China	AGS9, Kwanggyo and Dabeima, AGS9, Kwanggyo, Xuzhou 424 and Yanhuangyihao	Zhang et al. (1989)
	Jilin 21, Gongjiao8107-12, Gongjiao8045-5224-2	Sun et al. (1990)
	Tiefeng 18	Ma (1991)
	"Yanhuang 1," "Dongnong 47-1C"	Liao et al. (1993)
	Xudou 2 and Liao 81-5017	Liao et al. (1995)
	Zheng 772-9	Liang et al. (2000)
	Dongnong 81-43, Tie 6915	Zheng et al. (1998)
Japan	Nemashirazu, Harosoy, Tohoku 45, Tohoku 46, Tohoku 47, Tohoku 51, Tohoku 53,	Nagasawa and Iizuka (1977)
	"Suzuyutaka"	Hashimoto and Nagasawa (1986)
	Peking, PI 90763, PI 84751	Takahashi et al. (1987)
	"Yumeyutaka"	Kitamura et al. (1992)
	"Tousan 140," "Hourei"	Gunduz et al. (2002)
India	"Hardee," "Ankur," PK 327	Suteri (1986)
	PK 472, PK 262, PK 564, PK 327, PK 1042, PS 1225, JS 71-05, LSb 1, KHSb 2	Gupta and Chauhan (2005)
	"PS 1225"	Pushpendra et al. (2008)
Pakistan	"Crawford," "Chico," "Zane," "80-B-4007"	Akhtar et al. (1992)

interval of less than 100 kb on chromosome 2, and gene candidates were characterized (Saghai Maroof et al. 2010). In the case of the Rsc8 resistance gene, two genomic simple sequence repeat (SSR) markers BARCSOYSSR_02_0610 and BARCSOYSSR_02_0616 were identified that flank the two sides of RSC8. Sequence analysis of the soybean genome indicated that the interval between the two genomic SSR markers is 200 kb. Quantitative real-time polymerase chain reaction (QRT-PCR) analysis of the candidate genes determined that five genes (Glyma02 g13310, Glyma02 g13320, Glyma02 g13400, Glyma02 g13460, and Glyma02 g13470) are likely involved in soybean SMV resistance (Wang et al. 2011).

Marker-assisted selection (MAS) has been widely and successfully used in selecting disease-resistant plants by identifying genetic markers that select specific genes/alleles or combination of multiple resistance genes/alleles in plants

(Fjellstrom et al. 2004; Moose and Mumm 2008; Foolad and Panthee 2012). SSR and SNP are powerful tools in genome mapping, association studies, diversity analysis, and tagging of important genes in plant genomics (Giancola et al. 2006; Caicedo et al. 2007; Choi et al. 2007; Hyten et al. 2007; Saghai Maroof et al. 2010; Song et al. 2010; Suh et al. 2011). Molecular marker-assisted selection has proven to be a highly efficient strategy for selecting resistant lines (Mudge et al. 1997; Cregan et al. 1999; Meng et al. 2003). Recently, high-density simple sequence repeat (SSR) markers were developed based on the whole-genome sequence of soybeans (Song et al. 2010). Some of these markers that are tightly linked to R genes have been used to develop soybean lines resistant to SMV (Ma et al. 2010).

Transgenic soybean plants resistant to soybean mosaic virus (SMV) were obtained by transforming with the coat protein gene and the 3-UTR from SMV (Wang et al. 2001). Four insertion events were detected in a T0 plant obtained by using *Agrobacterium tumefaciens-mediated* transformation. Self-pollination of T0 progeny yielded four homozygous transgenic lines with a single insertion event or combinations of two insertion events in the T3 generation. A single coat protein gene transcript was detected in all four transgenic lines, and a virus coat protein was detected in three transgenic lines. Two transgenic lines were highly resistant to the virus. These constitute the first example of stable genetically engineered disease resistance in soybean. Pyramiding respective Rsv genes from different loci (Rsv1, Rsv3, and Rsv4) through marker-assisted selection (MAS) is an ideal method for creating durable and wide-spectrum resistance to all strains of SMV.

14.7 Management

Strategies for the management of viral diseases normally include control of vector population using insecticides, use of virus-free propagating material, appropriate cultural practices, and use of resistant cultivars.

Host-plant resistance is the most effective method of controlling this disease. Three independent loci, Rsv1, Rsv3, and Rsv4, conferring SMV resistance have been identified and mapped on different chromosomes (linkage groups [LG]) of soybean (Hayes et al. 2000; Jeong et al. 2002; Yu et al. 1994). Nine alleles have been reported at the Rsv1 locus: Rsv1 in PI 96983, Rsv1-h in "Suweon 97," Rsv1-k in "Kwanggyo," Rsv1-m in "Marshall," Rsv1-n in PI 507389 and V262, Rsv1-r in "Raiden," Rsv1-s in LR1, Rsv1-t in "Ogden," and Rsv1-y in "York" (Chen and Choi 2008).

The disease can be effectively managed through the deployment of single dominant resistance genes known as Rsv genes that confer resistance to different strains of SMV. Pyramiding respective Rsv genes from different loci (Rsv1, Rsv3, and Rsv4) through marker-assisted selection (MAS) is an ideal method for creating durable and wide spectrum. Results indicate that an Essex background or modifier genes from the donor source had effects on reactions of Rsv3 and Rsv4 genes, causing the isogenic lines to be more susceptible than the Rsv donor parents. Two-gene and three-gene isolines of Rsv1Rsv3, Rsv1RSv4 and Rsv1Rsv3Rsv4, acted in a complementary manner, conferring resistance against all strains of SMV, whereas isolines of Rsv3Rsv4 displayed a late susceptible reaction to selected SMV strains. Saghai Maroof et al. (2008) demonstrated with MAS and three near-isogenic lines, each containing a different SMV resistance gene, that pyramided lines can be generated in a straightforward manner into two- or three-gene-containing lines with high levels of resistance to SMV.

In India, cultivation of moderately resistant varieties such as PK 472, PK 262, PK 564, PK 327, PK 1042, PS 1225, JS 71-05, LSb 1, KHSb 2, etc. is recommended. Cultivation of 2–3 or more varieties and change of varieties at least after every two years help minimizing disease incidence.

Sowing of virus-free seeds collected from disease-free fields and elimination of affected plants early in the season followed by control of vector population by a minimum of two sprays of thiamethoxam 25 WG at 100 g/ha or methyl demeton at 800 ml/ha at 30 and 45 days after sowing also minimizes the spread of disease. Clean cultivation without weeds is also effective in checking the spreading of disease.

14.8 Conclusion

The soybean mosaic virus (SMV) is known to reduce the yield wherever soybean is grown. Resistance to SMV is controlled by single dominant genes at three distinct loci, *Rsv1*, *Rsv3*, and *Rsv4*. The mechanisms of resistance at the *Rsv3* and *Rsv4* loci were investigated by tracking virus accumulation and movement over time using leaf immunoprints. The mechanisms of *Rsv3* resistance include extreme resistance, hypersensitive response, or restriction to virus replication and movement, which are strain specific. The *Rsv4* gene was found to function in a non-strain-specific and non-necrotic manner. The mechanisms of *Rsv4* resistance involve restricting both cell-to-cell and long-distance movement of SMV. Cultivars containing two genes for resistance, *Rsv1* and *Rsv3* or *Rsv1* and *Rsv4*, were resistant to multiple strains of SMV tested and show great potential for gene pyramiding efforts to ensure a wider and more durable resistance to SMV in soybeans.

Rsv4 is an important resistance gene because it confers broad resistance to SMV and also because its mode of action, though not entirely understood, appears to be distinct from the hypersensitive response type of disease resistance exhibited by Rsv1 alleles (Ma et al. 1995). Because of its unique resistance nature, there is interest in pyramiding this gene with other resistance loci such as Rsv1 and Rsv3 to incorporate multiple lines of defense against SMV infection. The ability to pyramid resistance genes into a single cultivar is greatly expedited by the use of closely linked molecular markers. Since genes such as Rsv4 and Rsv1 can mask one another's presence, selecting lines that contain both genes is not always possible by simple phenotypic methods. By using marker-assisted selection, gene pyramiding becomes a valuable alternative for introducing multiple disease resistance genes.

References

Akhtar A, Sher H, Ali A, Hassan S (1992) Biological characterization of soybean mosaic virus. Sarhad J Agric 8:555–561

Almeida MRA, Almeida LA, Kiihl RAS, Domit LA (1995) Identification of a new strain of the Soybean mosaic virus in Brazil. Agric Biol Technol 38:1095–1100

Arif M, Hassan S (2002) Evaluation of resistance in soybean germplasm to Soybean mosaic potyvirus under field conditions. Online J Biol Sci 2:601–604

Babu M, Gagarinova AG, Brandle JE, Wang A (2008) Association of the transcriptional response of soybean plants with soybean mosaic virus systemic infection. J Gen Virol 89:1069–1080

Buss GR, Ma G, Chen P (1995) The gene for resistance in soybean mosaic virus (SMV) in the Raiden cultivar is not independent of *Rsv*1. In: Agronomy abstracts. ASA, Madison, p 91

Buss GR, Ma G, Kristipati S, Chen P, Tolin SA (1999) A new allele at the Rsv3 locus for resistance to Soybean mosaic virus. In: Kauffman HE (ed) Proceedings of the World soybean research conference VI, Chicago, IL. 4–7 August 1999. Superior Printing, Champaign. p. 490

Buzzell RI, Tu JC (1984) Inheritance of soybean resistance to soybean mosaic virus. J Heredity 75(1):82

Buzzell RI, Tu JC (1989) Inheritance of a soybean stem-tip necrosis reaction to soybean mosaic virus. J Heredity 80(5):400–401

CAB International (2003) Crop protection compendium. CAB International, Wallingford

Caicedo AL, Williamson SH, Hernandez RD (2007) Genome-wide patterns of nucleotide polymorphism in domesticated rice. PLoS Enetics 3:1745–1756

Celia VC, Khetarpal RK, Mishra A, Jain A, Gupta GK (2004) Seed transmission and detection of Soybean Mosaic Potyvirus in Soybean. Ind J Mycol Plant Pathol 34:86–87

Chen P, Choi CW (2007) Characterization of genetic interaction between soybean and soybean mosaic virus. In: Rao GP (ed) Molecular diagnosis of plant viruses. Studium Press, Houston, pp 389–422

Chen P, Choi CW (2008) Characterization of genetic interaction between soybean and soybean mosaic virus. In: Rao GP (ed) Molecular diagnosis of plant viruses. Studium Press, Houston, pp 389–422

Chen P, Buss GR, Roane CW, Tolin SA (1991) Allelism among genes for resistance to Soybean mosaic virus in strain-differential soybean cultivars. Crop Sci 31:305–309. doi:10.2135/cropsci1991.0011183X003100020015x

Chen P, Buss GR, Roane CW, Tolin SA (1994) Inheritance in soybean of resistant and necrotic reactions to soybean mosaic virus strains. Crop Sci 34(2):414–422

Chen P, Ma G, Buss GR, Gunduz I, Roane CW, Tolin SA (2001) Inheritance and allelism tests of Raiden soybean for resistance to Soybean mosaic virus. J Hered 92:51–55. doi:10.1093/jhered/92.1.51

Chen P, Buss GR, Tolin SA, Gunduz I, Cicek M (2002) A valuable gene in Suweon 97 soybean for resistance to Soybean mosaic virus. Crop Sci 42:333–337. doi:10.2135/cropsci2002.0333

Cho EK, Goodman RM (1979) Strains of soybean mosaic virus: Classification based on virulence in resistant soybean cultivars. Phytopathology 69:467–470

Cho EK, Goodman RM (1982) Evaluation of resistance in soybeans to *Soybean mosaic virus* strains. Crop Sci 22:1133–1136

Choi BK, Koo JM, Ahn HJ, Yum HJ, Choi CW, Ryu KH, Chen P, Tolin SA (2005) Emergence of *Rsv*-resistance breaking *Soybean mosaic virus* isolates from Korean soybean cultivars. Virus Res 112:42–51

Choi IY, Hyten DL, Matukumalli LK, Song Q, Chaky JM, Quigley CV, Chase K, Lark KG, Reiter RS, Yoon MS, Hwang EY, Yi SI, Young ND, Shoemaker RC, van Tassell CP, Specht JE, Cregan PB (2007) A soybean transcript map: Gene distribution, haplotype and single nucleotide polymorphism analysis. Genetics 176:685–696

Chung BY-W, Miller WA, Atkins JF, Firth AE (2008) An overlapping essential gene in the *Potyviridae*. Proc Natl Acad Sci U S A 105:5897–5902

Cregan PB, Jarvik T, Bush AL, Shoemaker RC, Lark KG, Kahler AL, Kaya N, VanToai TT, Lohnes DG, Chung J, Specht JE (1999) An integrated genetic linkage map of the soybean. Crop Sci 39:1464–1490

Domier LL, Steinlage TA, Hobbs HA, Yang Y, Herrera-Rodriguez G, Haudenshield JS, McCoppin NK, Hartman GL (2007) Similarities in seed and aphid transmission among *Soybean mosaic virus* isolates. Plant Dis 91:546–550

Edwardson JR, Christie RG (1986) Viruses infecting forage legumes, Florida agricultural experiment station monograph series, no. 14. Agricultural Experiment Stations, Institute of Food and Agricultural Sciences, University of Florida, Gainesville, p 432

Fjellstrom R, Conaway-Bormans CA, McClung AM, Marchetti MA, Shank AR, Park WD (2004) Development of DNA markers suitable for marker assisted selection of three Pi genes conferring resistance to multiple Pyricularia grisea pathotypes. Crop Sci 44:1790–1798

Foolad MR, Panthee DR (2012) Marker-assisted selection in tomato breeding. Crit Rev Plant Sci 31:93–123

Giancola S, McKhann HI, Berard A, Camilleri C, Durand S, Libeau P, Roux F, Rebound X, Gut IG, Brunel D (2006) Utilization of three high-throughput SNP genotyping methods, the GOOD assay, Amplifluor and Taqman, in diploid and polyploidy plants. Theor Appl Genet 112:1115–1124

Giesler LJ, Ghabrial SA, Hunt TE, Hill JH (2002) Bean pod mottle virus: A threat to U.S. soybean production. Plant Dis 96:1280–1289

Goodman RM, Oard JH (1980) Seed transmission and yield losses in tropical soybeans infected by soybean mosaic virus. Plant Dis 64:913–914

Gore MA, Hayes AJ, Jeong SC, Yu YG, Buss GR, Saghai Maroof MA (2002) Mapping tightly linked genes controlling potyvirus infection at the Rsv1 and Rpv1 region in soybean. Genome 45:592–599

Gunduz I, Buss GR, Ma G, Chen P, Tolin SA (2001) Genetic analysis of resistance to Soybean mosaic virus in OX670 and Harosoy soybean. Crop Sci 41:1785–1791. doi:10.2135/cropsci2001.1785

Gunduz I, Buss GR, Chen P, Tolin SA (2002) Characterization of SMV resistance genes in Tousan 140 and Hourei Soybean. Crop Sci 42:90–95

Gunduz I, Buss GR, Chen P, Tolin SA (2004) Genetic and phenotypic analysis of *Soybean mosaic virus* resistance in PI 88788 soybean. Phytopathology 94:687–692

Guo D, Zhi H, Wang Y, Gai J, Zhou X, Yang C (2005) Identification and distribution of strains of *Soybean mosaic virus* in middle and northern of Huang Huai Region of China. Soybean Sci 27:64–68

Gupta GK, Chauhan GS (2005) Symptoms, identification and management of soybean diseases, Technical Bulletin 10. National Research Centre for Soybean, Indore

Hashimoto K, Nagasawa T (1986) Breeding for virus resistance in soybeans in Tohoku National Agricultural Experiment Station. Misc Publ Tohoku Nat Agric Exp Stat 5:1–12

Hayes AJ, Saghai Maroof MA (2000) Targeted resistance gene mapping in soybean using modified AFLPs. Theor Appl Genet 100:1279–1283

Hayes AJ, Ma G, Buss GR, Saghai Maroof MA (2000) Molecular marker mapping of Rsv4, a gene conferring resistance to all known strains of soybean mosaic virus. Crop Sci 40(5):1434–1437

Hayes AJ, Jeong SC, Gore MA, Yu YG, Buss GR, Tolin SA, Saghai Maroof MA (2004) Recombination within a nucleotide-binding-site/leucine-rich-repeat gene cluster produces new variants conditioning resistance to soybean mosaic virus in soybeans. Genetics 166:493–503

Hill JH, Koval NC, Gaska JM, Grau CR (2007) Identification of field tolerance to bean pod mottle and soybean mosaic viruses in soybean. Crop Sci 47:212–218

Hong EH, Kim SD, Hwang YH, Lee YH, Moon YH, Park EG, Park RK (1987) A semi-dwarf, high density-adaptable and high yielding soybean variety: Paldalkong. Res Rep Rural Dev Admin Crops Korea Republic 29(1):203–209

Hong EH, Kim SD, Lee YH, Hwang YH, Park EH, Moon YH, Kim HS, Seong YG, Kim WH, Kim YH, Park RK (1991) A good quality, high yielding and wide cultivation new soybean variety "Mallikong". Res Rep Rural Dev Admin Upland and Industrial Crops 33(3):1–4

Hwang T, Moon J, Yu S, Yang K, Mohankumar S, Yu Y, Lee Y, Kim H, Kim H, Maroof M (2006) Application of comparative genomics in developing molecular markers tightly linked to the virus resistance gene *Rsv4* in soybean. Genome 49:380–388

Hyten DL, Hartman GL, Nelson RL, Frederick RD, Concibido VC, Narvel JM, Cregan PB (2007) Map location of the Rpp1 locus that confers resistance to soybean rust in soybean. Crop Sci 47:837–838

Jayaram C, Hill JH, Miller WA (1992) Complete nucleotide sequences of two soybean mosaic virus strains differentiated by response of soybean containing the *Rsv* resistance gene. J Gen Virol 73:2067–2077

Jeong SC, Saghai Maroof MA (2004) Detection and genotyping of SNPs tightly linked to two disease resistance loci, *Rsv1* and *Rsv3*, of soybean. Plant Breed 123:305–310

Jeong SC, Kristipati S, Hayes AJ, Maughan PJ, Noffsinger SL, Gunduz I, Buss GR, Saghai Maroof MA (2002) Genetic and sequence analysis of markers tightly linked to the soybean mosaic virus resistance gene, Rsv3. Crop Sci 42:265–270

Kendrick JB, Gardner MW (1924) Soybean mosaic: Seed transmission and effect on yield. J Agr Res 27:91–98

Kennedy BW, Cooper RL (1967) Association of virus infection with mottling of soybean seed coats. Phytopathology 57:35–37

Kiihl RAJ, Hartwig EE (1979) Inheritance of reaction to soybean mosaic virus in soybeans. Crop Sci 19:372–375

Kim SD, Hong EH, Seung YG, Kim YH, Lee YH, Hwang YH, Kim HS, Lee SH, Kim WH, Ryu YH, Park RK (1994) New soybean variety resistant to disease and lodging, with adapted high yielding "Sinpaldalkong 2". RDA J Agric Sci, Upland and Industrial Crops 36(1):153–157

Kitamura K, Ishimoto M, Kikuchi A, Kaizuma N (1992) A new soybean variety "Yumeyutaka" lacking both L-2 and L-3 of the lipoxygenase isozymes. Jpn J Breed 42(4):905–913

Koshimizu Y, Iizuka N (1963) Studies on soybean virus diseases in Japan. Tohoku Nat Agr Expt Sta Bul 27:1–103

Laguna IG, Rodriguez PP, Truol GA, Herrera P, Plopper LD (1999) Virus diseases of soybean in Argentina. In: Proceedings World soybean research conference, Chicago, Illinois, pp 635

Li Z, Bernard RL, Nelson RL (1998) A RAPD marker with a heterozygous form linked to the Rsv1 locus in soybean. In: Agronomy abstracts, Annual meetings USA, Baltimore, Maryland, p 157

Li Z, Nelson RL, Bernard RL, Peng Y (2004) Identification and sequence analysis of a RAPD marker tightly linked to the Rsv1 locus in soybean. Soybean Genet Newslet 31:1–12

Li K, Yang QH, Zhi HJ, Gai JY, Yu DY (2010) Identification and distribution of soybean mosaic virus strains in southern China. Plant Dis 94:351–357

Liang HZ, Li WD, Lu WG, Xu JJ, Wang TF, Liu PX (2000) Analysis of the value for breeding of Zheng 77249, a unique promising germplasm selected for cultivation in the Huanghe and Huaihe Valleys. Chin J Oil Crop Sci 22(2):10–12

Liao L, Zhao RL, Zhang ZQ, Wang JL (1993) Changes induced by SMV infection in some physiological and biochemical characters of soyabean varieties with different resistance. Oil Crops China 1:26–29

Liao L, Liu YZ, Sun DM, Tian PZ (1995) Top necrosis symptom of the soybean plants infected by soybean mosaic virus (SMV). Soybean Genet Newslet 22:61–65

Liao L, Chen P, Buss GR, Yang Q, Tolin SA (2002) Inheritance and allelism of resistance to *Soybean mosaic virus* in Zao18 Soybean from China. J Hered 93(5):447–452

Liao L, Chen P, Rajcan I, Buss GR, Tolin SA (2011) Genetic analysis of "8101" soybean containing three genes for resistance to soybean mosaic virus. Crop Sci 51:503–511

Lim S (1985) Resistance to soybean mosaic virus in soybeans. Phytopathology 75:199–201

Lima Neto VC, Costa AS (1975) Determination of resistance to common mosaic virus in soybean. Summa Phytopathol 1(3):221–230

Ma SM (1991) Evaluation of resistance of Chinese soyabean cultivars to soybean mosaic potyvirus. Soybean Sci 10(3):240–244

Ma G, Chen P, Buss GR, Tolin SA (1995) Genetic characteristic of two genes for resistance to soybean mosaic virus in PI 486355 soybean. Theor Appl Genet 91:907–914

Ma G, Chen P, Buss GR, Tolin SA (2002) Complementary action of two independent dominant genes in Columbia soybean for resistance to Soybean mosaic virus. J Hered 93:179–184. doi:10.1093/jhered/93.3.179

Ma G, Chen P, Buss GR, Tolin SA (2003) Genetic study of a lethal necrosis to Soybean mosaic virus in PI 507389 soybean. J Hered 94:205–211. doi:10.1093/jhered/esg059

Ma Y, Li HC, Wang DG, Liu N, Zhi HJ (2010) Molecular mapping and marker assisted selection of soybean mosaic virus resistance gene RSC12 in soybean. Legum Genomics Genet 8:41–46

Meng Z, Zaykin DV, Xu CF, Wagner M, Ehm MG (2003) Selection of genetic markers for association analyses, using linkage disequilibrium and haplotypes. Am J Hum Genet 73:115–130

Michelmore RW, Paran I, Kesseli RV (1991) Identification of markers linked to disease-resistance genes by bulked segregant analysis: A rapid method to detect markers in specific genomic regions by using segregating populations. Proc Natl Acad Sci U S A 88:9828–9832

Moon J-K, Jeong S-C, Van K, Maroof M, Saghai A, Lee S-H (2009) Marker-assisted identification of resistance genes to soybean mosaic virus in soybean lines. Euphytica 169:375–385. doi:10.1007/s10681-009-9970-z

Moose SP, Mumm RH (2008) Molecular Plant Breeding as the Foundation for 21st Century. Crop Improv Plant Physiol 147:969–977

Mudge J, Cregan PB, Kenworthy JP, Kenworthy WJ, Orf JH, Young ND (1997) Two micro satellite markers that flank the major soybean cyst nematode resistance locus. Crop Sci 37:1611–1615

Nagasawa T, Iizuka N (1977) On the breeding of soya beans for resistance to virus diseases. Bull Tohoku Nat Agric Exp Stat Tohoku Nogyo Shikenjo Kenkyu Hokoku 55:75–80

Pushpendra SK, Singh BV, Gupta MK (2008) Pant soybean 1225 – an improved variety of soybean with broad genetic base. Soybean Res 6:72–76

Ross JP (1968) Effect of single and double infections of soybean mosaic and bean pod mottle viruses on soybean yield and seed characteristics. Plant Dis Rptr 52:344–348

Ross JP (1969) Pathogenic variations among isolates of soybean mosaic virus. Phytopathology 59:829–832

Ross JP (1975) A newly recognized strain of soybean mosaic virus. Plant Dis Rptr 59(10):806–808

Ross JP (1977) Effect of aphid transmitted soybean mosaic virus on yields of closely related resistant and susceptible soybean lines. Crop Sci 17:869–872

Saghai Maroof MA, Jeong SC, Gunduz I, Tucker DM, Buss GR, Tolin SA (2008) Pyramiding of soybean mosaic virus resistance genes by marker-assisted selection. Crop Sci 48(2):517–526

Saghai Maroof MA, Tucker DM, Skoneczka JA, Bowman BC, Tripathy S, Tolin SA (2010) Fine mapping and candidate gene discovery of the soybean mosaic virus resistance gene, Rsv4. Plant Genome 3:14–22

Shakiba E, Chen P, Shi A, Li D, Dong D, Brye K (2013) Inheritance and allelic relationships of resistance genes for *Soybean mosaic virus* in 'Corsica' and 'Beeson' soybean. Crop Sci 53:1455–1463

Shi A, Chen P, Li DX, Zheng C, Hou A, Zhang B (2008) Genetic confirmation of 2 independent genes for resistance to *Soybean mosaic virus* in J05 soybean using SSR markers. J Heredity 99:598–603

Shi A, Chen P, Vierling R, Li D, Zheng C (2013) Identification of soybean mosaic virus resistance alleles in Jindou 1 soybean. Euphytica 192:181–187. doi:10.1007/s10681-012-0816-8

Shin DC, Park CK, Cho EG, Sung DK, Chang SD, Chung GS, Suh HS (1988) A new high yielding and disease resistant sprouting soybean variety "Eunhakong". Res Rep Rural Dev Admin Upland Industrial Crops 30(1):17–20

Sinclair JB, Shurtleff MC (eds) (1975) Compendium of soybean diseases. American Phytopathological Society, St. Paul

Song QJ, Jia GF, Zhu YL, Grant D, Nelson RT, Hwang EY, Hyten DL, Cregan PB (2010) Abundance of SSR motifs and development of candidate polymorphic SSR markers (BARCSOYSSR_ 1.0) in soybean. Crop Sci 50:1950–1960

Steinlage TA, Hill JH, Nutter FW Jr (2002) Temporal and spatial spread of *Soybean mosaic virus* (SMV) in soybeans transformed with the coat protein gene of SMV. Phytopathology 92:478–486

Suh SJ, Bowman BC, Jeong N, Yang K, Kastl C, Tolin SA, Saghai Maroof MA, Jeong SC (2011) The Rsv3 locus conferring resistance to Soybean Mosaic Virus is associated with a cluster of coiled-coil nucleotide-binding leucine-rich repeat genes. Plant Genome 4:55–64

Sun ZQ, Liu YZ, Sun DM, Liao L (1990) Inheritance of resistance to soybean mosaic virus strains 1, 2 and 3 in soybean (*Glycine max*). Oil Crops China 2:20–24

Suteri BD (1986) Resistance in soybean cultivars to two strains of soybean mosaic virus. Ind Phytopathol 39(4):578–580

Takahashi K, Nagasawa T, Tanaka T, Iizuka N (1987) Reaction of soyabean varieties to the strains of soyabean mosaic virus and soyabean stunt virus in Japan. Misc Publ Tohoku Nat Agric Exp Stat Japan 7:1–35

Thottapilly G, Rossel HW (1987) Viruses affecting soybean. In: Singh SR, Rachie KO, Dashiell KE (eds) Soybean for the tropics: research, production, and utilization. Wiley, Chichester, pp 53–68

Tu JC, Ford RE, Quinones SS (1970) Effects of soybean mosaic virus and/or bean pod mottle virus infection on soybean nodulation. Phytopathology 60:518–523

Urcuqui IS, Haenni AL, Bernardi F (2001) Potyvirus proteins: a wealth of functions. Virus Res 74:157–175

USDA (2013) United States Department of Agriculture foreign agricultural service. Available at http://www.fas.usda.gov/psdonline/psdreport.aspx?hidReportRetrievalName=BVSandhidReportRetrievalID=906andhidReportRetrievalTemplateID=1, Accessed on 18 July 2013

Vos P, Hogers R, Bleeker M, Reijans M, van de Lee T, Hornes M, Frijters A, Pot J, Peleman J, Kuiper M, Zabeau M (1995) AFLP: a new technique for DNA fingerprinting. Nucleic Acids Res 23:4407–4414

Wang Y, Nelson RL, Hu Y (1998) Genetic analysis of resistance to soybean mosaic virus in four soybean cultivars from China. Crop Sci 38:922–925

Wang X, Eggenberger AL, Jr N, Hill JH (2001) Pathogen-derived transgenic resistance to soybean mosaic virus in soybean. Mol Breed 8:119–127

Wang X, Gai J, Yu D (2003) Inheritance of resistance to the SMV strain group SC8 in Kefeng No. 1. Soybean Sci 22:190–192

Wang D, Ma Y, Yang Y, Liu N, Li C, Song Y, Zhi H (2011) Fine mapping and analyses of RSC8 resistance candidate genes to soybean mosaic virus in soybean. Theor Appl Genet 122:555–565. doi:10.1007/s00122-010-1469-4

Wrather A, Shannon G, Balardin R, Carregal L, Escobar R, Gupta GK, Ma Z, Morel W, Ploper D, Tenuta A (2010) Effect of diseases on soybean yield in the top eight producing countries in 2006. Online Plant Health Progress. doi:10.1094/PHP-2010-0125-01-RS

Yu YG, Saghai Maroof MA, Buss GR, Maughan PJ, Tolin SA (1994) RFLP and microsatellite mapping of a gene for soybean mosaic virus resistance. Phytopathology 84:60–64

Yu YG, Saghai Maroof MA, Buss GR (1996) Divergence and allelomorphic relationship of a soybean virus resistance gene based on tightly linked DNA microsatellite and RFLP markers. Theor Appl Genet 92(1):64–69

Zhang YD, Gai JY, Ma YH (1989) Inheritance of resistance to two local soybean mosaic virus strains SA and SC in soybeans. Acta Agrono Sinica 15(3):213–220

Zheng CM, Bing T, Gao FL, Wu ZP, Zheng CM, Bing T, Gao FL, Wu ZP (1998) Study on the changes of free amino acid and soluble protein in seed coat of soybeans with different resistance to seed coat mottling after infection with soybean mosaic virus. Acta Phytopathol Sinica 28(3):227–231

Zheng C, Chen P, Gergerich R (2005) Characterization of resistance to *Soybean mosaic virus* in diverse soybean germplasm. Crop Sci 45:2503–2509

Zheng C, Chen P, Gergerich R (2006) Genetic analysis of resistance to Soybean mosaic virus in J05 soybean. Crop Sci 97:429–437

Boerhaavia diffusa-Derived Antiviral Glycoprotein: A Novel, Eco-friendly Approach for the Management of Viral Diseases

15

L.P. Awasthi, S.P. Singh, and H.N. Verma

15.1 Introduction

Many higher plants are known to contain endogenous proteins that act as virus inhibitors (Hansen 1989; Chessin et al. 1995 and Verma et al. 1998). All these belong to a class of proteins called ribosome-inactivating proteins (RIPs). These proteins have been studied in *Phytolacca americana* (Irvin 1975), *Mirabilis jalapa* (Kubo et al. 1990), and *Trichosanthes kirilowii* (Yeung et al. 1988) and show antiviral activity when mixed with virus inoculum (Loebenstein 1972; Kumar et al. 1997). On the other hand, some virus inhibitors of plant origin have been reported to induce systemic resistance in non-treated parts of plants also and thereby preventing infection of viruses (Verma et al. 1979; Ponz and Broening 1986; Verma et al. 1995, 1996, 1998). One such glycoproteinaceous substance isolated from *B. diffusa* roots has prevented virus infection and multiplication in plants (Verma and Awasthi 1979: Verma et al. 1979). It has shown very high antiviral activity when mixed with viruses in vitro and provoked the plant system to produce new protein(s) in the treated plants which is the actual virus inhibitory agent (VIA) (Verma and Awasthi 1980). This glycoprotein induces antiviral state in the plants, through formation of a de novo synthesized protein, and perhaps is active in signaling the activation of defense mechanism in susceptible hosts.

High antiviral activity was found in the sap extracted from the leaves of host plants sprayed with *Boerhaavia diffusa* glycoprotein (Verma and Awasthi 1980; Awasthi and Kluge 1984). No such activity was detected in the sap from leaves of non-treated (control) plants. Obviously, sap from treated leaves contains some virus inhibitory agent (AVA)/protein which is absent in non-treated healthy plants. Induced antiviral agent (AVA) was active against serologically unrelated and morphologically different viruses, when mixed in vitro (Verma and Awasthi 1980). Isolation, biophysical, and chemical characteristics of induced antiviral agent(s) (AVA)/proteins from several host plants have been studied.

15.2 Characteristics of Viruses

Viruses are very small (submicroscopic) infectious particles (virions) composed of a protein coat and a nucleic acid core. They carry genetic information encoded in their nucleic acid, which typically specifies two or more proteins.

L.P. Awasthi (✉) • S.P. Singh
Department of Plant Pathology, N.D. University of Agriculture and Technology,
Kumarganj, Faizabad 224229, Uttar Pradesh, India
e-mail: lpawasthi14@gmail.com

H.N. Verma
Jaipur National University, Jaipur, India

L.P. Awasthi (ed.), *Recent Advances in the Diagnosis and Management of Plant Diseases*,
DOI 10.1007/978-81-322-2571-3_15

Translation of the genome (to produce proteins) or transcription and replication (to produce more nucleic acid) takes place within the host cell using host's biochemical "machinery." Viruses do not capture or store free energy and are not functionally active outside their host. A large number of phytopathogenic viruses infect a wide range of crops and cause great economic losses every year throughout the world.

15.3 Yield Losses Caused by Viruses

Disease is an alteration in one or more of the ordered sequential series of physiological processes culminating in a loss of coordination of energy utilization in a plant as a result of the continuous irritation from the presence or absence of some factors or agents. All types of living organisms including animals, plants, fungi, and bacteria are hosts for viruses, but most viruses infect only one type of host. Viruses cause many important plant diseases and are responsible for losses in crop yield and quality in all parts of the world. The worldwide losses caused by viral diseases are estimated at about US $ 220 billion per year.

15.4 How to Control

In recent years the most active areas of research for the control of viral diseases are: the breeding of resistant or immune cultivars by classic genetic procedures, the production of virus-free stocks of seed and vegetative propagules, the production of transgenic plants containing viral genes that confer resistance to the virus, the control strategies for viral diseases which consist of the removal of infected host, and the breeding for resistance and interruption of the disease cycle by measures such as vector control and screening of partially infected seed lots. Direct control measures including chemicals have been tried by many workers, but none of the chemicals could prevent or control the infection and spread of the viruses in fields. Not much success has been achieved for the control of viruses by managing their vectors through insecticides.

15.5 Why Not Insecticides

Insecticides may kill the insect vectors and prevent the spread of vector-borne viruses up to some extent. Many friendly insects which serve as pollinators/beneficial like honeybees, silk worm, etc. are also killed. Majority of vectors have developed resistance against insecticides. The agrochemicals are very costly; besides, they cause various human health hazards, including soil, water, and environmental pollution. After indiscriminate use, for longer time, most of the agrochemicals leave various types of residues in crop produce, soil, water, and environment.

15.6 What Is Alternative Approach?

Looking into all these facts, an eco-friendly approach has been tried by using extracts from different plant parts from a number of phanerogamic plants. These plants, viz., *Boerhaavia diffusa* (root extract), *Datura metel, Solanum melongena, Euphorbia hirta, Clerodendrum aculeatum, Mirabilis jalapa, Phytolacca americana, Azadirachta indica, Terminalia arjuna* (leaf extracts), and *Cuscuta reflexa* (filament extract), were found effective in preventing the infection and spread of many plant viruses. Out of these, the root extract from *Boerhaavia diffusa* plants was found very effective.

The glycoprotein isolated from *Boerhaavia diffusa* has shown broad spectrum and very high antiviral activity against isometric as well as anisometric viruses, both in hypersensitive and systemic hosts. This systemic resistance-inducing agent, when sprayed on to the host plants before virus infection, induced systemic resistance in whole of the plant (at treated as well as at remote site). The application of BD inhibitor onto basal or upper leaves of the hosts protected the plants against virus infection. These antiviral substances of plant origin interfering virus infection have been termed as VIAs.

15.7 *Boerhaavia diffusa*

The plant was named in honor of Herman Boerhaave, a famous Dutch physician of the eighteenth century (Chopra 1969). *Boerhaavia* is an herbaceous plant. Six species are found in India – *B. diffusa*, *B. chinensis*, *B. erecta*, *B. repens*, *B.rependa*, and *B. rubicunda* (Chopra 1969; CSIR 1988). The whole plant or its specific parts (leaves, stem, and roots) are known to have medicinal properties and have a long history of use by indigenous and tribal people in India. It has many ethnobotanical uses (the leaves are used as vegetable; the root juice is used to cure asthma, urinary disorders, leucorrhea, rheumatism, and encephalitis), and it is medicinally used in the traditional Ayurvedic system. Besides these, *B. diffusa* has potent antiviral efficacy against phytopathogenic viruses. Antiviral agent isolated from this plant was found to be a glycoprotein with a molecular weight of 16–20 kDa (Verma and Awasthi 1979).

15.8 Chemical Composition of *Boerhaavia diffusa*

The *Boerhaavia diffusa* plant contains a large number of such compounds as flavonoids, alkaloids, steroids, triterpenoids, lipids, lignins, carbohydrates, proteins, glycoproteins, and punarnavine (Agarwal and Dutt 1936; Basu et al. 1947; Surange and Pendse 1972; Ahmad and Hossain 1968). Hypoxanthine-9-L-arabinofuranosid, ursolic acid punarnavoside (liirodendrin), and a glycoprotein having a molecular weight of 16–20 kDa have been isolated and studied in detail for their biological activity (Mishra and Tiwari 1971; Jain and Khanna 1989; Verma et al. 1979).

15.9 Biological Activity

15.9.1 As Medicine in the Traditional System

B. diffusa plant has been widely used by indigenous tribes in the traditional system of medicine. The roots have been widely used for the treatment of dyspepsia, jaundice, enlargement of spleen, abdominal pain, abdominal tumors, and cancers (Kirtikar and Basu 1956).

15.9.2 As Medicine in the Ayurvedic System

The roots and leaves with flowers have been found to be highly potent in Ayurvedic medicine, and different parts of this plant were reported to have various medicinal properties (CSIR 1988).

15.9.3 Pharmacological and Clinical Properties

Pharmacological studies have demonstrated that *Punarnava* possesses punarnavoside, which exhibits a wide range of properties: diuretic, anti-inflammatory, antifibrinolytic, anticonvulsant, antibacterial, antistress agent, antihepatotoxic, antiasthmatic, antiscabies, and anti-urethritis.

15.10 Antiviral Activity of *Boerhaavia diffusa*

The roots of *B. diffusa* are a rich source of a basic protein, which has been used for inducing systemic resistance in many susceptible crops against commonly occurring viruses (Verma and Awasthi 1979, 1980; Verma et al. 1979; Awasthi et al. 1984, 1985). Maximum antiviral activity, in each case, was recorded with the aqueous extract of dried root powder when applied before virus inoculation. The active principle was purified and isolated by Verma et al. 1979. This protein or antiviral agent was active against tobacco mosaic virus in *Nicotiana glutinosa, Datura metel, Chenopodium amaranticolor*, and *Nicotiana tabacum* (Ky58 White Burley and NP31); sunn hemp rosette virus in *Cyamopsis tetragonoloba, Vigna unguiculata*, and *Crotalaria juncea*; and gomphrena mosaic virus in *Chenopodium amaranticolor, Vigna unguiculata*, and *Gomphrena globosa*, when applied a few hours (2–24 h) before virus inoculation by the respective inocula of viruses (Verma and Awasthi 1979; Awasthi et al. 1984).

Table 15.1 Biophysical characteristics of purified AVA

Treatment	Results
Dilution	Active up to 0.002 mg/ml
Heat sensitivity	Inactivated
Longevity	Active up to 6 months
Dialyzability	Non-dialyzable
Absorption on animal charcoal or celite	Adsorbed
Effect of trypsin	Inactivated
Effect of pronase	Inactivated
Effect of RNAs	Not affected
Effect of DNAs	Not affected
Nature	Proteinaceous

15.11 Nature of Antiviral Agent and Induction of Systemic Resistance

The antiviral agent was a basic glycoprotein (70–80 % protein and 8–13 % carbohydrate) with a molecular weight of 16–20 kDa as determined by gel filtration chromatography (Verma et al. 1979). Following application of systemic resistance-inducing protein, the susceptible healthy hosts produced a virus inhibitory agent (VIA). The VIA showed the characteristics of the protein and upon incubation with the virus, reduced infectivity of the viruses both in vitro and in vivo. The biophysical characteristics (Table 15.1) of induced VIA were found to be a basic protein. The glycoprotein occurring in *B. diffusa* roots functions as a signal molecule and is of great interest as it has a role in stimulating the defense systems of plants against viruses (Verma and Awasthi 1980 and Awasthi et al. 1987).

15.12 Mode of Action

It seems most probable that *B. diffusa* glycoprotein acts directly on virus (es) when incubated in vitro and decreases the virus infection by the aggregation of virus particles or coating or fracturing the virus particles. While in vivo glycoprotein prevents infection and multiplication of virus (es), as only a small number of particles were seen in electron micrographs of purified virus (es) samples taken from systemic hosts of different virus(es) separately. It is supposed therefore that glycoprotein acts indirectly through host by inducing the formation of some virus inhibitory substance in treated host plants, which is normally absent in non-treated control plants. Induction of systemic resistance was reversed by simultaneous application of transcription inhibitor actinomycin-D.

15.13 Preparations of BD Root Extract/Glycoprotein

Seeds and root cuttings of *Boerhaavia diffusa* were sown/planted on the roadside/barren land/orchards/unused waste land and on other areas not under cultivation of any crop. Roots of *Boerhaavia diffusa* were collected and allowed to dry under shade at room temperature. Dried roots were cut into small pieces, ground to powder in a Willey grinder, and stored at low temperature. The crude root extract was prepared by mixing water w/v (1 g/10 ml), the pulp was strained through two folds of cheese cloth, and the homogenate was clarified by centrifugation at 5000 g for 15 min. The glycoprotein was precipitated by adding an equal amount (v/v) of a saturated solution of ammonium sulfate to the supernatant. The precipitate obtained was dissolved in water; the solution was dialyzed against running water and used for experimental work.

15.14 Methods of Application

15.14.1 Seed Treatment

Seeds of crop varieties were soaked in *B. diffusa* root extract/partially purified glycoprotein for 2–6 h, depending on the crop, before sowing.

15.14.2 Spraying of Antiviral Agents

The first spray of *B. diffusa* root extracts (1:10)/partially purified glycoprotein was done after 7–15 days of germination of seeds, depending on the nature of crop. The second, third, fourth, and the subsequent sprays were done at weekly to fortnightly intervals, following the first spraying. In control plots, water was sprayed, instead of antiviral substance.

15.14.3 Management of Viral Diseases of Crops

Planned experiments, conducted at the University Research Farm and also on the farmers' fields on various locations for the management of papaya viruses in papaya; legume viruses in mung bean/urd bean, soybean, and cowpea; cucumber viruses in cucurbits (bottle gourd, water melon, and cucumber); tomato viruses in tomato; and potato viruses in potato, revealed that disease incidence, severity, and crop losses were significantly reduced/managed by the fortnightly sprayings of BD inhibitor.

15.15 Conclusion

The viral diseases are of immense importance considering the extensive damage and severe losses they cause to the crops. Because of their peculiar nature and characteristic association with hosts and vectors, no therapeutic method to completely control them has been found successful. However, certain preventive majors, if adopted suitably, can be of great help in avoiding viral diseases. The antiviral substances of plant origin may be used for the disease management system. Infection of several viral diseases could be prevented by the application of extracts/antiviral compounds from *Clerodendrum* spp., *B. diffusa*, *A. indica*, *Bougainvillea* spp., *Datura metel*, *Cuscuta reflexa*, etc.

Use of plant products in plant protection is a recent and an eco-friendly approach. Results obtained on the management of papaya viruses in papaya; legume viruses in mung bean/urd bean, soybean, and cowpea; cucumber viruses in cucurbits (bottle gourd, water melon, and cucumber); tomato viruses in tomato; and potato viruses in potato revealed that in each case of disease incidence, severity and crop losses were significantly reduced/managed by the weekly/fortnightly sprayings of BD root extract/partially purified glycoprotein. Antiviral agents from *B. diffusa*/different plants may provide acquired resistance against virus infection in plants.

The strategies need to be developed to prolong the effect of the inhibitors. BD root extract may be a source of possible prophylactic substance against the virus diseases of crops. Following treatment with antiviral substances/glycoprotein onto the leaves on which it is sprayed. the protective effect is observed systemically, resulting in the synthesis of some virus inhibitory/inactivating protein which is actually antiviral and defends the plants against virus infection. Because of the phytochemical nature, it has other advantages over chemicals. The SRIP is easily biodegradable, does not leave any residue, is non-phytotoxic, and is very cheap. *B. diffusa* glycoprotein creates an antiviral state in the host plants that persists for up to several days. If *B. diffusa* glycoprotein, applied at regular intervals, it could maintain a high nonspecific antiviral state for an extended period of time. There is no toxic effect on the hosts so that by proper time sequencing along with certain modifiers and spreaders antiviral state can be maintained throughout the life of many seasonal as well as annual host (crops) plants.

Unlike other plant pathogens, viruses are non-obligate parasites. Till date no antiviral chemical is available to control the viral diseases. Only prevention is possible as in case of many animal viruses, where vaccines are given, as fixed doses, as prophylactic preventive measure to protect the animals against the infection of viruses. A similar strategy for the immunization of host plants by various systemic resistance inducers can be practiced for preventing virus diseases in plants also. Here also the number of doses/treatments to protect the crop plants against the infection by the viruses under natural field conditions has been standardized. Management of viral diseases of crops using immunization concept and not the direct control concept may be workable for plants also. Treatments less than four to six, depending on the duration of crops, may not be much effective. Therefore, the number of treatments must be given as per requirement.

The antiviral effect of plant products cannot be compared with any chemical, because so far no standard antiviral drug is reported to prevent/control the infection of the viruses. It is a beginning, and a lot of work is needed to be done for the exploitation of wild/cultivated plants as source of antiviral agents for future, to develop eco-friendly approach to manage the viral diseases of crops.

References

Agarwal RR, Dutt SS (1936) Chemical examination of *Punarnava* or *Boerhaavia diffusa* Linn. II. Isolation of an alkaloid punarvine. Chem Abstr 30:3585

Ahmad K, Hossain A (1968) Isolation, synthesis and biological action of hypoxanthine-9-L-arabinofuranoside. J Agric Biol Sci 11:41

Awasthi LP, Kluge S (1984) Interferon like substances from plants regulating the resistance of hosts to virus diseases. In: VIth International congress of Virology. Sendai, Japan. September 7–9

Awasthi LP, Chowdhury B, Verma HN (1984) Prevention of plant virus disease by *Boerhaavia diffusa* inhibitor. Indian J Trop Plant Dis 2:41–44

Awasthi LP, Pathak SP, Gautam NC, Verma HN (1985) Control of virus diseases of vegetable crops by a glycoprotein isolated from *B. diffusa*. Indian J Plant Pathol 3:311–327

Awasthi LP, Kluge S, Verma HN (1987) Characteristics of an antiviral agent induced by *Boerhaavia diffusa* glycoprotein in host plants. Indian J Virol 3:156–169

Basu NK, Lal SB, Sharma SN (1947) Investigations on Indian medicinal plants. Q J Pharm Pharmacol 20:38–42

Chessin M, De Borde D, Zipf A (eds) (1995) Antiviral proteins in higher plants. CRC Press, Boca Raton

Chopra GL (1969) Angiosperms. Systematics and life cycle. S. Nagin & Co., Jalandhar, pp 361–365

CSIR (1988) The wealth of India: raw materials, vol VII B. CSIR, New Delhi, p 174

Hansen AJ (1989) Antiviral chemicals for plant disease control. Plant Sci 8:45–88

Irvin JD (1975) Purification and partial characterization of the antiviral protein from *Phytolacca americana* which inhibits eukaryotic protein synthesis. Arch Biochem Biophys 169(2):522

Jain GK, Khanna NM (1989) Punarnavoside: a new antifibrinolytic agent from *Boerhaavia diffusa* Linn. Indian J Chem 28(B):163–166

Kirtikar KR, Basu BD (1956) Indian medicinal plants, vol III, 2nd edn. Lalit Mohan Basu, Allahabad, pp 2045–2048

Kubo S, Ikeda T, Imaizumi S, Takanami Y, Mikami Y (1990) A potent plant virus inhibitor found in *Mirabilis Jalapa*. Ann Phytopathol Soc Jpn 56:481–487

Kumar D, Verma HN, Narendra T, Tewari KK (1997) Cloning and characterisation of a gene encoding an antiviral protein from *Clerodendrum aculeatum* L. Plant Mol Biol 33:745–751

Loebenstein G (1972) Localization and induced resistance in virus infected plants. Annu Rev Plant Physiol Plant Mol Biol 10:177–206

Mishra AN, Tiwari HP (1971) Constituents of the roots of *Boerhaavia diffusa*. Phytochemistry 10:3318

Ponz F, Broening G (1986) Mechanism of resistance to plant viruses. Annu Rev Plant Physiol Plant Mol Biol 24:355–381

Surange SR, Pendse GS (1972) Pharmacognostic study of roots of *Boerhaavia diffusa* Willd. (*punarnava*). J Res India Med 7:1

Verma HN, Awasthi LP (1979) Antiviral activity of *Boerhaavia diffusa* root extract and the physical properties of the virus inhibitor. Can J Bot 57:926–932

Verma HN, Awasthi LP (1980) Occurrence of a highly antiviral agent in plants treated with *Boerhaavia diffusa* inhibitor. Can J Bot 58:2141–2144

Verma HN, Varsha, Baranwal VK (1995) Agricultural role of endogenesis antiviral substances of plant origin. In: Chessin M, De Borde D, Zipf A (eds) Antiviral proteins in higher plants. CRC Press, Boca Raton, pp 23–37

Verma HN, Awasthi LP, Saxena KC (1979) Isolation of the virus inhibitor from the root extract of *Boerhaavia diffusa* inducing systemic resistance in plants. Can J Bot 57:1214–1217

Verma HN, Srivastava S, Varsha, Kumar D (1996) Induction of systemic resistance in plants against viruses by a basic protein from *Clerodendrum aculeatum* leaves. Phytopathology 86:485–492

Verma HN, Baranwal VK, Srivastava S (1998) Antiviral Substances of plant origin. In: Hadidi A, Khetrapal RK, Koganezawa H (eds) Plant virus disease control. APS Press (The American Phytopathological Society), St. Paul, pp 154–162

Yeung HW, Li WW, Feng Z, Barbieri L, Stripe F (1988) Trichosanthin α – momorcharin and ß – momorcharin: identity of abortifacient and ribosome-inactivating proteins. Int J Pept Protein Res 31:265–268

Nematode Diseases of Crops in India

16

Matiyar Rahaman Khan

16.1 Introduction to Nematodes

Nematode is a widely recognized group of a distinct phylum Nematoda or Nemata under superphylum Ecdysozoa (all moulting animal group). Early members of phylum Nematoda are thought to have evolved in marine habitats during the Cambrian period (600–550 million years ago). The fossil records of nematodes are extremely rare. A nematode fossil of *Palaeonema phyticum* was traced in association with *Aglaophyton major* (a free-sporing land plant) in the early Devonian period (nearly 416–396 million years ago). More and older fossils from water bear or marine tardigrades (animal closely related to nematode) were found, and its members arose in the mid-Cambrian. This gives an indirect evidence for the marine origin of the phylum Nematoda during the Cambrian period. Among other related pseudocoelomates (phyla: Rotifera, Gastrotricha, Priapulida, Kinorhyncha), the closest relative of Nematoda is Nematomorpha (horsehair worms). Nematodes can inhabit in marine, freshwater and terrestrial environments. They are the most abundant (approximately 0.5–10 million) multicellular animals on earth. Mostly their size is small (plant parasites and free livings, 0.25 mm to 12 mm; animal parasites 5 cm to 10 m) and they have a colourless body; thereby, they are inconspicuous organisms. As a pathogen, nematode is popularly known as the causal agent of a number of human, animal and plant diseases. Human beings are frequently infected with pinworm (*Enterobius vermicularis*, mostly affected children and cause of itching in anal area), giant worm (*Ascaris lumbricoides*, intestinal parasite which causes anaemia and intestinal blockage), hookworms (*Necator americanus* and *Ancylostoma duodenale*), filarial worm (*Wuchereria bancrofti* and *Onchocerca volvulus*, cause of elephantiasis and river blindness, respectively) and guinea worm (*Dracunculus medinensis* and *Brugia malayi*, cause of filariasis). Some animal parasitic nematodes also affect livestock such as sheep (*Haemonchus contortus*), dog (*Dioctophyma renale*) and chicken (*Ascaridia galli*).

Nematodes also parasitize on insects such as grasshopper (*Mermis nigrescens*) and mosquitoes (*Romanomermis culicivorax*), and some become entomopathogenic (such as *Steinernema* and *Heterorhabditis* in bacterial symbiotic association with *Xenorhabdus* and *Photorhabdus* spp., respectively) on many insect-pests of crops. Among the plant parasitic nematodes (PPNs), the most economically important ones are root-knot (*Meloidogyne*), cyst (*Heterodera* and *Globodera*), seed gall (*Anguina tritici*), reniform

M.R. Khan (✉)
Department of Agricultural Entomology, Bidhan Chandra Krishi Viswavidyalaya,
Kalyani, Nadia 741235, West Bengal, India
e-mail: drmrkhanbckv@gmail.com

L.P. Awasthi (ed.), *Recent Advances in the Diagnosis and Management of Plant Diseases*,
DOI 10.1007/978-81-322-2571-3_16

(*Rotylenchulus reniformis*), citrus (*Tylenchulus semipenetrans*), burrowing (*Radopholus similis*), stem and bulb (*Ditylenchus* spp.), foliar (*Aphelenchoides* spp.), lesion (*Pratylenchus* spp.) and pinewood nematodes (*Bursaphelenchus xylophilus*). Nematode infestation accounts for about 14 % global loss amounting to ~$100 billion dollars annually (Chitwood 2003). In fact, PPNs constitute a small fraction (~15 %) of the described nematodes (~30,000 species), the majority being non-parasitic and free-living organisms that play essential roles in terrestrial and sediment food webs. Free-living nematodes constitute about 40 % of the described species, which mostly feed on bacteria (bacterivores), fungi (fungivores), algae, protozoan and diatom (unicellular eukaryote feeders) and other nematodes (predator), while animal parasites (or pathogens) of vertebrate and invertebrate (carnivores) and plant parasites (herbivores) share 44 % and 15 %, respectively, of the described species.

16.2 General Characteristics of Nematodes

Nematode displays conserved and simple body plan (Fig. 16.1). Generally, the nematodes (*Greek* words *nema* (thread) and *oides* (form)) are pseudocoelomates (absence of mesodermal lining) and wormlike similar to unsegmented animals popularly known as roundworm (circular in cross section), eelworm (eel-like animal), nemas, threadworm (thread-like), etc. and generally lack external appendages. They are typically vermiform (with tapered ends), thread-like and fusiform (spindle shaped) and covered with a usually translucent, flexible, acellular cuticle secreted by an underlying cellular hypodermis. Most nematodes are sexually dimorphic (dioecious). Although they are generally oviparous, some are viviparous or ovoviviparous. The tail (post-anal portion) varies in shape from broadly rounded to filiform or intermediate forms, and it may differ between the sexes and even developmental stages. In transverse section, they are vermiform and circular. They are triploblastic (comprise of three germ layers: ectoderm, mesoderm and endoderm) and pseudocoelomic and unsegmented (segmentation does not go below the hypodermal layer) animals and have bilateral (in sagittal section, two halves are mirror image of each other) and symmetrical (being radially symmetric in the region of the head and pharynx) body. Their body consists of two tubes – the outer tube (body wall) composed of the cuticle, hypodermis and somatic musculature (longitudinal and spindle shaped) and the inner tube composed of the alimentary system separated by pseudocoelomic cavity. They do not possess circular body muscles and their movement is accomplished by contraction and relaxation of longitudinal muscles. Their body cuticle is highly flexible and the body shape is maintained through pressure from body fluid which is analogous to hydrostatic skeleton.

Nematodes have complete digestive system (alimentary system), which consists of the pharynx (so-called oesophagus) which has a lumen (triradiate in cross section) and the intestine which is a simple tube that is usually of single cell thickness on the peritoneal side and internally lined with microvilli. They lack the circulatory and respiratory systems. The nervous system mainly consists of circumcentric nerve rings. In female, the reproductive organs are in the middle to posterior of the nematode. Nematode species often have both male and female and often reproduce asexually by parthenogenesis. The reproductive organs are sometimes used as attributes for identification because the number of ovaries and the position of the vulva in the female nematode's body are easily seen under the light microscope. Male nematodes can be easily identified by the presence of spicules. Spicules are copulatory structures that are used during mating to guide the sperm into the vagina of the female nematode. They may have one or two testes which open into the cloaca. Many species have gubernaculums that guide the spicules during copulation. The females have one or two gonads which open separately through a vulva. Some nematodes are hermaphroditic (produce both sperm and ova at different times during ontogeny). The excretory (secretory–excretory) system can be glandular (in adenophorea) or tubuloglandular (secernentea) and opens through a short or

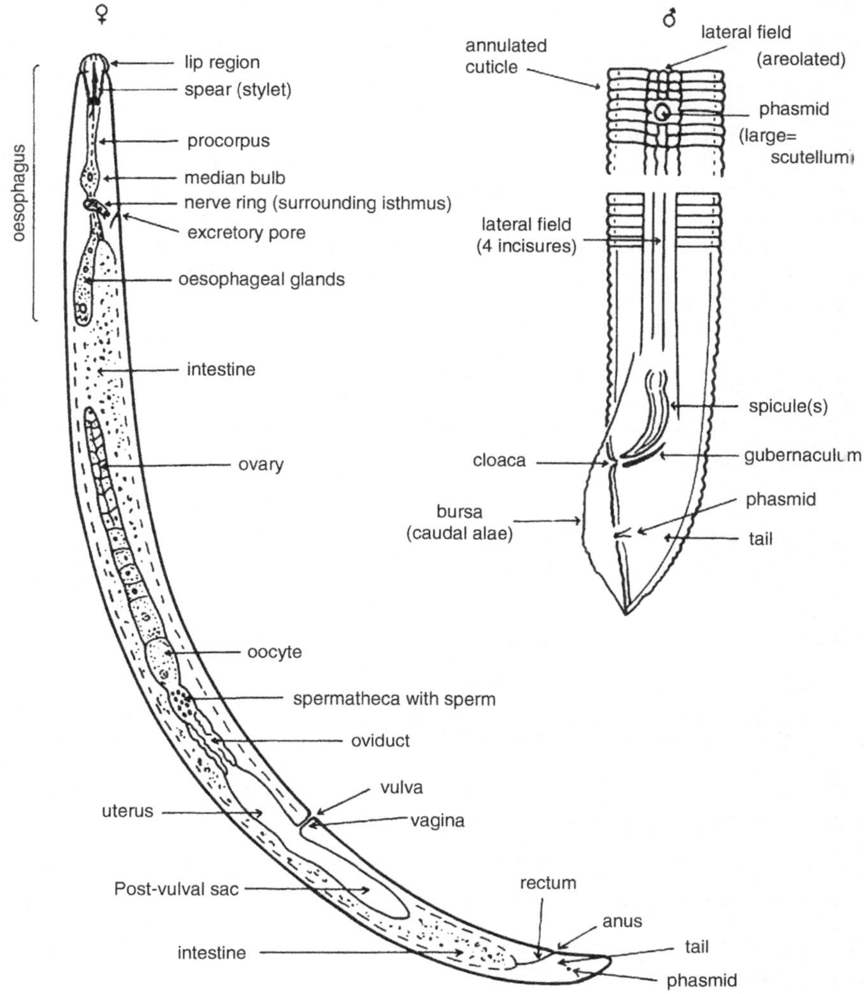

Fig. 16.1 Morphology of *Tylenchorhynchus* (From O'Brien and Stirling 1991)

long cuticularized terminal duct connected to the S–E pore. All the natural body openings (vulva, anus, S–E pore and cloaca) are ventral. The mouth opening is always terminal in the anterior end. Sensory organs like amphids, deirids, phasmids, cephalids, caudalids, hemizonid and hemizonion on the nematode body at different positions are either in pair or singly. In contrast to all other nematodes, some marine species have eye spots that enable them to detect light in their environment (e.g. *Draconema*). A synapomorphy of the Nematoda is the presence of noncontractile axon-like myoneural processes or extensions that run from the contractile or body portion of muscle cells to the neural junctions of the nerve cords. They have four juvenile stages (designated as J1, J2, J3 and J4) between the egg and the adult (in Dasgupta 1988, 1998).

The plant-feeding nematodes are distributed in four distinct orders, viz. Tylenchida, Aphelenchida, Dorylaimida and Triplonchida (Table 16.1). Their size generally ranges from 250 μm to 12 mm in length and about 15–35 μm in width (Fig. 16.2).

Table 16.1 Differentiating characteristics of four plant parasitic nematodes orders

Character	Tylenchida	Aphelenchida	Dorylaimida	Triplonchida
Stylet	Stomatostylet	Stomatostylet	Odontostylet	Onchiostylet
Pharynx	Three parts: corpus, isthmus and basal bulb	Three parts: corpus, isthmus and basal bulb	Two parts: distinct corpus and post corpus	Two parts: corpus gradually broadening into posterior part
Pharyngeal gland number	3 glands	3 glands	3–5 glands	5 glands
Position of gland opening	Dorsal gland opens near the spear base	Dorsal gland opens in metacorpus, anterior to valve	All glands open to body of gland	All glands open to body of gland
Metacorpus	Width <75 % of body width	Large, width almost (>75 %) of body width	Absent	Absent
Isthmus	Present	Absent in Aphelenchoididae	Absent	Absent
Pharyngeal gland	In a bulb or overlapping lobe	In a bulb (Paraphelenchidae) or overlapping lobe	In posterior bulboid expansion	In posterior bulboid expansion

Manzaniilla-Lopez and Hunt (2009)

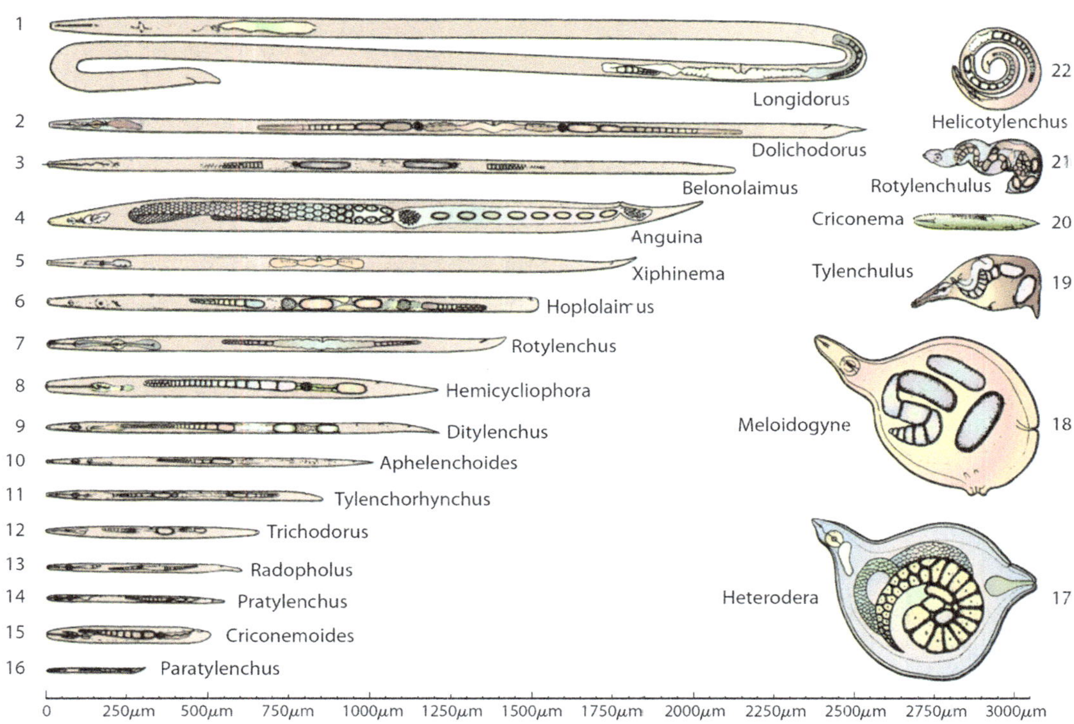

Fig. 16.2 Relative size of major genera of plant parasitic nematodes (Source: G.N, Agrios 1987)

An outline classification of plant parasitic nematodes is given below:

Phylum: Nematoda Potts, 1932
Class: Secernentea (Chromadorea Inglis, 1983)
Subclass: Diplogasteria
Order:

- **Tylenchida** Orley, 1880
- **Aphelenchida**

Class: Adenophorea (Enoplea Inglis, 1983)

Subclass: Enoplia Pearse, 1942
Order:
- **Dorylaimida** Pearse, 1942
- **Triplonchida** Cobb, 1920

Order: Tylenchida Orley, 1880
Suborder: Tylenchina Thorne, 1949
Superfamily: Tylenchoidea
Family: Tylenchidae (*Tylenchus*, *Filenchus*, *Coslenchus*)
- Tylenchorhynchidae (*Tylenchorhynchus*, *Merlinius*)
- Dolichodoridae (*Dolichodorus*)
- Belonolaimidae (*Belonolaimus*)
- Pratylenchidae (*Pratylenchus*, *Radopholus*, *Hirschmanniella*)
- Hoplolaimidae (*Hoplolaimus*, *Rotylenchulus*, *Helicotylenchus*, *Scutellonema*, *Aorolaimus*, *Peltamigratus*, *Rotylenchus*)
- Psilenchidae (*Psilenchus*)
- Tylodoridae (*Tylodorus*)

Superfamily: Atylenchoidea
Family: Atylenchidae (*Eutylenchus*, *Atylenchus*)
Superfamily: Heteroderoidea
Family: Heteroderidae (*Heterodera*, *Globodera*, *Cactodera*, *Punctodera*, *Meloidogyne*)
- Nacobbidae (*Nacobbus*)

Superfamily: Criconematoidea Taylor, 1936 (1914)
Family: Criconematidae (*Criconema*, *Criconemoides*, *Hemicycliophora*, *Hemicriconemoides*)
- Paratylenchidae (*Paratylenchus*)
- Tylenchulidae (*Tylenchulus*, *Trophotylenchulus*, *Sphaeronema*)
- Tylenchocriconematidae (*Tylenchocriconema*)

Superfamily: Neotylenchoidea Thorne, 1941
Family: Neotylenchidae (*Neotylenchus*)
- Paurodontidae (*Paurodontus*)
- Nothotylenchidae (*Nothotylenchus*)
- Ecphyadophoridae (*Ecphyadophora*)

Superfamily: Sphaerularioidea Lubbock, 1861
Family: Anguinidae (*Anguina*, *Ditylenchus*, *Subanguina*)
- Halenchidae (*Halenchus*)

Order : Aphelenchida Siddiqi, 1980
Suborder: Aphelenchina Geraert, 1966
Superfamily: Aphelenchoidea Fuchs, 1937 (Thorne, 1949)
Family: Aphelenchidae (*Aphelenchus*)
Superfamily: Aphelenchoidoidea Skarbilovich, 1947 (Siddiqi, 1980)
Family: Aphelenchoididae (*Aphelenchoides*, *Bursaphelenchus*)
Order: Dorylaimida Pearse, 1942
Suborder: Dorylaimina Pearse, 1942
Superfamily: Dorylaimoidea de Man, 1876 (Thorne, 1934)
Family: Longidoridae Thorne, 1935 (Meyl, 1961)
(*Longidorus*, *Paralongidorus*, *Xiphinema*)
Order: Triplonchida Cobb, 1920
Suborder: Diptherophorina Coomans & Loof, 1970
Superfamily: Diptherophoroidea Micoletzky, 1922
Family: Trichodoridae (*Trichodorus*, *Paratrichodorus*)

16.3 Types of Plant Parasitism in Nematode

Plant parasitic nematodes are broadly grouped into three categories, viz. ectoparasites, semi-endoparasites and endoparasites, based on their parasitic habits on plants.

Ectoparasitic nematodes are the most primitive group and feed externally by inserting their stylet (spear-like organ) into the cells of root hairs and root tips. They are migratory browsers or grazers (e.g. *Tylenchorhynchus*, *Belonolaimus*, *Trichodorus*, *Paratrichodorus*, etc.) that frequently move over the root surface, and many others are specialized sedentary (sessile) ectoparasites (*Hemicycliophora*, *Paratylenchus*, *Xiphinema*, *Longidorus*, etc.) and adapted to feed sedentarily on deeper cells by penetrating their long stylet for a longer period of time. The cells fed by them are sometimes modified in the feeding sites.

Nematodes that invade the host tissues are considered as endoparasites. Endoparasites are further grouped into two as migratory and sedentary. Migratory endoparasites (e.g. *Pratylenchus*, *Radopholus*, *Hirschmanniella*, *Bursaphelenchus*, etc.) invade roots, stems, leaves, rhizomes, bulbs, etc. and cause extensive damage during their migration through the cortex and other tissues leading to the formation of cavity and lesions on the necrotic tissues, while sedentary endoparasites (e.g. *Meloidogyne*, *Heterodera*, *Globodera*, etc.) are the most advanced parasites among all plant parasitic nematodes causing economic damage to the crops. Invasion of plant tissues by sedentary endoparasites and their oesophageal gland cell secretion through the stylet induce the most sophisticated cellular changes to establish their feeding sites.

Some relatively highly adaptive plant parasitic nematodes like *Rotylenchulus* (commonly known as reniform nematode) and *Tylenchulus* (citrus nematode) are considered as semi-endoparasites. They feed with their head and anterior end of the body embedded in host tissue and undergo cellular changes at their feeding sites.

16.4 Plant–Nematode Interactions

Plant parasitic nematodes are obligate parasite and they exclusively feed on the cytoplasm of living plant cells. Many nematodes feed briefly and ectoparasitically on root hairs or epidermal cells. Some others invade roots and feed as sedentary or migratory end parasites. Among the economically important plant parasitic nematode species, root-knot (*Meloidogyne*) and cyst nematodes (*Heterodera* and *Globodera*) are the primary model of host parasitic relationships for their economic significance. In fact, sedentary endoparasites have evolved sophisticated mechanism to establish feeding sites through transformation of host cells to ensure their survival. The feeding process involves the capability of nematode to regulate the plants' gene in their favour and elaborate transformation of multinucleate cells for enhanced metabolic activity. The group has complex but disruptive interactions with their hosts. There are both similarities and differences between the groups, most importantly in their interactions with their hosts. Juveniles of root-knot nematodes enter the root near the root tip and move intercellularly (consequently no direct damage) down to the cortex and migrate to vascular cylinder towards the root tip. Upon reaching the vascular cylinder, they establish feeding sites in the differentiation zone of the root by repeated nuclear division without cytokinesis of plant cell. This gives rise to the formation of multinucleate 'giant cell'. This giant cell is characterized by the enlarged (hypertrophied), dense cytoplasm and increased number of nuclei (up to 100) with high metabolic activity. The plant cells at the feeding site and adjoining cells also divide (hyperplasia) and swell (hypertrophy), leading to the formation of the characteristic outward manifestation of root gall or root-knot. The juveniles feed on the cytoplasm of giant cells through their protrusible stylet and moult thrice (J2-J3-J4-adult) and develop into pear-/melon-shaped females which lay eggs in a protective gelatinous matrix. Most common tropical root-knot species, viz. *M. incognita*, *M. javanica* and *M. arenaria*, reproduce through mitotic parthenogenesis, but sexual reproduction also occurs in some species. Further root-knot nematode species are polyphagous with wide hosts range, and some species have narrow hosts.

In contrast to root-knot nematodes, juveniles of cyst nematodes penetrate roots and reach the vascular cylinder by piercing and destroying the cell wall with their stylet on their way. They establish their feeding site at the vascular cylinder through injection of stylet secretions and induce multinucleate *syncytium*. The formation of syncytium involves breakdown of the cell walls and fusion of hundreds of cells at the initial feeding sites and surrounding cells. A syncytium consists of enlarged nuclei with large nucleolus, a dense cytoplasm and pronounced cytoplasmic streaming. Similar to root-knot nematodes, infective juveniles feed on nutrient-rich syncytial cells and undergo three consecutive moults inside the root to develop into egg-laying mature females. Their reproduction is exclusively sexual, and upon fertilization, the female body becomes full of eggs. After the death of the female, its body cuticle transforms into a hard, protective, persistent and tanned structure containing both eggs and hatched juveniles, termed as *cyst*. Majority of cyst nematodes are host specific (*Globodera* spp., *H. glycines*) and have narrow host range requiring host root diffusate for recognition of their hosts. Like potato cyst nematodes (*G. rostochiensis* and *G. pallida*), attack on potato and related Solanaceae crops and hatching of eggs and emergence of infective juveniles depend on the release of potato root diffusates.

The juveniles of both the nematode groups become sedentary at their feeding sites close to or within the vascular cylinder (Fig. 16.2). Induction of cellular changes around their head occurs through oesophageal gland secretion via stylet during feeding processes to enhance their nutritional requirement. The nuclei within syncytium and giant cell become polyploid with transfer cell-like structures. These changes are primarily to increase the metabolic capacity of feeding sites (metabolic sink) to meet constant supply of nutrients for the nematodes.

The nematode oesophageal gland secretions play a vital role to initiate these complex interactions. All plant parasitic nematodes secrete substances from glands through their hollow, protrusible stylet from the head's end. All these secretions emanate from two sub-ventral and one dorsal gland cells located at the basal bulbs of oesophagus. Both the glands participate in the host infection process and formation of feeding cells. The sub-ventral glands are believed to be involved for early stages of parasitism and dorsal gland for the development of feeding sites. Mostly, the nematode oesophageal secretions are thought to regulate the plant's genes and establish feeding sites.

There is signal exchange between nematode and plant cell just to trigger the several molecular events for the transformation of feeding cells, though these processes are not yet resolved. The signal for induction for cellular changes comes from the nematode, and therefore, secretions emanate from sub-ventral gland cells and injected through stylet into plant cell for initiation parasitism.

The infection of nematode on plant initiates complex changes in plant gene expression. A large number of genes induced by the infection of nematode are likely to contribute in order to establish parasitic interactions. The genes that are induced in the defence response are found to upregulate upon infection of both root-knot and cyst nematode species. Root-knot and cyst nematodes are known to upregulate the host endoglucanase and polygalacturonase genes after infection of plant cells and expression of endoglucanase are linked with the formation of giant cell and syncytium.

The gene encoding cell-wall-degrading enzymes (β-1,4 endoglucanase or cellulase) is known in root-knot (*Meloidogyne incognita*) and cyst nematodes (*Heterodera schachtii*, *H. glycines*, *Globodera rostochiensis*, *G. tabacum*). Other cell-wall-degrading enzymes like pectate lyase and polygalacturonase along with their encoding genes are known to exist in plant parasitic nematodes (e.g. pine nematode, *Bursaphelenchus xylophilus*). In most of the cases, their transcripts are found in sub-ventral glands and these encoded enzymes appear to be involved in softening the cell wall for easy migration of nematodes through root tissue.

Several nematode gene products (cellulase, hemicellulase, pectinase, expansin) were also identified from sub-ventral gland cells (remain active in early stage of parasitism) secretions, and these genes (cell-wall-degrading proteins) show strong sequence similarity to the genes encoded by microorganisms (bacteria). Therefore, it is speculated that these genes might have been transferred horizontally into plant parasitic nematodes from soil microbes. Further, the presence of almost similar gene product (venom allergen-like proteins ~ vaps also found *Ancylostoma caninum*) in the secretions of plant and animal parasitic nematodes indicated that they could have a general role in parasitism (Williamson and Gleason 2003) (Fig. 16.3).

16.5 Nature of Nematode Problems

PPNs can be detrimental to crop growth and development depending on population density and host susceptibility. Nematode can feed on all

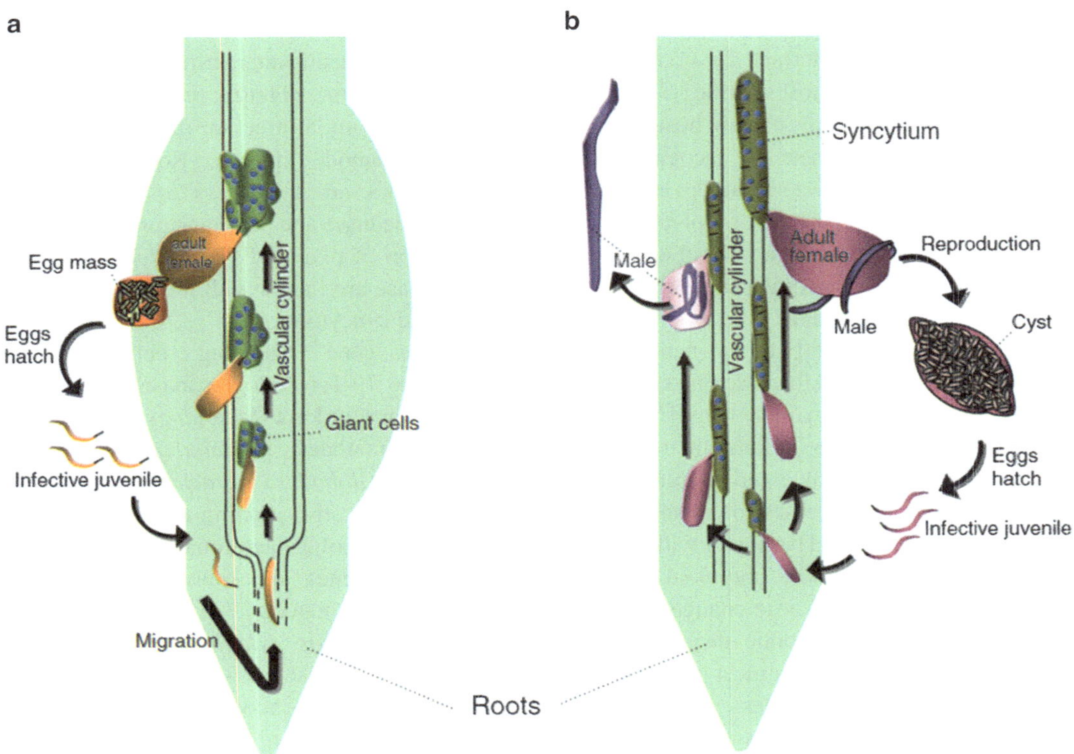

Fig. 16.3 Life cycle of (**a**) root-knot nematode and (**b**) cyst nematode (Abad and Williamson 2010)

parts of plants including roots, leaves, flowers, stem and seeds. Generally, they feed on the host tissues with the help of their protrusible stylet causing plant injury, and due to feeding and secretion, they modify the host cell into specialized nutritive cells such as multinucleate giant cell, syncytium or nurse cells to ensure permanent feeding. Some other nematodes induce gall formation on the plant's root, leaf and seed. Lesions develop due to the feeding on plant tissues, resulting in cell death and subsequent discolourations. Nematode infection on root reduces the ability of the root to uptake water and nutrients. Damaged plants are easily attacked by various soil pathogens like bacteria and fungi and develop a disease complex/syndrome. The aetiology of those diseases caused by the organisms involved is difficult to determine. Several nematodes belonging to the orders Dorylaimida and Triplonchida serve as a vector of plant viruses. Thus, nematode functions as a plant pathogen, predisposing agent and vector of plant viruses. PPNs are also known to interfere with the activity of beneficial nitrogen fixing *Rhizobium* bacteria in leguminous crops. Infestation of root-knot nematodes in pulse crops causes reduction of *Rhizobium* nodulation (Ali et al. 2010).

16.6 Field Diagnosis of Nematode Problems in Crop Plants

16.6.1 Aboveground Field Symptoms

(a) Stunting, unthrifty and patchy growth, yellowing (chlorosis) of leaves, reduction of foliage (e.g. *M. incognita* in betelvine) and progressive decline of plants (*Pratylenchus loosi* in tea, *Radopholus similis* in banana), poor yield, etc.

(b) Declines: slow decline (e.g. *Tylenchulus semipenetrans*), spreading decline (*Radopholus similis*) of citrus, slow decline (slow wilt or yellows disease) of pepper and grape, pomegranate decline (*M. incognita*)
(c) Leaf with angular spots and necrosis (*Aphelenchoides ritzemabosi*, *A. fragariae*), white tip (*A. besseyi* in rice)
(d) Daytime wilting of plant (*Meloidogyne* spp. in tomato and brinjal)
(e) Reduction in number of tillers in rice (e.g. *Heterodera oryzicola*, *Hirschmanniella* spp.)
(f) Deformed or distorted foliage (crinkling, curling and twisting), ear head, panicles (e.g. *Ditylenchus angustus*, *D. dipsaci*, *Aphelenchoides* sp., *Anguina tritici*, *Heterodera avenae*) and flowers (floral malady on tuberose due to infestation of *A. besseyi*) and grain discolouration (e.g. *A. besseyi* in rice)
(g) Formation of seed gall (e.g. *Anguina tritici* induces ear-cockle of wheat), stem gall or leaf gall (e.g. *Anguina australis* on *Ehrharta longiflora*) and galls on peanut shells (*M. arenaria*)
(h) Blackhead toppling and yellows disease (*R. similis* in banana and pepper)
(i) Replant problems (peach replant (*Pratylenchus penetrans*) and citrus replant (*T. semipenetrans*) along with other pathogens)

16.6.2 Belowground Symptoms

(a) Root knot or root gall (*Meloidogyne* spp. on vegetables), swelling of root tip (e.g. *M. graminicola* on rice), small galls (*Ditylenchus radicicola* on oat, rye, wheat etc., *Hemicycliophora arenaria* on citrus, *Xiphinema diversicaudatum* on rose), forking, branching (*Meloidogyne* spp. on carrot) or stubbed root and coarse/abbreviated root (*Trichodorus* and *Paratrichodorus* spp. on onion, corn etc. and *M. hapla*)
(b) Root lesions or brown/black spots and root discolouration (*Pratylenchus*, *Radopholus*, *Hirschmanniella*, *Bursaphelenchus citrophilus* – red ring disease of coconut)
(c) Lesion on roots/rhizomes (e.g. *Pratylenchus* sp., *Helicotylenchus multicinctus*, *Belonolaimus longicaudatus*) or pegs/pods (*Kalahasty malady* (*Tylenchorhynchus brevilineatus*) on groundnut) or deformation of tubers (e.g. *M. javanica* on potato)
(d) 'Dirty root' – soil stick to root surface of citrus (e.g. *Tylenchulus semipenetrans* in citrus)
(e) Thick root (e.g. *Meloidogyne* sp. in passion fruit, coffee)
(f) Curly tip (fish hook) galls – *Xiphinema* spp. on rose, plum, celery, etc. and *M. graminicola* on rice seedlings
(g) Dry tuber rot (*Ditylenchus destructor* in potato, *Scutellonema bradys* on yam)

16.7 Nematode Interactions with Other Pathogens

PPNs independently can cause plant diseases. They interact indirectly with other microorganisms (fungi, bacteria, viruses) leading to disease complexes. As an obligate parasite, nematode by themselves rarely kills the plants to ensure their own survival. The role played by the nematodes often is described as *incitant* (attack healthy tissues, thereby forming infection courts for other organisms), *aggravator* (release metabolic byproducts capable of killing host cell directly and/or predispose unaffected cells to invasion by other organisms and/or stimulate the growth of harmful organisms), *vector* (carry pathogens into healthy host tissue without involving themselves in the aetiology of the disease), *predisposer* (set conditions favourable for other organisms for the development of disease in the healthy host tissues) and *pathogen* (cause disease by themselves).

16.8 Nematode–Fungus Interactions

The involvement of nematodes in disease complexes was reported for the first time by Atkinson in (1892) on cotton wilt (*Fusarium oxysporum* f.sp. *vasinfectum*) – a severe wilt was observed in

Table 16.2 Nematode–fungus disease complexes (Back et al.2002)

Nematode	Pathogen	Crop	Reference
Meloidogyne javanica	*Fusarium oxysporum* f.sp. *lentis*	Lentil	De et al. (2001)
M. incognita	*Thielaviopsis basicola*	Cotton	Walker et al. (1999, 2000) Wheeler et al.(2000)
M. incognita	*Rhizoctonia solani*	Peanut	Abdel-Momen and Starr (1998)
M. incognita	*Rhizoctonia solani*	Tomato	Arya and Saxena (1999)
M. arabicida	*Fusarium oxysporum*	Coffee	Bertrand et al. (2000)
Heterodera glycines	*Phytophthora sojae*	Soybean	Kaitany et al. (2000)
H. glycines	*Fusarium solani*	Soybean	Rupe et al. (1999) Sugawara et al. (1997)
Globodera rostochiensis	*Rhizoctonia solani*	Potato	Back et al. (2000)
Pratylenchus thornei	*Fusarium oxysporum* f.sp. *ciceri*	Chickpea	Castillo et al. (1998)
P. thornei	*Rhizoctonia solani*	Chickpea	Bhatt and Vadhera (1997)
P. neglectus	*Verticillium dahliae*	Potato	Hafez et al. (1999)
P. penetrans	*Verticillium dahliae*	Mint	Johnson and Santo (2001)
R. reniformis	*Fusarium oxysporum* f.sp. *pisi*	Pea	Vats and Dalal (1997)
R. reniformis	*Phytophthora palmivora*	Betelvine	Jonathan et al. (1997)

the presence of root-knot nematode. There are numerous reports (Table 16.2) of nematode–fungus disease complex.

Mechanism of interactions of plant parasitic nematode with fungus is primarily through enhancement of pathogenicity mechanism of fungus. A nematode-infected plant is altered physiologically and the host becomes susceptible to pathogens. Most synergistic interactions between these endoparasitic nematode and *Fusarium* occur after 3–4 weeks of nematode infection. It is reported that RKNs cause increased leakage of electrolyte from galled tissues which activate the resting spores of pathogenic fungi. The galled roots exude more carbohydrates, protein, amino acids, etc., and these invite rapid fungal colonization on root. Further, mechanical injury made by nematodes is also responsible for the breakdown of resistance in some crop cultivars against fungal pathogen. Nematode can vector fungal propagules both internally and externally on the body surface. This was evident for ear-cockle nematode (*Anguina tritici*) which carries (a vector) the spores of *Dilophospora alopecuri* for initiation of infection on aerial parts of the cereals. Fungi also suppress PPNs through predation, parasitization, toxin production and several other ways. *Fusarium solani* reduces the population of cyst of *Heterodera zeae* on maize (Lal et al. 1982). *F. solani*, *F. oxysporum*, *Paecilomyces lilacinus*, *Pochonia chlamydosporia*, etc. parasitize the eggs of *Meloidogyne* spp. Some of these fungi are being used for control of PPNs.

16.9 Nematode–Bacteria Interaction

The PPN may predispose plants to bacterial disease. Bacteria can easily enter into plants through wounds created by the nematode injury. Nematodes also transmit bacterial propagules externally on their body surface or internally within their gut. Hunger (1901) showed for the first time that tomato plants are readily attached by *Pseudomonas solanacearum* in nematode-

Table 16.3 Nematode–bacteria interaction

Nematode	Bacteria	Host	Disease	Reference
Anguina tritici	*Corynebacterium michiganense* pv. *tritici*	Wheat	Tundu disease	Gupta and Swarup (1972)
Aphelenchoides ritzemabosi or A. fragariae	*Corynebacterium fascians*	Strawberry	Cauliflower disease	Crosse and Pitcher (1952)
Meloidogyne hapla	*Pseudomonas solanacearum*	Tomato	Wilt	Libman et al. (1964)
Meloidogyne incognita	*Pseudomonas solanacearum*	Eggplant	Wilt	Reddy et al. (1979)
Meloidogyne incognita	*Pseudomonas solanacearum*	Tobacco	Wilt	Lucas et al. (1955)

infested soil. A few cases of nematode–bacterial disease complexes are known (Table 16.3). RKN increases the frequency and severity of the bacterial wilt of tobacco (*Ralstonia solanacearum*), bacterial wilt of alfalfa (*Clavibacter michiganensis* sub sp. *insidiosus*) and bacterial scab of gladiolus (*Pseudomonas marginata*). The role of the nematode for most of the cases is to provide the bacteria with an infection court and/or easy access of bacteria into host through wounding. The combined infection of *Clavibacter tritici* (*Corynebacterium michiganense* pv. *tritici*) and *Anguina tritici* produces yellow ear rot or *tundu* disease of wheat (Gupta and Swarup 1968, 1972). *A. tritici* alone causes ear-cockle of wheat, while *C. tritici* alone is incapable of causing yellow ear rot disease. Simultaneous infection of *Ralstonia solanacearum*, *Macrophomina phaseolina* and *Meloidogyne incognita* on jute results in wilt symptom, popularly known as 'Hooghly wilt' in West Bengal.

Bacteria are also capable of parasitizing nematodes, and this has been exploited for managing crop problems due to nematodes. Two most popular bacteria pathogens of nematodes are *Pasteuria penetrans* and *Pseudomonas fluorescens*. The use of bacterial antagonist *P. fluorescens* to control root-knot nematode species, *M. incognita*, *M. javanica* and *M. graminicola*, is suggested in vegetables and rice.

16.10 Nematode–Virus Interaction

Nematode as vector of plant viruses is now well documented. Hewit et al. (1958) proved the role of the *Xiphinema index* for successful transmission of grapevine fanleaf virus in the vineyards in California. This was the first report of reliable proof of transmission of a plant virus by a plant parasitic nematode. For a successful transmission of plant viruses by the nematode species, the steps involve acquisition, retention, dissociation and inoculation. Generally, virus particles attached on the cuticular lining of the feeding apparatus and the anterior part of the alimentary tract (oesophagus) of nematode. During the feeding process, ingested virus particles pass through the stylet (feeding apparatus) and oesophageal lumen (internally lined by cuticle) into the alimentary canal leaving a small portion retained (adsorbed) on the wall. Both the retention and dissociation of virus particles depend on the compatibility of the virus (coat protein) and the vector, the nematode (position of oesophageal gland opening, gland secretion and its flow, pH, chemical nature of lumen and feeding apparatus, etc.). Usually, nematodes may be infective from 2 to 4 months and even longer and can transmit the viruses after feeding on infected plants from 1 h to 4 days. Both juvenile and adults are capable of acquiring and transmitting viruses. All the plant parasitic nematodes ingest the virus particles during feeding on plants, but successful virus transmission occurs in nematode species belonging to Order Dorylaimida (*Xiphinema*, *Longidorus* and *Paralongidorus* spp.) and Triplonchida (*Trichodorus* and *Paratrichodorus* spp.).

The site of retention for virus particles varies among the nematode genera, for example, virus particles attached on the inner surface of odontostyle in *Longidorus* and *Paratrichodorus* spp., on cuticle limning and lumen of odontophore of *Xiphinema* spp. and on the cuticle lining along the entire length of the oesophagus lumen in *Trichodorus* and *Paratrichodorus*.

The plant viruses transmitted by the nematodes are broadly grouped as below:

16.10.1 NEPO Viruses

Nematodetransmitted particles are 25–30 nm in size, polyhedral in shape and mostly transmitted by *Xiphinema* and Longidorus spp. The viruses belonging to this group are *grapevine* fanleaf virus (*Xiphinema index*), tomato ringspot virus (*Xiphinema americanum*) and raspberry ringspot Scottish strain (*Longidorus elongatus*).

16.10.2 TOBRA or NETU Viruses

Nematodetransmitted particles are 180–210 nm and 45–115 nm in size depending on virus isolates and rod-shaped (tubular) viruses, mostly transmitted by *Trichodorus* and *Paratrichodorus* spp. The viruses transmitted are tobacco rattle (*Trichodorus similis*), pea early browning virus (*Paratrichodorus anemones*), pepper ringspot virus (*Paralongidorus maximus*), etc.

16.11 Major Nematode Problems of Crops in India

Some economically important PPN problems of different crops in India have been documented (Khan et al. 2010), and their management options are briefly discussed as follows.

16.11.1 Root-Knot Nematodes (*Meloidogyne* spp.)

Root-knot nematodes (RKNs, *Meloidogyne* spp.) are global menace to crop production (Sasser 1980). It has a very wide distribution and causes serious damage to crops particularly in vegetables. The average yield losses in the world are believed to be about 5 % and could be more in the developing countries of tropic and subtropic (Taylor and Sasser 1978). Considering the universal significance of RKNs, an International *Meloidogyne* Project (IMP) was operated (1975–1984) with its headquarter at North Carolina State University, USA, and its collaborating centres were in many developing countries of the tropics and subtropics. Worldwide more than 97 known species of RKN have been recorded and only 14 known species of *Meloidogyne* are recorded in India. Four species of RKNs, viz., *Meloidogyne indica* (citrus, Delhi), *M. lucknowica* (from sponge gourd, UP), *M. triticooryzae* (rice and wheat, Delhi) and *M. piperi* (pepper, Kerala), have been described from India. Various insect-pests, diseases and weeds are inflicting damages to vegetable crops. RKNs (*Meloidogyne* spp.) are one of the potential constraints for cultivation of vegetables particularly in the developing countries of tropics and subtropics. Vegetable crops harbour large number of plant parasitic nematodes, but RKN is the most damaging one. It affects the crop directly and indirectly by interaction with various soil-borne fungi, bacteria and viruses. The most predominant species of RKNs are *Meloidogyne incognita*, *M. javanica*, *M. arenaria* and *M. hapla*. All the species of RKNs produce a characteristic 'root gall 'or 'knotted root symptom' which could be easily recognized by the naked eye. There is hardly any vegetable crop which is not attacked by the RKNs. Therefore, it has widely been considered as a limiting factor for cultivation of vegetables. The lack of awareness among the farmers about the nematode problems and nonavailability of suitable package of practices to extension workers for managing the RKNs are the major hindrance for protecting the vegetable crops from the nematodes. Chemical approach of nematode management is no doubt effective, but high doses of nematicides required for managing nematodes are neither economical nor environmentally safe.

The infection of RKN produces characteristic disease symptoms on the belowground root system popularly known as 'root gall' or 'knotted roots'. Different sizes of galls are induced depending on their host and the species of the nematode involved. On cucurbits, the nematode induces large galls, whereas in chilli, a small size of galls is produced. Usually the infection of *M. hapla* produces small galls as compared to *M. incognita* and *M. javanica*. The size of galls also differs with the level of infection as in the case of heavy infection, large size or multiple galls or secondary galls develop. Besides galling, forking of taproot in carrot and tubercle on potato tubers is also noticed. Aboveground symptoms are non-specific in nature. Infected plants exhibit symptoms of general mineral deficiency, yellowing, stunting, wilting during hotter part of the day, chlorosis, premature shedding of leaves and poor look of plants resulting in low yield. The nematodes are also involved in interaction with other soil-borne fungi, bacteria and viruses and cause serious damage to crops. The interaction of RKNs is known in many vegetables, fibre, pulses and plantation crops. However, the most common problems are the breakdown of disease resistance and wilting of healthy plants. The most common interaction of RKN with *Ralstonia (Pseudomonas) solanacearum* is causing "pseudomonas wilt" in tomato, brinjal and potato.

16.12 Nematode Management Options

RKNs are polyphagous in nature, having high reproductive potential, and have acquired a unique mechanism of survival strategy through laying their eggs in protective gelatinous matrix. Management of root-knot nematode is not an easy task under intensive crop cultivation system. Therefore, the idea of keeping the nematode population below the economic damage level by adopting different available tactics is advised to the growers. The young tender seedlings of various crops are very much vulnerable to attack by nematode, while the older plants achieve some degree of tolerance. Considering the farmer's suitability, clean cultivation practices of vegetable crops could be suggested for managing root-knot nematodes:

16.12.1 Cultural Practices

Cultural practices are the most effective and economical means of managing insect-pests and disease including nematode problems.

- Two to three summer ploughings (20 cm deep) during the months of May to June at the interval of 15 days expose nematodes, weeds, pathogen propagules and hibernating stages of insect-pests to sunrays.
- Intercropping with antagonistic plants like marigold (*Tagetes* spp.) reduces soil population of many soil nematodes including RKNs. Incorporation of such crop in cropping system either as intercrop or alternative crop may be considered whenever feasible.
- Crop rotation with resistant varieties or non-host crops like mustard, sesame, maize, wheat, etc. is useful in bringing down soil nematode populations below the damage threshold level.
- Application of organic manure, farmyard manures (FYM), at 18–20 t/ha reduces nematode population through their multiple actions: releasing of toxic substances, enhancing of crop tolerance and encouraging of soil microbial antagonists either alone or simultaneously.

16.12.2 Resistant Varieties

Plant resistance plays an important role in the integrated management of RKN diseases. However, availability of resistant varieties of vegetable crops is very few in number, and many of them are not acceptable to the farmers for their suitability.

Some of the resistance varieties that exhibited resistance or tolerance reaction to RKNs are as given below:

Tomato	SL-120, Hisar Lalit, PNR-7, Hisar N-1, Hisar N-2, Hisar N-3, NT-3, NT-8 NT-12, Ronita, Patriot, PAU-5, Mangla and Karnataka Hybrids
Chilli	Pusa Jwala, CAP-63, CA-2057, Sindhuri, NP-46 A, Mohini, SP-26, P-6-3, K-235
Brinjal	Giant of Banaras, Black Beauty, Gola, Gachha Baigan, Pbr-91-2, IC-95-13, HOE-101, Red Wonder
Cowpea	Barasati mutant, 82-IB, C-152, IHR-29-5, GAU-1
Pea	B-58, C-50
Potato	Kufri Dewa
Okra	Kanki local green, Harichickni, Vaishali Badher
Pumpkin	Dasna, Jaipuri
Watermelon	Shahjanpuri
Ridge gourd	Panipati, Meerut special

Modified after Anonymous (1988) and Parvatha Reddy (2008)

16.12.3 Chemical Control

Chemical control with the application of nematicides is one of the most effective means of nematode management. However, most of the effective nematicides have been withdrawn from the world market. At present, a few insecticides having nematicidal property are available to the farmers, but because of their high doses required to manage nematode, they become cost-ineffective and leave high pesticide residues to the harvested crops. Despite their inherent drawbacks, chemical nematicides could be applied judiciously so that the doses and costs are reduced drastically. The application of nematicides through bare-root dip treatment, seed treatment and nursery bed treatment has been proven to be effective to protect the young seedlings from nematode attack.

16.12.4 Nursery Bed Treatment

In most of the cases, the infection is carried through infested seedlings from nursery bed. The damage caused by RKN to the root system of tender seedlings is more harmful than to older plants. The application of nematicides to nursery bed helps to raise nematode-free seedlings. Moreover, it reduces the dose of nematicides and cost substantially. The soil application of carbofuran (Furadan 3G) at 0.3 g a.i./m^2 is sufficient for producing nematode-free seedlings of many transplanted vegetable crops. The treatment of nursery bed with sebuphos (Rugby 20WP) or carbofuran (Furadan 3G) or benfurocarb (Oncol 50WP) at 0.3 or 0.6 g a.i./m^2 at the time of sowing reduces the infestation of RKNs.

16.12.5 Bare-Root Dip Treatment

The seedlings of many transplanted vegetable crops can be dipped in systemic nematicides like oxamyl, prophos and dimethoate at 500–1000 ppm for 6 h to denematize the roots. These practices will further ensure to protect the root system of tender seedlings from early attack of nematodes. The seedlings of transplanted vegetables like brinjal, tomato, chilli and planting materials of pointed gourd treated with carbosulfan (Marshal 25EC) at 500 ppm for 6 h provided effective control against RKNs. Nursery bed treatment with carbofuran at 0.3 g a.i./m^2 and with carbosulfan 25EC at 500 pm in seedlings before transplanting effectively manages *M. incognita* and enhances crop yield in vegetables. Bare-root dip treatment of tomato and brinjal seedlings with Zolone (Phosalone 35EC) or monocrotophos (Monocil 36SL) or carbosulfan (Marshal 25ST or DS) at 0.05 % reduces RKN infestation.

16.12.6 Seed Treatment

The practice of seed soaking and seed dressing is an important prophylactic measure which gives adequate initial protection to the young seedlings of tomato, brinjal, okra, chilli, etc. The most commonly used systemic nematicides, viz. fenamiphos, isofenphos, carbosulfan, etc., are used at 2–3 % w/w. Seed dressing with carbosulfan (Marshal 25ST) at 3 % w/w is quite effective for managing RKN in okra, bottle gourd, pointed gourd, bitter gourd and jute. Seed soaking with dimethoate, carbosulfan (Marshal 25EC), etc. can be adopted for providing better crop with early protection against nematode.

16.12.7 Field Application

Field application of carbofuran 3G at 2 kg a.i./ha in tomato, brinjal and okra reduces nematode population and increases yields.

16.12.8 Biological Control

16.12.8.1 Biological Control and Nematode-Suppressive Soils

Despite its several limitations, biological control of RKN is a cost-effective and eco-friendly method. As a component of integrated nematode management, biological suppression of RKN is well known. There has been some success in controlling *Meloidogyne* spp. in the fields with *Arthrobotrys robustus* (Royal 300), *A. irregularis* (Royal 350) and *Paecilomyces lilacinus* (Biocon, PL Plus, Paecil, Nemachek). Several biocontrol agents (BCA) have been exploited against this, but only a few BCAs, viz. *Paecilomyces lilacinus*, *Pochonia chlamydosporia*, *Trichoderma viride*, *Pseudomonas fluorescens* and *Pasteuria penetrans*, have been found effective and showed promising for managing nematode. The current prospects of *Trichoderma* for management of PPNs have been documented (Haseeb and Khan 2012). Currently, a few commercial formulations of *P. lilacinus* (Bionematon, Yorker), *Trichoderma viride* and *T. harzianum* (Tricho guard, Bioderma, Ecoderma), *Myrothecium verrucaria* (DiTera) and *Pasteuria penetrans* (Pasutsuria 50WP) are available in the market for controlling root-knot and other nematodes. Integration by vine dipping of pointed gourd in monocrotophos 36SL at 0.1 % followed by soil inoculation of *T. viride* at 10 g/pit at planting and 40 days after planting results in the reduction of root galling caused by *M. incognita* and more fruit yield (Khan et al. 2009). *P. lilacinus* formulations (at 15 kg/ha) enriched (at 1 kg mixed with 25 kg FYM followed by 7 days incubation) with FYM are quite effective to manage RKN (*M. incognita*) infestation in pointed gourd (Dey et al. 2012).

16.13 Integrated Approach for RKN Management

Individual method of nematode control has either proved ineffective or uneconomical approach against RKNs. Therefore, integration of various suitable tactics may be an eco-friendly, economically viable and practically feasible approach for managing nematode problems in crops. The adoption of deep summer ploughing during the summer period at fortnightly interval along with organic matter application followed by planting with nematode-free seedlings is a feasible approach to reduce nematode population. Similarly, farmers with their available resources could follow integration of cultural, biological and chemical methods and resistant varieties in suitable combination for each crop cultivation system. Soil solarization/summer ploughing alone as well as in combination with carbofuran 3G at 2 kg a.i./ha has been found effective against nematodes infesting brinjal, chilli and tomato. Seedlings raised in solarized nursery beds treated with carbofuran at 0.3 g a.i/ m^2 integrated with the application of neem cake at 5q/ha give better check against nematodes infesting vegetables.

16.14 Wheat Seed Gall Nematode (*Anguina tritici*)

The nematode is one of the most serious pests of wheat in some parts of the country. It is the first known plant parasitic nematode to be described in scientific literature in 1743. The nematode alone causes *ear-cockle* disease in wheat, and in association with the bacterium *Clavibacter tritici*, it produces *yellow ear rot* or *tundu* disease. The host range of this nematode is very few in number and wheat is considered as the most suitable host. Although the control of this nematode is simple and easy as compared to other plant

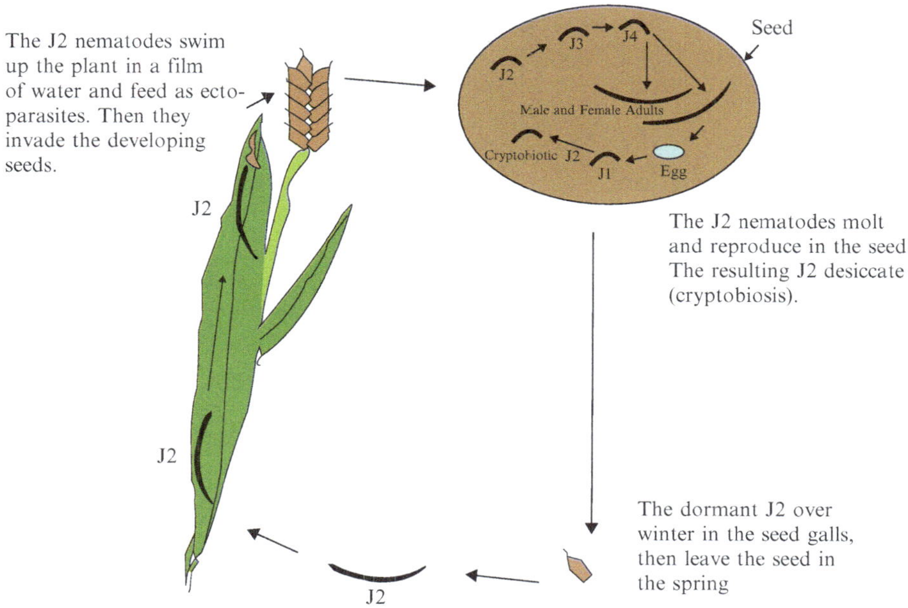

Fig. 16.4 Life cycle of *Anguina tritici*, the seed gall nematode (Source: http://www.apsnet.org/edcenter/intropp/PathogenGroups)

parasitic nematodes, it is still troublesome in many wheat-growing parts of Rajasthan, UP, Bihar and Madhya Pradesh particularly in tribal belts where tons of wheat grains are wasted every year (Anon 1995–2001). The wheat seed galls are the primary source of dissemination for nematode either as seed mixture of cockle. The second-stage juvenile (J2) of *A. tritici* is the survival stage and remains viable in anhydrobiotic state inside the cockle for several years (up to 30 years). After sowing the seed galls (cockles) come in contact with soil moisture and become soft leading to the release of a large number of J2s. These J2s infect the growing point of seedlings as ectoparasite, causing leaf distortions, and reach the inflorescence due to the natural growth of seedling. The nematode enters the floral primordia and becomes endoparasite, and eventually the floral primordia converts into blackened seed gall (cockle).

The initial visible symptom is enlargement of basal stems near the soil surface at the stage of 20–25-day-old seedlings. Generally the infested plant exhibits more number of tillers and grows fast as compared to the healthy ones. Twisting, curling and crinkling of leaves and stunted growth are common symptoms in the early stage of plant growth and the affected ears are typically swollen and broader with a few awns on the glumes. The cockled ears contain initially green galls; later in each spikelet, 1–5 galls can be seen (Fig. 16.4).

16.15 Yellow Ear Rot or Tundu Disease

The early stage of plant growth exhibits similar disease symptoms as that of ear-cockle disease. The yellow ear rot disease is primarily caused by a bacterium, *Clavibacter tritici*, only in the presence of nematode, *A. tritici*. Under humid climatic conditions, the characteristic symptoms appear with the production of bright yellowish bacterial slime on the leaf surface which can be seen trickling down the ears. During dry weather, these slimes become hard and diseased spikes are generally distorted, stunted and narrower than the healthy ones with the grains partially or completely converted by bacterial mass. Generally low temperature and high humidity are favourable for *tundu* disease and just reverse conditions – high temperature and low humidity are conducive for ear-cockle symptom.

16.15.1 Management Options

Nematode is easy to manage because the gall is the only source of ear-cockle and *tundu* disease. Both physical and mechanical methods are successful for eradicating nematode from several developed countries of the world; however, in India, nematode is still becoming a problem probably due to poor awareness and failure of national campaign against the dreadful disease of wheat.

16.15.2 Physical Methods

- Hot water treatment of wheat seed lots at 54–56 °C for 10–20 min.
- Water floatation of seed galls in 5–10 % salt solution for 5–15 min. Seeds containing nematode gall will float in water surface and can be collected and discarded.

16.15.3 Mechanical Methods

Fanning or winnowing is an effective method to remove the galls from seed lots. Sieving/screening is a common practice and the most successful method for the eradication of seed gall nematodes from many countries, though complete removal of gall is not possible with this method because some large-sized galls remain on the sieves.

16.16 Cereal Cyst Nematode (*Heterodera avenae*)

Infestation of nematode which caused *molya* disease was recorded for the first time from the Sikar district of Rajasthan in India (Vasudeva 1958), and subsequently, it is known to occur in the major wheat-growing states, viz. Punjab, Haryana, Uttar Pradesh, Delhi, Himachal Pradesh, Gujarat, Jammu and Kashmir and Madhya Pradesh of India (Kaushal et al. 2001; Kanwar and Bajaj 2010). The nematode is mainly confined to the family Poaceae except *Senebiera pinnatifida* (Cruciferae), most importantly barley, wheat, oats, rye and triticale.

Heterodera avenae complex group of closely related species *H. filipjevi* and *H. ustinovi* is the causal organism of a serious disease popularly known as *molya* of wheat and barley. All the three species are known to exist in India. However, *H. avenae* complex group is prevalent in India with several biotypes. There were 11 biotypes of *H. avenae* known worldwide based on international assortment of wheat, oat and barley cultivars. In India, the occurrence of five biotypes from Rajasthan and Haryana (Mathur et al. 1974) and biotype 3 is most common. While Swarup et al. (1979) reported only two biotypes from India- Rajasthan and Haryana populations are one biotype and the Punjab population is different biotype.

The nematode-infested fields exhibit patchy growth with stunted and yellowish plant. Infested plants show thin narrow leaves, reduced tillering, fewer leaves and small size of ear heads with reduced number of grains. The roots of nematode-attacked plants appear bushy and bunchy due to the emergence of rootlets at the site of infection, and slight swelling of the root tips may be encountered. The aboveground symptoms are often confused with the general deficiency symptoms. However, the presence of cysts is the only confirmatory evidence for nematode infection.

The second-stage juveniles (J2) infect the growing tips of roots and upon feeding develop specialized syncytial cells for their growth and development. After 3 months, the juveniles achieve lemon-shaped sedentary female which is found to attach with roots. Eggs are laid inside the female body, and after the death of the female, the body cuticle transforms into brown cyst. These cysts are dislodged from the root in the soil and become the inoculum for the next season. The activity of the nematode perfectly matched with the growing season of wheat crop grown in the *rabi* season (October to March/April) in India. During off-season (April to Sept), eggs and J2 undergo diapauses induced by the rising of temperature during the end of April and remain dormant inside the cyst in the soil. In India, nematode completes only one generation during winter (October to March). The duration of its life cycle depends on the temperature and development of white females. Under Indian conditions, December-sown crop may take 8–9 weeks and February-sown crop 5 weeks on wheat (Bajaj and Kanwar 2005) (Fig. 16.5).

16.16.1 Management Options

16.16.1.1 Cultural Practices

This nematode can be managed by introducing crop rotation with non-hosts like sarson, toria, raya, taramira, gram, berseem, carrot, coriander, etc. with wheat. Deep summer ploughing (2–3) at an interval of 10–15 days during hot summer months is helpful to reduce soil nematode populations. Growing wheat cultivar C-306 as trap crop early in October, before growing main crop, proves effective. However, growing resistant varieties as the wheat cultivar Rak Molya Rodhall (Raj MR1) and barley cultivars Rajkiran, C-164 and BH-72 is most effective to control the nematodes. However, the duration of rotation with keeping the field free from weed hosts is most important for effective suppression of the nematode population.

16.16.1.2 Chemical Control

Field application of carbofuran (Furadan 3G) at 1.0–1.5 kg a.i./ha has been found effective (Kaushal et al. 2001; Kanwar and Bajaj 2010).

16.16.2 Integrated Management

Integration of different tactics was found economical against cereal cyst nematode. Early sowing in the month of November along with field application of carbofuran at 2 kga.i./ha is quite effective in increasing yield and reducing cyst population in soil. Natural nematode-suppressive soils against the cereal cyst nematode (CCN, *Heterodera avenae*) are known and widespread in northern Europe. The practice of nematode-suppressive soils, which are often associated with monoculture of a susceptible host (Kerry and Crump 1998) is a common crop rotation practice in nematode management. The

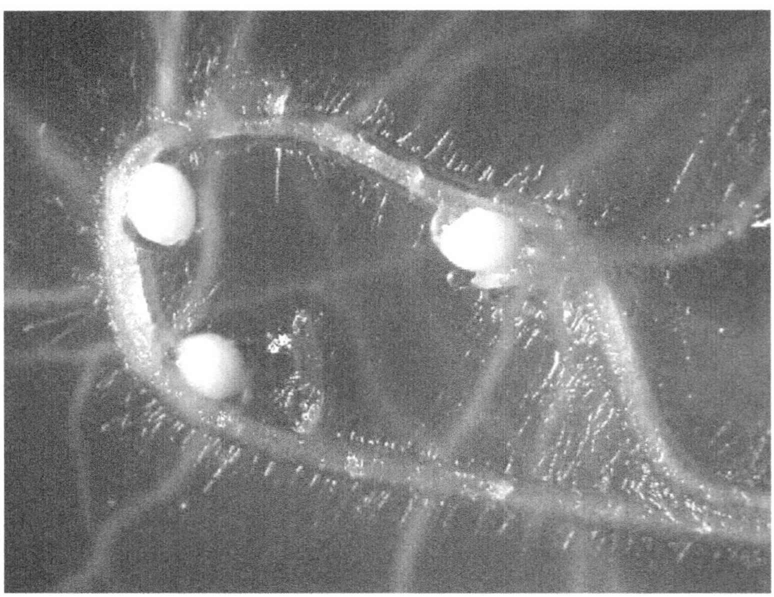

Fig. 16.5 Life cycle of cereal cyst nematode (*Heterodera avenae*) (Source: http://www.sardi.sa.gov.au)

main nematode-suppressing agents in these suppressive soils are *P. chlamydosporia* and *Nematophthora gynophyla,* and after 8 years of growing wheat cultivar, the CCN population declined to non-detectable levels (Kerry 1982).

16.17 Potato Cyst Nematode (*Globodera rostochiensis, G. pallida*)

Potato cyst nematode (PCN) is one of the serious nematode pests of potato in southern states like Tamil Nadu, Karnataka and Kerala of India. The nematode is popularly known as *golden nematode* and is considered as one of the major crop protection problems of the world. It causes around 9 % loss of global potato, amounting to about 40 million tons (Krishna Prasad 1995). Jones (1961) detected this nematode from a field at Vizianagaram state farm in Ootacamund, Nilgiri Hill, Tamil Nadu. Considering the importance of potato cyst nematode in the country, the government of Tamil Nadu imposed the Destructive Insects and Pests Act 1919 (DIP Act 1919) in 1971 to contain the nematode in the Nilgiri Hill.

The disease symptoms are not specific at low infestation of nematode, but the symptoms become prominent at high density of soil nematode population in soil. The symptoms include the following: small patches in poorly growing plants, temporary wilt of plants during daytime, stunted growth of plants, unhealthy yellowish foliage and poor root systems, reduction in the number and size of tubers and gradual reduction in the production of potato yield over the years. The nematode is primarily confined to the family Solanaceae and depends on the host root diffusates which induce the hatching of second-stage juveniles from eggs. The life cycle is completed in about 5 weeks, and three generations are completed in a cropping season. Unlike other cyst nematodes, the cysts containing eggs are protected and remain viable for several years in soil even in the absence of potato.

Five pathotypes (Ro1, Ro2, Ro3, Ro4 and Ro5) of *G. rostochiensis* and three pathotypes (Pa1, Pa3 and Pa3) are known to exist worldwide. Pathotypes Ro2 and Ro5 of *G. rostochiensis* and Pa_1 and Pa_3 of *G. pallida* are known to be prevalent in India. The life cycle of the potato cyst nematode is shown in Fig. 16.6.

The spread of PCN occurs through soil particles adhering to tubers, farm implements, gunny bags, farmers' feet, etc. However, irrigation water or rain water running down the hill slope carries the cyst from the infested fields to the uninfested fields.

16.17.1 Management Options

16.17.1.1 Cultural Practices

Growing non-host and non-solanaceous vegetable crops like cabbage, cauliflower, beet roots, carrots, garlic, radish, turnips, etc reduce soil nematode population. One of the most effective rotations, potato–cabbage–carrot, is commonly practised by the farmers of Nilgiri Hills. Growing the nematode-resistant variety Kufri Swarna is popular among the growers in Nilgiri areas (Anonymous 1985).

16.17.1.2 Biological Control

Application of talc-based formulation of *Paecilomyces lilacinus* and *Pochonia chlamydosporia* with nematicide showed promising results (Krishna Prasad 2006).

16.17.1.3 Chemical Control

The use of carbofuran 3G at 2 kg a.i./ha is effective in reducing nematode population as well as in increasing potato yield in Nilgiri conditions.

16.17.1.4 Integrated Approach

Crop rotation with cabbage or carrot, intercropping with bean or wheat followed by fodder oats or raising radish or French bean for a short period proved economical for growing potato. Soil application of carbofuran at 2.0 kg a.i./ha is a local practice of Nilgiri farmers for bringing down PCN population in soil (Krishna Prasad 2006).

16.17.1.5 Nematode Pests of Rice in India

More than 300 nematode species of 35 genera are known to parasitize rice. However, a dozen of nem-

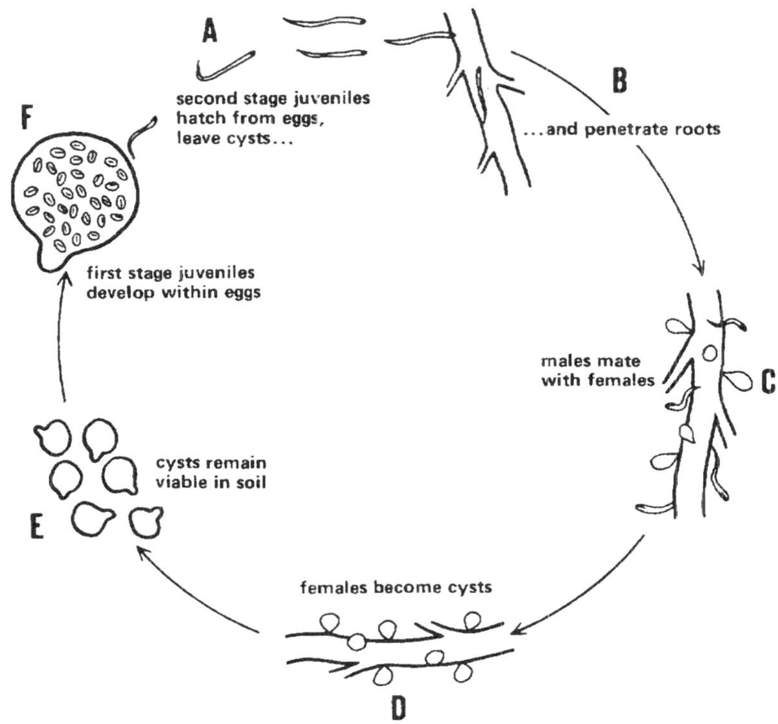

Fig. 16.6 Life cycle of potato cyst nematode (Franco 1986)

atode species is inflicting economic damage to rice crop grown under different situations. Rice root-knot nematode (*Meloidogyne graminicola*) is one of the major rice nematodes emerging as a potential threat to rice cultivation in India. Stem nematodes (*Ditylenchus angustus*) and white tip nematodes (*Aphelenchoides besseyi*) are aboveground nematode parasites and induce conspicuous '*ufra* disease' and 'white tip disease' symptoms in rice, respectively. Both of these nematode species are seed-borne in nature and have quarantine significance. Rice root nematodes (both species *Hirschmanniella oryzae* and *H. mucronata*) and rice cyst nematodes (*Heterodera oryzicola*) are also economically important and later have a limited distribution in Kerala infesting both rice and banana.

16.18 Rice Root-Knot Nematode (*Meloidogyne graminicola*)

Rice root-knot nematode (*Meloidogyne graminicola*), sedentary endoparasite of rice root, is a well-established nematode pest of rainfed upland rice. It poses serious problems in *boro* and *kahrif* nursery particularly in sandy loam or recent alluvial soils of West Bengal and other rice-growing regions of India. It is also becoming a problem in transplanted rice grown in waterlogged conditions. The widespread occurrence of *M. graminicola* has been found in Assam, West Bengal, Gujarat, Orissa, Karnataka, Tripura and other states of India (Fig. 16.7).

The aboveground symptoms are nonspecific in nature as yellowing, stunting of foliage, delayed flowering by 10–15 days and reduced number of tillers. The presence of characteristic 'hook-shaped' or 'ring-like' root gall on the root tip of growing rice seedlings is the confirmatory evidence for the association of this nematode. Galls produced by the nematode induce growth of lateral rootlets and root hairs. The yield loss due to *M. graminicola* has been estimated to be between 16 and 32 % in upland rice and in severe cases go up to 64 % (Phukan 1995). A complete failure of *boro* rice nursery in 'Simurali' in the district of Nadia, West Bengal, was reported (Anonymous 2001).

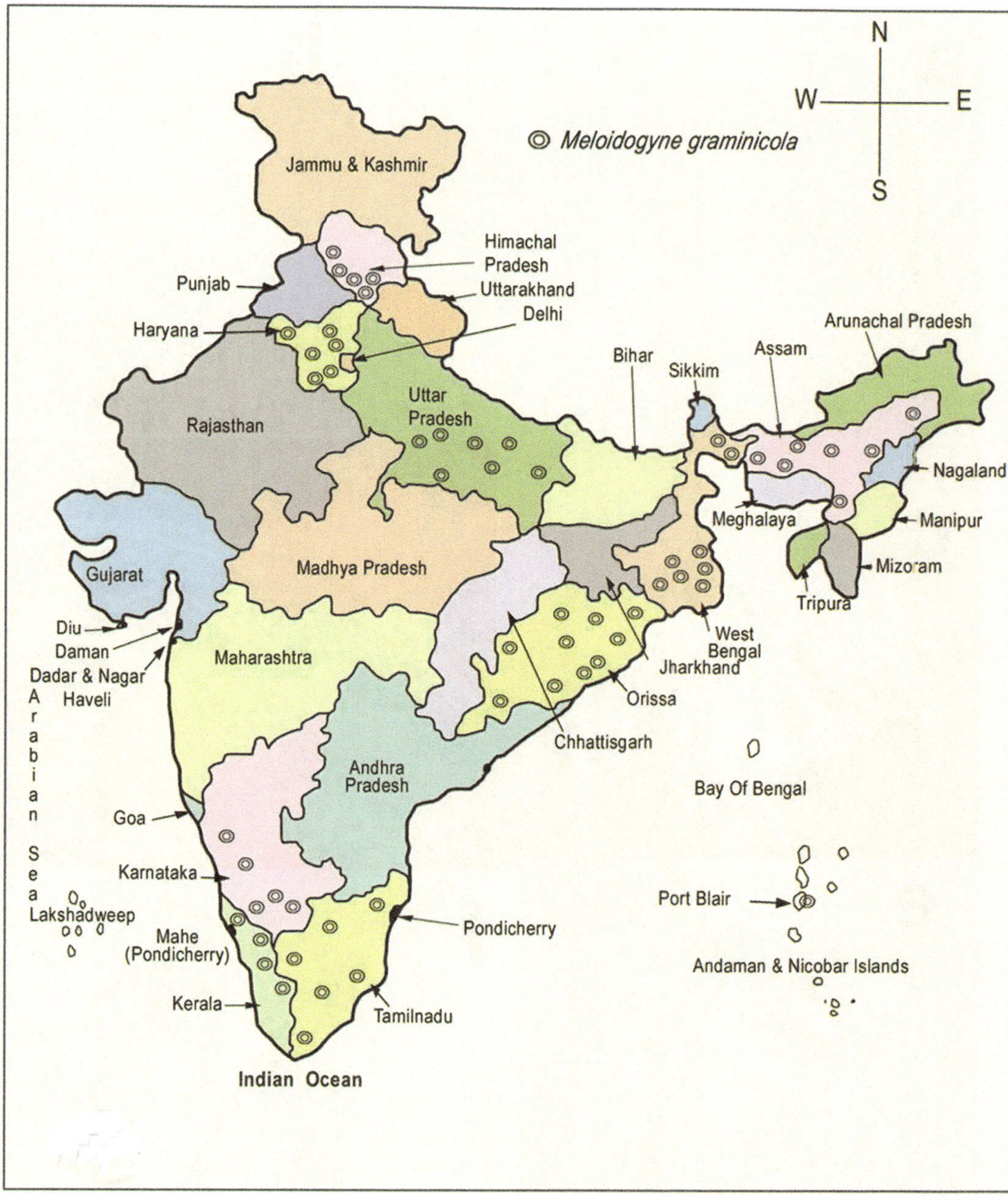

Fig. 16.7 Distribution of rice root-knot nematode, *Meloidogyne graminicola* infecting rice in India

After harvesting of rice, the nematode may survive in egg stage in soil or continue to reproduce on various weeds (Fig. 16.8). Female often remains concealed within root tissue and eggs are laid in cortical tissues and hatched juveniles reinfect the same roots. It completes life cycle within 19 days at 22–29 °C in upland rice.

16.18.1 Management Strategies

Based on extensive research, suitable eco-friendly management technologies suggested for the root-knot nematode-infesting rice in India (Jain et al. 2012) are given below:

Fig. 16.8 Rice root-knot nematode-infested (**a**) rice field, (**b**) rice nursery, (**c**) rice root, (**d**) *Echinochloa colonum* root, (**e**) *Brachiaria ramosa* root, (**f**) *Cyperus rotundus* root

- Use of non-host crop rotation (groundnut, mustard, black gram and potato) or fallows ideally for two sessions can help in bringing down the populations of rice root-knot nematode.
- Use of resistant genotypes such as ARC–12620, INRC–2002 and CR–94–CCRP–51 which have shown resistance against *M. graminicola* can be taken up in the affected areas.
- Nursery bed treatment with carbofuran at 0.3 g a.i./m^2 and its field application at 1 kg a.i./ha 40 days after transplanting reduced the nematode populations.
- Soaking of rice seeds in 0.1 % solution of carbosulfan for 12 h adversely affects egg mass production of *M. graminicola*, reduces root galling due to rice root-knot nematode and enhances rice yield.
- Incorporation of neem cake at 100 g/m^2 in the nursery area 15 days prior to the sowing checks the build-up of rice root-knot nematode.
- Summer solarization of nursery beds of rice for 15 days under tropical and subtropical conditions followed by nursery bed treatment with carbofuran at 0.3 g a.i./m^2 + main field application of carbofuran at 1 kg a.i./ha 40 days after transplanting is recommended against *M. graminicola*-infesting rice.
- Application of bioagent *Pseudomonas fluorescens* at 20 g/m^2 as nursery bed treatment reduces nematode populations.

16.19 Rice Root Nematode (*Hirschmanniella* spp.)

Rice root nematode (*Hirschmanniella* spp.), a migratory endoparasite of roots, occurs predominantly in rice soil. The nematode species are unique and ecologically adapted migratory endoparasites of rice and cause yield losses to the extent of 19 % in rice in West Bengal (Ahmad et al. 1984). Two species, viz. *Hirschmanniella* oryzae and *H. mucronata*, are economically important and widely distributed in rice-growing parts of India, and the occurrence of *H. gracilis* is doubtful in India (per. Comm. Dr. M.R. Siddiqi). The juveniles and adults penetrate through the entire length of roots and feed on cortical cells, leading to the formation of channels or cavities in the roots. Its feeding sometimes extends to the central vascular regions. The infected roots exhibit water-soaked brown lesions which are mostly spindle shaped. The physiological function of infected plants is disrupted and plant growth reduced. The aboveground symptoms are nonspecific as stunted growth, leaf chlorosis, reduced tillering and delayed flowering.

The population of *Hirschmanniella* species was found to be at maximum during the active growth phases of rice. The population build-up of this nematode increases after transplanting of rice up to 80 days (Singh and Jain 1995) and declines when the roots of rice start degenerating.

Rice root nematode survives better in poorly drained clay and heavy soils. It can survive even in high temperature of May to June (35–45 °C) as well as low temperature of December to January (8–12 °C) in the North Indian conditions (Mathur and Prasad 1973). Their survival in soil is much longer than in roots in flooded soils. *H. oryzae* can survive more than 12 months in wet soils. A number of weeds found in rice fields serve as alternative host for this nematode. In West Bengal, the nematode can survive in the months of summer in the absence of any crop under laterite soil conditions. Under rice–wheat cropping system, the nematode maintains a very high population, though wheat is not the host for the nematode. The spread of this nematode occurs mainly through irrigation water, flood water, soil adhering to farm implements, field workers and root of rice seedlings.

16.19.1 Management Options

Direct seeding of rice has been found to be more vulnerable to attack by this nematode compared to transplanted crop (Singh and Jain 1995).

Cultural Practices

- Early planting of rice in the month of June or middle of July
- Use of organic amendments such as mustard cake or neem cake at 220–240 kg/ha
- Balanced NPK fertilization
- Crop rotation with wheat, linseed, potato, cauliflower, mustard and gram in *rabi* season
- Deep dry summer ploughings
- Weeding during standing rice and in the absence of the crop
- Growing resistant varieties/cultivars: TKM-9, CR-142-3-2, CR-52, N-136 and W-136
- *Sesbania rostrata* can be used as trap crop for *H. oryzae*

Chemical Approaches

- Nursery bed treatment with carbofuran 3G or phorate 10 G at 1.0 kg a.i./ha followed by 1.0 kg a.i./ha at 7 and 50 days after transplanting
- Bare-root dip treatment with chlorpyriphos/carbosulfan 25 EC/monocrotophos 36 EC at 0.1–0.2 % for 20–30 min
- Seed soaking with carbosulfan 25EC or isofenphos at 0.2 % for 6 h

16.20 Rice Stem Nematode (*Ditylenchus angustus*)

Rice stem nematode is usually problematic in rice grown in deep-water situations. The vernacular name 'stem nematode' is derived from the stem inhabiting nature of the nematode. It is an obligate parasite and serious pest of rice causing popular disease symptom referred to as '*ufra* disease'. In Bangladesh and some areas of India, 100 % yield loss has been recorded due to severe attack of *D. angustus* (Table 16.4). The nematode has been found prevalent in Malda, Murshidabad, Hooghly, 24-Parganas (north and south), Jalpaiguri, Coochbehar and West Dinajpur districts of West Bengal and Sibsagar, Jorhat, Morigaon, Sonitpur, Barpeta and Dhubri in Assam (Phukan 1995).

Table 16.4 Yield loss in rice due to *Ditylenchus angustus* in South and Southeast Asia

Country/states	Yield loss (%)	Sources
India	40–100	Chakraborti et al. (1985)
Assam	30–100	Anonymous (1986)
UP	5–50	Singh (1953)
Maharashtra	–	Present, no further details
Orissa	–	Present no further details
West Bengal	10–30	Rao et al. (1986); Pal (1970)
Bangladesh	50–100	Catling et al. (1978); Mondal and Miah (1987); Rahman et al. (1994)
Indonesia	–	Present, no further details
Malaysia	–	Present, no further details
Myanmar	–	Present, no further details
Philippines	–	Present, no further details
Thailand	10–90	Hashioka (1963)
Vietnam	20–100	Kinh and Phuang (1981)

The symptom produced by rice stem nematode is popularly known as *ufra* or *dakpora* disease. *Ufra* symptoms appear in patches and subsequently spread to the entire field. The nematode attack at vegetative stage results in yellowing or whitish pattern on the leaf sheath (Fig. 16.9a) and the margin becomes corrugated. In due course of time, the splash pattern turns into brownish stains and the stem and internode become black. Twisting of leaf and leaf sheath is a commonly found symptom. Sometimes infested nodes give bushy appearance due to branching. The *ufra* symptoms (Fig. 16.9b) may be grouped as:

Swollen or *thor ufra*	Panicle does not come out; it remains enclosed within the leaf sheath and infected portion tending to branch
Pucca or ripe *ufra*	Panicle emerges partially and bears filled grains at the tip only

The primary sources of *D. angustus* infection are rice stubbles, straw, wild rice and weeds found in rice fields. The nematode can overwinter through the quiescent state (fourth-stage juvenile) which remains viable up to 15 months. They live in coiled anhydrobiotic state in grains (Prasad

Fig. 16.9 (**a**) *Ufra* symptoms at vegetative stage of rice. (**b**) Ufra disease symptoms of rice (Source: Jorhat Centre, AAU, India)

and Varaprasad 2001); dried plant parts are left in the fields and reinfect the crop in the next season.

16.20.1 Management Options

The management of *ufra* nematode in rice is relatively difficult for the deep-water rice-growing conditions. Cultural practices include adoption of resistant cultivars (IR63142-J8-B-2-1, Rayada 16-06, Rayada 16-07, Rayada 16-011, Rayada 16-017, CN- 540, NC-493, Jalamagna, etc.), summer ploughing, destruction of rice stubbles and early maturing rice cultivars like Padmapani, Rayada and Digha, following crop rotation with non-hosts as jute, mustard, etc. Soil application of carbofuran 3G at 0.75 kg a.i./ha followed by spraying of standing crop with carbosulfan 25EC or triazophos 40EC at 0.2 % controls the disease.

16.21 White Tip Nematodes (*Aphelenchoides besseyi*)

The leaf and bud nematode *Aphelenchoides besseyi* (Christie 1942) is a specialized parasite attacking aerial parts of its natural host, rice. Though rice is the most suitable host of this nematode, it can infect tuberose, onion, soybean, sugar cane, oat, millets, orchids, etc. The most characteristic disease symptom is 'white tip' in rice leaf produced by this nematode, hence the common name of the nematode, 'white tip nematode'. Dastur (1936) reported for the first time in India the occurrence of *A. besseyi* on rice from Madhya Pradesh, then the Central Province. The occurrence of *A. besseyi* in rice is widespread in India (Fig. 16.10); however, it is more frequently encountered in rice from some states like Gujarat, Tamil Nadu, Madhya Pradesh Andhra Pradesh and West Bengal. In West Bengal, the nematode is also widespread (Das and Khan 2007) and posing a major problem in the rice fields adjacent to tuberose fields (Khan 2001). On an average, the nematode inflicts yield losses 10–60 %. Infestation of white tip nematode in rice has been extensively reviewed (Khan et al. 2010).

It is easy to detect the presence of nematode within the rice seeds. In field, the initial appearance of symptoms includes an up to 5 cm leaf tip that becomes pale yellow or whitish at tillering stage and the subsequently dry leaves. These symptoms are found for a short period in the plant. The tip of the flag leaf is often twisted which may obstruct the emergence of panicles. Infested panicles are shorter and lighter in weight as compared to healthy panicles.

A. besseyi survives in pre-adult and adult stages (quiescent state) beneath the hull of rice

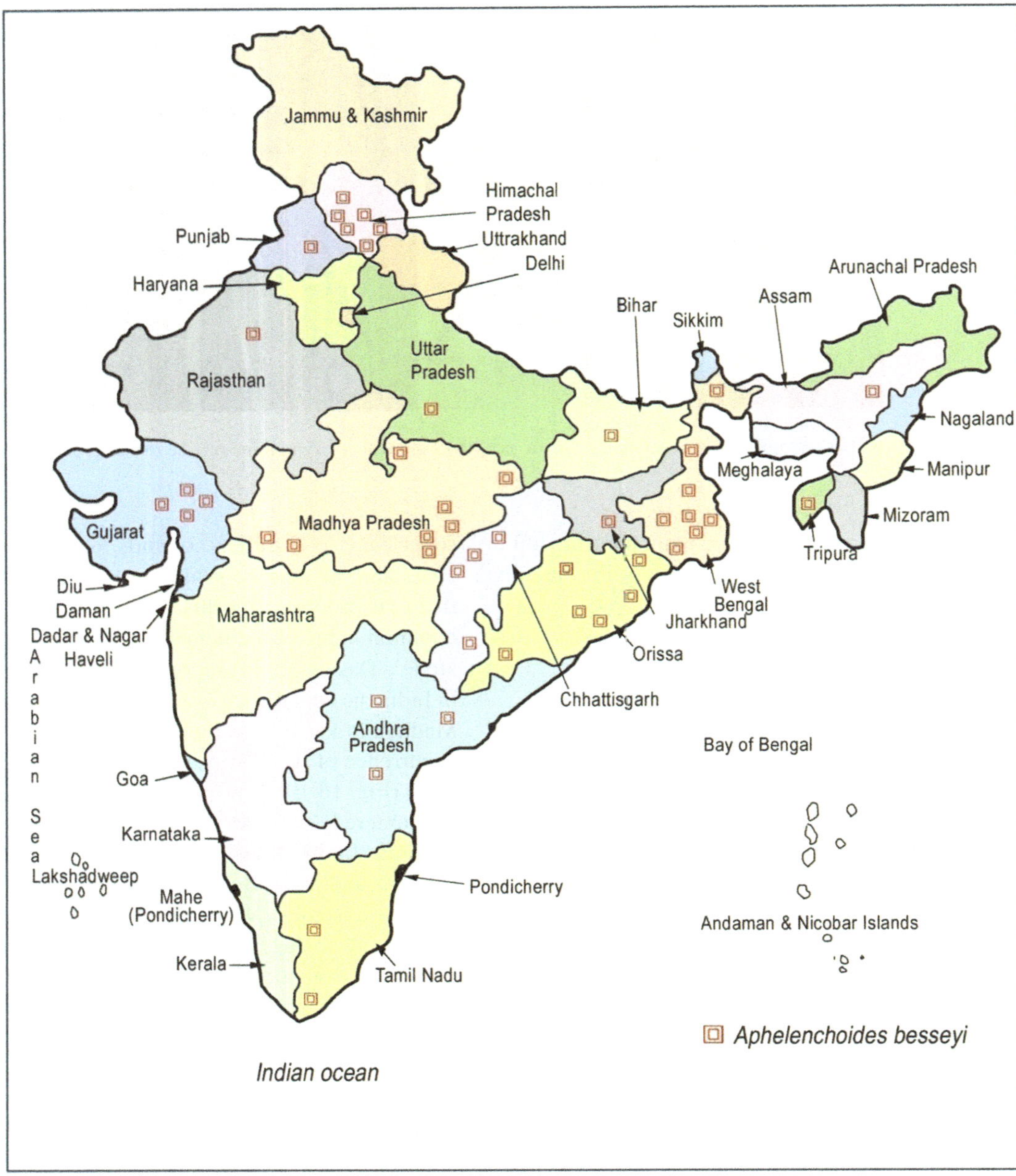

Fig. 16.10 White tip nematode (*Aphelenchoides besseyi*) infecting rice in different states of India

kernel and does not survive in soil after rice plants have been harvested. Infected seeds or the presence of other alternative hosts helps nematode to survive up to the next crop. They usually remain in coiled anhydrobiotic state in rice between lemma and palea up to 3 years. The infected rice seed is the only means for rapid spread of *A. besseyi*. It also spreads through irrigation water or flood water. Female lays eggs on rice plants. All the developmental stages occur in rice plants. The life cycle of *A. besseyi* is completed within 2 weeks, and therefore, several generations are completed within a cropping season.

Fig. 16.11 White tip disease symptoms of rice (Courtesy: T. Nishizawa and web-source)

16.21.1 Management Options

The nematode problem in rice could be managed by using healthy nematode-free seeds or disinfecting seeds by presoaking followed by hot water treatment at 52–54 °C for 10–15 min on area-wide basis. An infected seed can also be made nematode-free by following a simple method of spreading the rice seed on the concrete floor on bright sunny days at least 4 h for six consecutive days which kills nematodes inside the grain. Alternatively, seed soaking in water (1:2 ratio) overnight followed by adding two volumes of boiling water for 10 min and then drying of seeds in shade also proved effective to denematize the seeds. Seed treatment with carbosulfan 25EC at 0.1 % for 12 h is also recommended. In standing crop, timely spraying of crop with monocrotophos/cartap hydrochloride/oxamyl/benomyl at least twice at 15 days interval is useful for controlling the white tip disease in rice Fig. 16.11.

16.22 Rice Cyst Nematode (*Heterodera oryzicola*)

Rice cyst nematode (*Heterodera oryzicola*), sedentary endoparasite of rice, is considered as a serious pest of rice in Kerala. It was also known to occur in Burdwan and Bankura districts of West Bengal (Rao 1985). This nematode is one of the serious pests of rice and banana in Kerala (Kuriyan 1995), and its further occurrence in Karnataka and Goa was also reported (Prasad 2002).

The browning of roots and chlorosis of leaves, growth retardation, early flowering of plants by 10–15 days and partial filling of grains are the typical symptoms of *H. oryzicola* infection. No gall develops on rice roots. The presence of brown cyst on rice root confirms this nematode infestation. One life cycle is completed in 30 days and 12 generations may occur in a year. Yield losses could be as high as 38 % due to the attack of nematodes (Rao 1978). The nematodes spread through infested seedlings, irrigation water or farm implements.

16.22.1 Management Options

The nematode can be successfully managed by (i) soaking of seeds with fenamiphos at 0.02 % for 6 h, (ii) soil application of carbofuran or phorate 1 kg a.i./ha at 7 and 50 days after transplanting, (iii) growing of resistant varieties like Lalnakanda and CR143-2-2 and (iv) regulatory measures on the movement of banana rhizomes as well as rice seedlings from infested areas.

16.23 Burrowing Nematode (*Radopholus similis*)

This nematode is an internationally quarantined pest and is capable of parasitizing many fruits, spices and plantation crops. It is known to cause a serious disease of black pepper popularly referred to as 'pepper yellows' in Indonesia, 'slow wilt' in India and 'spreading decline' in Florida. Many economically important crops like banana, citrus, betelvine, coconut, areca nut, black pepper, ginger, etc. are seriously affected by the nematode. The nematode is a main problem in southern states like Kerala, Karnataka and Andhra Pradesh (Parvatha Reddy and Singh 1980). Recently, it has spread to the states like Orissa, Manipur, Maharashtra, Madhya Pradesh and Himachal Pradesh and Gujarat possibly through indiscriminate movement of planting materials of banana, ginger and turmeric rhizomes.

Nematode-infested banana plants exhibit *toppling disease* at bearing stage, premature defoliation, poor plant vigour and finally reduction of bunch size and weight. Lesion develops on roots and there is subsequent rotting and decaying of tender roots due to the involvement of other soil microorganisms. The coconut plants attacked by *R. similis* show general "decline" symptoms like yellowing, stunting, decrease in leaf size and button shedding, which resulted in low yield. In black pepper, slow growth of vines, yellowing of leaves, rapid panicle declines followed by severe dieback and death of vines are common symptoms. The roots of coffee, areca nut and betelvine are seriously attacked and develop characteristic lesions and root decay and rotting. The nematode is a migratory endoparasite of root and feeds on the succulent tissues of feeder roots. Due to their intracellular movement, the nematode destroys the cells and forms burrows or cavity inside the root. All the developmental stages are capable of feeding the roots. Eggs are laid in root tissues and newly hatched juveniles start feeding and develop inside the roots. Therefore, the entire life cycle may be completed within the roots. The total life cycle is completed in 20–25 days. In India, only the banana race of *R. similis* is prevalent.

16.23.1 Management Options

The nematode is difficult to manage due to its endoparasitic nature and wide host range.

Preventive Measures

- Planting materials (suckers/rhizomes/seedlings) should be free from nematode.
- Any occurrence of discolouration on the rhizome should be removed and treated with Bordeaux mixture or nematicides.
- Hot water treatment of rhizomes at 50–55 °C for 20–25 min could be done to denematize the planting materials.
- Raising of coconut, areca nut and black pepper seedlings from nematode-free nursery bed.
- Before planting, sun-drying of banana rhizomes is also effective to reduce nematode population.

Curative Measures

- Application of carbofuran 6 g a.i. or phorate 3 g a.i./plant 2–3 times in a year is effective to reduce the nematode in coconut and banana.
- Application of neem cake at 400 g/plant, with the first dose at planting and second dose after 4 months, increases bunch weight and the yield of banana.
- Avoiding susceptible crops as intercrop.
- Intercropping with *Crotalaria juncea* reduces *R. similis* population.
- Bioagents like *P. lilacinus*, *Glomus fasciculatum* and *Pasteuria penetrans* are promising agents against *R. similis*.

16.23.2 Resistant/Tolerant Varieties

Banana	Kadali, Pedalimoongil, Kunnan, Pey Kunnan, Pisang Seriby
Areca nut	Resistant: VTL-11x VTL-17 (Sundararaju and Koshy 1988)
Tolerant:	Indonesia-6 (VTL-11), Mahuva-B, Andaman-5 (VTL-29e)

Coconut	Kenthali, Klappawangi, Hybrid Java Giant (JG) X Kulasekharam Dwarf Yellow (KDY), KDY X JG, Java Tall X Malayan Yellow Dwarf, San Ramon X Gangabondan (Sosamma et al. 1980)

16.24 Citrus Nematode (*Tylenchulus semipenetrans*)

Citrus nematode is found in all the citrus-growing areas of the country and is widely recognized as an economically important pest of citrus. It is one of the causal factors for the *slow decline* in citrus, which is characterized by a general reduction in tree growth, lack of vigour, yellowing of foliage and small size of fruits. The nematode is a semi-endoparasite of citrus root. It causes symptoms that are often non-descriptive and difficult to diagnose. The nematode is often unnoticed in the seedlings in nursery which causes widespread distribution. The presence of nematode is best confirmed by microscopic observations of soil and root samples. The female nematodes and their gelatinous matrix-containing eggs adhere to soil particles and give the roots a dirty appearance which is not easily washed off. The most serious effects of the nematode on the growth and yield of citrus are usually encountered when new seedlings are planted on old orchard. This condition is known as 'citrus replant problem'. The young tree grows slowly and fruiting is delayed. This*slow decline* in infested trees implies a general deterioration of citrus trees beginning with the production of smaller and fewer fruits. The extent of decline in mature trees is related to their vigour, tolerance to nematode and the degree of infection.

T. semipenetrans feeds on the surface layers of roots causing discolouration and necrosis. A young female penetrates the deeper root tissues and establishes a feeding site around the head. The feeding site comprising 16 cortical cells is referred to as 'nurse cell'. The posterior part of mature female body remains outside and eggs are laid in a gelatinous matrix outside the host tissue. The life cycle of this nematode is completed within 6–8 weeks under optimum temperature at 25–31 °C.

16.24.1 Management Options

Prophylactic Measures

- Seedlings in nursery beds should be free from nematode infection.
- Previously infested citrus orchard should be either avoided or fumigated to kill any nematode population in soil.
- Runoff water from adjacent infested citrus orchard should be prevented.
- Clean equipment/implements for cultural practices should be used because movement of any adhering soil particles from one place to another in orchard may disseminate the nematode.

It is difficult to eliminate the nematode once it is established in orchard. Therefore, regular monitoring is essential to prevent the nematode from reaching damaging threshold levels. Usually the nematode at low population level (500/g of feeding root) is not damaging to the crop, but at high population (4000/g root), it causes devastating damage to citrus plants (Nickle 1991).

Curative Measures

- Application of oil cakes of neem, mahua, groundnut, etc. at 1.0 kg/plant can reduce nematode populations.
- Combined use of neem cake at 1 kg/plant along with carbofuran 3G (Furadan) 1.0 kg a.i./ha is also effective.
- Use of biocontrol agent like *Paecilomyces lilacinus* at 4 g/plant along with carbofuran 3G (Furadan) 1 kg a.i./ha gives good results (Parvatha Reddy and Nagesh 1995).
- Use of resistant 'trifoliate' citrus stock may be an efficient method to check the nematode.

16.24.1.1 Cultural Practices

Removal of old feeder roots before the start of growth flush followed by application of FYM helps to reduce nematode population in soil.

16.25 Tuberose Foliar Nematode (*Aphelenchoides besseyi*)

The foliar nematode *A. besseyi* in Hawaii (Holtzmann 1968), in West Bengal (Chakraborti and Ghosh 1993; Khan et al. 1999) and Orissa of India (Khan 2006a, b) and Mekong Delta of Vietnam (Cuc and Pilon 2007) is causing severe damage to tuberose. *A. besseyi* is now a serious problem in tuberose in West Bengal, and the disease 'floral malady' induced by the nematode species is recorded from Ranaghat areas of Nadia district of West Bengal (Chakraborti and Ghosh 1993). This nematode species is also a major problem in the Mekong Delta of Vietnam where the tuberose crop is severely damaged or completely destroyed (Cuc and Pilon 2007). It is now a major limiting factor for cultivation of tuberose in Ranaghat and Haringhata regions of Nadia, Rajarhat of 24-Parganas (north), Bangaon of 24-Parganas (south) and some pockets of Howrah and Midnapore districts of West Bengal. The high population of *A. besseyi* is also recorded from fields of Kolaghat, Panskura-II, of Midnapore district. The 'Calcutta Single' cultivar of tuberose is most vulnerable to nematode damage as compared to 'Double' cultivar. The nematode species cause severe infestation of tuberose and induce malformed flowers (Khan et al. 1999; Khan 2001). The population of *A. besseyi* causing 'white tip disease' in rice is the same population infecting tuberose and causing floral disease (Khan 2001). The foliar nematode is now a key nematode pest of tuberose, and it causes yield losses as high as 59 % in West Bengal (Pathak and Khan 2009). Infestation of foliar nematode is a potential threat for cultivation of tuberose in West Bengal and Orissa of India and Mekong Delta of Vietnam.

16.25.1 Symptoms

The disease incidence of foliar nematode in tuberose (cv. Calcutta Single and Calcutta Double) is estimated on a 0–4 scale: (0) (no malady symptoms on flower stalk), (1) (distortions at the basal part of flower stalk but exhibit flower bloom), (2) the entire flower stalk exhibits distortions, but few flowers bloom at the tip), (3) the entire flower stalk is distorted but no flower bloom) and (4) (complete sterility of flower stalk or blind head) (Khan 2004). Cuc et al. (2010) also proposed to estimate disease severity of foliar nematode into four categories: DS1 (normal plant height, slightly short panicle compared to healthy plants, some flowers fail to open), DS2 (short plant height, short panicle, many flowers fail to open), DS3 (very short height, very short panicle, many flowers fail to open) and DS4 (very short height, very short panicle, flower buds fail to develop and severe browning of stems, leaves and buds). Early symptoms are rather restricted to small water-soaked spots near the midrib. These spots enlarge along the midrib, causing elongated, black, greasy spots measuring from 2 to 6 in. long and finally causing the leaf to bend, wilt and dry. Nematodes are usually found in great abundance at the margin of the spots. However, in Indian conditions, the diseased flower is characterized as a floral malady of tuberose (Chakraborti and Ghosh 1993). The foliar nematode-infected tuberose flower stalk initially appears rough, stalk becomes crinkled, stunted and finally distorted (Plate 1) and in severe cases flower buds failed to bloom. These brown streaks appear on leaf bracts and petals and subsequently develop rusty brown spots. The severely infected flower stalk becomes rotten and brittle over drying, even get blind. The nematode *A. besseyi* remains in masses forming 'nematode wool' which could be easily recovered from dark brown spots (Khan and Pal 2001). The ovary contains a large number of nematodes. This nematode is more serious during the rainy season generally from July to September and percent loss of second year crop occurs particularly in tuberose cv. Calcutta Single. However, in the cv. Calcutta Double, 30–40 % flower stalk rendered unsaleable and individual flower stalk contained up to 45,000 nematodes (Khan 2004, 2006b).

16.25.2 Bionomics

The nematode is a foliar ecto- and endoparasite on tuberose. *Aphelenchoides besseyi* primarily disseminates through infested bulbs, dry plant parts and runoff and irrigation water from one field to other fields. The nematode survives in coiled anhydrobiotic condition (quiescent pre-adult and adult stages) in the scaly leaves outside the bulbs. The nematode can also survive in all stages (eggs, juveniles and adults) on the flower stalk of tuberose. It takes 10–12 days to complete life cycle at 30 °C (Khan and Ghosh 2009). The nematode can also survive in the dried scaly leaves, stems and flowers for more than 25 months; however, they cannot survive in soil under field conditions (Khan 2004; Khan and Ghosh 2011a). The nematode is a serious problem in areas where rice and tuberose are cultivated in the cropping sequence. The fungi *Fusarium oxysporum* and *Alternaria alternata* are associated with the floral malady in tuberose (Chakraborti and Ghosh 1993); however, they do not play any role in the development of typical foliar disease in tuberose (Khan 2004).

The peak period of multiplication occurs during March to August under West Bengal conditions, but least multiplication occurs during winter months. The temperature and RH influenced the nematode infestation in tuberose. The nematode maintains high population (1084–2768 per flower stalk) during July to November (at temperature of 25–33 °C) and the lowest population (87 per stalk) in the month of January (at temperature of 10–26 °C) in tuberose (Khan 2004). The canopy temperature of nematode-infested plant is higher than that of healthy crop, and morning dew deposition is significantly correlated with nematode infestation. Further, dew deposition at 7 days lag phase is considered as a good indicator for predicting nematode population in tuberose (Chowdhury et al. 2010).

16.25.3 Management

The foliar disease in tuberose was controlled with nuvacron (monocrotophos) at 0.15 % and metacid (parathion-methyl) at 0.15 % (Chakraborti 1995; Khan et al. 2006, 2008). The monocrotophos 36SL is highly effective against *A. besseyi* in rice (Kumar and Sivakumar 1998). The field efficacies of carbosulfan 25EC at 750 ppm, cartap hydrochloride at 700 ppm, NSKE at 10,000 ppm and *Pseudomonas fluorescens* at 2 g/l of water are in decreasing order; however, in terms of cost–benefit ratio, the order of preference is *P. fluorescens* at 2 g/l of water > NSKE at 10,000 ppm at 2 ml/l of water > carbosulfan 25 EC at 750 ppm > cartap hydrochloride 50 SP at 700 ppm (Pathak and Khan 2010). Carbosulfan 25EC at 0.1 % (Chakraborti 1995), cartap hydrochloride at 1.5 g/l of water and NeemAzal 1 % at 2 ml/l (a neem-based formulation) are also effective as bulb treatment for disinfection of *A. besseyi* (Khan et al. 2009). The foliar application of pongamia oil (nomite) and NeemAzal 1 % (azadirachtin) along with bulb treatment and foliar spraying with benomyl, carbosulfan, metacid and monocrotophos are also effective against the nematodes (Khan et al. 2006, 2008).

The hot water treatment of bulb is the most effective for managing the foliar nematode (Khan et al. 2005a). The thermal death point of *A. besseyi* is 48 ± 2 °C for 5 min, while the tolerance of the bulbs is 50 ± 2 °C for 30 min for germination to temperature (Khan et al. 2005a; Khan and Ghosh 2011b). The presoaking of bulbs overnight followed by hot water treatment (HWT) at 50 °C for 30 min + dipping of bulbs in monocrotophos 36SL in 500 ppm for 6 h + two sprayings with monocrotophos 36SL at 500 ppm in the first, second and third years of crop with three sprayings with monocrotophos 36SL at 500 ppm at 15 days interval effectively reduced foliar disease, suppressed nematode population and yielded quality flowers (Khan et al. 2005). Integration of bulb treatment (presoaking overnight followed by HWT at 50 °C for 20 min.) and 2–3 foliar spraying with monocrotophos 36SL at 0.15 % at 30 days interval effectively controlled (Khan and Ghosh 2009). The HWT consisting of soaking bulbs in water for 30 min at 57 °C is the most efficacious method to protect tuberose from *A. besseyi* and to produce healthy flowers in a cost-effective manner in Mekong Delta of Vietnam (Cuc et al. 2010).

Among the varieties of tuberose tested (Shringer, Calcutta Single, Calcutta Double, Hyderabad Single, Hyderabad Double, Suvasini,

Vaibhav, Prajwal and Phule Rajani), Prajwal and Shringer are tolerant to *A. besseyi* (Khan and Ghosh 2007; 2009). The suggested management strategies against foliar nematode in tuberose (Khan 2006a, b; Khan et al. 2012) are given below:

- The planting material (bulbs) should be soaked overnight either in plain water or in 5 % neem seed kernel extract (home preparation from locally available neem), or bulbs are dipped in monocrotophos 36SL at 500 ppm for 6 h.
- After sprouting of the bulb, 3–4 sprayings with monocrotophos 36 SL at 500 ppm (use sticker) at 15–20-day interval should be given.
- In the second and third year crop, sprayings with monocrotophos 36 SL at 500 ppm at 15–20-day interval starting from the month of April to May onwards reduce the nematode infestation.
- Clean cultivation of tuberose and any infested parts of plants found in the field should be removed and burnt on inspection.
- Grow tolerant cultivars of tuberose such as Prajwal, Phule Rajani and Shringer.

16.26 Chrysanthemum Foliar Nematode (*Aphelenchoides ritzemabosi*)

The foliar nematode (*Aphelenchoides ritzemabosi*) is popularly known as 'chrysanthemum foliar nematode', and the disease could be easily diagnosed after recovering the nematode from the infested leaves of plants. Upon infection, chrysanthemum plant usually exhibits elongated to angular spots. The infested plants show fewer leaves that are very often lower, are in varying stages of dryness. and have finally shed off. Usually, the infested plants are poor in vigour, have smaller leaves with shrunken appearance and bear fewer undersized flowers. Several ornamental plants like *Zinnia elegans* (Ahmed et al. 1966), *Salvia splendens* and species of chrysanthemum, dahlia and aster grown in Shimla, India, suffered from foliar nematode damage with sickly and stunted appearance during the rainy season (Gill 1981). The leaves of such plants exhibited small yellowish white spots which later gradually turned into dark brown. The discolouration at its initial stages of attack was mostly confined within leaf veins, but with the progress of the disease, two or more adjoining spots merged to give bigger appearance.

All these plants, *Zinnia elegans*, *Salvia splendens*, *Aster* sp. and *Dahlia* sp., are attacked by *A. ritzemabosi* with varying levels of leaf infection (Table 16.5). In *Z. elegans*, in addition to numerous small rectangular spots, spots of triangular to pentagonal shapes also appear, while in *S. splendens*, the number of spots is less but the size is almost double that of *Z. elegans* and *Dahlia*. *Aster* species also exhibit the symptoms similar to *S. splendens*, but the intensity of nematode infection is mild (Gill 1981). The severity of infestation in nature is given below.

Zinnia plants show heavy infestation and sometimes suffer more severe damage as compared to chrysanthemum depending upon the prevailing conditions. The development of disease symptoms and disease severity on chrysanthemum primarily depends on host varieties and nematode races (Gill 1981). He suggested the following approaches for effective management of the nematode:

- Keep the leaves dry to avoid the leaf injury by this nematode under glass house conditions.
- Collect cuttings for the propagation of plants from the tips of the tallest branches and not from the growth near the base of the plants to reduce the nematode incidence.

Table 16.5 Severity of infestation due to *A. ritzemabosi* in five hosts (Gill 1981)

Crop species	Family	% infection in leaves
Zinnia elegans	Compositae	++++
Salvia splendens	Labiatae	+++
Chrysanthemum sp.	Compositae	++
Aster sp.	Compositae	+
Dahlia sp.	Compositae	+

+=up to 15 %; ++=26–50 %; +++=51–75 %; ++++=76–100 % infestation

- Thorough cleaning and burning of all possible infested plant materials.
- The infested cutting can be denematized by submerging in hot water at 50 °C for 5 min or at 44.4 °C for 30 min.
- Application of phorate or aldicarb 10G at 1.5 kg a.i./ha as a basal dressing at the time of transplanting gives good control of this nematode, whereas methyl parathion or chlorpyriphos at 0.05 % as foliar spray at fortnightly intervals gives good protection to the foliage and blooms against the nematode attack.

16.27 Reniform Nematode (*Rotylenchulus reniformis*)

Reniform nematode is the next most important nematode pest of transplanted vegetable crops. The first report of *R. reniformis* feeding on cowpea was made by Linford and Oliviera (1940). About 10 valid species of *Rotylenchulus* are known to exist over the world (Gaur and Perry 1991). *R. reniformis* is the most prevalent species and attacks over 150 plant species of 50 families (Khan 2005). In India, this nematode was first recorded by Das (1960), Siddiqi (1961) and Seshadri and Sivakumar (1963), and now it is known to occur in almost all states of India. The adult female of reniform nematode (*R. reniformis*) is kidney shaped, obligate and sedentary semi-endoparasite. In West Bengal, *R. reniformis* is associated with banana, many vegetables, tuberose, tea, pulses, fruits, betelvine, etc. It is associated with old-age banana decline.

16.27.1 Hosts

Many important crops are attacked by the reniform nematodes. Among the vegetables, tomato, brinjal, okra, potato, cowpea, sweet potato, beans and onion are the ones mostly affected with the infestation of the nematode species. In India, tomato, brinjal, onion, pointed gourd, etc. have been reported as hosts for this nematode (Jain 1992).

16.27.2 Races/Biotypes

Dasgupta and Seshadri (1971) established the occurrence of two distinct races/biotypes designated as race A and race B in *R. reniformis*. Race A can complete its life cycle on cowpea, castor and cotton while race B on cowpea only. Prasadrao and Ganguly (1996) reported the occurrence of four physiological races based upon the reproduction capabilities on cotton, pearl millet, cowpea, castor and mustard.

16.27.3 Crop Loss

Reniform nematode is known for its pathogenic behaviour in vegetable crops. The yield losses due to *R. reniformis* in tomato ranged between 21 and 49 %. *R. reniformis*-infested pointed gourd showed slower growth, reduced size, thinner veins, reduced fruit size and significantly lower yield (Nath et al. 1976).

16.27.4 Symptoms

Symptoms of damage to crops are not specific. It feeds on cortical tissue, phloem and pericycles, causing necrosis on roots of many crops. Root discolouration, leaf shedding and malformation of fruits and seeds also occur. *R. reniformis* also interacts with other pathogens like *Fusarium* spp., *Verticillium* spp., *Sclerotium rolfsii* and *Rhizoctonia solani* leading to disease complexes. It can also parasitize the bacterial nodules.

16.27.5 Survival Mechanism

Nematode survives in air-dried soil for a long time. Retention of moulted cuticles of the previous stages is a unique adaptation for survival of the nematode in soil. Individual young females, males and fourth-stage juveniles survive in a coiled anhydrobiotic state within encrusted cuticles in soil. Survival increases when decreasing soil moisture starts. Alternate drying and wetting of soil result in a sharp decline of population density of the nematode.

16.27.6 Biology

The first moult occurs within eggs and eggs are hatched in water independent of root exudates. Juveniles develop to pre-adult stage without any host tissue to feed on, quickly completing three superimposed moultings. The young female is the only infective stage. Upon infection, the young female orients itself perpendicularly to the longitudinal axis of roots leaving its posterior part outside the root. After establishing the feeding site, it assumes a kidney shape with its greater part protruding outside the root. Egg laying starts within 7–10 days after invasion and eggs are laid into a gelatinous matrix secreted by six specialized cells around the vagina. Each egg mass contains 30–200 eggs. It takes 3–4 weeks to complete life cycle depending upon temperature and host suitability. The life cycle of *R. reniformis* on cotton is depicted in Fig. 16.12.

16.27.7 Management Options

16.27.7.1 Non-chemical Approaches

- Crop rotation with non-host crops like mustard, maize, sugar cane and marigold
- Growing susceptible crops in winter seasons in multiple cropping systems
- Application of oil cakes like neem, karanj, mustard, FYM, etc. have been found promising.
- Deep summer ploughings for two to three times during the hot months (May to June)
- Irrigation between ploughings results in alternate drying and wetting, which may stimulate exsheathment of *R. reniformis* young female
- Good crop cultivation practices like field preparation, fertilizer application and moisture management
- Resistant varieties: Cowpea (Pusa Phalguni, C-152, RC-48), papaya (Solo, Washington, Coorg Honey Dew), onion (Evergreen), chickpea (BG-425, BG-426, BG-434, BG-268, BG-273), chilli (Pusa Jawala) (in Gaur and Perry 1991).

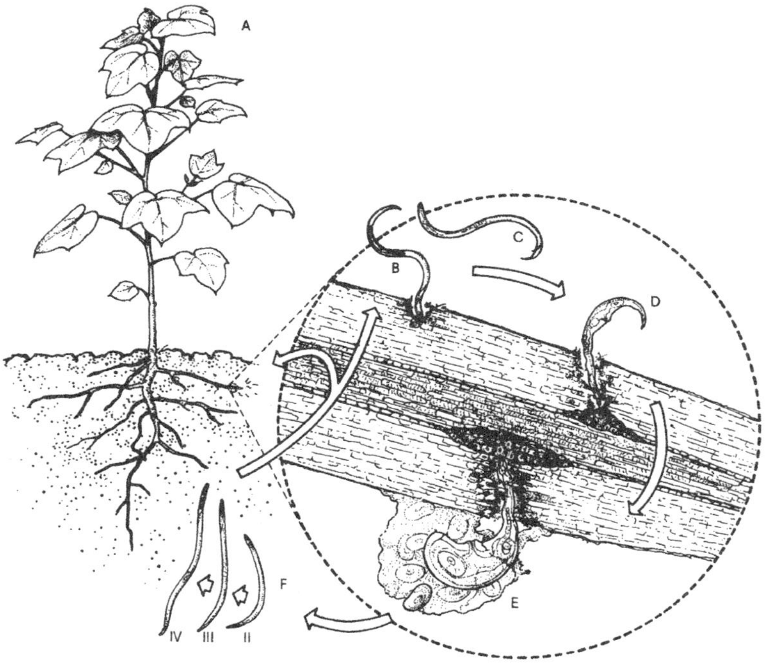

Fig. 16.12 Life cycle of *Rotylenchulus reniformis* on cotton (Robinson et al. 1997)

16.27.7.2 Chemical Approaches

Seed dressing or seed soaking, bare dip of clone/planting materials and nursery treatment are effective methods of nematicide application to protect the young tender stage of crops. Soil application of carbofuran (Furadan 3G) at 2 kg a.i./ha gives good control but may not be economical and eco-friendly for many crops. However, judicious use of nematicides may be adopted by restricting their use at nursery bed, seed treatments, bare dipping of vine cuttings and pit application particularly for transplanted crops.

16.27.7.3 Integrate Approach

Deep ploughings (20 cm) followed by fallowing for 1 month or fallowing for 1 month after weeding or integration of aldicarb application at 0.8 kg a.i./ha at sowing have been found to be effective (Sivakumar et al. 1973). It has been observed that combined application of organic amendments and nematicides helps in reducing the incidence of *R. reniformis* infecting tomato (Badra and Mohamed 1979). Incorporation of non-host *Capsicum annuum* and application of carbofuran/aldicarb/phorate/disulfoton at 1.0 kg a.i./ha, 15 days after transplanting of tomato, proved 89 % reduction of nematode population (Sivakumar et al. 1979).

16.28 Lesion Nematodes (*Pratylenchus* spp.)

More than 68 valid species of *Pratylenchus* are known worldwide, and it has a very wide host range (>400 host plant species). This nematode occupies the third position after root-knot and cyst nematodes for its economic impact on many crops. Lesion nematodes (*Pratylenchus* spp.) are migratory endoparasites of the root. The vernacular name 'root lesion nematode' is derived from the necrotic discoloured patches (lesion) which develop on roots. W.B. Mountain in the 1950s first demonstrated conclusively that lesion nematodes are pathogens of plants and are capable of producing root lesions.

The crops of economic importance attacked by the nematode include wheat, maize, cotton, potato, rice, coffee, banana, tea, vegetables, ornamentals and fruits. Some nematode species like *Pratylenchus thornei* in wheat, soybean, chickpea, sunflower and opium, *P. zeae* in maize, *P. indicus* in rice, *P. loosi* in tea (Dasgupta et al. 2010), *P. coffeae* in coffee and banana and *P. pratensis* and *P. vulnus* in fruits are serious problems.

The aboveground symptoms caused by the nematode are non-descriptive. The nematode usually infects roots, rhizomes, pods, tubers, etc. Having penetrated into the roots, they multiply in large numbers. All the stages (juveniles and adult) of this nematode are infective. The nematode takes 4–8 weeks to complete a life cycle (Fig. 16.13). Symptoms of nematode infestation often go unnoticed and initial infection appears as spotty along the root surface. The infected plant's root exhibits dark red brown lesions caused by necrosis of the invaded cells. Root lesion is the most characteristic symptom. The lesions initially appear as small, elongate, water-soaked spots that later turn brown to black. Loss of primary roots, pruning or decay of roots, reduced size of blossoms and shrinking of grains are also associated with the nematodes. Several secondary soil-borne fungi and bacteria are also involved in the rotting and decay of roots, and thus normal functioning of infested roots is heavily impaired. The interaction of *P. penetrans* and Verticillium wilt fungi (*V. dahliae*) induces a disease complex known as 'potato early dying syndrome'. This syndrome can lead to premature vine death and severe yield losses that can become a limiting factor in potato production. The nematode also infects potato tubers and causes a scabby appearance (depending upon species) with sunken lesions or dark, wart-like bumps that turn purple on tubers in storage (Davis and MacGuidwin 2000).

16.28.1 Management Options

- Summer ploughing of field reduces nematode populations.
- Seed dressing at 2–3 % (w/w) or soaking of pulse crops with carbosulfan at 0.1 % reduces nematode infestation.

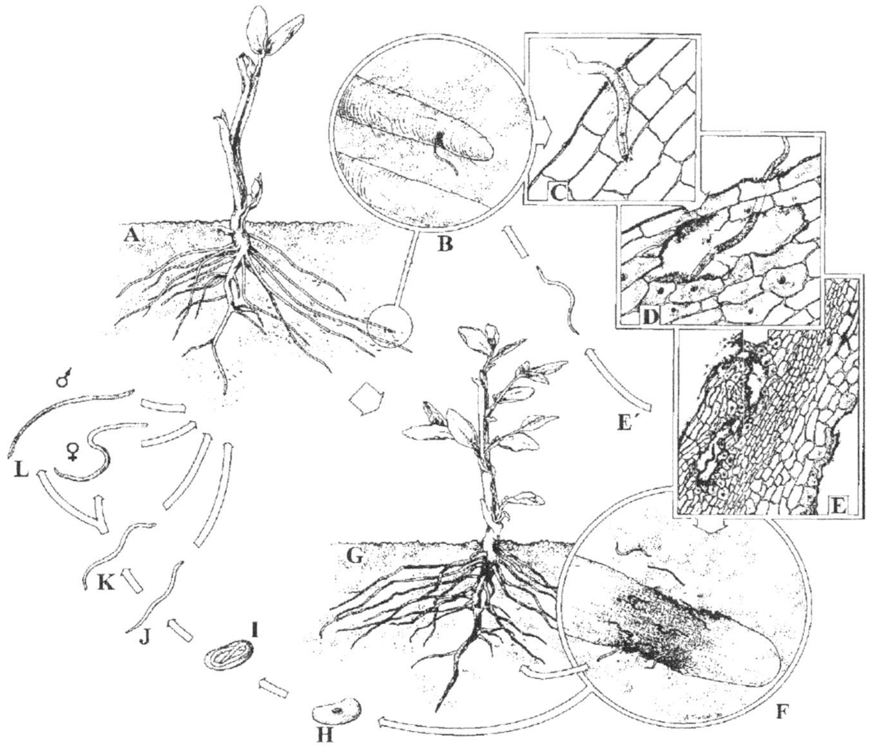

Fig. 16.13 Life cycle of lesion nematode (*Pratylenchus* spp.) (Vovlas and Troccoli 1990)

- Application of carbofuran (Furadan 3G) at 1 kg a.i./ha at sowing reduces crop damage caused by soil nematodes.
- Growing antagonistic crops like marigold (*Tagetes patula*) cv. Harmony in autumn after main crop or in between rows of main crop.
- Hot water treatment of bulbs, corms, tubers and fleshy roots can kill the dormant nematodes inside the root.

16.29 Pigeon Pea Cyst Nematode (*Heterodera cajani*)

Pigeon pea cyst nematode (*Heterodera cajani*) is the only species of cyst nematode that parasitizes a large number of leguminous crops. This nematode species was reported and described for the first time on pigeon pea from India (Koshy 1967), and now it is prevalent and gaining importance in almost all pigeon pea-growing states in India and accounting yield losses as high as 30 %. Some populations of this nematode are also known to attack sesame. The species is distinctive for having large egg sac (almost double of its cyst size). The nematode completes its life cycle (one generation) in 16 days at 29 °C and in 45–80 days at 18–25 °C (Koshy and Swarup 1971). It can also reproduce parthenogenetically, though it is a bisexual species. During a cropping season, it can quickly multiply and build up a huge population. At the seedling stage of the plant, pearl-like or lemon-shaped white female can be found attached with the roots. Cysts, the resting structure, contain the eggs derived from the dead female body and persist in soil for several years. The infective stage (J2) emerges from the cysts, penetrates the host roots and establishes a specialized feeding site (syncytium) in the stele. They later develop into swollen females, which retain the eggs. Infected plants show yellowing, stunting, poor vigour and pod formation. Several crops like pigeon pea, cowpea, mungbean, moth bean, cluster bean, garden pea, soybean, black gram and sesame are seriously damaged by the nematode.

16.29.1 Management Options

- Summer ploughing of fields during hot months.
- Crop rotation with non-host crops for 2–3 years.
- Soil application of carbofuran (Furadan 3G) at 1–2 kg a.i/ha along with soil amendments (neem cake, mustard cakes) at 5 q/ha.
- Bioagent like *Pasteuria penetrans* (Pasutsuria 50WP) may be used to suppress soil population.

16.30 Ectoparasitic Nematodes

Several ectoparasitic nematodes are emerging as the new problems of crops. In spite of being numerous in soil and attaining much adaptive biological features of survival and multiplication as compared to endoparasites, they are regularly ignored as nematode pest. Many genera like *Tylenchorhynchus*, *Hoplolaimus*, *Helicotylenchus*, *Paratylenchus*, *Hemicriconemoides*, *Hemicycliophora*, *Criconemoides*, *Xiphinema*, *Longidorus*, *Trichodorus*, etc. are prevalent in the rhizosphere of agricultural, horticultural and forest crops, and their pathogenic potential has been proven and their damaging nature is well documented. However, ectoparasitic nematodes are given least attention because the symptoms they induced on plant are not easily convincing and are often confused with other soil problems and soil pathogens. The role played by this group of nematodes is much more dangerous, particularly when they interact with other soil microorganisms, thereby making the plants vulnerable to weak pathogens.

In modern agriculture, intensive and extensive cultivation of the same crops, changes in cultivation practices like high yielding varieties, poor organic nutrition in soil, indiscriminate use of agrochemicals, etc. resulted in shift of pest status. For instance, *Tylenchorhynchus brevilineatus*, a polyphagy ectoparasite of diverse crops, has appeared as a serious problem as *Kalahasti malady* of groundnut in Andhra Pradesh accounts for 20–60 % yield losses (Reddy et al. 1984). This nematode is also reported to cause concern in tobacco in Gujarat. There are many instances wherein *T. brassicae* in cabbage and cauliflower, *Hoplolaimus indicus* in rice and jute, *Helicotylenchus multicinctus* in banana, *Paratylenchus* and *Criconemoides* in apple and peach, *Xiphinema basiri* and *Hemicriconemoides* in citrus and grape and *Paralongidorus* in sal are emerging as nematode problems (Khan and Ganguly 1995) in the changing agricultural scenario in India.

16.30.1 Management Options

Ectoparasitic nematodes are very much vulnerable to summer ploughing which expose and break the life cycle of many pathogens including nematodes. The manipulation of cultural practices like clean cultivation, crop rotation, organic amendments, etc. will reduce their soil populations.

16.31 Future Approaches and Conclusion

PPNs constitute one of the major limiting factors for cultivation of crops. The changes in agricultural situations have tremendous effects on the emergence of new nematode problems in India. The recent outbreaks of *M. graminicola* in the Mandya district of Karnataka, West Bengal, Orissa and Assam; *floral malady* (*Aphelenchoides besseyi*) in tuberose in West Bengal and Orissa; *Kalahasty malady* (*Tylenchorhynchus brevilineatus*) in groundnut in Andhra Pradesh (AP); *Meloidogyne indica* in kagzi lime and Bt cotton in Gujarat; *M. incognita* in pomegranate in Maharashtra, Karnataka, Gujarat and Andhra Pradesh; and *Pratylenchus thornei* in wheat, maize, chickpea and soybean are few examples of current threats and serious concerns. However, PPNs are neglected pests of crops and considered as of low priority for crop production and protection in India. The economic significance of nematodes in agriculture is very often underestimated, and their damage potential is not recognized by the plant protection specialists, scientists and administrators. As an important component of integrated pest management, nematode pathogens cannot be ignored and they could be tackled with the intelligent planning of nematode-suppressive crop sequences, summer ploughing, organic manuring, clean cultivation, adjusting sowing time, water and irrigation management and sensible use of nematicides. The increasing concerns on the ill effects of chemical pesticides on the environment have driven recent research interests on the use of several alternative strategies like botanicals, biopesticides and cropping system research for management of insect–pest diseases including nematodes. Some of the successes have been obtained for managing plant parasitic nematodes with neem-based formulations, fungal formulation of *P. lilacinus* and *Trichoderma* spp. and bacterial formulations of *Pasteuria penetrans* and *Pseudomonas fluorescens* strain. However, wide adoption of bioagents in field scale is still a fantasy rather than reality because of their inconsistent efficacy in different agro-ecological situations. Integration of more than one option can be explored on the basis of their compatibility, economic viability and availability to the farmers. Developing holistic approaches for managing field problems including nematodes inducing disease complexes in concert with other pathogens like fungi, bacteria and viruses is the major challenge ahead for the greater benefit of the growers.

References

Abad P, Williamson V (2010) Pant nematode interaction: A sophisticated dialogue. Adv Bot Res 53:148–192

Abdel-Momen SM, Starr JL (1998) *Meloidogyne javanica–Rhizoctonia solani* disease complex of peanut. Fundam Appl Nematol 21:611–616

Agrios GN (1987) Plant pathology, 2nd edn. Academic, New York

Ahmad N, Das PK, Baqri QH (1984) Evaluation of yield losses in rice due to *Hirschmanniella gracilis* (de Man 1888) Luc and Goodey, 1963 (Tylenchida: Nematoda) at Hooghly (West Bengal). Bull Zool Surv India 5:91

Ahmed S, Singh S, Misra SP (1966) *Aphelenchoides ritzemabosi* on *Zinnia elegans* at Simla. Ind Phytopathol 19:225–226

Ali SS, Naimuddin, Ali M (2010) Nematode infestation in pulse crops. In: Khan MR, Jairajpuri MS (eds) Nematode infestation, Part-I: Food crops. The National Academy of Sciences, Allahabad, India, pp 288–325, 325 p

Anonymous (1985) Extn Bull. No.21 (E). Central Potato Research Institute, Shimla, India, 4 pp

Anonymous (1986) All India coordinated research project on plant parasitic nematodes with integrated approaches and their control. Agricultural Research Institute, New Delhi, India

Anonymous (1988) Nematode management in horticultural crops, Technical bulletin No.4. Indian Institute of Horticultural Research, ICAR, Bengaluru, India, 52 p

Anonymous (1995–2001) Quinquennial review team report of All India Coordinated Research Project on Plant Parasitic Nematodes, Project Coordinating Cell, Division of Nematology, Indian Agricultural Research Institute, New Delhi, India

Anonymous (2001) Annual report of all India coordinated research project on plant parasitic nematodes. Directorate of Research, Bidhan Chandra Krishi Viswavidyalaya, Kalyani

Arya R, Saxena SK (1999) Influence of certain rhizosphere fungi together with *Rhizoctonia solani* and *Meloidogyne incognita* on germination of 'Pusa Ruby' tomato seeds. Ind Phytopathol 52:121–126

Atkinson GF (1892) Some diseases of cotton. Bull Alabama Agric Exp St 41:61–65

Back MA, Jenkinson P, Haydock PPJ (2000) The interaction between potato cyst nematodes and Rhizoctonia solani diseases in potatoes. In: Proceedings of the Brighton crop protection conference, pests and diseases. British Crop Protection Council, Farnham, UK, pp 503–506.

Back MA, Haydock PPJ, Jenkinson P (2002) Disease complexes involving plant parasitic nematodes and soil borne pathogens. Plant Pathol 51:683–697

Badra T, Mohamed MI (1979) Influence of combination of organic managements and nematicides on tomato infected with *Rotylenchulus reniformis* and associated populations of predacious and saprophytic arthropods and microphagous nematodes. Ind J Nematol 19:101–110

Bajaj HK, Kanwar RS (2005) Parasitization of maize by Heterodera avenae and *H. filipjevi*. Nematol Medit 33:203–207

Bertrand B, Nunez C, Sarah JL (2000) Disease complex in coffee involving *Meloidogyne arabicida* and *Fusarium oxysporum*. Plant Pathol 49:383–388

Bhatt J, Vadhera I (1997) Histopathological studies on cohabitation of *Pratylenchus thornei* and *Rhizoctonia bataticola* on chickpea (*Cicer arietinum* L.). Adv Pl Sci 10:33–38

Castillo P, Mora-Rodriguez MP, Navas-Cortes JA, Jimenez-Diaz RM (1998) Interactions of *Pratylenchus thornei* and *Fusarium oxysporum* f.sp. *ciceri* on chickpea. Phytopathology 88:828–836

Catling HD, Alam S, Miah SA (1978) Assessing losses in deep water rice due to insects and diseases in Bangladesh. Exp Agr 14:1–11

Chakraborti HS (1995) Chemical control of floral malady of *Polianthes tuberosa* L. J Mycopathol Res 33:25–28

Chakraborti HS, Ghosh SC (1993) Studies on the floral malady of *Polianthes tuberosa* L. in West Bengal. J Mycopathol Res 31(2):109–115

Chakraborti HS, Nayak DK, Pal A (1985) *Ufra* incidence in summer rice in West Bengal. Int Rice Res Newlet 10:15–16

Chitwood DJ (2003) Research on plant-parasitic nematode biology conducted by the United States Department of Agriculture-Agricultural Research Service. Pest Manag Sci 59:748–753

Chowdhury S, Ghosh S, Nanda MK, Khan MR (2010) Effect of foliar nematode, *Aphelenchoides besseyi* infestation on microclimate of tuberose. In: Pandey V (ed) Agrometeorological services for farmers. Anand Agricultural University, Anand, pp 252–259, 336p

Crosse JE, Pitcher RS (1952) Studies in the relationship of eelworm and bacteria to certain plant diseases. I. The etiology of strawberry cauliflower disease. Ann Appl Biol 39:475–484

Cuc NTT, Pilon M (2007) An *Aphelenchoides* sp. nematode parasitic of *Polianthes tuberosa* in the Mekong Delta. J Nematol 39:248–257

Cuc NTT, Thanh Son N, Trung TM, van Trang N, Dang LM, Pilon M (2010) Hot water treatment prevents *Aphelenchoides besseyi* damage to *Polianthes tuberosa* crops in the Mekong Delta of Vietnam. Crop Protec 30:1–4

Das VM (1960) Studies on the nematode parasites of plants in Hyderabad (Andhra Pradesh, India). Z Parasitenk 19:553–605

Das TK, Khan MR (2007) Occurrence and distribution of white tip nematode, *Aphelenchoides Besseyi* in West Bengal, India. Ind J Nematol 27(1):94–97

Dasgupta MK (1988) Principles of plant pathology. Allied Publishers Limited, Chennai, 1040 p

Dasgupta MK (1998) Phytonematology. Naya Prakash, Kolkata, 846 p

Dasgupta DR, Seshadri AR (1971) Races of the reniform nematode. Rotylenchulus reniformis Linford and Olivera. 1940. Ind J Nematol 1:21–24

Dasgupta MK, Mukheriee B, Bhattacharya C, Khan MR, Ghosh S (2010) Nematode infestation in tea. In: Khan MR, Jairajpuri MS (eds) Nematode infestation, Part-II: Industrial crops. The National Academy of Sciences, Allahabad, India, pp 213–255, 464 pp

Dastur JF (1936) A nematode disease of rice in Central Provinces. Proc Ind Acad Sci 4:108–122

Davis EL, MacGuidwin AE (2000) Lesion nematode disease. Plant Health Instructor. doi: 10.1094/PHI-I-2000-1030-02

De RK, Ali SS, Dwivedi RP (2001) Effect of interaction between *Fusarium oxysporum* f.sp. *lentis* and *Meloidogyne javanica* on lentil. Ind J Pulses Res 14:71–73

Dey P, Khan MR, Pal S, Solanki RD (2012) Managing root knot nematode (*Meloidogyne incognita*) problem by *Paecilomyces lilacinus* in pointed gourd (*Trichosanthes dioica*) – A success story, pp. 189. In: Khan MR, Jha S, Sen C, Banerjee H, Biswas T (eds) Book of abstracts on international symposium on plant health and climate change issues, 7–9 December, 2012, Kalyani, p 234

Franco J (1986) Potato cyst nematode, *Globodera* sp. Tech Infor Bull 9:19, CIP, Lima, Peru

Gaur HS, Perry RN (1991) The biology and control of the plant parasitic nematode, *Rotylenchulus reniformis*. In: Evans K (ed) Agricultural zoological review, vol 4. Intercept Ltd., Hants, England, pp 177–211

Gill JS (1981) Fighting the foliar nematode of chrysanthemum. Ind Hort 25:21–28

Gupta P, Swarup G (1968) On the ear-cockle and yellow rot disease of wheat. 1. Symptoms and histopathology. Ind Phytopathol 21:318–323

Gupta P, Swarup G (1972) Ear cockle and yellow rot disease of wheat: nematode bacterial association. Nematologica 18:320–324

Hafez SL, Al-Rehiayani S, Thornton M, Sundararaj P (1999) Differentiation of two geographically isolated populations of *Pratylenchus neglectus* based on their parasitism of potato and interaction with *Verticillium dahliae*. Nematropica 29:25–36

Haseeb A, Khan MR (2012) Prospects of *Trichoderma* in nematode management. In: Nagesh M, Rajkumar, Bhumannavar BS, Krishna Kumar NK (eds) Status and prospects for enhancing the uptake of antagonistic organisms for nematode management in India. NBAII (ICAR), Bangalore, pp 42–56, 187 p

Hashioka Y (1963) The rice stem nematode, Ditylenchus angustus in Thailand. FAO Plant Protec Bull 11:97–102

Hewit WB, Raski DJ, Goheen AC (1958) Nematode vector of soil borne fanleaf virus of grapevines. Phytopathology 48:568–595

Holtzmann OV (1968) A foliar disease of tuberose caused by *Aphelenchoides besseyi*. Plant Dis Rep 52:56

Hunger FWT (1901) Een bacterie-ziekte den tomaat. Mededlingen Plantentuin. Batavia 48:4–57

Jain RK (1992) Nematode pests of vegetable crops. In: Bhatti DS, Walia RK (eds) Nematode pests of crops. CBS Pub. & Distributors, New Delhi, pp 77–97, 381 p

Jain RK, Khan MR, Kumar V (2012) Rice root knot nematode (*Meloidogyne graminicola*) infestation in rice. Arch Phytopathol Plant Prot 45(6):635–645

Johnson DA, Santo GS (2001) Development of wilt in mint in response to infection by two pathotypes of

Verticillium dahliae and co-infection by *Pratylenchus penetrans*. Plant Dis 85:1189–1192

Jonathan EI, Gajendran G, Arulmozhiyan R (1997) Interaction of *Rotylenchulus reniformis* and *Phytophthora palmivora* on betelvine. Nematol Medit 25:191–194

Jones FGW (1961) The potato root eelworm *Heterodera rostochiensis* Woll. in India. Curr Sci 30:187

Kaitany R, Melakeberhan H, Bird GW, Safir G (2000) Association of *Phytophthora sojae* with *Heterodera glycines* and nutrient stressed soybeans. Nematropica 30:193–199

Kanwar RS, Bajaj HK (2010) Cereal cyst nematode infestation in wheat. In: Khan MR, Jairajpuri MS (eds) Nematode infestation, Part-I: Food crops. The National Academy of Sciences, Allahabad, India, pp 192–217, 325 p

Kaushal KK, Sharma GL, Paruthi IJ (2001) Nematode diseases of wheat and barley and their management. In: National congress on centenary of nematology in India – Appraisal and future plans, 5–7 December, 2001, Division of Nematology, Indian Agricultural Research Institute, New Delhi, India, pp 23–24.

Kerry BR (1982) The decline of *Heterodera avenae* populations. EPPO Bull 12:491–496

Kerry BR, Crump DH (1998) The dynamics of the decline of the cereal cyst nematode, Heterodera avenae, in four soils under intensive cereal production. Fundam Appl Nematol 21:617–625

Khan MR (2001) White-tip nematode, *Aphelenchoides besseyi* in rice-tuberose cropping system. In: Proceedings of National seminar on frontiers of crop management, 1–3 February, 2001, Visva-Bharati: Sriniketan, India, pp 142–144

Khan MR (2004) Observation on the foliar nematode, *Aphelenchoides besseyi* infecting tuberose in West Bengal. Ann Pl Protec Sci 12(1):106–109

Khan MR (2005) Hosts and Non-hosts of *Rotylenchulus reniformis* Linford& Oliveira, 1940 – A critical review. Env Ecol 23(1):124–140

Khan MR (2006a) Foliar nematode (*Aphelenchoides besseyi*)-a threat for cultivation of tuberose. AAPP Newslet 1(1):3

Khan MR (2006b) Current options for managing nematode pest of crops in India. In: Mohilal N, Gambhir RK (eds) Plant nematology in India. Parasitology Lab., Department of Life Sciences, Manipur University, Manipur, pp 16–50, 178 p

Khan MR (2010) White tip nematode infestation in rice. In: Khan MR, Jairajpuri MS (eds) Nematode infestation, Part-I: Food crops. The National Academy of Sciences, Allahabad, India, pp 140–170, 325 p

Khan E, Ganguly S (1995) Ectoparasites as emerging nematode problems of crops in the changing agricultural scenario. In: Lokeswar RR (ed) National Seminar on changing pest situation in the current agricultural scenario of India, Publication and Information Division, Indian Council of Agricultural Research, New Delhi, India, pp 444–453

Khan MR, Ghosh S (2007) Evaluation of tuberose cultivars against foliar nematode, *Aphelenchoides besseyi*. In: National symposium on nematology in 21st century: Emerging paradigms held during 22–23 November, 2007 at AAU, Assam, India, p 77

Khan MR, Ghosh S (2009) Control of foliar nematode, *Aphelenchoides besseyi* in tuberose (*Polianthes tuberosa* L.) from West Bengal, India. In: International symposium on nematodes in tropical ecosystems, 17–21 August, Hanoi, Vietnam, p 22

Khan MR, Ghosh S (2011a) Survival and population dynamics of foliar nematode, *Aphelenchoides besseyi* infecting tuberose in West Bengal, India. Ind J Nematol 41(1):47–51

Khan MR, Ghosh S (2011b) Bioefficacy of some pesticides against *Aphelenchoides besseyi* and sensitivity of tuberose bulbs to hot-water. Ind J Nematol 41(1):29–33

Khan MR, Pal AK (2001) Plant parasitic nematodes associated with tuberose (*Polianthes tuberosa* L.) in West Bengal. Ann Plant Protec Sci 9(2):357–359

Khan MR, Karmakar SG, Maiti CS (1999) Investigation on severe infestation by *Aphelenchoides besseyi* in tuberose (*Polianthes tuberosa*). In: National seminar on nematological research in India, Challenges and preparedness for the new millennium, C. S. Azad University of Agriculture and Technology, Kanpur, India, 17 December, 1999. Nematological Society of India, p 20

Khan MR, Shit S, Pal AK, Biswas B (2005) Integrated approach for management of foliar nematode, *Aphelenchoides besseyi* infecting tuberose in West Bengal, India. Int J Nematol 15:30–34

Khan MR, Shit S, Pal AK, Biswas B (2006) Management of foliar nematode, *Aphelenchoides besseyi* infecting tuberose in West Bengal, India. Ind J Nematol 36(1):44–47

Khan MR, Pathak B, Ghosh S, Ghoshdostidar K (2008) Management approaches of tuberose foliar nematode, *Aphelenchoides besseyi* in West Bengal, India. Ind J Nematol 38(1):25–29

Khan MR, Bhattacharya I, Chattopadhyay SB, Ghosh S (2009) Integrated approach for managing root knot nematode (*Meloidogyne incognita*) in pointed gourd (*Trichosanthes dioica* Roxb.). Ind J Nematol 39(1):25–28

Khan MR, Jain RK, Singh RV, Pramanik A (2010) Economically important plant parasitic nematodes distribution-atlas. Directorate of Information and Publications of Agriculture, Indian council of Agricultural Research, Krishi Anusandhan Bhavan, New Delhi, 145 p

Khan MR, Nagesh M, Khan MR (2012) Nematode infestations in ornamental crops. In: Khan MR, Jairajpuri MS (eds) Nematode infestation, Part-II: Horticultural crops. The National Academy of Sciences, Allahabad, pp 398–442, 604 p

Kinh DN, Phuang CD (1981) Reaction of some deep water rice varieties of *Ditylenchus angustus*. Int Rice Res Newslet 14:32–33

Koshy PK (1967) A new species of *Heterodera* from India. Ind Phytopathol 20:272–274

Koshy PK, Swarup G (1971) Investigations on the life history of the pigeon pea cyst nematode, *Heterodera cajani*. Ind J Nematol I:44–51

Krishna Prasad KS (1995) Nematode problems of potato. In: Swarup G, Dasgupta DR, Gill JS (eds) Nematode pest management – An appraisal of ecofriendly approaches. Nematological Society of India, New Delhi, pp 177–185

Krishna Prasad KS (2006) Central Potato Research Institute (ICAR), Shimla, India, Technical Bulletin No. 77, 20 pp

Kumar S, Sivakumar CV (1998) Management of rice white tip nematode, *Aphelenchoides besseyi*. Ind J Nematol 29:35–37

Kuriyan KJ (1995) Nematode problems of rice In India. III. Rice cyst nematode and white tip nematode. In: Swarup G, Dasgupta DR, Gill JS (eds) Nematode pest management – An appraisal of ecofriendly approaches. Nematological Society of India, New Delhi, pp 151–155

Lal A, Mathur VK, Agarwal PC (1982) Studies on the effect of *Fusarium solani* parasitism on *Heterodera zeae*. Nematologica 28:447–450

Libman G, Leach JG, Adama RE (1964) Role of certain plant parasitic nematodes in infection of tomato by *Pseudomonas solanacearum*. Phytopathilogy 54:151–153

Lucas GB, Sasser JN, Kelman A (1955) The relationship of root knot nematodes to Granville wilt resistance in tobacco. Phytopathology 45:537–540

Mathur VK, Prasad SK (1973) Survival and host range of the rice root nematode, *Hirschmanniella oryzae*. Ind J Nematol 3:88–93

Mathur BN, Arya HC, Mathur RL, Handa DK (1974) The occurrence of biotypes of cereal cyst nematode (*Heterodera avenae*) in the light soils of Rajasthan and Haryana, India. Nematologica 20:19–26

Manzaniilla-Lopez RH, Hunt DJ (2009) Tropical plant and soil nematodes: diversity and interactions. In: Tropical biology and conservation management, vol. VI, Encyclopedia of Life Support Systems (EOLSS). Eolss Publishers, Paris, 197pp. http://www.eolss.net

Mondal AH, Miah SA (1987) Control of ufra of rice by seedling treatment. Bang J Plant Path 3:25–30

Nath RP, Haider MG, Akhter SW, Prasad H (1976) Studies on the nematodes of vegetables in Bihar-II: combined effect of *Meloidogyne incognita* and *Ozonium texanum* var *parasiticum* on germination of eggplant. Ind J Nematol 6:175–177

Nickle WR (1991) Manual of agricultural nematology. Marcel Dekker Inc., New York, 995 pp

O'Brien PC, Stirling GR (1991) Plant nematology for practical nematologists. Queensland department of primary industries information series Q190023

Pal AK (1970) Survey of nematode infection of paddy and successful treatment with Hexadrin. Proc Ind Sci Cong Assoc 57:296

Parvatha Reddy P (2008) Diseases of Horticultural crops: nematode problems and their management. Scientific Publishers (India), Jodhpur, 379 pp

Parvatha Reddy P, Nagesh M (1995) Nematode problems of ornamental crops in India. In: Swarup G, Dasgupta DR, Gill JS (eds) Nematodes pest management an appraisal of ecofriendly approach. Nematological Society of India, New Delhi, pp 244–251, 300 pp

Parvatha Reddy P, Singh DB (1980) Plant parasitic nematodes associated with banana production technology. Tamil Nadu Agricultural University, Coimbatore, pp 161–163

Pathak B, Khan MR (2009) Yield loss potential of foliar nematode, *Aphelenchoides besseyi in tuberose*. Ind J Nematol 39(1):38–40

Pathak B, Khan MR (2010) Comparative efficacy of chemical, botanical and biological agents against foliar nematode, *Aphelenchoides besseyi* infecting tuberose in West Bengal, India. Ind J Nematol 40(1):83–87

Phukan PN (1995) Nematode problems of rice crops in India. IV. Stem nematode and rice root knot nematode. In: Swarup G, Dasgupta DR, Gill JS (eds) Nematode pest management – An appraisal of ecofriendly approaches. Nematological Society of India, New Delhi, pp 156–160

Prasad JS (2002) Nematode diversity in rice based cropping system vis-à-vis sustainable agriculture. In: Singh RV, Pankaj SC, Dhawan SC, Gaur HS (eds) Proceedings of National Seminar on Biodiversity and Management of nematodes in Cropping systems for sustainable agriculture. Nematological Society of India, IARI, New Delhi, pp 3–6

Prasad JS, Varaprasad KS (2001) Ufra nematode, *Ditylenchus angustus* is seed borne! Crop Protec 2(1):75–76

Prasadrao GMV, Ganguly S (1996) Host preference of six geographical isolates of reniform nematode, *Rotylenchulus reniformis*. Ind J Nematol 26:19–22

Rahman ML, Mondal AH, Bakr MA (1994) Widespread *ufra* disease incidence in different rice ecosystems in Bangladesh. Int Rice Res Newlett 19:27–28

Rao YS (1978) National Symposium on increasing rice yield in *kharif*, CRRI, Cuttack, Orissa, 8–11 February 1978

Rao YS (1985) Nematodes. In: Padmanabhan SY (ed) Rice in India, ICAR, Monograph, pp 591–615

Rao YS, Prasad JS, Panwar MS (1986) Nematode problems of rice: crop loss, symptomatology and management. In: Swarup G, Dasgupta DR (eds) Plant parasitic nematodes of India problem and progress., pp 279–299, 497 p

Reddy DDR, Subramanyam P, Sankar Reddy GH, Raja Reddy C, Siva Rao DV (1984) A nematode disease of peanut caused by *Tylenchorhynchus brevilineatus*. Plant Dis 68:526–529

Reddy PP, Singh DB, Ramkishun (1979) Effect of root-knot nematodes on the susceptibility of Pusa purple cluster brinjal to bacterial wilt. Curr Sci 48:915–916

Robinson AF, Inserra RN, Caswell-Chen EP, Volvas N, Troccoli A (1997) *Rotylenchulus reniformis*: identification, distribution, host range and crop plant resistance. Nematropica 27(2):127–180

Rupe JC, Robbins RT, Becton CM, Sabbe WA, Gbur EE (1999) Vertical and temporal distribution of *Fusarium solani* and *Heterodera glycines* in fields with sudden death syndrome of soybean. Soil Biol Biochem 31:245–251

Sasser JN (1980) Root knot nematodes: a global menace to crop production. Plant Dis 64(1):36–41

Seshadri AR, Sivakumar CV (1963) A preliminary note on the reniform nematode (*Rotylenchulus reniformis*) Linford and Oliveira, 1940 on a number of cultivated crops in South India. Madras Agric J 50:134–137

Siddiqi MR (1961) Occurrence of the citrus nematode, *Tylenchulus semipenetrans* Cobb, 1913 and the reniform nematode, *Rotylenchulus reniformis* in India. Proc 48 Ind Sci Congr 3:504

Singh B (1953) Some important diseases of paddy. Agr Anim Hus Ind 3:27–30

Singh I, Jain RK (1995) Nematode problems of rice in India, I. Rice root nematode. In: Swarup G, Dasgupta DR, Gill JS (eds) Nematode pest management – An appraisal of ecofriendly approaches. Nematological Society of India, New Delhi, pp 140–150

Sivakumar CV, Kuppuswamy S, Meerzainuddin M (1973) Evaluation of carbofuran seed treatment for the control of root knot on okra. Ind J Nematol 3:71–72

Sivakumar CV, Lakshmanan PL, Balasubramanian M (1979) Interaction of a non-host plant (*Capsicum annuum* L.) and certain systemic nematicides in population decline of *Rotylenchus reniformis Rev.* Nematol 2:165–168

Sosamma VK, Koshy PK, Bhaskara Rao EVV (1980) Susceptibility of coconut cultivars and hybrids to *Radopholus similis*. Ind J Nematol 10:250–252

Sugawara K, Kobayashi K, Ogoshi A (1997) Influence of the soybean cyst nematode (*Heterodera glycines*) on the incidence of brown stem rot in soybean and adzuki bean. Soil Biol Biochem 29:1491–1498

Sundararaju P, Koshy PK (1988) Separate and combined effects of Radopholus similis and Cylindrocarpon obtusisporum on arecanut seedlings. Indian J Nematol 17(2):301–305

Swarup G, Sethi CL, Seshadri AR, Kaushal KK (1979) On the biotypes of Heterodera avenae, the causal organism of 'Molya' disease of wheat and barley in India. Ind J Nematol 9:164–168

Taylor AL, Sasser JN (1978) Experimental and agronomic use of nematicides, International Meloidogyne Project, North Carolina State University, Raleigh, N.C. 27607

Vasudeva RS (1958). Scientific report of the Indian Agricultural Research Institute for the year ending, 30 June 1958

Vats R, Dalal MR (1997) Interaction between *Rotylenchulus reniformis* and *Fusarium oxysporum* f.sp. *pisi* on pea (*Pisum sativum* L.). Ann Biol 13:239–242

Vovlas N, Troccoli A (1990) Histopathology of broad bean roots infected by the lesion nematode, *Pratylenchus penetrans*. Nematol Medit 18:239–242

Walker NR, Kirkpatrick TL, Rothrock CS (1999) Effect of temperature on and histopathology of the interaction between *Meloidogyne incognita* and *Thielaviopsis basicola* on cotton. Phytopathology 89:613–617

Walker NR, Kirkpatrick TL, Rothrock CS (2000) Influence of *M.incognita* and *Thielaviopsis basicola* populations on early-season disease development and cotton growth. Plant Dis 84:449–453

Wheeler TA, Hake KD, Dever JK (2000) Survey of *Meloidogyne incognita* and *Thielaviopsis basicola*: their impact on cotton fruiting and producers management choices in infested fields. J Nematol 32:576–583

Williamson VM, Gleason CA (2003) Plant–nematode interactions. Curr Opinion Plant Biol 6:327–333

17 Recent Advances in Diagnosis and Management of Diseases of Vegetable Crops in India

K.K. Pandey

17.1 Introduction

Vegetables play a vital role in providing nutritional security to human health as they are rich source of essential nutrients, vitamins, minerals, antioxidants, and other growth-promoting phytochemicals. Shifting from a non-vegetarian diet to vegetarian diet, global recognition of importance of the vegetables for human health, and their medicinal and nutritional value have contributed to a steady upward trend in vegetable production. The total vegetable production of India during 2005–2006 was 113.5 million tons and made a quantum leap in vegetable production, securing second position in the world. China is ranked first in the world and approximately produces 302 million tons of vegetables. The average productivity of our country has increased from 10.5 t/ha in 1991–1992 to 15.8 t/ha in 2005–2006. In spite of these achievements, the productivity per unit area of vegetable crops is much lower as compared to developed countries because of different reasons, the most important being the prevalence of diseases.

The biotic factors like fungi, bacteria, viruses, phytoplasma, and nematodes are responsible for causing diseases in vegetable crops. The abiotic factors like sunscald, frost injury, and nutritional disorders are also indirectly associated with the vegetable diseases (Pandey and Pandey 2002). Vegetable crops are more susceptible to diseases as compared to cereals. A very high percentage of water, relatively high metabolic activity, and inherent short self-life of vegetables make them highly perishable commodities. These characteristics propose a great problem in the post-harvest management. It has been estimated that a total of 65 million tons of preharvest losses of vegetables (10.5 %) are due to disease alone (Snowden 1991). In preharvest loss, the quality of products was affected by indirect physiological interference of the pathogen or it was directly destroyed by disease before harvesting. The extent of preharvest losses is much higher in developing countries as compared to developed countries, and maximum loss of 44 million tons has been reported from developing countries. Post-harvest losses have been estimated to be about 30 % of the total vegetable production of country.

Disease pressure in the standing crop from the seedling stage to harvest and spoilage caused by microorganisms during transit, storage, and marketing are the main constraints in total vegetable production. Among pathogens, fungi are the main reason for the extensive damage of vegetable crops especially in tropical and subtropical countries. Introduction of hybrid cultivars, intensive monoculture of vegetables, off-season vegetable growing practices, and WTO agreements has contributed significantly to increase

K.K. Pandey (✉)
Plant Pathology, Indian Grassland and Fodder Research Institute, Jhansi 284003, India
e-mail: kkpiivr@gmail.com; headseedcpri@gmail.com

L.P. Awasthi (ed.), *Recent Advances in the Diagnosis and Management of Plant Diseases*,
DOI 10.1007/978-81-322-2571-3_17

the disease pressure and prolong the survival period of pathogens.

Presently, the control of vegetable diseases is mainly based on the application of agrochemicals. Despite the proven efficiency of agrochemicals in the prevention and spread of the diseases, their high cost makes the vegetable production very expensive for the marginal farmers of the agricultural-based developing countries like India. Moreover, chemical residues remaining in the vegetables form an important health hazard. The successful management of any disease involves its detailed study regarding symptoms, causal agents, disease cycle, and epidemiology which lead to integrated disease management strategies.

17.2 Nursery Diseases of Vegetable Crops

Damping-off is the most important disease of tomato, brinjal, chili, onion, cabbage, and cauliflower. Seed rotting, bacterial blights, downy mildew, black rot, and nematode infestation are other diseases of nursery crops. The tender radical and plumule may be killed by several seed- and soilborne fungi in the nursery bed. The preemergence damping-off and seed rotting are the common cause of a lesser amount of seed emergence and many time confused with poor vigor of seeds. Postemergence damping-off results infection in seedlings near the soil line after emergence and gives wire-stem symptoms. The tissues become soft, water-soaked, and weak causing the seedlings to fall over and later they die. The infected seedling sometimes appears green but next morning collapses and is unable to stand further. Damping-off is caused by several species of soil- and seed-borne fungi where *Rhizoctonia*, *Pythium*, *Fusarium*, *Phytophthora*, *Phomopsis*, *Sclerotium*, *Colletotrichum*, *Phoma*, and *Alternaria* are associated with these vegetable crops. The losses by damping-off were recorded 80 % in tomato, 60–75 % in onion, 60–75 % in early cauliflower, 40 % in chili, and 20–25 % in brinjal (Pandey et al. 2004).

Bacterial leaf spot is the second most important seedling disease of tomato, chili, and capsicum. It is caused by *Xanthomonas campestris* pv. *vesicatoria*. Symptoms appear as very minute, dark spots on all aboveground parts. Leaves turn yellow, blighted, and fall down leaving only naked stem.

Black rot is the third most important disease of nursery prevalent in cauliflower, cabbage, Chinese cabbage, Brussels sprouts, knol-khol, and broccoli. The symptoms appear as rotting of cotyledons and brown to black, soft discoloration on seedlings. The marginal yellowing of juvenile leaves of seedlings was observed at later stage. Sometimes entire seedling becomes yellow and dries. This is seed-borne disease and is caused by *Xanthomonas campestris* pv. *campestris*. Nursery diseases of vegetable crops can be managed successfully by the following integrated approach.

17.2.1 Soil Solarization

Soil disinfestations by the use of white transparent polythene in hot summer are one of the most effective approaches for management of soilborne diseases of nursery beds. The main objective of soil solarization is to eliminate pathogens, weeds, toxic metabolites, biotic agents, and insects. The major soil pathogens like *Pythium* spp. *Phytophthora* spp., *Fusarium* spp., *Sclerotium rolfsii*, *Macrophomina phaseolina*, *Verticillium dahliae*, *Pseudomonas* sp., *Streptomyces scabies*, and *Meloidogyne* spp. are effectively reduced in nursery beds. The extent of bacterial blight reduction was recorded 55.8 % in tomato seedling due to solarization (Pandey and Pandey 2004). It also reduces the population of viable sclerotia of *S. rolfsii* and *Sclerotinia* spp. Availability of soil nutrients particularly phosphorus and potash is increased in solarized bed. Incorporation of organic matter and residues of cruciferous plants in nursery beds before polythene mulching reduces the infection of cabbage yellows, *Pythium* spp., *S. rolfsii*, and *Rhizoctonia solani*.

17.2.2 Biological Control

Treatments of soil and seed with bioagents can manage nursery diseases effectively. Seed treatment is preferred in direct sown crops, while soil treatment and seed treatment are both feasible in nursery crops. Effective utilization of biocontrol agent in soil is only possible when it is supplemented with decomposed organic amendments and green manuring. Congenial moisture and protection from extreme weather are very desirable. Proper rhizosphere competence maximizes the effectiveness of a particular bioagent in soil system. Therefore, the use of resident antagonists is always beneficial. Biological seed treatment can be done by priming, seed coating, or dry powder treatment of seed depending upon the nature of biocontrol agents. Usually for seed treatments, 6–10 g *Trichoderma* is used for one kilogram of seed, but spore concentration should be in between 10^6 and 10^9 cfu/ml. Similarly 10–25 g *Trichoderma* powder should be applied per square meter nursery area depending upon the soil type and organic matter status of the soil. Seed treatment by *Trichoderma* sorghum formulation at 1 % along with soil application at 10 g/m^2 area is also very effective in tomato, brinjal, and chili. Soil application of *Trichoderma* at 10 g along with 100 g neem cake/m^2 nursery area is effective for control of nursery diseases as well as nematodes. *Trichoderma* spp. can be integrated with fungicides like propineb, captan, azoxystrobin, copper hydroxide, and pencycuron at lower doses (200–400 ppm) during seed treatment (Pandey et al. 2006).

17.2.3 Chemical Control

Nursery bed should be treated with 0.3 % solution of captan in such a manner that 5 l of the solution uniformly drenched in one square meter of area. Seeds must be treated with carbendazim at 0.25 % before sowing. After complete germination of seeds, copper oxychloride is sprayed at 0.3 % followed by spray of streptocycline at 100 ppm thoroughly to cover seedlings and its root zone. It should be repeated after 15 days of the first spray.

17.2.4 Cultural Method

Nursery bed should be properly drained and pulverized. Soil texture of nursery bed should be light and porous by adding equal proportion of sand and compost. Seed should be sown in lines at proper distance to avoid overcrowding. Provide proper aeration and light to check the pathogen infection. Protect germinating seeds and juvenile seedlings from heavy rain and scorching sunlight through agronet, muslin cloth, and portable low tunnel polythene arch. Most of the insect vector for viral diseases can be managed by cage of muslin cloth or fine agronet.

17.3 Fungal Diseases of Vegetable Crops

17.3.1 Alternaria Blight of Tomato

The term early blight is commonly used for the potato crop because of its appearance early in cropping stage than late blight. It is not true in tomato where early blight always appears late in the season of tomato crop after appearance of late blight. The same pathogen also attacks on tomato and gives blight appearance. However, tomato is infected by two species of *Alternaria*. The name Alternaria blight is scientifically more appropriate in tomato than early blight. Alternaria blight of tomato commonly known as early blight is caused by *Alternaria solani* with on average of 45 % severity every year in North India. This is the most destructive disease and prevalent all over the country. *A. alternata* f.sp. *lycopersici* is another species causing leaf blight in vegetative stage, where irregular leaf blotch appears from margins of leaves and has prominent yellowing on their outer margin due to host-specific AAL toxin. Symptoms of early blight appear on all aboveground parts of the plant. Scattered, brown leaf spots with concentric rings are formed. The spots coalesced and enlarged in humid weather. Girdling by brown lesions at the base of seedling and on main stem and branches of the plant appears in later stage. The stem lesions are usually restricted to one side of a stem and become elongated and sunken. Primary infection also occurs on blossoms. Fruit symptoms initiate from calyx and pedicels. Gradually it progresses on apical portion of the fruit (stem end) as a circular lesion and radiates from the attachment between calyx and fruit. Fruit spots are dark brown, depressed, firm, and with distinct continuous rings. The yield loss up to 80–86 % is reported in experimental field. The pathogen is soil- as well as seed-borne. The main crop and late-sown hybrid tomato are more severely damaged by the disease during January to April. Distinct pathogenic and cultural variability in the strains of *A. solani* of tomato was observed in our country. The cross-inoculation test of potato and tomato isolate of *A. solani* has easily infected both the hosts which clearly showed lack of host specificity by the pathogen.

Crop rotation with non-solanaceous host is essential for an effective reduction of the inoculum. Healthy seeds should be collected from disease-free fruits. Summer plowing is advised to increase the desiccation of pathogen and infected plant parts. Field sanitation must be adopted as a regular practice by plucking the lower leaves and burning of infected crop debris. Minimize relative humidity in plant canopy for preventing the infection. Seed should be treated by captan at 0.25 % before sowing. Two sprays of chlorothalonil at 0.2 % at 8-day interval are effective against the disease, but spray must be started soon after the infection on the floral part. Tomato cultivars CLN-2071-C, CLN-2070-A, BSS-174, and DTH-7 having resistance expressed as slow blighting against four pathogenic isolates of *A. solani* were selected for cultivation in disease-prone areas (Pandey et al. 2003). Keinath et al. (1996) reported application of fungicides based on TOMCAST forecasting model for effective and economic spray schedule.

17.3.2 Late Blight of Tomato

The disease is more common in northwest states. It also appears in Uttar Pradesh and Bihar in the months of December–January. Late blight is caused by *Phytophthora infestans*. The yield loss was recorded 70–80 % during 1997 in Uttar Pradesh (Pandey and Pandey 2003a). Symptoms on leaves, twigs, and fruits are more common. Blight appears on foliage as light-pale green, water-soaked dead area. The lesions are wet in morning and become dried and shriveled in daylight. It enlarges rapidly until entire leaves are killed and exposing all the fruits for infection. In moist and cloudy weather, a downy white fungus develops near the outer margin of lesion on underside of the leaves. Green and ripe fruit affected by the blight appears as greenish brown and internal tissues become spongy with foul odor. In addition to tomato, the same pathogen also attacks on potato but not on brinjal and pepper.

Always use healthy and certified seeds collected from disease-free area. Infected crop debris and fruits must be collected from the field and burnt. The balanced nutrients may be maintained in the crop to minimize the disease incidence. Preventive sprays of mancozeb at 0.25 % at 5–7 days provide good control in cloudy, cold, and drizzling weather. One spray of coordinated product of metalaxyl plus mancozeb at 0.2 % is very effective when applied soon after infection, but spray should not be repeated. Staking of plant reduces late blight diseases of tomato.

17.3.3 Gray Leaf Spot of Tomato

The disease has become very severe in Gangetic plains since last few years during the month of September to November. The disease is caused by *Stemphylium solani*. Disease symptoms appeared as irregular, small necrotic, brown spots mostly on leaves. The spots spread very fast on all the leaves and blight appearance of the plant is observed within few days. Later on infected leaves defoliate and naked plant is left over in the field. The management schedule for gray leaf spot is the same as described in the Alternaria blight.

17.3.4 Septoria Leaf Spot of Tomato

The disease is mainly prevalent in winter season and is caused by *Septoria lycopersici*. The plant may be attacked at any stage of its growth. The disease is characterized by numerous, small, gray, circular leaf spots having dark border. A few black pin-point-size pycnidia may be seen within the spot. The fungus survives in plant residue from diseased plant and also on the seed. Wet weather conditions favor fungus activity. Temperature between 10 and 21 °C coupled with high relative humidity is most congenial for the disease development. Hyaline, septate, and long pycnidiospores produced abundantly in pycnidia are the source of secondary infection.

For control of the disease, destroy infected plant residue in the field. Seed treatment with carbendazim at 0.2.5 % is useful in checking seed-borne infection. Effective control of the disease can be obtained with the spray of carbendazim at 0.1 % at 10 days of interval. The spray also prevents infection in seed crop of tomato resulted in disease-free seed production.

17.3.5 White Rot of Vegetable Crops

The disease is common in tomato, brinjal, chili, capsicum, cauliflower, cabbage, winter bottle gourd, carrot, pea, French bean, and dolichos bean. White rot is also known as sclerotinia rot. The disease is caused by ubiquitous fungus *Sclerotinia sclerotiorum*. Yield loss up to 50 % has been recorded in pea. Symptoms are observed as wet and soft rot of the tissues. Very soon white fungus growth is observed on the rotten portion. Later embedded sclerotia in white mycelium are formed on the infected portion as well as inner portion of stem, pith, and fruits. These sclerotia fall down in the soil and survive for several years. These sclerotia germinate and develop apothecia containing numerous ascospores. Primary infection is always initiated from ascospore infecting floral parts followed by falling of infected petals on adjoining tissues. Infected seeds in pea and French bean are converted into sclerotia and become concomitant mixture during threshing. Low temperature between 8 and 15 °C coupled with high soil moisture, foggy weather, and high relative humidity is essential for infection and disease development.

All infected part must be carefully collected and burnt before drying of the plants. Deep summer plowing and low land paddy cultivation with continuous stagnation of water reduce the sclerotial population. Close planting and dense canopy of plant should be avoided. Field should be weed free. Alternate spray of carbendazim at 0.1 % and mancozeb at 0.25 % at 7–10 days of interval is found effective. Fungicidal spray must be started at early bloom or flowering stage. Seed cleaning is essential to remove the sclerotial mixture during seed processing and storage. Development of host resistance is practically not possible against this necrotrophic pathogen. However, Pandey et al. (2002c) reported moderately resistant reaction under challenge inoculation in Lerica and Rishi cultivar of tomato.

17.3.6 Fruit Rots of Tomato

Tomato is severely damaged by several types of fruit rots during fruit setting to ripening stage in the field. *Myrothecium roridum* causes water-soaked rotting with white to black sporodochial concentric rings over it. *Rhizoctonia solani-infected* fruit rot appears as papery, light white, soft rot with crakes and mycelium. *Sclerotium rolfsii*-infected fruits are always visible with abundant white mycelium and mustard grain-like sclerotia. Buck eye fruit rot caused by *Phytophthora parasitica* (*P. nicotianae*) is the most devastating disease of tomato. The symptoms mostly on green fruits appeared as grayish brown and water-soaked with concentric lesions, and later on fruits become mummified. Colletotrichum fruit rot was first reported from Maharashtra followed by Uttar Pradesh. The pathogen was identified as *Colletotrichum capsici* (Pandey 2006). Symptoms appear as brown, dry rotting of ripe fruits with numerous acervuli especially in hybrid tomato. Black circular dot-like acervuli are found below the skin, and the pathogen deeply penetrates in the fruit to infect seeds and become internally seed-borne.

The strategies for management of all the fruit rot diseases of tomato are similar. Avoid the contact of fruits from soil by staking of plant and mulching. Provide proper drainage in the field. Green manuring along with soil application of *Trichoderma* at 5 kg/ha colonized in 100 kg FYM is very effective in checking most of the fruit rot. Collect affected fruits and burn them to reduce primary inoculum. Seeds should be treated by carbendazim at 0.25 % just before sowing.

17.3.7 Collar Rot of Vegetable Crops

Collar rot is a very common disease in all vegetable crops including elephant foot yam. The disease is also known as southern blight. It is caused by *Sclerotium rolfsii*, an necrotrophic pathogen with broad host range. The yield loss was recorded 30 % in elephant foot yam as direct mortality of the plants (Pandey and Pandey 1996). Collar rot is becoming a serious problem in many hybrids of tomato and the incidence was recorded as high as 40 % (Pandey and Pandey 2003a). The disease generally occurs as a sudden wilt of individual plants scattered in the field. Initially wet rotting of bark near collar region without foliage yellowing is observed. Rotting covers entire bark of the plant near collar region. Characteristic symptoms are observed as white, fungus growth on affected portion as well as on contact soil. Gradually this hyphal mat is converted into small, mustard-like sclerotia that survive in the soil. However, the sclerotia survived for short period in the lack of host.

Crop rotation with low land paddy and other cereals reduces the sclerotial population from the soil. Field should be cleaned by uprooting and burning of all the infected plants. Fallowing, plowing, and irrigating the field in summer season followed by further plow during this period reduce the soil inoculum as well as weed population. Green manuring in July followed by soil application of *Trichoderma* at 5 kg/ha soon after plowing of the sunhemp. Seedling root dip by *Trichoderma* at 1 % suspension may be given during transplanting. Remove all the weeds from field. Drench *Trichoderma* at 1 % suspension after 20 days of sowing to reduce early infection. Drenching of copper oxychloride at 0.3 % near the collar region in evening is very effective. However, a commercial neem-based formulation Eco-Neem (0.5 %) was also found effective for the eco-friendly management of tuber crop like elephant foot yam (Pandey and Pandey 1996).

17.3.8 Phomopsis Blight of Brinjal

This disease is the most devastating disease of brinjal and prevalent throughout the country. It is caused by *Phomopsis vexans* and the perfect stage of the pathogen is known as *Diaporthe vexans*. Incidence of Phomopsis blight was recorded as high as 70 % in seed production block with cent-percent yield loss due to complete rotting of most of the fruits from entire field (Pandey and Pandey 2003a). Infected seeds also cause damping-off in the nursery. Leaves of seedlings show distinct circular, light brown spots mostly on lower leaves. Center of the spots is slightly lighter in color. Old spots are having numerous black dots like pycnidia. These spots are papery, often crack, and sometimes have shot hole. Almost similar symptoms are observed on the leaves of transplanted crop in main field. Sometime irregular, big lesions are also observed on leaves, particularly in cool and humid weather. Stem lesions are started from basal part of nodal portion as brown dry rot with constriction. This leads to drying of few twigs or partial wilting of the some portion of plant. Old and dried twigs show several dots like erumpent pycnidia on the bark. Pale to light brown sunken spots develop on the old fruits. Individual spots expand and coalesce to cover entire fruit and later the fruit mummified. Fruit rot is soft, spongy, and with rhythmic growth. Mature, prominent, and erumpent black pycnidia are found on inner circle of rot, while submerged conspicuous and immature pycnidia are found on the outer ring of rot with concentric arrangement. The seeds of infected fruit are colonized by the pathogen and disease becomes internal seed-borne.

The use of disease-free healthy seeds is the most important. Treat the seed with carbendazim at 0.25 % to eliminate the seed inoculum. Crop rotation with cereals at least for 3 years is very effective. Infected crop residue and fallen mummified fruits should be collected and burnt. Spray carbendazim at 0.1 % after 15 days of transplanting and also during flowering and fruit setting stage. In seed crop, first harvest of immature fruits is done for consumption and seeds are extracted from subsequent fruits. KS-233 and Ramnagar Giant were found moderately resistant to Phomopsis fruit rot and can be used in endemic area (Pandey et al. 2002a).

17.3.9 Anthracnose of Pepper

Anthracnose of pepper is also known as ripe fruit rot and is caused by *Colletotrichum capsici*. Anthracnose (fruit rot) is mainly caused by *Colletotrichum capsici* in India. However, *C. gloeosporioides* is also reported in Karnataka and recently identified from Chhattisgarh also. The pathogen is causing 23 % yield loss in red chili as fruit and seed rot (Pandey and Pandey 2003a). Symptoms of anthracnose on red fruits are more commonly observed than green fruits. Pre- and postemergence damping-off was also observed due to seed-borne inoculum. Papery, sunken, elongated, soft lesions were observed on green fruits. Infection on flower and small juvenile fruits results in premature drop. Characteristics symptoms of anthracnose are not expressed until the fruit becomes fully mature and red. Small, irregular, sunken, light brown lesions are observed on mature fruit. Acervuli formation starts from the center of lesion and moved toward periphery on red fruits. Arrangement of acervuli is observed in concentric ring fashion. The red fruits appear healthy during harvest but show symptoms of the disease during drying. Maximum infection is observed in first fruiting. Fruit rot extends to the seed cavity making it internally seed-borne pathogen.

The use of disease-free clean seeds and crop rotation are important elements in disease management. Screening and shorting of diseased fruits must be done after complete drying of the fruits. Seeds should be treated with carbendazim at 0.25 % before sowing. Seedling should be sprayed by carbendazim at 0.1 % before transplanting. Foliar spray of carbendazim at 0.1 % followed by copper oxychloride at 0.3 % at flowering stage is effective to manage the disease. Consume green fruits of first harvest and seed should be collected from subsequent fruit setting.

17.3.10 Choanephora Dieback of Pepper

The disease is prevalent throughout North India in post-rainy season from September to November. Symptoms often initiate on tender twigs from the apical growing point of the plant and flower buds as necrosis and withering. Infected tissue turns dark brown and progresses from tip to downward. Rapid downward killing of leaves and twigs with wet rotting and silvery appearance is the characteristic symptoms. *Choanephora capsici*-rotted leaves and twigs show grayish black spore producing stalks and sporangial structure. Later on the infected leaves defoliated and dry rot, bare twigs remain attached with the plants giving typical dieback symptoms. *Choanephora cucurbitarum* is also reported to cause this disease on *Capsicum* spp. (Black et al. 1991).

Infected twigs along with healthy part should be cut in the morning, collected carefully, and burnt. Avoid apical injury during transplanting and also at flowering stage. Foliar spray of mancozeb at 0.25 % followed by copper oxychloride at 0.3 % should be given in the month of October when dew drop started to fall.

17.3.11 Ascochyta Blight of Beans

The disease is becoming a serious problem in pea, cowpea, and French bean. The associated pathogen is *Ascochyta phaseoli* with beans while *Ascochyta pisi* on pea. *A. pisarum* was reported on pea from Himachal Pradesh (Paul and Rathour 1998). However, four species of Ascochyta, viz., *A. pinodes*, *A. pisi*, *A. pinodella*, and *A. pisarum*, are reported to be associated which cause different types of symptoms including leaf and pod spot, blight, and foot rot on pea (Gupta and Paul 2001). The fungus is seed-borne as well as soil-borne. The fungus produces circular to oval spots on leaves which are dark brown with white to light brown centers. The infected lesions may desiccate and fall to produce a shot hole. Lesions may also coalesce and cause defoliation. Stem lesions are circular to elongate. Symptoms on pods are mostly circular, dark colored, and sunken. Severely affected pods become shrivel and infected seeds were reduced in size. The fungus is seed-borne as well as soilborne. The disease is favored by high moisture level and relatively low temperature (20–22 °C).

Effective cultural operations involving crop rotation and destruction of diseased plant refused in the field help in reducing the primary inoculum. As the disease is seed-borne, the use of healthy and disease-free seeds is recommended for planting. It is essential to treat the seeds by carbendazim at 0.25 % to eliminate the pathogen inoculum from seed.

17.3.12 Anthracnose of Beans

Anthracnose is a major problem in French bean, cowpea, and dolichos bean caused by *Colletotrichum lindemuthianum*. All aerial parts of the plant are susceptible to pathogen. The disease is characterized by black sunken lesions on pods, cotyledons, and stems. Pinkish spore masses may appear in these lesions during favorable conditions. The infected pods get shriveled prematurely and the seed size is reduced. The infection may extend to the seed and become seed-borne. Stems are also infected in severely affected plants. The fungus can survive from one season to another in infected plant debris as well as in diseased seed. Optimum range of temperature and humidity for disease development is 18–27 °C and 90–100 %, respectively. Disease incidence increases with the age of the plant.

The seed must be collected from healthy pods. Crop rotation and destruction of infected debris are useful in reducing the source of primary inoculum. Treatment of seed with carbendazim at 0.25 % seed is most effective to reduce the seed-borne inoculum. Foliar spray of carbendazim at 0.1 % followed by mancozeb at 0.25 % should be given soon after initiation of visible symptoms on the crop.

17.3.13 Ashy Stem Blight of Cowpea

This disease is most severe in cowpea and is caused by fungus *Macrophomina phaseolina*. Disease symptoms appeared mostly in late kharif season as conspicuous, reddish brown to light black sunken cankers. The lesions spread rapidly extending to the entire stem and killing the plant. Infection on older plants results in yellowing and drooping of foliage. Drying of intermittent branches and fine twigs was also observed. The disease is seed-borne and has a very wide host range. Warm weather conditions (30–34 °C) are favorable for disease incidence.

Excessive soil moisture should be avoided. Disease-free healthy seeds should be used. Treatment of seed with carbendazim at 0.25 % is useful. Infected crop debris must be burnt after harvesting.

17.3.14 Angular Leaf Spot of French Bean

This disease is very common in French bean and is caused by fungus *Isariopsis griseola*. The pathogen is also known as *Phaeoisariopsis griseola* (Sacc.) Ferr. Leaf spots are formed usually on the lower leaves of plant. The spots are reddish brown to dark brown with a distinct margin. The lesions are usually confined between leaf veins. Severely spotted leaves senescence prematurely and drop off. Lesions on pods have reddish brown centers with black border. These spots coalesce covering the entire pod. Stem lesions are dark and elongate. The disease severity is increased in wet weather. The fungus can survive for 2–3 years in crop debris of diseased plants in soil.

In order to minimize disease incidence, crop rotation should be followed. Infected plant debris is burnt. For effective disease control, treatment of seed with carbendazim at 0.25 % as seed dressing and foliar spray of carbendazim at 0.1 % at 10 days interval are recommended.

17.3.15 Anthracnose of Cucurbits

Bottle gourd is the most severely infected by the anthracnose. Disease symptoms are found on all aboveground plant parts. Emerging seedling from infected seeds has small circular lesions on epicotyls, cotyledons, and leaves. This disease is caused by *Colletotrichum orbiculare* (*C. lagenarium*). Symptoms on leaves are observed as water-soaked, small, yellowish angular spots that enlarge and turn to brown. The necrotic portion dries and shatters. Elongated water-soaked, sunken lesions appear on stem. Light yellow to brown discoloration of the stem lesions is due to abundant sporulation. Severe incidence resulted infection of juvenile fruits as small, sunken, light brown, cracked spots. Young infected fruits cankered and drop. Red gummy exudates appear from the cankers developed on fruits and stems. The fungus can survive from one season to another in infected plant debris as well as in diseased seed. Optimum range of temperature and humidity for disease development is 18–27 °C and 90–100 %, respectively. Disease incidence increases with the age of the plant and frequent rainfall. Pathogen perpetuates in the infected crop debris in the soil, weeds, and on the seeds.

Crop rotation and destruction of infected debris are useful in checking the infection. The crop should be grown on bower system to avoid soil contact. Maintain proper drainage in the field. Seed production should be preferably carried out in summer season because summer crop is often free from pathogen. Seed treatment with carbendazim at 0.25 % is most effective against the seed-borne infection. Spray carbendazim at 0.1 % or chlorothalonil at 0.2 % soon after infection.

17.3.16 Downy Mildew of Cucurbits

Disease appears on all the cucurbits. However, it is a severe problem in cucumber, bitter gourd, bottle gourd, sponge gourd, ridge gourd, pointed gourd, and muskmelon. The disease is caused by *Pseudoperonospora cubensis*. Symptoms appear as irregular, numerous, small, yellow, and angular areas surrounded by green tissues scattered all over the leaf lamina. It appears just like in definite mosaic pattern particularly in cucumber. The yellow areas are angular and bounded by veins. Symptoms on bottle gourd are light brown while grayish brown and diffused on bitter gourd without yellowing. In high humid weather, faint white downy growth of fungus is observed on lower side of the leaves. Later on infected fruit dries and falls down. The pathogen survives as mycelium, because oospore formation is not common. The pathogen can infect the plants at temperatures between 10 and 30 °C. Five pathotypes on different cucurbits have been distinguished based on high compatibility with specific hosts (Bains and Prakash 1985) in India. Muskmelon grown in plastic house is severely infected by the disease.

Crop should be grown with wide spacing in well-drained soil. Air movement and sunlight exposure help in checking the disease initiation and development. Bower system of cropping reduces the disease incidence. Field sanitation is done by burning crop debris to reduce the inoculum. Seed production should be preferably carried out in summer season to obtain disease-free seed. Protective spray of mancozeb at 0.25 % at weekly interval gives good control without any fear of resistance development. In severe case, one spray of metalaxyl plus mancozeb or cymoxanil plus mancozeb at 0.2 % is given. Metalaxyl-resistant strains of the pathogen can be controlled by dimethomorph. Bitter gourd lines NIC-12285 and VRBT-39 were moderately resistant in challenge inoculation (Pandey et al. 2005).

17.3.17 Powdery Mildew of Cucurbits

Powdery mildew is prevalent in all the states and appears on most of the cucurbits. However, it is a severe problem on bottle gourd, bitter gourd, cucumber, sponge gourd, and pumpkin. This is more severe in winter season and green house crops. Two pathogens are associated with this disease, i.e., *Sphaerotheca fuliginea* and *Erysiphe cichoracearum*. Symptoms first appear on lower surface of leaves followed by the upper surface followed by all foliar part as white to dull white, floury, powdery growth. In case of *S. fuliginea* infection, lesions become rusty brown, whereas in case of *E. cichoracearum*, these are white (Sirdhana and Chaudhari 1972). *Leveillula taurica* has also been associated with powdery mildew symptoms. The white growth quickly covers most of the leaf surface and leads to heavy reduction in photosynthesis area. Plant may wither and die. Growth of plant and fruits seized drastically reduced. Transpiration rate increased considerably which leads to premature drying of leaves. Two distinct races of *E. cichoracearum* were established on muskmelon. Physiological specialization has been reported in *S. fuliginea* with the existence of four races (0, 1, 2, and 3) on different cucurbits in India (Khan and Sharma 1993).

Foliar sprays of penconazole at 0.05 % or tridemorph 0.1 % or carbendazim 0.1 % are very effective to control the disease, but their continuous spray should be avoided. Alternate spray of these systemic fungicides and wettable sulfur at 0.2 % may be given to check the development of resistant strains of the pathogen. Use tolerant line of cucumber like Poinsett, Palmetto, and Ashley. Field sanitation by removing weed hosts and infected crop debris reduces the inoculums.

17.3.18 Gummy Stem Blight of Cucurbits

This disease was earlier considered of minor importance but now becoming severe since the last one decade in muskmelon, bottle gourd, ridge gourd, and sponge gourd. Introduction of hybrids spreads this disease in all cucurbit-growing area because majority of the hybrids are susceptible to the disease. The disease is caused by *Phoma cucurbitacearum* (anamorph) and its teleomorph is *Didymella bryoniae*. Initially water-soaked areas are observed on the stem near soil line. Later translucent gum-like exudates released from the affected portion is deposited over it. Black dot-like pycnidia are also observed on the affected bark. The pathogen also infects leaves and side branches. Large, irregular, blotchy, brown lesions were observed on the leaves. Most of the lesions initiated from margin of the leaves and rapidly increased, engulfing entire leaves.

Deep summer plowing, crop rotation with cereals, and field sanitation are essential to reduce the inoculum. Avoid growing of susceptible hybrids. Maintain proper drainage and aeration in the field. Avoid injury near soil line during intercultural operation and through soil insects. Treat the seed with carbendazim at 0.25 %. Drenching of carbendazim at 0.1 % near collar region is very effective, but economics should be considered depending on commercial value of the crop.

17.3.19 Fruit Rots of Cucurbits

Fruit rot is a common problem during kharif season in most of the cucurbits. *Pythium* and *Phytophthora* are major pathogens for this rotting. *Phytophthora cinnamomi* is causing rotting of fruits, vines, and leaves of pointed gourd. Whole crop is completely blighted within 3–4 days. Pythium fruit rot is also called as cottony leak and characterized by water-soaked, soft rotting of the fruits starting from the lower portion, which is close to soil contact. White, fluffy, mycelium growth is observed on the affected portion in early morning in most of the gourds. Rhizoctonia fruit rot is most severe in muskmelon showing rotting and cracking of infected portion of fruit. Phomopsis fruit rot on ash gourd caused by *Phomopsis cucurbitae* was first time observed during cropping season of year 2000 and is a new record in the world (Pandey and Pandey 2003b). Symptoms are mostly observed on matured fruits as comparatively dry rotting with characteristic pycnidia over it. Generally lower portion of fruits touching soil surface is affected.

Avoid contact of the fruits from soil either by staking of plant or providing suitable mulch below the fruits. Adopt bower system of cultivation which is very effective for the management of most of the diseases and good commercial yield. Provide proper drainage in the field. Green manuring followed by soil application of *Trichoderma* at 5 kg/ha is very effective in checking most of the fruit rots. Collect all the affected fruits and burn them to reduce primary inoculum.

17.3.20 Leaf Spots of Cucurbits

The associated pathogens with this disease are *Cercospora citrullina*, *Phoma cucurbitacearum*

(*Didymella bryoniae* teleomorph), *Alternaria cucumerina*, and *Corynespora melonis*. Several types of leaf spots occurred on different cucurbits. Often these leaf spot diseases are more pronounced at maturity stage. Phoma leaf blight was severely observed on ash gourd, ridge gourd, and bottle gourd in Gangetic plains during 1998. Thereafter, severity of the disease is increasing every year, engulfing most of the gourd and spreading throughout the cucurbits growing area. Exotic hybrids are more susceptible to the pathogen. Papery with rhythmic large spots followed by shot hole is observed on the leaves. Black small dot-like fruiting structures are also observed on the old spots. White fungus growth is clearly visible on outer margins of the spots in morning periods. Circular spots with variable size are observed on the leaf lamina in Alternaria infection while small, light brown to dark brown with white center in *Cercospora*.

Field sanitation, selection of healthy seeds, and crop rotation reduce disease incidence. Seed production should be preferably carried out in summer season because summer crop is often free from disease. Foliar sprays of contact fungicide like mancozeb at 0.25 % alternated with one spray of hexaconazole at 0.05 % is very effective. Cucurbits particularly gourds should be grown on bower system.

17.3.21 Alternaria Leaf Spot of Cruciferous Vegetables

Alternaria brassicae and *A. brassicicola* in which *A. brassicae* is most common in India. Alternaria leaf spot usually appears in early stage of plant growth in cauliflower while in late stage in cabbage. *Alternaria raphani* is a serious pathogen of radish but is also observed in cauliflower. The disease is prevalent during January to March. All foliar parts are affected by the pathogen. Alternaria leaf spots are restricted to lower leaves only and do not cause economic loss in the crop grown for vegetable purpose. However, it causes severe loss in seed crop of cauliflower. Symptoms appear as circular light brown spots on leaves. Concentric rings are visible on the spots. Black sporulation is observed in humid weather. Inflorescence and siliqua are severely infected in seed crop. Brown discoloration of individual florets and flower clusters occur in cauliflower. Pericarp of seed is severely infected by the pathogen. Cotyledons of seed are also infected resulting internally seed-borne nature of the pathogen. Infected seeds act as a primary source of infection. Cabbage infections confined mostly on outer leaves of heads.

Detach all the infected lower leaves in morning and burn it with help of dry straw. Always use healthy and disease-free seeds. Seed should be treated with captan at 0.25 % seed. Spray chlorothalonil at 0.2 % along with sticker at 0.1 % in evening hour. Spray mancozeb at 0.25 % along with sticker during silique formation stage to prevent seed infection. Fungicide should be applied in the evening of same day when the infected leaves have to be detached in morning. Cauliflower lines, namely, Hazipur-4, Deep Malika, Suryamukhi, Kathmandu local, and Aghani-1, were found resistant in artificial inoculation conditions (Pandey et al. 2002b).

17.3.22 Black Leg of Cruciferous Vegetables

The disease is prevalent in all the vegetables of this group but more importantly in cabbage and cauliflower. It is caused by most commonly occurring anamorphic stage of the pathogen called *Phoma lingam*. The teleomorph *Leptosphaeria maculans* is not common in nature and sometimes produced in infected crop debris left in the field. Symptoms appear at any stage of growth from seedling to maturity. Papery, grayish white, round lesions appeared on leaves. Lesions were also observed on seed stalks and silique becoming seed-borne. Light brown depressed and cankerous lesions can be seen near the base of stem. Black, linear streaks in the affected stem when split vertically. Small, black dot-like, numerous pycnidia appeared on the lesions. The infected plants finally either topple down or wilt. The pathogen survives in seed as well as crop debris left in the soil.

Hot water treatment of seeds at 50 °C for 30 min is effective to control the spread of the disease. Long crop rotation of at least for 3 years with cereal and millet is recommended. Field sanitation by uprooting and burning of all the infected plants is advised. Deep summer plowing should be done to reduce the soil inoculum. Seed production should be carried out in either in disease-free area or under the strict supervision of pathologists.

17.4 Bacterial Diseases of Vegetable Crops

17.4.1 Bacterial Wilt of Solanaceous Vegetables

Bacterial wilt of solanaceous vegetables is the most destructive and widespread disease in tropical and subtropical states of India. Bacterial wilt is a very destructive disease for tomato, brinjal, chilies, capsicum, and potato. The disease has been recorded on several hundred species belonging to more than 50 plant families. Considering the economic importance, wide host range, and worldwide distribution, bacterial wilt is ranked as one of the most important diseases to be caused by phytopathogenic bacteria. The losses have been reported in brinjal and tomato as high as 80 and 90 %, respectively, under Indian conditions (Rao 1976). The symptoms are characterized by the sudden wilting and death of infected plants. The petiole of older leaves droops down and the leaves showed pale yellowing symptoms accompanied by stunting of whole plant. Excessive development of adventitious roots also takes place in tomato. There is typical browning of vascular tissues of roots, stems, or tubers. Whitish bacterial exudates come out from cross sections of infected plants. The name of bacterium has been changed several times but now accepted as *Ralstonia solanacearum*. The species is divided into five races, five biovars, and two RFLP groups (Chakrabarti et al. 2001). The bacterium is rod-shaped flagellate having 1–4 polar flagella and gram-negative. The size is variable being influenced by the growth conditions. The length is usually 1.5–3 times greater than the width. Extremely short rods frequently occurring in pairs and measuring 0.5–0.6 × 0.8–1.2 μm occur in the infected tissues.

The *R. solanacearum* bacterium is mainly soilborne. It is also seed-borne in nature (Shakya 1993). The pathogen survives in infected plant debris in wild host plants and weeds. Bacteria enter in plants through wounds occurring in roots due to cultural practices, nematodes, wind, rain splashes, hailstorms, and insects. The bacteria enter in xylem vessels and the intercellular spaces of the parenchyma cells and dissolve the cell walls. The slimy masses of bacterial cells along with their polysaccharides cause occlusion and clogging of xylem vessels. Phenol oxidases cause oxidation of phenolic compounds into quinones and polymerize to form melanin resulting vascular browning. Relatively high soil moisture and soil temperature (25–35 °C) favor the disease. Acidic soil has more severe problem of bacterial wilt. Spontaneous reversion of an avirulent phenotype of *R. solanacearum* to virulent wild types has been reported in Indian isolates (Shekhawat et al. 1992).

The bacterial wilt is a very difficult disease to manage due to variable nature of pathogen, wide host range, and soilborne. Reduction on soil inoculum is possible with chemicals, soil amendments, crop rotation, summer fallowing, escaping the disease, growing the crop in new fields, resistant cultivars, and biocontrol. Chemical control has been tried by antibiotics, chloropicrin, and bleaching powder but neither economical nor feasible. *Pseudomonas fluorescens* and *Bacillus polymyxa* have been found to delay the development and reduce the incidence of bacterial wilt. Reduced nitrogen application may also increase susceptibility to solanaceous plants to wilt incited by *R. solanacearum*. Resistance to bacterial wilt in tomato is available but the quality and quantity attributes is not satisfactory. High level of resistance to bacterial wilt has been identified in India in brinjal variety, namely, Arka Nidhi, Arka Keshav, Arka Neelkantha, BB-1, BB-44, BB-49, EP-49, and Surya (Sadashiva et al. 1993). Crop rotation with non-solanaceous hosts is effective against this disease. Green manuring with cruciferous crops and soil application of bacterial antagonist reduces the level of bacterial wilt inoculum in soil. The bacterial wilt of acidic soil can be managed effectively by increasing the soil pH up to neutral through soil application of lime.

17.4.2 Bacterial Blight and Speck of Tomato

Bacterial blight of tomato is caused by *Xanthomonas campestris* pv. *vesicatoria*. The disease is prevalent throughout the country. Small, dark, water-soaked spots appear on leaflets and stems of seedlings as well as transplants. Numerous spots coalesced and cause severe burning of the leaves. Kharif crop of tomato is severely affected by the bacteria with an average incidence of 35–40 % every year (Pandey and Pandey 2003a). Yield loss is recorded more than 40 % in early sown tomato. Disease incidence continues from July to October. Prominent yellow halo is observed around the spot. The most conspicuous and damaging phase is infection of immature fruits. Small, slightly raised, and corky spots appear on exposed surfaces of the green and turning-red fruits.

Bacterial speck is caused by *Pseudomonas syringae* pv. *tomato*. Bacterial speck is prevalent from November to January and favored by cool, moist weather having temperature range between15 and 25 °C. It is characterized by absence of chlorotic halo around the spot. Small black spots appear on leaf, petiole, and pedicel while irregular, elongated lesions can be seen on stem. The spots are necrotic, circular to roughly circular measuring 1.0 mm in diameter. As the fruit approaches the pink stage, the tissue around the spot retains its green color longer than normal. Generally, disease is observed when fruits start ripening.

The bacteria survive mainly through seeds, crop debris, and several weed hosts. Disease appeared in rainy season and continued up to fruit initiation stage. The disease is prevalent in warm humid weather when RH is more than 90 % and temperature is between 25 and 35 °C. A low temperature less than 17 °C limited the entrance and multiplication of *X. campestris* pv. *vesicatoria* in the intercellular space of pepper leaves. Injury created by high-pressure fungicidal sprays also enhances the infection of *X. campestris* pv. *vesicatoria* in tomato. In general, injuries caused by other animate agencies like nematodes, insects, and mites increase the susceptibility of vegetable crops by providing wounds for entry of bacteria.

Seed dipping in streptocycline solution at100 ppm or hot water treatment at 50 °C for 30 min is effective. Summer plowing is recommended to desiccate the bacteria and infected host. Soil solarization of nursery bed is beneficial to minimize infection. Rotate nursery seedbed and main field with cereal crops. Seed should be collected from disease-free plants. One spray of streptocycline at 150–200 ppm followed by one spray of kasugamycin at 0.2 % in afternoon is recommended to minimize infection. One spray of copper oxychloride at 0.3 % should be given after 15 days of antibiotic application.

17.4.3 Bacterial Wilt of Cucurbitaceous Vegetables

Bacterial wilt of cucurbits is sporadic in India and affects many cultivated and wild species of the family Cucurbitaceae. Cucumber is the most susceptible host followed by muskmelon, squash and pumpkin, while watermelon is least susceptible. Bacterial wilt of cucurbits caused by *Erwinia tracheiphila* is a capsulated bacterium having four to eight peritrichous flagella. The first signs of wilt appeared as drooping of leaves, which become flaccid in sunny weather. Soon after dropping, all the leaves and entire plant are wilted. Affected stem becomes soft and macerated but later on plant dries up permanently. Wilted stems release viscid, sticky, bacterial matrix exudates from the vascular bundles after cutting. Droplets of white bacterial ooze appear on the cut surface of the infected stem. This bacterial ooze can form delicate threads that may be extended for several centimeters.

This is a vascular pathogen and differs from other *Erwinia* spp. because it is unable to degrade middle lamella. The bacteria overwinter in the bodies of adult cucumber beetles particularly red striped and spotted beetle. Primary infection is initiated when beetles feed on the young leaves or cotyledons of cucurbits plants causing deep wounds. The bacteria are deposited in such wounds with the feces of the insects. This is the only means of natural infection known and the bacterium is completely dependent upon the insect for its survival. After infection, the bacteria swim through the droplets of sap present in the wound. It enters through wound sap in the xylem vessels and moves rapidly to all parts of the plant. Penetration of the host through stomata does not take place. The bacteria present in the vessels of infected plants die within 2 months after the dead plants dry up. The survival period of these bacteria is very short in infected crop debris.

Summer plowing of soil should be carried to expose all the stages of beetles. Soil application of neem cake should be carried out at initial infestation stage of beetles. Control of cucumber beetles at high infestation level may be only possible by applying granular insecticides like carbofuran in surrounding soil of the plants or in the pits followed by dusting and foliar spray of carbaryl. Restricted use of exotic hybrids in the endemic areas is helpful to reduce the pathogen pressure.

17.4.4 Angular Leaf Spot of Cucurbits

Angular leaf spot occurs all over the world on cucumber, gherkin, muskmelon, pumpkin, squash, watermelon, and ash gourd. The disease manifests as small water-soaked spots on leaves, stem, blossoms, and fruits. The numerous spots on leaves are angular due to delimitation by veins, become tan colored, and may enlarge to irregular shapes giving the leaves as blighted or scorched appearance. On squash and watermelon, these are brown and black, respectively, with a yellow halo. The bacterial ooze forms a crust on them and dead tissues may tear off leaving shot holes. The spots on fruits are small, circular, brown to black, and superficial. The causal agent *Pseudomonas syringae* pv. *lachrymans* is a gram-negative, aerobic, straight, rot-shaped bacterium with 1–5 polar flagella.

Infested soil and contaminated seeds transmit the disease. The pathogen can survive for 2 years in the soil or debris from diseased plants. The bacteria can spread through irrigation water and rain splashes and can colonize the buds on plants. It can remain there for 2–3 weeks. The bacteria prefers cooler climate, and the temperature range of 20–24 °C is optimum for disease development, but it may continue up to 37 °C.

Field sanitation and crop rotation reduces the disease incidence. To eliminate the risk of seedborne infection, the seed should be obtained from disease-free dry areas. Hot water treatment of seed is effective to kill the pathogen, but it may affect germination. Spraying of antibiotics like kasugamycin at 0.2 % and streptocycline at 150 ppm in afternoon gave good control of *Pseudomonas syringae* pv. *lachrymans* on cucumber. Copper fungicides are useful, but care should be taken not to spray at high temperature or to young plants.

17.4.5 Bacterial Blights of Leguminous Vegetables

The bacterial pathogens associated with beans are common blight caused by *Xanthomonas campestris* pv. *phaseoli*, fuscous blight by *X. phaseoli* var. *fuscans*, and halo blight by *Pseudomonas syringae* pv. *phaseolicola*. Common blight is the major disease of bean particularly in cowpea. Disease appear in the month of July to September in rainy season. Symptoms of common blight are first seen as small translucent, water-soaked spots on leaves followed by on pods. Bigger blotch symptoms appear on the leaves in severe case of common blight particularly in rainy season. Small, sunken, reddish brown lesions appeared on pods making it shriveled, leading to premature death. Vascular discoloration may become evident in the severe incidence. Halo blight has most of the symptoms of common blight. A chlorotic halo appears around the individual spots in halo blight. The areas surrounding lesion remain green even on pod maturity. Kishun (1988) reported that on cultivars Contender and Arka Komal of French bean, the disease intensity varied from 4 to 71 % and yield loss of green pods estimated up to 84 %. Bacterial blight of pea is widely distributed throughout India and 80–100 % incidence was recorded in Kullu Valley (Sood 1989). The disease is caused by *Pseudomonas syringae* pv. *pisi*. Another bacterium, *P. syringae* pv. *syringae*, has been reported to infest on vegetable pea in temperate region. Spots on leaflets are round, oval, or irregular, 2–5 mm in diameter, reddish brown with translucent center, and a darker brown margin. Chocolate-brown linear streaks are observed on the stem and petiole. Sepals are readily infected which causes drooping of blossom and juvenile pods. The pods are chocolate brown, thin, twisted, and shriveled. Bacterial blight of cluster bean caused by *X. axonopodis* pv. *cyamopsis* is a widespread in guar growing regions of India and is severe in rainy season crop. The disease appears as both leaf spot and blight. The leaf spots are interveinal, round, and water-soaked in appearance and later coalesce resulting in blight phase. Leaf infection advances systemically through the petiole to stem, producing black, longitudinal streaks leading to cracking and breaking of the stem.

The bacterium overwinters in the infected seeds and crop debris but seed is the main source of primary inoculum and an effective means of dissemination. The bacterium is a warm-temperature pathogen and causes greatest damage to the plants at 28–32 °C. High humidity and high rainfall increase the disease spread. Secondary spread of disease occurs by windblown rain splashes, contact between leaves, irrigation water, and insects. Cool, moist weather favors the disease, while warm and dry weather retards the disease. The extent of disease spread depends upon the frequency of rainy periods and is positively correlated. If enough rains occur, within the span of a week, the primary infection from a tiny amount of infected seed can spread over large areas. Most of the bacterial diseases in leguminous vegetables are seed-borne. Young plant foliage was more susceptible then old plants.

Crop rotation of about 2–3 years is essential to dispose of overwintering bacteria and infected crop debris. Field sanitation by collection of all the infected leaves and crop debris followed by burning it reduces the inoculum load in soil. It is very essential because bacteria survive for longer period in stubble leftover in field. The use of clean and disease-free seed is the most effective control measure of bacterial blights in bean and peas. The use of tolerant varieties for each of the blight is desirable. Seed-borne inoculum is effectively checked by streptocycline seed treatment at 250 ppm for 2 h. Proper cultural practices like wide spacing between plants, removal of weeds, and drainage should be maintained in the field to avoid water stagnation. Sowing time should be adjusted in such a way that fruiting stage should not coincide with heavy rainy period.

17.4.6 Black Rot of Cruciferous Vegetables

Black rot of cole crops is prevalent in all agroclimatic regions of India. It occurs worldwide and attacks all cultivated brassicas and numerous cruciferous weeds. The disease is caused by the bacterium *Xanthomonas campestris* pv. *campestris*. Symptom initiates along the margins of leaves as V-shaped with chlorotic lesions and progresses in the direction of midrib. However, such symptoms are not necessary on cabbage and appear from any side or center of the leave. In highly susceptible varieties, irregular marginal yellowing and burning of the leaves appeared. Severe infection of black rot is resulting complete burning of crop in cauliflower before the curd formation. Irregular round yellowing are formed after hailstorm and insect feeding in the center of leaves. Some of the veins and veinlets within chlorotic area turn black in severe case. Black vascular scar is observed on any detached infected leaves, midribs, and veins. Yellowing of leaves was observed from lower portion of midvein in severe infection.

The bacterium is transmitted primarily through seed. The pathogen is also survived for not more than a year in diseased plant left in the field. Disease severity increases rapidly whenever hailstorm is taken place in cropping season. Cruciferous weeds are also source for perpetuation of bacteria in field. The bacteria usually enter the cotyledons through stomata and pass to the young leaves and progress systematically throughout the plant system. The foliage infection and secondary transmission is through water pores, insect injury, infested soil, hail storm, cultural practices, seedlings, and movement of farmers in the field particularly during morning time. Black rot development is more rapid between 20 and 30 °C temperatures with high humidity and sufficient dew. The incubation period varies between 7 and 20 days depending upon temperature.

Always collect seeds from disease-free plants. Seed treatment with hot water at 50 °C for 30 min or seed dipping in 100 ppm streptocycline solution for 30 min is effective in eliminating the seed-borne bacterial inoculum. Crop rotation should be carried with non-cruciferous crops. Mulches of grass and straw can also be used for this purpose. Detach the lower infected leaves in afternoon when dew and bacterial ooze dried up from the leaves and then burn it. Use antagonistic bacteria in the soil. Spraying of antibiotic like streptocycline at 150–200 ppm or kasugamycin at 0.2 % along with sticker at 0.1 % at 10–15 days of interval gives good control, but spray must be given after detachment of leaves. Sprays of copper oxychloride at 0.3 % at 10 days of interval reduce the spread of disease. Mixture of streptocycline 100 ppm and copper oxychloride at 0.3 % with sticker at 0.1 % gives good control of the disease. Pusa Mukta variety of cabbage was reported as resistant to black rot (Kalda et al. 1991). MR-1 is a very good variety of cabbage having multiple-disease resistance including black rot developed at Katrina by Dr. K.S. Kapoor. Hsb-19 strain of *Bacillus* was reported an effective and safe means to suppress this important bacterial disease in *Brassica* and got 83.15 % disease control over control (Jalali and Parashar (1995). Cauliflower lines Kunwari-18, Kataki-7, and BT-10-2 were found moderately resistant after artificial screening and can be used in black-rot-prone area (Pandey et al. 2003).

17.4.7 Bacterial Soft Rot of Vegetables

Bacterial soft rot is the most important disease after post-harvest mainly during transportation and storage. Vegetable having more water content gets easily infected by soft rot bacteria. The first symptom on affected tissue is small, water-soaked lesion, which enlarges very fast engulfing entire plant part. The affected area becomes soft, mucilaginous, water-soaked, translucent, and depressed. Leafy vegetables are more susceptible to the disease. The cauliflower curds are severely infected during hail storming, snow, and rain fall in temperate climate as well as plains. Cabbageheads start rotting from stump region and engulf entire head. Bacterial-soft-rot-infected vegetables completely collapsed and decomposing tissues give repulsive bad odors. The causal bacteria is *Erwinia carotovora* pv. *carotovora*. The bacterium is straight rod shaped, measuring 0.5–1.0 × 1.0–3.0 μm, facultative anaerobe, capsulated, and motile by several peritrichous flagella.

The pathogen enters in the host through different types of injury caused by insects, snow, hail, wind, sunburning, and other extreme environmental conditions; mechanical, physical, and diseased lesions; and natural openings. The pathogen survives in infected plant parts, soil, stored tubers, and planting materials. The bacterium is a typical week facultative anaerobe. Bacteria multiply on the exudates released by the wounded cells. The bacteria have strong potential to produce pectolytic and cellulolytic enzymes that result in degradation of middle lamella and plasmolysis of cells. Infection is favored by high humidity and optimum temperature for disease development. Previously occurring diseases like black rot and *Rhizoctonia* and *Pythium* infections predispose for further attack of this disease.

This disease can be only managed by checking all kinds of injury. Spraying of plantomycin or streptocycline at 300 ppm in combination with copper oxychloride at 0.2 % at fortnightly interval is required soon after the disease initiation in standing crops. However, it is not possible in harvested products, so the best method is to avoid injury, proper packaging, free air movement, low-temperature transportation, and delicate handling. Dipping of harvested vegetables in a solution of sodium hypochlorite containing 50 ppm of available chlorine is recommended for certain vegetables. The fields to be planted must be free from the previous year's plant debris. Plants subjected to curd rot infection should be kept free of surface moisture by planting in well-drained areas, by providing sufficient plant-to-plant distance for adequate air circulation and by avoiding overhead irrigation. Chemical sprays for curd rot are effective at initial stage of infection. Chemical control in cauliflower and cabbage should be only recommended in seed crops.

References

Reference for Section 17.3

Bains SS, Prakash V (1985) Susceptibility of different cucurbits to *Pseudocercospora cubensis* under natural and artificial epiphytotic conditions. Indian Phytopath 31:42–46

Black LL, Green SK, Hartman HL, Poulos JM (1991) Pepper diseases: a field guide. Asian Vegetable Research and Development Center, Taiwan, p 20

Gupta VK, Paul YS (2001) Diseases of vegetable crops. Kalyani Publishers, New Delhi, p 47

Keinath AP, DuBose VB, Rathwell PJ (1996) Efficacy and economics of three fungicide application schedule for early blight control of yield of fresh market tomato. Plant Dis 80:1277–1282

Khan MW, Sharma GK (1993) Races of *Sphaerotheca fuliginea* infecting cucurbits in India and host differentials for their identification. Indian Phytopath 46:236–242

Pandey KK (2006) *Colletotrichum capsici*: a new fruit rot pathogen of tomato in Utter Pradesh. J Mycol Plant Pathol 36(1):104–105

Pandey KK, Pandey PK (1996) Symptomatology, epidemiology and chemical control of collar rot of elephant foot yam. Veg Sci 23(2):171–175

Pandey PK, Pandey KK (2002) Emerging strategies to control fungal diseases in vegetables. In: Khachatourians GG, Arora DK (eds) Applied mycology & biotechnology, vol 2. Elsevier Science International Publication, Amsterdam, pp 197–217

Pandey KK, Pandey PK (2003a) Survey and surveillance of vegetable growing areas for prevalence of major diseases. Veg Sci 30(2):128–134

Pandey KK, Pandey PK (2003b) First report of some new fungal diseases on waxgourd (*Benincasa hispida*) in the world from India. J Mycol Plant Pathol 33(3):439–441.

Pandey KK, Pandey PK (2004) Effect of soil solarization on disease management in vegetable nursery. J Mycol Plant Pathol 34(2):398–401

Pandey KK, Pandey PK, Kalloo G, Chaurasia SNS (2002a) Phomopsis blight in brinjal and sources of resistance. Indian Phytopath 55(4):507–509

Pandey KK, Pandey PK, Singh B (2002b) Slow blight response of early group of cauliflower (*Brassica oleracea* var. *botrytis* subvar *cauliflora)* to alternaria blight (*Alternaria brassicae*) under artificial inoculation conditions. Indian J Agric Sci 72(11):682–684

Pandey KK, Pandey PK, Kalloo G (2002c) A novel inoculation technique to screen tomato cultivars for Sclerotinia rot resistance. Ann Appl Biol Test Agrochem Cultivars 23:14–15

Pandey KK, Pandey PK, Kalloo G, Banerjee MK (2003) Resistance to early blight of tomato with respect to various parameters of disease epidemics. J Gen Plant Pathol 69(6):364–371

Pandey KK, Pandey PK, Mishra KK (2004) Integrated disease management of vegetable nursery. Veg Sci 31(1):45–50

Pandey KK, Ram D, Pandey PK, Rai M (2005) Artificial screening technique and resistance to downy mildew disease in bitter gourd. Veg Sci 32(1):107–108

Pandey KK, Pandey PK, Mishra KK (2006) Bioefficacy of fungicides against different bioagents for tolerance level and fungistatic behaviour. Indian Phytopath 59(1):68–71

Paul YS, Rathour R (1998) New blight of pea from Himachal Pradesh. Indian Phytopath 51:197–198

Snowden AL (1991) Post harvest diseases of fruits and vegetables, vol. I, II. Wolfe Scientific Ltd., London

Sirdhana BS, Chaudhari SL (1972) Occurrence of *Erysiphe cichoracearum* and *Sphaerotheca fuliginea* in Udaipur, Rajasthan. Indian J Mycol Plant Pathol 2:76–79

References for Section 17.4

Chakrabarti SK, Gadewar AV, Shekhawat GS (2001) Regulation of virulence in *Ralstonia solanacearum* and its probable exploitation in manipulating host resistance. In: Nagarajan S, Singh DP (eds) Role of resistance in intensive agriculture. Kalyani Publishers, Ludhiana, pp 37–52

Jalali I, Parashar RD (1995) Biocontrol of *Xanthomonas campestris* pv. *campestris* in *Brassica juncea* with phylloplane antagonist. Plant Dis Res 10:145–147

Kalda TS, Tewari RN, Singh R, Singh N, Verma TS (1991) Evolution of Pusa Mukta a resistant variety to black rot disease in cabbage. Indian J Hortic 48:247–250

Kishun R (1988) Losses in yield of french beans affected with common blight. In: Gnanamanickam SS, Mahadevan AJ (eds) Advances in research on plant pathogenic bacteria. Proceedings of the national symposium on phytobacteriology, University of Madras, India, 14–15 March, 1986, pp 29–31

Pandey KK, Pandey PK (2003) Survey and surveillance of vegetable growing areas for prevalence of major diseases. Veg Sci 30(2):128–134

Pandey KK, Pandey PK, Singh B (2003) Artificial screening for black rot resistance based on different disease parameter in early cauliflower. Mycobiology 31(3):173–178

Rao MVB (1976) Bacterial wilt of tomato and eggplant in India. In: Sequeira L, Kelman A (eds) Proceedings of the first international planning conference and workshop on the ecology and control of bacterial wilt caused by *Pseudomonas solanacearum*, Raleigh, NC, 18–24, July, 1976, pp 92–94

Sadashiva AT, Deshpande AA, Reddy KM, Singh R (1993) New sources of resistance to bacterial wilt in eggplant. Capsicum Eggplant Newsl 12:94–96

Shakya DD (1993) Occurrence of *Pseudomonas solanacearum* is tomato seeds imported into Nepal. In: Hartman GL, Hayward AC (eds) Bacterial wilt. Proceedings of international conference held at Kaohsiung, Taiwan, 28–31 October 1992, ACIAR proceedings no. 45:371–372

Shekhawat GS, Gadewar AV, Chakrabarti SK (1992) Spontaneous phenotypic reversion from afluidal to fluidal state in strain of *Pseudomonas solanacearum*. ACIAR Bacterial Wilt Newsl 8:5–9

Sood AK (1989) Outbreak of bacterial blight of peas in Kullu Valley. Indian Phytopath 42:322

18 Novel Detection Techniques for Plant Pathogens and Their Application in Disease Management

Sandeep Jain, T.S. Thind, P.S. Sekhon, and Amarjit Singh

Precise disease diagnosis and pathogen detection are of utmost importance, as without this ability we can neither hope to understand the disease nor in many cases control it. Detection deals with establishing the presence of a particular target organism within a sample, including symptomless individuals. Diagnosis relates to the identification of the nature and cause of the disease problem and thus deals with plants showing the symptoms. All aspects of plant disease epidemiology, from disease spread to estimation of yield losses, require the ability to identify the pathogen. Implementation of plant disease regulations through quarantine also requires the ability to diagnose plant disease. Increasing globalization and the international trade in plants and plant products also poses threat of inadvertent introduction of exotic pests and pathogens. Hence, importance of diagnostic tests has increased manyfold in modern era. In addition to detecting new invasive species, rapid and accurate diagnostic tests are required to monitor the emergence of novel variants of well-established pathogens. Improved techniques are vital to safeguard food security in the face of threatening diseases emerging as a consequence of climate change or other environmental shifts, or due to new agricultural practices. Detection of pathogens may be carried out either by employing conventional or relatively modern methods.

18.1 Conventional Tools

Conventional methods for identifying fungal plant pathogens are basically descriptive which rely on the interpretation of visual symptoms, isolations, culturing, and laboratory identification of the pathogen. The accuracy and reliability of these methods depend largely on the experience and skill of the person. Diagnosis requiring culturing can be time-consuming and can be impractical when rapid results are required. Many plant pathogens are indeed difficult to identify and require extensive taxonomical knowledge. All these factors put together may complicate timely disease management decisions.

Traditionally, for virus detection it was essential to perform time-consuming indexing or to culture for one or more days at a certain temperature on the appropriate medium in order to identify bacterial colonies using biochemical and physiological tests. For large number of samples, this process was obviously not suitable. Many times certain pathogens remain latent in plant tissue without showing any symptom. In these cases, visual inspection fails in detecting the presence of pathogen, with potential risk in quarantine prevention. Closely related organisms

S. Jain (✉) • T.S. Thind • P.S. Sekhon • A. Singh
Department of Plant Pathology, Punjab Agricultural University, Ludhiana 141004, India
e-mail: sandeepjain@pau.edu

L.P. Awasthi (ed.), *Recent Advances in the Diagnosis and Management of Plant Diseases*,
DOI 10.1007/978-81-322-2571-3_18

may be difficult to discriminate on the basis of morphological characters alone. Sometimes it is also difficult and time-consuming to discriminate between populations of the same pathogen that have specific properties, e.g., fungicide resistance, toxin production, or differences in virulence.

18.2 Modern Tools

Newer methods that are increasingly being applied to the diagnosis of plant pathogens include immunological methods, DNA/RNA probe technology, and polymerase chain reaction (PCR) amplification of nucleic acid sequences. These techniques have several potential advantages over conventional diagnostic methods in that they are more accurate, are faster, and can be used by personnel with no specialized taxonomical expertise (Ward et al. 2004). Even more important, these techniques allow detection of non-culturable microorganisms. Furthermore, molecular identification techniques are useful in revealing new diseases of unknown etiology. These technologies not only reveal the presence of pathogen but could even be used for accurate quantification of its biomass. The comparative analysis of genomic sequences allows the phylogenetic reconstruction of the pathogen relationships at different taxonomic levels.

Molecular detection has undergone some major breakthroughs over the last four decades which have helped to develop and apply these techniques for various purposes. The advent of antibody-based detection, in particular monoclonal antibodies and enzyme-linked immunosorbent assay, was the first very important breakthrough. This approach was an important turning point in virology and bacteriology because pathogens could be identified and detected much more rapidly. Afterward, DNA-based technologies came, such as the polymerase chain reaction (PCR) which has the capability to amplify the original target DNA several million-fold. Nobel prizes were awarded to J.F. Kohler and C. Milstein in 1984 and to K. Mullis in 1993 for the development of monoclonal antibodies and for the amplification of nucleic acid sequences by polymerase chain reaction (PCR), respectively.

18.3 Serological Detection

Since the introduction of ELISA (enzyme-linked immunosorbent assay) for plant virus detection (Clark and Adams 1977), serological assays have increasingly become the mainstay of plant certification schemes. Now it is used to detect several plant pathogens including bacteria, fungi, and phytoplasmas. It has the advantage that it enables handling large numbers of samples and requires only one antiserum prepared in a single animal species (Ahoonmanesh et al. 1990). However, it has the disadvantage that it necessitates the preparation of a different antibody-enzyme conjugate for each virus to be tested. Furthermore, it is highly strain specific and hence can sometimes fail to detect serologically distinguishable yet closely related strains of the same virus (Koenig 1978).

The serologic detection is based on the reaction between the antigen and its self-homologue antibody. There are several assays revealing the antigen-antibody reaction in vitro. The first tests were direct ones, developed on agar medium, in which the reaction was revealed by a precipitation band visible by a simple visual inspection. A significant improvement of the techniques was achieved through the introduction of ELISA assay, allowing simultaneous testing of large number of samples with low cost and commercial scale applications. The end point of such ELISA tests using 96-well immunoplates is a color change that can be read electronically with spectrophotometer or assessed with naked eye. At first only polyclonal antiserum was available, but now monoclonal antibody-based antisera are more commonly used because of their increased specificity. Monoclonal antibodies (MCAs) recognize only a single epitope and therefore differentiate between related pathogens. However, specificity can also pose problem where all the strains of pathogen need to be detected. This potential shortcoming is often overcome by preparing a "cocktail" of several MCAs each specific to particular strain.

18.4 Dipsticks and Immunoblotting

Sticks onto which specific MCAs have been bounded are dipped into test specimen, allowing ELISA reaction to be carried on in situ. Such flexible plastic dipsticks are quick and easy to use in the field to test sap squeezed onto coated surface. Any viruses present are first entangled and then detected following incubation with antibody-enzyme conjugate. If specified virus is present, dipstick becomes colored, whereas if free of virus, it remains colorless. ELISA reactions can be carried out in a similar way on membranes, where a drop containing the specific MCA has already absorbed as "dot," onto which a drop of test sample is later added then blotted. The technique is similar to ELISA except that the plant extracts are spotted onto a membrane rather than using a microtiter plate as the solid support matrix.

Cahill and Hardham (1994) developed dipstick technique for the detection of *Phytophthora cinnamomi* zoospores in soil. Using this method, it was possible to detect 40 zoospores ml^{-1} in less than 45 min. Plant pathogens have been detected in situ by squashing plant tissues directly onto membranes followed by immunodetection, that is, tissue printing or squash blot systems (Gwinn et al. 1991; Shine and Comstock 1993).

The disadvantages of these techniques are possible interference of sap components with the subsequent diagnostic reactions. Sometimes the color of the sap will prevent weak positive reactions from being observed, and the results cannot be readily quantified. Nevertheless, their sensitivity, the relatively short time required to assay large numbers of samples, the need for minimum laboratory facilities for the assay, the ability to store blotted membranes for extended periods, and low costs favor these as useful diagnostic techniques. The other advantage is that the samples can be blotted onto the membranes right in the field and such membranes can be carried or shipped by mail for further processing at a central location either within the country or in a different country.

18.5 Lateral Flow Devices

Recently, there has been great progress in the development of very rapid diagnostic tests that are also simple and do not require any special equipment or knowledge. Most of these use the lateral flow assay. Lateral flow device kits are based on existing technology similar to a pregnancy test kit. They use specific antibodies and the tests give results in a couple of minutes. Lateral flow devices (LFDs) have been designed for several plant viruses and bacteria (Danks and Barker 2000). Although detection specificity is very high when using the appropriate monoclonal antibodies, the sensitivity of these methods, although good for virus detection, is relatively low for bacteria, and they are more appropriate for analyzing plants with symptoms.

18.6 Electron Microscopy

Electron microscopy (EM) provides very useful information on the morphology of the pathogen including virus particles and is commonly used for detection when EM facilities are readily available. Electron microscopes rely on beam of electrons produced from hot filament, rather than light as the source of illumination, and their power of resolution is consequently very much greater as it is not limited by wavelength of light. Electron microscopes are of two types: transmission electron microscopes (TEMs) and scanning electron microscopes (SEMs). While TEM is analogous to a compound electron microscope, an SEM is more similar to binocular stereoscopic microscope. Filamentous and rod-shaped viruses such as potyviruses, potexviruses, and tobamoviruses can more readily be differentiated in negatively stained leaf-dip preparations. Viruses that occur in low concentrations in plant sap are not easily seen unless the virus in the test material is concentrated before visualization. The efficiency of virus visualization can be improved in combination with serology. As EM is labor intensive and expensive, it cannot often be used for the rapid processing of multiple samples. Many agricultural research institutions cannot afford to have an electron microscope facility due to the prohibitively high costs involved in installation and maintenance of the facility.

18.7 Immunosorbent Electron Microscopy (ISEM)

ISEM combines the specificity of serological assays with the visualization capabilities of the EM. The principle of immunosorbent electron microscopy (ISEM) is the selective trapping of plant viruses onto electron microscope grids precoated with a specific antiserum. Electron microscope grids coated with carbon are first exposed to antibodies. The grid is then exposed to the sap of an infected host plant. Virus particles are selectively "trapped" on antibody-coated grids with little contaminating host-plant material. After staining, the virus particles adsorbed to the antibodies can clearly be seen under TEM. This method is at least 5000 times more sensitive than conventional transmission electron microscopy. Hence, the technique is more effective when virus particles are infrequent, and there is no need to fix, embed, and cut sections which means that it is much quicker and easier.

18.8 Nucleic Acid-Based Methods

Nucleic acid-based methods, based on amplification or hybridization, have revolutionized the way of pathogen detection and identification because these are sensitive and specific that allows genetic relationships to be determined. Specificity is directly related both to design of primers or probes and to the amplification or hybridization protocols. All viable propagules contain the entire nucleic acid complement of the organism. The presence of the nucleic acid sequence is not altered by the development or by the response to environment or by the host.

18.9 Hybridization Formats

The nucleic acid is bound to nylon and nitrocellulose filters. DNA must be denatured prior to binding. This is normally achieved by treatment with alkali. There are two general formats: dot blots and Southern blotting. When an extract or cell sap believed to contain the target nucleic acid is dotted onto the filter, this is referred to as dot blot. When hybridized to a number of probes, the pattern of hybridization can identify the pathogens present and the intensity of signal can give some idea of the quantity. The advantages of dot blot are its simplicity, speed, and number of samples that can be processed simultaneously. However, a dot blot can give information if specific probes are available.

If a specific probe is not available, or if nonspecific hybridization is suspected, then it is necessary to perform a Southern blot. To do this it is necessary to prepare reasonably pure DNA from culture of the pathogen. Then, DNA is digested with one or more restriction enzymes. The digest is then run on agarose gel before being blotted onto the filter. The cost and time required increases. However, identification criterion is raised to include not just hybridization of the probe but hybridization of specific band sizes. Nonspecific hybridization, recognizable as a smear, can be ignored.

18.10 Polymerase Chain Reaction (PCR)

PCR is extremely sensitive and popular technique used for the detection and identification of plant pathogens. PCR allows the amplification of millions of copies of specific DNA sequences by repeated cycles of denaturation, polymerization, and elongation at different temperatures using specific oligonucleotides (primers), deoxyribonucleotide triphosphates (dNTPs), and a thermostable *Taq* DNA. The amplified DNA fragments are visualized by agarose gel electrophoresis or alternatively by colorimetric or fluorometric assays. The presence of a specific DNA band of the expected size indicates the presence of the target pathogen in the sample. The main advantages of

PCR techniques include high sensitivity, specificity, and reliability. PCR detection is also usually completed more rapidly than culture techniques (Harmon et al. 2003; Zhang et al. 2006). Even RNA-based viruses can be detected via PCR, by first reverse transcribing the RNA into DNA (Schaad et al. 2003). Moreover, it is not necessary to isolate the pathogen from the infected material reducing the diagnosis time from weeks to hours and allowing the detection and identification of non-culturable pathogens. This characteristic has been especially useful in the analysis of symptomless plants. It is fairly expensive and requires minimal skill to perform. Detection of pathogen in a given sample by PCR does not depend only on the performance of PCR assay but also on the efficiency of the procedure employed to extract the nucleic acid from the plant materials as inhibitors that are present in the extract of nucleic acids reduce the sensitivity of detection.

18.11 BIO-PCR

Culturing for a short time can be combined with PCR detection (BIO-PCR) to increase the quantity of pathogen present and to ensure that only viable microorganisms are detected (Schaad et al. 1999). Target cell enrichment followed by PCR or BIO-PCR improves the efficiency and sensitivity of PCR by allowing target pathogen population to increase in a pre-enrichment phase, before DNA extraction and PCR. Selective pre-enrichment increases pathogen population relative to nontarget microorganisms and results in high quantity of target DNA, which ultimately results in sensitivity. Additionally, during incubation and enrichment on artificial media, inhibitory compounds are adsorbed or diluted during cell harvest and do not interfere with DNA amplification.

18.12 RT-PCR

Since PCR can only amplify double-stranded templates such as DNA, RNA should be converted to DNA (called complementary DNA or cDNA) prior to use in a PCR-based assay. Conventional PCR is applicable directly to DNA plant viruses; however, for diagnosis of plant viruses with RNA genomes, the RNA target has to be converted to a complementary DNA (cDNA) copy by reverse transcription before PCR is begun. The cDNA provides a suitable DNA target for subsequent amplification. During the initial cycles of PCR, a complementary strand of DNA will be synthesized from the cDNA template, and thereafter, the reaction will proceed as for double-stranded DNA described above. This process of amplification is called reverse transcription-polymerase chain reaction (RT-PCR). On the completion of the reaction, the amplified DNA can be analyzed by agarose gel electrophoresis.

Since detectable DNA may be obtained from dead cells, hence, starting from RNA may more accurately reflect viable pathogen propagules. This type of assay is commonly used in the detection of plant viruses, most of which have RNA genomes (Waterhouse and Chu 1995).

18.13 Immunocapture (IC)-PCR

A technique that combines the technical advantages of PCR with the practical advantages of ELISA, called immunocapture (IC)-PCR, was developed for the detection of several different plant viruses. In this assay, the virus particles are first "concentrated" by trapping onto a solid surface (either microcentrifuge tube or ELISA plate) using virus-specific antibodies. The trapped virus particles are disrupted and the released viral nucleic acid are amplified by RT-PCR. This results in greater sensitivity, and problems encountered with RNA extraction are minimized and inhibitors of RT-PCR washed away prior to amplification. Thus, IC-PCR is a very useful alternative for RT-PCR in virus detection from plant material and insect vectors.

18.14 Real-Time PCR

Real-time PCR allows fast, accurate detection and quantification of plant pathogens in an automated reaction. This technology differs from conventional PCR by monitoring products online while they accumulate at each reaction cycle in a closed tube format, without the need of post-reaction processing such as gel electrophoresis. As a consequence, real-time PCR is generally faster than conventional PCR, enabling high-throughput analyses. Typically, DNA amplification is monitored each cycle based on the emission of fluorescence.

In addition to simplifying quantification, real-time PCR has a number of other advantages over conventional PCR. It is faster and a higher throughput is possible. It can be more specific than conventional PCR, if a specific probe is used in addition to the two specific primers. The high specificity of the probes used may detect single nucleotide polymorphisms (SNPs) – specificity is determined by a single base pair change in the DNA, as is often the case, for example, when distinguishing between fungicide resistant and sensitive isolates (Ward et al. 2004).

18.15 Multiplex PCR

Multiplex PCR allows the simultaneous and sensitive detection of different DNA or RNA targets in a single reaction. PCR detection protocols can be designed to verify the presence of more than one pathogen in plant material by looking for common specific sequences in two or more of them or to detect related viruses or bacteria on multiple hosts. Multiplex PCR is useful in plant pathology because different bacteria and/or RNA viruses frequently infect a plant material.

18.16 Loop-Mediated Isothermal Amplification (LAMP)

Notomi et al. (2000) revealed a novel means of DNA amplification which was highly specific, efficient, and rapid and which required only one temperature and no thermal cycler known as loop-mediated isothermal amplification (LAMP). Similar to PCR, this technique uses primers to detect and amplify sequences of target DNA. Unlike PCR, a LAMP reaction does not require a thermocycler to carry the DNA amplification through specific temperature phases. Instead, the reaction is done isothermally (typically at 65 °C) in a heat block or water bath.

18.17 Microarray Technology

Microarrays are generally composed of thousands of specific probes spotted onto a solid surface (usually nylon or glass). Up to 30,000 DNA probes (gene sequences) can be arrayed onto a single chip. Each probe is complementary to a specific DNA sequence (genes, ITS, ribosomal DNA), and hybridization with the labeled complementary sequence provides a signal that can be detected and analyzed. For the diagnostic laboratory, microarrays offer potential as a generic method for detecting large numbers of known pathogens in a single test. Although there is great potential for microarray technology in the diagnosis of plant diseases, the practical development of this application is still in progress.

18.18 Some Applications of Modern Methods in Disease Management

18.18.1 On-Site Diagnosis and Large-Scale Field Detection

Simple commercial methods for rapid detection are required for testing large numbers of samples by non-experienced technicians. For this purpose, tissue print-ELISA and lateral flow devices are helpful for detecting several plant viruses and bacteria. Tissue print-ELISA has also demonstrated potential in detection of viruses in plant tissues when imprinted on nitrocellulose membranes (without the need for extract preparation), for example, CTV.

18.18.2 Portable Real-Time PCR

Nucleic acid-based hybridization formats like dot blot or closely related techniques are generally used in large-scale analysis plans to identify spreading patterns of diseases in the field and in quarantine or certification schemes. In these cases sampling may be carried out in the field, using a single drop of plant sap, immobilized on a membrane pretreated with NaOH/EDTA or using a stem fresh cut or even whole leaf print. Then, the material is shipped to the competent laboratory, where nucleic acid hybridization is performed using probes.

18.19 Detection of Seed-Borne Pathogens

Detection of fungal pathogens in seeds and other propagative material by conventional methods involving their isolation and cultivation in suitable medium has been not only time-consuming but also difficult. ELISA formats allow sensitive and specific detection of several pathogens associated with seed. The loose smut disease of barley caused by *Ustilago nuda* is internally seed-borne. A DAS-ELISA has been applied successfully to test naturally infected barley seeds. The results are comparable with conventional seed embryo test. BIO-PCR has been successfully used for the detection of seed-borne pathogens causing halo blight of beans (*Pseudomonas syringae* subsp. *phaseolicola*), bacterial ring rot of potato (*Clavibacter michiganensis* subsp. *sepidonicum*), and black rot of carrot (*Alternaria radicina*) (Walcot 2003).

DNA chips or microarrays may also be applied to test seeds for pathogens. Currently, few DNA-chip seed detection assays have been developed. However, it is envisioned that this technology will be more widely employed for routine seed testing in the future.

18.20 Detection of Soilborne Pathogens

The largest group of soilborne pathogens for which immunologically based assays have been developed are the oomycetes. Commercial kits for the detection of Phytophthora species have been developed and have found increasing use in agriculture (Papavizas et al. 1981; Miller et al. 1994). These kits use DAS-ELISA and polyclonal antibodies and come in either a 96-well plate format or as a field assay that relies on a flow-through membrane system. The advantage of the latter is an assay time of 10 min and an easily observed color end point.

Inhibition of PCR amplification by substances such as phenolics and humic acids within soils had made it almost impossible to amplify low quantities of DNA from soils directly, but recent advances in techniques for the extraction and purification of DNA from soil have meant that this problem can be overcome. Protocols that have been successful in reducing inhibition of amplification generally involve grinding of soil samples with detergent followed by purification of the extract with polyvinylpyrrolidone (PVP) or gel chromatography. There has been considerable interest in the development of DNA identification techniques for the oomycetes like *Phytophthora*.

18.21 Monitoring of Airborne Inoculum

Airborne inoculum plays an important role in the spread of some plant diseases, and monitoring it could be used as an alternative to direct measurements of disease in the crop, when assessing disease risk. Detection of airborne inoculum, traditionally based on trapping of spores combined with microscopy, has now been adapted to PCR methods. PCR-based methods can detect and quantify biological material in air samples. However, more work needs to be done to test the methods under field conditions and determine the sensitivities of detection. Serological methods are also potentially useful for detecting and quantifying airborne bioparticles.

Kennedy et al. (2000) described a new microtiter immunospore trapping device (MTIST device) that uses a suction system to directly trap air particulates by impaction in microtiter wells. This device can be used for rapid detection and immunoquantification of ascospores of *Mycosphaerella brassicicola* and conidia of

Botrytis cinerea by an enzyme-linked immunosorbent assay (ELISA) under controlled environmental conditions.

18.22 Mycotoxin-Producing Fungi

Some phytopathogenic and food spoilage fungi produce mycotoxins, and detection of these is of great concern, both from a food safety and food trade standpoint. The main genera involved are plant pathogenic *Fusarium*, *Penicillium*, and *Aspergillus*, which are predominantly food spoilage fungi. Many of these fungi are difficult to classify and identify by conventional morphology-based techniques. These factors have prompted the development of various DNA-based and immunological techniq for identification of the mycotoxigenic fungi themselves, the mycotoxins they produce, and the genes involved in mycotoxin biosynthesis (Paterson 2006).

18.23 Fungicide Resistance Monitoring

To make the best use of available fungicides, the resistance status of the pathogen population should be known. Conventional screening methods can be slow and costly, especially for obligate pathogens that cannot be grown on artificial media. DNA-based methods targeted at specific resistance genes offer rapid, cost-effective alternatives. PCR linked with allele-specific probes, PCR-restriction fragment length polymorphisms, and allelle-specific PCR have been used to detect fungicide resistance in different pathogens. For example, real-time PCR assay has been to quantify strobilurin resistance in *Blumeria graminis* f. sp. *tritici* populations (Fraaije et al. 2002).

18.24 Quarantine Detection

International and transborder movements of plant germplasm are necessary to satisfy increasing demands for food, access to germplasm with desirable traits, and access to newly developed or identified varieties for evaluation or inclusion in commercial production systems. Unregulated movement of plant germplasm could result in introduction of new species of pathogens which may cause devastation of crops.

Diagnostic techniques used in a quarantine program should have adaptability for screening a large number of samples. The techniques should be relatively simple and robust, with minimal risk of cross contamination or false-positive results. Serology-based techniques such as ELISA have met these needs very successfully. Some molecular diagnostic techniques have been modified into an ELISA-type format. Serology has been combined with nucleic acid-based technologies in techniques such as immunocapture PCR. LAMP and real-time PCR assay have been developed to detect regulated plant pathogens to increase diagnostic capability. For example, *Xylella fastidiosa* causing Pierce's disease of grapevine, citrus veinal chlorosis, almond leaf scorch, and phony peach is a regulated plant pathogen in many parts of the world which is detected using these techniques (Ward et al. 2010). With the availability of genomic sequences of pathogens and the rapid development of microarray technology, urgent need is felt to set up this technology and apply it to detection of quarantine pathogens.

References

Ahoonmanesh A, Hajimorad MR, Ingham BJ, Francki RIB (1990) Indirect double antibody sandwich ELISA for detecting alfalfa mosaic virus in aphids after short probes on infected plants. J Virol Methods 30:271–2821

Cahill DM, Hardham AR (1994) Exploitation of zoospore taxis in the development of a novel dipstick immunoassay for the specific detection of *Phytophthora cinnamomi*. Phytopathology 84:193–200

Clark MF, Adams AN (1977) Characteristics of the microplate method of enzyme linked immuno-sorbent assay for the detection of plant viruses. J Gen Virol 34:475–483

Danks C, Barker I (2000) On-site detection of plant pathogens using lateral-flow devices. EPPO Bull 30:421–426

Fraaije BA, Butters JA, Coelho JM, Jones DR, Hollomon DW (2002) Following the dynamics of strobilurin resistance in Blumeria graminis f.sp. tritici using quantitative allele-specific real-time PCR measurements with the fluorescent dye SYBR Green *I*. Plant Pathol 51:45–54

Gwinn KD, Collins-Shephard MH, Reddick BB (1991) Tissue print-immunoblot, an accurate method for the detection of *Acremonium coenophialum* in tall fescue. Phytopathology 81:747–748

Harmon PF, Dunkle LD, Latin R (2003) A rapid PCR-based method for the detection of Magnaporthe oryzae from infected perennial ryegrass. Plant Dis 87:1072–1076

Kennedy R, Wakeham AJ, Byrne KG, Meyer UM, Dewey FM (2000) A new method to monitor airborne inoculum of the fungal plant pathogens *Mycosphaerella brassicicola* and *Botrytis cinerea*. Appl Environ Microbiol 66(7):2996–3003

Koenig R (1978) ELISA in the study of homologous and heterologous reactions of plant viruses. J Gen Virol 40:309–318

Miller SA, Bhat RG, Schmitthenner AF (1994) Detection of Phytophthora capsici in pepper and cucurbit crops in Ohio with two commercial immunoassay kits. Plant Dis 78:1042–1046

Notomi T, Okayama H, Masubuchi H, Yonekawa T, Watanabe K, Amino N, Hase T (2000) Loop-mediated isothermal amplification of DNA. Nucleic Acids Res 28:e63

Papavizas GC, Bowers JH, Johnston SA (1981) Selective isolation of Phytophthora capsici from soils. Phytopathology 71:129–133

Paterson RM (2006) Identification and quantification of mycotoxigenic fungi by PCR. Process Biochem 41:1467–1474

Schaad NW, Berthier-Schaad Y, Sechler A, Knorr D (1999) Detection of *Clavibacter michiganensis* subsp. *sepedonicus* in potato tubers by BIO-PCR and an automated real-time fluorescence detection system. Plant Dis 83:1095–1100

Schaad NW, Frederick RD, Shaw J, Schneider WL, Hickson R, Petrillo M, Luster DG (2003) Advances in molecular-based diagnostics in meeting crop biosecurity and phytosanitary issues. Annu Rev Phytopathol 41:305–324

Shine JM, Comstock JC (1993) Digital image analysis system for determining tissue-blot immunoassay results for ratoon stunting disease of sugarcane. Plant Dis 77:511–513

Walcot RR (2003) Detection of seedborne pathogens. Hortic Technol 13(1):40–47

Ward E, Foster SJ, Fraaije BA, Mccartney HA (2004) Plant pathogen diagnostics: immunological and nucleic acid-based approaches. Ann Appl Biol 145:1–16

Ward L, Harper S, Clover G (2010) Development of a LAMP assay for *Xylella fastidiosa*. MAF Biosecurity New Zealand technical paper No: 2010/14

Waterhouse PM, Chu PWG (1995) Nucleic acid based approaches to plant virus and viroid diagnostics. In: Skerritt JH, Appels R (eds) New diagnostics in crop sciences, Biotechnology in agriculture No. 13. CAB International, Wallingford, UK, pp 195–214

Zhang ZG, Li YQ, Fan H, Wang YC, Zheng XB (2006) Molecular detection of *Phytophthora capsici* in infected plant tissues, soil and water. Plant Pathol 55:770–775

19 Molecular Tools and Techniques for Detection and Diagnosis of Plant Pathogens

Pramila Pandey, N.S. Pandey, Md. Shamim, Deepti Srivastava, D.K. Dwivedi, L.P. Awasthi, and K.N. Singh

Plant pathogenic bacteria, phytoplasmas, viruses, and viroids cause destructive, extensive, and inexpensively important diseases in a very extensive range of plant species worldwide (Agrios 2001; Janse 2007). The damage is often sufficient to cause significant yield losses in cultivated plants (Scortichini 1995; Cambra et al. 2006). The two main effects on agriculture are decreased production and, in a less direct way, the need of implementation of expensive management and control procedures and strategies. In addition, efficient registered products for the chemical control of bacteria are lacking, and there is no chemical control available for viruses. Consequently, preventive measures to avoid planting of contaminated material are of the highest importance in the context of an integrated approach to control. Among such procedures, testing of planting material for pathogen-free status is an important, although not exclusive, method for controlling bacterial, fungal, and viral diseases of plants (Martin et al. 2000; Janse and Wenneker 2002). As many plant pathogens remain latent in the planting material, and in very low numbers, methods of high sensitivity, specificity, and reliability are required. Communal institutions and the agro-food industry used to control the hygienic superiority of seeds, fruits, and plant material by microbiologically testing plant pathogen. These methods were often expensive and time consuming, and some of them were not sensitive and specific enough. In addition, biological indexing cannot be applied on the large scale required. Diagnosis of the pathogenic diseases based on the symptoms induced, host range, and reactions on differential host plant species or crop cultivars has been accomplished. Biochemical and physiological characteristics like utilization of carbon sources using the Biolog system have also been used as the basis for identification of bacterial species. However, these tests require long time and often provide inconsistent results which have to be confirmed by other tests.

P. Pandey (✉) • D. Srivastava • D.K. Dwivedi
K.N. Singh
Department of Plant Molecular Biology and Genetic Engineering, N.D. University of Agriculture and Technology, Kumarganj, Faizabad 224229, Uttar Pradesh, India
e-mail: pramila28@gmail.com

N.S. Pandey
Department of Botany, National P.G. College, Lucknow 224001, Uttar Pradesh, India

Md. Shamim
Department of Molecular Biology and Genetic Engineering, Bihar Agricultural University, Sabour Bhagalpur-813210, Bihar, India

L.P. Awasthi
Department of Plant Pathology, N.D. University of Agriculture and Technology, Kumarganj, Faizabad 224229, Uttar Pradesh, India

L.P. Awasthi (ed.), *Recent Advances in the Diagnosis and Management of Plant Diseases*,
DOI 10.1007/978-81-322-2571-3_19

The prevention measures demand pathogen detection methods of high sensitivity, specificity, and reliability because many phytopathogenic bacteria and viruses can remain latent in "subclinical infections," and/or in low numbers, and/or in some special physiological states in propagative plant material and in other reservoirs (Grey and Steck 2001; Janse et al. 2004; Biosca et al. 2006). Accurate detection of phytopathogenic organisms is crucial for virtually all aspects of plant pathology, from basic research on the biology of pathogens to the control of the diseases they cause. Moreover, the need for rapid techniques of high accuracy is especially necessary for quarantine pathogens because the risk of the disease and the spread of the inoculum must be reduced to nearly zero (Lopez et al. 2003). Here, we present the state of the art of different morphological, biochemical, serological, and molecular detections of plant pathogenic bacteria, phytoplasmas, fungi, and viruses. Pathogenic detection refers to the presence of a particular target organism in plant tissues, vectors, plant products, or environmental samples, with emphasis on symptomless plants, whereas diagnosis is related to the identification of the nature and cause of a disease in plants showing symptoms (Jeffers and Aldwinckle 1987; Shurtleff and Averre 1997; Fillhart et al. 1998; Lopez et al. 2006). In spite of the great advances in sensitivity and specificity of the available techniques and protocols, there is an evident lack of information on the epidemiology of most diseases, on the sources of inoculum, and on the secreted life of pathogens in soil and further reservoirs. Additionally, the distribution of most bacteria and viruses is not homogeneous in the plant and even less in the plot, orchard, or nursery, and there is an urgent need for studies on sampling methodology and sample processing. The battery of available techniques and probes for detection of plant pathogens has increased considerably over the last few years. In addition to time benefits, there is a great advantage in terms of specificity when using serological techniques with specific monoclonal or recombinant antibodies or PCR with specific primers, as they allow the detection of plant pathogenic bacteria and viruses even camouflaged by a high number of other microorganisms. The open question that we will try to answer is left hanging in the air: Are molecular methods solving the challenges of the high sensitivity, specificity, and accuracy posed by detection of plant pathogens?

19.1 Host Range and Symptomatology of Diseases

The main limitation in the plant disease management is the lack of rapid, accurate, and reliable means of pathogen. Traditionally, the most predominant techniques used to identify plant pathogens relied upon culture-based morphological approaches. The major limitations of these methods, however, are the reliance on the ability of the organism to be cultured, the time consuming and laborious nature, and the requirement for extensive taxonomical knowledge, all together often complicating timely disease management decisions.

Although the primary host of the pathogen may be known, its host range may not, despite the fact that such information may be vital for epidemiological studies. Host range may also be a valuable aid to identification. In particular, some fungal and bacterial pathogens are given trinomials, the third name designating the host. For example, isolates of *Fusarium oxysporum* are classified into many formae speciales according to the hosts that they infect; *F. oxysporum* f. sp. *lycopersici* infects tomatoes and *F. oxysporum* f. sp. *cubense* infects bananas, causing Panama disease. Similarly, isolates of *Xanthomonas campestris* are classified into many pathovars: *X. campestris* pv. *citri* causes citrus canker, whereas *X. campestris* pv. *malvacearum* infects cotton. However, it is in the determination of physiological races (often referred to simply as races) of an organism that host-range studies are of particular importance and were, until recently, the only way in which a race could be identified. Physiological races, although morphologically indistinguishable, differ in their virulence toward a range of genotypes of the host plant, causing severe symptoms on some but not others. For example, some isolates of *Xanthomonas campestris* pv. *vesicatoria* cause bacterial spot of tomato, and others cause a similar disease in pepper; these have been designated the tomato and pepper races, or the A and B groups, respectively. More recent work has classified the tomato race or A group as *Xanthomonas axonopodis* pv. *vesicatoria* and the pepper race or B group as *X. campestris* pv. *vesicatoria*. However, the pathogens may be differentiated still further by their virulence or virulence for different genotypes of these hosts. Originally, two tomato races were recognized, one of which, T1, induced the hypersensitive response (a resistant reaction) on the tomato genotype Hawaii 7998 while the other, T_2, was compatible with this genotype (i.e., the plant was susceptible) and did not induce the hypersensitive response.

When the pathogen is international in distribution, such as stem rust of wheat, caused by *Puccinia graminis* f. sp. *tritici*, an agreed set of cultivars is usually used, but these should be supplemented with local ones. A further refinement, well illustrated by *P. graminis* f. sp. *tritici*, is the classification of symptoms according to severity. With the increased knowledge of the complementary genetic systems of such obligate pathogens and their hosts, it is now possible to test isolates of the pathogen against cultivars with known resistance genes. The results from such tests quickly show if the resistance gene is effective or not, and if this information is obtained early enough, it may influence the choice of cultivars planted subsequently. Many plant pathogenic fungi are spread by microscopic airborne spores (inoculums). The detection of such inoculums is potentially useful in disease forecasting and management, as it gives an early warning of the risk of infection. Conventional methods for measuring spore concentrations in the air rely on trapping followed by microscopy to identify and enumerate fungal spores. In Koch's postulate, the aim is to show a consistent association of symptoms of a disease with a suspected causal organism. Usually, diseases are first noticed only when symptoms have become severe since earlier and more slight effects, which may be discovered later, are liable to be missed initially. Disease symptoms in plants are diverse and range from biochemical perturbations of a few cells to death of the whole plant.

Symptoms caused by other groups of plant pathogens are usually less easy to attribute to specific organisms since, although the symptoms caused may be obvious, the pathogen itself is likely to be microscopic or submicroscopic in size and may be completely embedded in the tissues of the host. In particular, diseases caused by viruses are especially prone to erroneous diagnosis when this is made entirely on the basis of symptoms (Damsteegt 1997). As Bock (1982) pointed out, errors may result from a failure to distinguish between a "new disease" and the virus causing it which may or may not be "new" as well as naming viruses as "new" on insufficient evidence.

Plant virologists have made considerable use of indicator plants which often react to mechanical inoculation with distinctive local lesions. Sap from plants showing symptoms is rubbed onto the leaf of the indicator plant, often with an abrasive, and any symptoms that develop are observed. Choice of the indicator plant is important as, for example, the work of Van Dijk et al. (1987) has shown. They examined accessions of Australian *Nicotiana* species and found that susceptibility and sensitivity were more common in the sections *Acuminatae*, *Bigelovianae*, and *Suaveolentes* than in other sections of the genus.

Indicator plants have also been used to detect the presence of viruses in insect vectors. Allen and Matteoni (1991) tested a range of plants, including three species of tobacco, as monitors for the detection of tomato spotted wilt virus (TSWV) in western flower thrips (*Frankliniella occidentalis*). Petunia was found to sustain the greatest number of feeding wounds, the highest percentage of infected plants, and the greatest number of viral lesions. Once an informed guess as to the nature of the causal agent of a disease has been made from the symptoms caused, confirmation should be sought by application of one or more of the techniques described below.

19.2 Morphology of the Causal Organism

Parasitic angiosperms and some fungal pathogens of plants are sufficiently large and distinct for a diagnosis to be made on the basis of morphology seen with the naked eye. The sporophores of *Ganoderma boninense*, a fungal pathogen of African oil palm (*Elaeis guineen-*

sis), coconut (*Cocos nucifera*), and betel nut palm (*Areca catechu*), are also distinctive. Other pathogens will probably require the aid of the light microscope or the higher magnifications obtainable with the scanning or transmission electron microscope. Fungi often prove difficult to identify on the basis of morphology as seen under the light microscope since, when cultured on agar media, they frequently refuse to yield anything more interesting than wefts of mycelium. Various culture treatments can sometimes induce these species to sporulate, but even then, a skilled mycologist is often required to make a definite identification. Moreover, their morphology may be influenced by the medium. For example, Adaskaveg and Hartin (1997) found that the morphology of conidia of isolates of *Colletotrichum acutatum* varied according to whether they were grown on potato dextrose agar, pea straw agar, or almond fruit. Even if additional morphological information from transmission and Stereoscan electron microscopy is available, a positive identification usually requires data from the supplementary techniques.

19.3 Techniques Used for the Detection and Diagnosis of Plant Pathogens

19.3.1 Detection Methods of Pathogens in the Earlier Period

19.3.1.1 Selective Media

Some plant pathogenic organisms may be cultured in vitro. For these, placing infected material on a suitable medium and inspecting the resulting colonies of the pathogen may be sufficient for a positive identification. However, the choice of medium is important. All media are selective and considerable effort has been made for some pathogens or groups of pathogens to adapt substrates so that only the organisms of interest will grow. The media which was selective for the five genera of plant pathogenic bacteria recognized as: *Agrobacterium*, *Corynebacterium*, *Erwinia*, *Pseudomonas*, and *Xanthomonas*. Both their selectivity and plating efficiency have been improved by further work. For example, Fatmi and Schaad (1988) and Fatmi et al. (1991) published a semi-selective medium for the isolation of *Clavibacter* (formerly *Corynebacterium*) *michiganense* subsp. *michiganense* from tomato seed which, besides nutrients, contained nalidixic acid, potassium tellurite, and cycloheximide. The medium gave recovery rates of 85–132 % when compared with nutrient-broth yeast-extract medium and allowed the detection of a single contaminated seed containing 50 colony-forming units in samples of 10,000 seeds. Similarly, Gitaitis et al. (1997) have published a diagnostic medium which was effective for the semi-selective isolation and enumeration of *Pseudomonas viridiflava*, the causal agent of bacterial streak and rot of onion. The medium contained tartrate as a carbon source and the antibiotics bacitracin, vancomycin, cycloheximide, novobiocin, and penicillin G. Selectivity was enhanced by incubation at 58 °C rather than higher temperatures. Many media selective for particular fungi have been published. Here, the problem is often to suppress fast-growing, saprophytic fungi and bacteria associated with diseased plant material, which would otherwise swamp the slower-growing pathogens, while allowing these to grow out of the infected plant material. For example, Manandhar et al. (1995) found that a semi-selective medium consisting of one-quarter-strength potato dextrose agar and seven antibiotics, fenarimol, vinclozolin, chloramphenicol, erythromycin, iprodione, neomycin, and tetracycline, was significantly less inhibitory to Colletotrichum capsicum and *C. gloeosporioides* than to *Alternaria* spp. and *Fusarium* spp. As a result, *C. gloeosporioides* was detected more frequently in pepper seeds than when they were placed on moist filter paper, the normal means of detection. Similarly, Duffy and Weller (1994) supplemented diluted potato dextrose agar with rifampicin and tolclofos-methyl for isolation of *Gaeumannomyces graminis* var. *tritici*, the causal agent of take-all, an important disease of wheat. Isolation of this organism is difficult since roots of the plant often harbor several other pathogenic fungi. Moreover, lesions caused by *G. graminis* var. *tritici* are rapidly invaded by many opportunistic soil fungi, which displace the pathogen. Another difficulty with the identification of *G. graminis* var. *tritici* is that it does not sporulate in culture, necessitating pathogenicity and molecular tests. However, the medium of Duffy and Weller (1994) was also of value in identification since the fungus altered its orange color, imparted by rifampicin, to purple.

19.3.1.2 Biochemical Methods

All organisms have distinctive biochemical features, and these can be used for diagnosis. Some characteristics are shared by large groups, while, at the opposite end of the scale, others are unique to individual populations. The characterization of pathogen is therefore paramount in determining the taxonomic level to which an organism defined.

Bacteria have long been identified by their metabolic functions such as their ability to metabolize certain substrates and, more recently, by analysis of their fatty acid profiles. Additionally, soluble protein analysis by gel electrophoresis has been adopted both for bacteria and fungi. All methods rely upon gene expression, and since this may be regulated by environmental factors, care has to be taken to standardize these.

19.3.1.3 Substrate Metabolism

Goor et al. (1984) were among the first to investigate the applicability of the "galleries" of biochemical tests contained in the API (*appareils et procédés d'identification*) systems to the identification of phytopathogenic strains of *Erwinia* and *Pseudomonas*. Those useful for distinguishing strains of *Erwinia* were API 20E, API 50CHE, and the oxidase, enzyme, and aminopeptidase systems. *Pseudomonas* strains were differentiated using the API auxanographic systems 50CH, 50AO, and 50AA. Biolog is an alternative to the API system and gave better differentiation of 204 bacterial pathogens associated with a sheath rot complex and grain discoloration of rice in the Philippines (Cottyn 1996). Using this system, Cottyn (1996) found that all the reported strains of *Pseudomonas fuscovaginae* were positive for the production of 2-ketogluconate, but strains of *Acidovorax avenae* and *Burkholderia glumae* were negative. In contrast, *B. glumae* was positive for the production of acid from inositol but negative for the production of 2-ketogluconate, and *A. avenae* was negative for both these reactions.

De Laat et al. (1994) identified the causal agent of bacterial leaf rot of a species of aloe (*Aloe vera*) as *Erwinia chrysanthemi* biovar 3 on the basis of its ability or failure to metabolize a number of substrates as well as its agglutination by an antiserum prepared against a defined strain of the organism. Similarly, Pernezny (1995) were able to define the bacterial species causing a severe outbreak of bacterial spot in lettuce fields in Florida as *Xanthomonas campestris* on the basis of substrate utilization, the pathovar being defined as vitians by its fatty acid profile. In some instances organisms may be identified by their production of unusual metabolites. For example, strains of *Aspergillus flavus* that were aflatoxigenic were recognized by their production of volatile C15H24 compounds such as alpha-gurjunene, trans-caryophyllene, and cadinene. These compounds were not produced by nontoxigenic strains (Zeringue et al. 1993).

19.3.1.4 Fatty Acid Profiles(FAME Analysis)

Identification of bacterial pathogens of plants by fatty acid methyl ester analysis is usually performed on pure cultures of the organism. About 40 mg of wet cells is saponified and methylated. The fatty acid methyl esters (FAME) are extracted in an ether–hexane mixture and analyzed by gas chromatography. Areas of the resulting peaks on the chromatograms are calculated and compared with profiles of known reference strains by computer programs (Roy 1988). For example, the organism responsible for an outbreak of bacterial spot of lettuce was defined as *Xanthomonas campestris* pv. *vitians* as the fatty acid profiles of the strains collected from the field matched this pathovar most closely (Pernezny 1995). In a more extensive study, Wells et al. (1994) were able to differentiate the five species of *Erwinia* of the *amylovora* group as well as the four species of the *herbicola* group.

19.3.1.5 Protein Analysis

Electrophoresis of soluble proteins from plant pathogens often gives rise to complex patterns, and these can be used for identification purposes. Instead of using a general protein stain, such as Coomassie Blue, a particular protein dye, which, for example, might have enzymic activity, may be revealed by appropriate staining methods. MacNish et al. (1994), by staining for pectic enzymes, were able to place 4250 Australian isolates of *Rhizoctonia solani* in 10 groups which they termed zymograms. However, a major disadvantage of the technique is that, compared with immunological or PCR methods, relatively large quantities (50–100 mg) of the pathogen must be grown in order to obtain the enzymes either from culture filtrates or from extracts of the organism itself (Bonde et al. 1993).

In contrast, peptide profiling may be performed on comparatively small amounts of pathogen, and this technique has been used to define virus strains. Kittipakorn et al. (1993) investigated virus isolates from groundnut which caused a wide variety of symptoms varying from mild mottle to systemic necrosis and stunting. Virion proteins (150–190 mg) were isolated, digested with trypsin, and separated by high-performance liquid chromatography (HPLC). Despite the disparate symptoms, all were found to have identical HPLC profiles and were identified as strains of peanut stripe potyvirus.

19.3.1.6 Serological Techniques

Serological techniques were initially used for virus identification but are now applied to several of the other groups of plant pathogenic organisms. Serological techniques rely on the specificity of antigen–antibody binding and a sensitive method for detection of the resulting complex. They have been used for the detection of plant viruses for many years, but more recently, they have been applied to the diagnosis of fungi, *Oomycetes*, *Phytoplasmas*, and bacteria as well. In order to obtain reliable results, the antibody preparation used, whether polyclonal or monoclonal, must have the required specificity and high affinity for an appropriate epitope of the pathogen. Polyclonal antibody preparations are obtained by injecting an animal with antigen and, as their name implies, contain a population of different antibodies. These are likely to be heterogeneous with regard to the antigen with which they bind, the affinity of the binding, and their suitability for the attachment of reagents used to detect them. Purification of the antigen to a high level will eliminate some of this variation. In contrast monoclonal antibodies, although more difficult to obtain, may be screened for specificity to a single epitope and will provide a stable source of antibody since the hybridoma may be kept in perpetuity by culturing and storage in the frozen state. In some instances the extreme specificity of a monoclonal antibody can be a disadvantage since it may not detect all members of a taxonomic group.

19.3.1.7 Nucleic Acid Techniques

The genome of every organism consists of a unique sequence of nucleotides, but the degree of conservation of different domains is variable. Highly conserved regions are found among many organisms which are distantly related, whereas some regions such as satellite DNA sequences, which are repetitive and usually non-transcribed, are very variable. In consequence, a considerable degree of discrimination is possible. For example, Piotte (1994, 1995) found two highly reiterate but different satellite DNA sequences with monomeric units of 295 bp and 169 bp in the nematodes, *Meloidogyne incognita* and *M. hapla*, respectively. The monomers were cloned and used as probes. That from *M. incognita* was nonspecific since it hybridized with all populations belonging to *M. incognita*, *M. arenaria*, and *M. javanica*. In contrast, the monomer from *M. hapla* was specific and, because of its variability, could be used to differentiate three populations of this species as well as to distinguish it from sympatric populations of *M. chitwoodi* and *M. incognita*. Moreover, because of the high reiteration – about 15,000 monomers per haploid genome – identification of the organism was possible by direct hybridization with as little material as one squashed female nematode, obviating the need to extract DNA (Bertaccini 1990).

A further advantage of nucleic acid analyses is that, unlike the techniques described so far, they are not dependent on gene expression and therefore are independent of environmental factors. In the past few years, a large number of techniques have been published, many of which are extremely ingenious and which are revolutionizing the detection and assay of plant pathogens. In order to give some coherence to this rapidly developing field, the techniques are described roughly in order of specificity.

19.3.1.8 Gel Fractionation

Nucleic acid fragments are separated according to size by gel electrophoresis. Agarose gels are normally used for separating fragments which are 300 nucleotides in length or more and polyacrylamide gels for shorter lengths. The resulting bands are visualized by staining with ethidium bromide, which is often included in the gel, and viewing under ultraviolet light. Usually, the technique is used in conjunction with other procedures such as cutting extracted DNA with restriction enzymes or amplifying specific domains by the polymerase chain reaction. However, Hodgson et al. (1998) were able to obtain preliminary identification of a viroid associated with the lethal Tinangaja disease of coconuts present on the island of Guam by two-dimensional polyacrylamide gel electrophoresis of RNA extracted from palms showing symptoms.

19.3.1.9 Nucleic Acid Probes and Polymerase Chain Reaction

Probes are single-stranded DNA or RNA molecules that have been labeled with a reporter molecule such as a radioactive isotope, an enzyme, or a fluorescent dye. They bind to complementary DNA/RNA sequences on the target samples which can then be detected in various ways, depending on the type of reporter molecule used (Strange 1993). The blackening of X-ray film can be used to detect radioactivity or light (chemiluminescence) emitted directly or indirectly by the bound probe. Detection of a color change, e.g., on nylon membranes or in microtiter plates, can also be used for enzyme-mediated probing methods. Detection using fluorogenic probes in real-time PCR is discussed later. Probes can be generated from double-stranded genomic, cloned, or PCR-derived DNA by heat or alkali treatment. Alternatively, short synthetic single-stranded oligonucleotides, usually around 20–40 bases long, can be used. Prior to the introduction of PCR, nucleic acid-based diagnostics generally involved the use of probes, and they continued to be used widely as an alternative to PCR for identification until the mid-1990s (Rasmussen and Scheffer 1988a, b; Rasmussen and Reeves 1992; Ward and Gray 1992; Mutasa et al. 1993; Schots et al. 1994). However, in recent years most molecular diagnostic assays have used PCR because of its greater sensitivity, simplicity, and speed and because it is possible to use much simpler protocols to produce the nucleic acid samples for detection. However, probes are now increasingly used in conjunction with PCR to produce diagnostic protocols that are more sensitive, more specific, or simpler than PCR alone (Williams et al. 2001; Knoll et al. 2002).

PCR, a method for rapidly synthesizing (amplifying) millions of copies of specific DNA sequences, is the most important technique used in diagnostics. In this method, the two strands of the DNA are first separated by heating to 95 °C. The temperature is then reduced to 40–65 °C to allow binding of two primers, one at each end of the region to be amplified. The primers are short pieces of single-stranded DNA which bind specifically to the target DNA by complementary base pairing. The primers give the PCR its specificity. A second strand of the DNA is then generated (usually at 72 °C), extending from the primers, using a thermostable DNA polymerase and deoxynucleoside triphosphates (dNTPs). At the end of this first cycle of the PCR reaction, the number of copies of target DNA sequence has effectively doubled. The whole process is then repeated many times, so that after about 25 such cycles, taking a few hours or less, millions of copies of the sequence have been produced. The procedures are very simple requiring only the mixing of a few ingredients (DNA, buffer, dNTPs, primers, and DNA polymerase) and placing them in a PCR machine, which is essentially a computer-controlled heating block that can be programmed to switch between the different temperatures required. The presence of the amplified DNA is usually checked, by agarose gel electrophoresis, but alternative detection formats include using colorimetric assays (Beck et al. 1996; Mutasa et al. 1996) or fluorometric assays (Fraaije et al. 1999) in a microtiter plate format. PCR is a very sensitive technique, only small amounts of DNA are needed (e.g., from a single spore), and often this can be prepared fairly simply. Sometimes, nested PCR is used to improve the sensitivity and/or specificity of the assay. This involves two consecutive PCR reactions, the second one using primers that recognize a region within the PCR product amplified by the first set. PCR products can also be detected with a probe (Mutasa et al. 1995). This can improve the sensitivity and specificity of the assay, particularly when the amplified product may not be sufficient to be seen on an agarose gel. Another approach is to use immunocapture PCR (IC-PCR). Antibodies immobilized on the surface of a microtiter plate or microcentrifuge tube are used to capture the pathogen, which is then detected using PCR. IC-PCR can improve the sensitivity and specificity of the assay and reduce problems with inhibitors in the sample. PCR has also been used to quantify the amount of pathogen DNA present in a sample, but, using conventional PCR, this is not straightforward. Standard PCR is not inherently quantitative. It is relatively easy to quantify the amount of PCR product generated, but it is difficult to relate this to the concentration of the target DNA originally present. Because the reaction rate in PCR is exponential, small differences in the amount of starting material result in large differences in the amount of product formed (Wang et al. 1989). Also, as the PCR proceeds toward the final cycles, the reagents are eventually used up and the amplification rate is consequently no longer exponential; thus, the final amount of product is no longer proportional to the amount of original template present. A method, called competitive PCR, has been developed to quantify target DNA using standard PCR technology (Nicholson et al. 1996, 1998). A single primer pair, in the same reaction, is used to simultaneously amplify target DNA and a known amount of competitor DNA (a piece of DNA with annealing sites for the primers but producing a differently sized product). The amount of target DNA can be estimated by comparing the relative amounts of target and competitor product produced (Gilliland et al. 1990). However, a carefully constructed competitor target is needed for each PCR assay, serial dilutions are required to ensure a reliable ratio of target to competitor product, and care must still be taken to ensure that the PCR reaction is not permitted to proceed beyond the exponential phase.

19.3.2 Detection Methods of Plant Pathogens at the Moment

Currently, the detection of several phytopathogenic viruses, phytoplasmas, fungi, and bacteria responsible for plant diseases is altering, active, and developing worldwide where reputable protocols can be modified or optimized only months after having been developed. In the past, protocols were established and used routinely for many years, which is not the case today. There is a propensity to use polyphasic approaches to detection including conventional, serological, and molecular techniques and to validate them in ring tests. Diagnostic protocols for detection of the 23 EU quarantine viruses, bacteria, fungi, nematodes, and insects have recently been set up and validated by ring tests in the DIAGPRO project financed by the "Standard, Measurements and Testing" program of the EU. They are intended to form the basis of improved detection of pathogens in plant material and are available through the web page of the Central Science Laboratory (http://www.csl.gov.uk/ science/organ/ph/diagpro/). The European and Mediterranean Plant Protection Organization (EPPO) is also publishing diagnostic protocols for the most important quarantine and non-quarantine organisms (http://www.eppo.org). Currently, automation and electronic data management are vital to increase the productivity and efficiency of routine analysis for detection of plant pathogenic bacteria and viruses (Lopez et al. 2003). Furthermore, rigorous quality control systems have had to be introduced into the laboratory because standards of accreditation and certification are increasing to control not only the outcome of laboratory testing but also the actual process of carrying out the tests. Several genomes from causal agents of plant diseases, both viral and bacterial, have been completely sequenced and more are under way (wit.integratedgenomics.com/GOLD). Based on their analysis, new specific sequences could be used to design detection probes for different pathogens (Van Sluys et al. 2002). The sequences of complete genomes in GenBank are available through NCBI (www.ncbi.nlm.nih.gov/Entrez/) and other databases. In the subsequent sections, a number of significant developments and improvements in serological and molecular techniques are described.

19.3.2.1 Serological Detection Techniques

Monoclonal and Recombinant Antibodies

The specificity of detection of pathogens by well-known serological techniques such as IF and ELISA has improved greatly with the use of specific monoclonal and recombinant antibodies. Both allow the selection of specific target epitopes to avoid false positives. Specificity problems are frequent when analyzing bacteria in plant material, soil, water, etc., due to the large quantities of other microorganisms present in the sample and to the difficulties in obtaining polyclonal antibodies specific for the target pathogen without cross-reaction with other microorganisms. Commercial monoclonal antibodies for detection of plant pathogenic viruses are now available from several companies. Several companies have developed a wide range of kits for rapid and/or in situ bacterial detection using polyclonal or monoclonal antibodies, based on slide agglutination, IF, and ELISA. The sensitivity of those based on IF allows detection of latent bacterial infections. The kits based on agglutination or ELISA are appropriate only for diagnosing plants already showing symptoms. Although the use of recombinant antibodies for diagnosis is only at the research level, it has a promising future. Single-chain Fv (scFv) technology (Bird et al. 1988) allows the cloning of variable (V) antibody genes, linking them to a flexible peptide as a single-chain Fv. These constructs are of great interest in plant pathology because they can be expressed in bacteria (Roberts 1996) as soluble proteins, fused with the capsid proteins of filamentous phages (Winter et al. 1994), and expressed in plants. However, their applications for diagnostic purposes are still scarce. In plant pathology, serological detection has been widely used, and some recombinant constructs have been produced, but only a few of them have been applied to routine ELISA tests.

Advances in molecular immunology have allowed the development of specific recombinant monoclonal antibodies. One such reagent has been used to detect *Ralstonia solanacearum* (formerly *Pseudomonas solanacearum*) race 3, a pathogen of potato, tomato, and eggplant for which many countries have quarantine regulations (Griep et al. 1998). The antibody is specific to the lipopolysaccharide of the bacterium and holds promise for being sufficiently reliable to obviate the need for time-consuming bioassays, which are normally required to provide confirmation of positives detected by polyclonal antibodies. Antibody binding may be recorded and quantified by a variety of techniques, but enzyme-linked immunosorbent assay (ELISA) in one of its variants is the one that is used predominantly. The most common of these are double-antibody sandwich ELISA (DAS-ELISA) and indirect ELISA (I-ELISA).

In DAS-ELISA, the antibody is bound to a solid support, such as the wells of a microtiter test plate and the test sample, containing the antigen of interest, added. This is "sandwiched" by adding another antibody which has been conjugated with an enzyme such as alkaline phosphatase. When substrate for the enzyme is added such as p-nitrophenyl phosphate, a colored reaction product is obtained. Normally, the intensity of the color is determined spectrophotometrically to give a measure of the amount of antigen, using a plate reader.

In I-ELISA, the antibody is bound to a solid support, as in DAS-ELISA, and this binds the antigen. However, the bound antigen is recognized by an antibody raised in a second animal, and this second antibody is recognized by a further antibody, which is conjugated to an appropriate enzyme. The advantage of I-ELISA is that the antibody used to recognize the antigen of interest is not constrained by conjugation with an enzyme and is likely to bind more completely. For this reason it may be less strain specific than DAS-ELISA. Another variant of the technique involves trapping the antigen to the F(ab')2 fragments of specific IgG on a solid support and detecting it by intact specific immunoglobulin (IgG) to which enzyme has been conjugated in the Fc region (Miller and Martin 1988). Dot ELISA and dipsticks are two further developments of the ELISA technique. In both, antibodies are bound to a membrane and the membrane is incubated with a preparation suspected to contain the antigen. After washing, the membranes are treated with enzyme-conjugated antibody and incubated with the enzyme substrate. The membrane is then dried and the intensity of color measured, for example, with a handheld reflectometer.

Cahill and Hardham (1994) describe a development of the dipstick technique which they combined with baiting to detect *Phytophthora cinnamomi* from soil (for examples of the damage done by this pathogen). Soil samples (20 g) were flooded with distilled water and baited with cotyledons from *Eucalyptus sieberi*. After incubation for 3 days, the samples were cold shocked at 48 °C for 20–30 min to induce the release of zoospores and three dipsticks were floated, membrane side downwards, on the water for 1.5 h. Some of the released zoospores encysted on the dipsticks and were detected immunologically as follows.

The dipsticks were treated with a murine monoclonal antibody specific for an antigen of*P. cinnamomi* located on the cyst periphery; the antibody was recognized by an anti-mouse antibody conjugated to alkaline phosphatase, and the conjugate was visualized by treatment of the dipsticks with a mixture of 4-chloro-2-methylbenzene-diazonium salt and naphthol AS-MX phosphate to give a red insoluble dye, which was easily visible under a handheld lens or dissecting microscope.

Other techniques for detecting bound antigens include gold-conjugated antibody, antibody conjugated with fluorescent dyes, and, for biotinylated antibodies, avidin conjugates, since the glycoprotein avidin binds strongly to biotin. Milne et al. (1995) used immunogold labeling successfully to detect three unrelated phytoplasmas in plant tissue both before and after embedding. Jones (1994) used fluorescence labeling to detect *Erwinia carotovora* subsp. *atroseptica* in seed potato. This organism is responsible for blackleg in cool temperate regions, and the incidence of the disease is directly related to the incidence of contamination of the seed potato, the threshold for disease development being about 103 cells per tuber. Juice was extracted from the peel of potatoes and placed in the wells of 24-well tissue-culture plates together with an agarized medium containing polygalacturonic acid. After incubation for 48 h, the agar in the wells was dried to a thin film in a hot-air oven at 50 °C and stained with fluorescein isothiocyanate conjugated with antiserum specific to the bacteria. The films were washed before photographing under a fluorescence microscope, and colonies of the bacterium were enumerated from the negatives using a commercial imaging system. Immunofluorescence was also used by Chittaranjan and de Boer (1997) to detect *Xanthomonas campestris* pv. *pelargonii* in greenhouse nutrient solution. A murine monoclonal antibody, 2H5, was used, and binding was detected by an anti-mouse antibody conjugated with indocarbocyanine.

19.3.2.2 Flow Cytometry

Flow cytometry is a technique for rapid identification of cells or other particles as they pass individually through a sensor in a liquid stream. Bacterial cells are identified by fluorescent dyes conjugated to specific antibodies and detected electronically using a fluorescence-activated cell sorter, which measures several cellular parameters based on light scattering and fluorescence. Multiparameter analysis includes cell sizing, fluorescence imaging, and gating out, or elimination of unwanted background associated with dead cells and debris (Alvarez 2001; Davey and Kell 1996). Flow cytometry has excellent potential as a research tool and possibly for routine use in seed health testing and other fields. Several parameters can be analyzed simultaneously, including total particle count, distinction between living and dead cells, and differentiation of target and nontarget bacterial populations associated with seeds or other plant material. This technique has also been adapted to the analysis of viability, metabolic state, and antigenic markers of bacteria. Fluorescent markers based on membrane integrity can be selected to assess the viability of cells by staining dead and live cells in different colors. This procedure can be combined with specific antibody staining using antibodies labeled with R-phycoerythrin. The cost for instrumentation is currently a major disadvantage that will be resolved when less-expensive models become available.

19.3.2.3 Molecular Detection Techniques

Molecular techniques based on hybridization or amplification, and especially on PCR, have been developed for the most important plant pathogenic viruses and bacteria. Although PCR can reach high sensitivity and specificity, its introduction for routine detection has been hampered by a lack of robustness (Van der Wolf et al. 2001). The failure of PCR amplification to correctly diagnose infected and noninfected plant material has been reported in different comparative assays. PCR, in different formats, is the most widely used molecular technique for detection of bacteria and viruses. Less-employed techniques based on hybridization are of interest for specific requirements. Their main advantages are specificity and rapidity. Specificity is directly related both to the design of the primers or probes and to the amplification or hybridization protocols. Furthermore, the possibility of designing a multiplex PCR saves time and reagent costs compared with monospecific PCR, which requires several reactions for the same number of tests (Mullis and Faloona 1987; Singh et al. 1996; Becker and Manz 1999). Colorimetric detection of PCR products, on membranes or in microtiter plates, has been employed successfully, increasing sensitivity and facilitating interpretation of results for the use of the technique in routine analyses (Bertolini et al. 2001; Chandelier et al. 2001). Although amplicon hybridization requires more time and manipulation of samples than gel electrophoresis using ethidium bromide staining (Landgraf et al. 1991), its great recompense favors its application for indexing programs. In addition, the procedure avoids the use of the lethal ethidium bromide, and it is possible to immobilize hundreds of samples in a single membrane. The design of other internal probes would allow subtyping of isolates where essential.

19.3.2.4 Fluorescence In Situ Hybridization

Fluorescence in situ hybridization (FISH) is a technique applied for bacterial detection that combines the simplicity of microscopy observation and the specificity of hybridization (Volkhard et al. 2000). Its use in detection of plant pathogenic bacteria is recent (Wullings et al. 1998) and is dependent on the hybridization of DNA probes to species-specific regions of bacterial ribosomes. They are particularly suitable as diagnostic targets because ribosomal RNA contains functional sequences that are common to all species but also sequences that are very specific to individual species, and FISH only needs to recognize this specific information. The probes hybridize with a three-dimensional protein/RNA structure not only with a linear sequence of RNA. The sensitivity of the FISH technique is equivalent to that of amplification technologies, and, in theory, FISH can detect single cells. This high sensitivity is the result of the high affinity and selectivity of DNA probes because FISH takes place under very stringent hybridization conditions, where a difference of one nucleotide in a 15–20-oligonucleotide probe is sufficient to discriminate binding. Furthermore, FISH maintains the structural integrity of the microorganism, confining the reagents in one small vessel, and one probe will bind to each of the 1–5·104 ribosomes inside. This extremely high signal is responsible for the theoretical ability of the technique to achieve single-cell sensitivity.

19.3.2.5 Co-operational PCR

A new PCR concept of high sensitivity for the amplification of viral RNA or bacterial targets from plant material has recently been described (Olmos et al. 2002). The method has been patented as Co-PCR (Spanish patent P20002613; 31 October 2000). The Co-PCR (co-operational amplification) technique can be performed easily in a simple reaction based on the simultaneous action of four or three primers. The reaction condition consists of simultaneous reverse transcription of two different fragments from the same target, one internal to the other; the production of four amplicons by the combination of the two pairs of primers, one pair external to the other; and the co-operational action of amplicons for the production of the largest fragment. The Co-PCR technique has been used successfully, both in metal block and capillary air thermal cyclers, for the detection of plant RNA viruses [Cherry leaf roll virus (CLRV), Strawberry latent ringspot virus (SLRSV), Cucumber mosaic virus (CMV), Plum pox virus (PPV), and Citrus tristeza virus (CTV)]. Coupled with colorimetric detection, the sensitivity observed is at least 100 times higher than that achieved with Real-Time Polymerase Chain Reaction (RT-PCR) and is similar to that of nested RT-PCR. Co-PCR usually produces the largest amplicon, in contrast to nested PCR, which requires two sequential reactions and obtains the smallest fragment. Metal block and capillary air thermal cyclers have been employed for the detection of some plant RNA viruses from different genera, and to a bacterium, but by using only three primers (Caruso et al. 2003) which shows the possibilities of this new approach. The low amount of reagents (ten times less than in conventional PCR) probably increases susceptibility to inhibitors, requiring prior RNA extraction for sensitive virus detection. However, this step was not necessary when analyzing the presence of *R. solanacearum* in water.

19.3.2.6 Multiplex PCR

Multiplex PCR allows the simultaneous and sensitive detection of different DNA or RNA targets in a single reaction (James 1999; Nassuth et al. 2000; Williams et al. 1999). On the other hand, PCR detection protocols can be designed to verify the presence of more than one pathogen in plant material by looking for common specific sequences in two or more of them or to detect related viruses or bacteria on multiple hosts (Louws et al. 1999). Multiplex PCR is useful in plant pathology because different bacteria and/or RNA viruses frequently infect a single host, and consequently, sensitive detection is needed for the propagation of pathogen-free plant material. There are several examples in plant pathology of simultaneous detection of several targets, and the amplification by multiplex PCR of two or three plant viruses has been reported (Russo et al. 1999; Saade et al. 2000). Nevertheless, there are still only a few examples in which more than three plant viruses were amplified in a single PCR-based assay (Bariana et al. 1994; Nassuth et al. 2000; Okuda and Hanada 2001), probably due to the technical difficulties of designing a reaction involving many compatible primers. One of them is the simultaneous detection of the six major characterized viruses described in olive trees, which belong to four different genera: Cucumovirus (CMV), Nepovirus [CLRV, SLRSV, and Arabis mosaic virus (ArMV)], Necrovirus [Olive latent virus-1 (OLV-1)], and Oleavirus (Olive latent virus-2) (Bertolini et al. 2001). This includes accurate design of six primer pairs for one-step RT-PCR amplification in a single closed tube and specific probes, enabling the detection of all major viruses described in olive trees, which are problematic for RNA extraction. A few plant pathogens have been detected using this approach. Multiplex PCR can be used where there is a need to detect several pathogens simultaneously. The assay uses several PCR primers in the same reaction, which saves time and reduces costs. However, multiplex PCR assays can be difficult to develop. The products from the different targets need to be of different sizes to ensure that they can be distinguished from one another, and they must all be amplified efficiently using the same PCR conditions (Henegariu et al. 1997). Multiplex PCR has been used to develop an assay for the detection of different foliar wheat pathogens (*Septoria tritici, Stagonospora nodo-*

rum, *Puccinia striiformis*, and *Puccinia recondita* in wheat) (Fraaije et al. 2001).

19.3.2.7 Multiplex Nested RT-PCR

A multiplex nested RT-PCR in a single closed tube has been developed for simultaneous and sensitive detection of the viruses CMV, CLRV, SLRSV, and ArMV and the bacterium *Pseudomonas savastanoi* pv. *savastanoi* from olive plants (Bertolini et al. 2003) using 20 compatible primers in a compartmentalized tube. This newly developed method combines the advantages of multiplex RT-PCR with the sensitivity and reliability of nested RT-PCR carried out in a single closed tube. It enables the simultaneous detection of several viral RNA and bacterial DNA targets in a single analysis, performed with woody plants. It also saves time and reagent costs because it can be performed in a single reaction, although accurate design of compatible primers is needed. The compartmentalization of a single Eppendorf tube with a pipette tip (Olmos et al. 1999; Olmos et al. 2003) allowed multiplex PCR and nested PCR to be combined effectively. During the first amplification reaction, there is no interference of the external with internal primers because they are physically separated from the initial reaction cocktail. Once the multiplex RT-PCR ends, the internal primers are mixed with the products of the first reaction before proceeding to the nested multiplex. Because the concentration of internal primers is very high compared with that of the external primers (which will also have been consumed by the first amplification), the nested multiplex can be performed with minimal interference. Consequently, sensitivity is increased at least 100-fold over that of multiplex RT-PCR for the detection of viruses. Furthermore, the sensitivity achieved by multiplex nested RT-PCR for the bacterium *P. savastanoi* pv. *savastanoi* is similar to that reached by applying monospecific nested PCR after an enrichment step, which was demonstrated to be 100-fold more sensitive than conventional PCR. This multiplex nested RT-PCR has been coupled with colorimetric detection, allowing the discrimination of amplicons of similar size, which would require additional monospecific analysis if gel visualizations only were employed.

19.3.2.8 Real-Time PCR

The process of quantifying target DNA has recently been simplified considerably with the advent of real-time PCR. This method avoids the usual need for post-reaction processing, as the amplified products are detected by a built-in fluorimeter as they accumulate. This is done by using nonspecific DNA-binding dyes (e.g., SYBR Green) or fluorescent probes that are specific to the target DNA. The principle underlying real-time PCR is that the larger the amount of target DNA present in the sample being tested, the quicker the reaction progresses and enters the exponential phase of amplification. The amount of PCR amplicon produced at each cycle is measured, using the fluorescent dyes or probes, and for each sample tested, the cycle threshold (Ct) is calculated. This is the cycle number at which a statistically significant increase in fluorescence is detected. The Ct increases with decreasing amounts of target DNA. A calibration curve relating Ct to known amounts of target DNA is constructed and used to quantify the amount of initial target DNA in an unknown sample. Software supplied with real-time PCR machines is used to rapidly analyze the results. Using DNA-binding dyes to detect the PCR product is a simpler and cheaper approach than using specific probes, but the assays are more prone to error since any nonspecific PCR products, and primer–dimer artifacts, can generate a signal. It is therefore very important to use highly specific primers and optimal amplification conditions. Using specific probes, such as TaqMan oligonucleotide probes, has the advantage of reducing signals due to mispriming or primer–dimer formation. TaqMan probes contain a fluorescent reporter dye with a quencher dye in close proximity to reduce the fluorescent signal. During PCR, the fluorogenic probe binds to its target sequence, within the DNA fragment being amplified. As the PCR process continues, the 5′ nuclease activity of the DNA polymerase cleaves the probe, thus separating the reporter and quencher dyes and causing an increase in fluorescence proportional to the amount of PCR product present. Other types of fluorescent probes and primers that utilize the properties of reporter and quencher dyes have also been used in real-time PCR assays. These include molecular beacons. It is possible to detect several targets simultaneously by using probes with different fluorescent reporter dyes. Such assays can be used to quantify different organisms, polymorphisms, or single point mutations. Real-time PCR methods are not yet widely used for plant pathogen diagnostics, but many assays have already been developed for detection of fungi, bacteria, viruses, and viroids (Fraaije et al. 2001, 2002; Schaad and Frederick 2002). In addition to simplifying quantification, real-time PCR has a number of other advantages over conventional PCR. It is faster and a higher throughput is possible. Postreaction processing is unnecessary, eliminating the risk of carry-over contamination. It can be more specific than conventional PCR if a specific probe is used in addition to the two specific primers. The high specificity of the probes used means that it is good for detecting single nucleotide polymorphisms (SNPs). These are where the specificity is determined by a single base-pair change in the DNA, as is often the case (e.g., when distinguishing between fungicide-resistant and fungicide-sensitive isolates). It is sometimes difficult to design conventional PCR methods that will detect such small differences reliably.

19.3.2.9 DNA Microarrays

DNA microarrays or biochips are made of a surface on which multiple capture probes are linked, each one being specific for a DNA or RNA sequence of the targets. Their purpose is the detection of numerous sequences in a single assay. Various supports are currently in use for the elaboration of microarrays, including glass, nylon, and different polymers. Up to 30,000 DNA probes (gene sequences) can be arrayed onto a single chip. The probes arrayed can be PCR products amplified to high concentrations or relatively short (30–50 bp) oligonucleotide probes. Once arrayed, the chip can be exposed to fluorescently labeled DNA/RNA from the sample to be tested (Lopez et al. 2003). The detection system uses one or several fluorophores that can be read with laser technology to reveal the targets present in the sample. Extraction of nucleic acids from the sample, labeling, and hybridization can be achieved with standard laboratory facilities. Another possibility is to use the nanochip technology developed by Nanogen (San Diego, USA), based on the combination of microelectronics with microarray technology in a solid support covered with streptavidin to increase the power of union with biotin-labeled DNA. The theoretically possible detection of multiple targets is problematic in practice, especially when trying to simultaneously detect bacteria and viruses. Until now, the protocols used in the different laboratories involved in developing microarrays for detection of phytopathogens require a prior step of PCR amplification and reach low levels of sensitivity. Consequently, their use for routine detection is still far from being common, in contrast with their widespread use for functional genomics studies. Microarrays can generate fast results for several pathogens, but their cost is still very high. They also generate significant amounts of data requiring expert interpretation. It is likely that microarrays will follow a path similar to that of PCR, which spent several years as a research tool before being used in diagnostics.

19.3.2.10 Target Sequences Used in Diagnostic Assay Development

The first stage in the development of a diagnostic assay is to select the nucleic acid sequences to be used to identify the organism. There are some generally applicable techniques that can be used for bacteria and fungi, but viruses usually need different approaches. Their genomes are relatively small, and in many cases data on complete virus genomes are available in the sequence databases. Probably the most common target for virus diagnostics is the coat protein gene, but any part of the genome could be suitable depending on how much sequence data is available from related viruses in the same region of the genome. As most viruses have RNA as their genetic material, there are usually extra steps involved in the assay because the RNA is usually converted to a DNA copy prior to detection. There are two main approaches that can be used to select target sequences for use in nucleic acid-based diagnostics. The first involves targeting particular known genes, and the second uses randomly selected DNA fragments. In the first strategy, the same known gene would be isolated and sequenced from target and nontarget pathogens. Regions of the sequence that are different would then be identified and used to design primers for PCR. The most commonly used target for bacteria and fungi is the DNA encoding the ribosomal RNA genes (ribosomal DNA or rDNA: Gurtler and Stanisich 1996). There are several reasons why rDNA has been such a widely used target for diagnostic development. It is present in many copies in each cell, and this allows for very sensitive detection. The genes are present in all organisms so are universally applicable. They include some regions that are highly conserved such as the 5.8S gene and others that are highly variable such as the internal transcribed spacer (ITS) regions. The conserved regions allow consensus, universal primers to be designed that can isolate the region from a wide range of organisms, e.g., all fungi, or all bacteria so that they can be studied further. The variable regions provide scope for discriminating between related organisms. The number of rDNA sequences in the publicly available databases is far greater than for any other region of DNA. The availability of so many sequences for comparison means that it is possible to check the specificity of primers by screening computer databases (i.e., by in silico testing) before testing the primers in the laboratory. Other targets for detecting fungi are beta-tubulin genes, which have also been used to detect fungicide resistance (Fraaije et al. 1999), and mating type genes. In bacteria, a variety of other targets have been used to design specific PCR assays (reviewed recently by Louws et al. 1999). These include plasmid DNA and genes associated with pathogenicity.

Detection of pathogen in plants is essential to ensure safe and sustainable agriculture production. The techniques available have evolved significantly in the last few years to achieve rapid and reliable detection of pathogens, extraction of the target from the sample being important for optimizing detection. Plant pathogens infect various crops from seedling to heading stages in the field conditions as well as after harvest and during storage time. To improve the sensitivity of techniques for pathogen detection, a prior enrichment step in liquid or solid medium is advised. Many techniques are available for pathogen identification. They include substrate metabolism, fatty acid profiles (FAME analysis), protein analysis, and serological techniques, usually as one of the formats of the enzyme-linked immunosorbent assays (ELISA). Serological and molecular techniques are currently the most appropriate technique when high numbers of samples are available to be analyzed. Specific monoclonal and/or recombinant antibodies are available for many plant pathogens and have contributed to the specificity of serological detection. Molecular detection can be optimized through the automatic purification of nucleic acids from pathogens by column chromatographic techniques or robotics technique. Polymerase chain reaction (PCR) and new variants of PCR, such as simple or multiplex nested PCR in a single closed tube, co-operative PCR, and real-time monitoring of amplicons or quantitative PCR, allow high sensitivity in the detection of one or several pathogens in a single assay. The latest development in the analysis of nucleic acids is microarray technology, but it requires generic DNA/RNA extraction and pre-amplification methods to increase high level of detection sensitivity. The advances in research that will result from the sequencing of many plant pathogen genomes, especially now in the era of proteomics, represent a new foundation of information for the future enlargement of sensitive and specific recognition techniques for these microorganisms. Abundant detection methodologies now exist, but regardless of the approach, important questions need to be answered prior to their inclusion into experiments These include sensitivity, accuracy, robustness, frequency of testing, and cost. Despite many novel technologies being accessible, challenges remain to recognize as yet unculturable pathogens, to detect cryptic species, and to characterize the assemblage and diversity of fungal communities in different environments without bias. There is always a need to characterize plant pathogen rapidly and accurately. No one knows how many plant pathogen species exist, but sequencing of environmental DNA may improve the accuracy of current estimates. Next-generation sequencing and pyrosequencing approaches will also provide promising ways of enlarging the scope of molecular detection studies. However, new technologies, integrated with other conventional tools, which they should accompany, not replace, provide useful information in the thoughtful prevention of plant disorder.

References

Adaskaveg JE, Hartin RJ (1997) Characterization of *Colletotrichum acutatum* isolates causing anthracnose of almond and peach in California. Phytopathology 87:979–987

Agrios GN (2001) Fitopatologia, 2nd edn. Uteha Noriega Editores, Mexico

Allen WR, Matteoni JA (1991) Petunia as an indicator plant for use by growers to monitor for thrips carrying the tomato spotted wilt virus in greenhouses. Plant Dis 75:78–82

Alvarez AM (2001) Differentiation of bacterial populations in seed extracts by flow cytometry. In: de Boer SH (ed) Plant pathogenic bacteria. Kluwer, Dordrecht, pp 393–396

Bariana HS, Shannon AL, Chu PWG, Waterhouse PM (1994) Detection of five seedborne legume viruses in one sensitive multiplex polymerase chain reaction test. Phytopathology 84:1201–1205

Beck JJ, Ligon JM, Etienne L, Binder A (1996) Detection of crop fungal pathogens by polymerase chain reaction technology. In: Marshall G (ed) Diagnostics in crop production, BCPC symposium proceedings No 65. British Crop Protection Council, Farnham, pp 111–118

Becker H, Manz A (1999) Microsystem technology in chemistry and life science. Springer Desktop Editions in Chemistry, vol 194. Springer, Berlin/Heidelberg/New York

Bertaccini A (1990) Detection of chrysanthemum yellows mycoplasmalike organism by dot hybridization and Southern blot analysis. Plant Dis 74:40–43

Bertolini E, Olmos A, Martınez MC, Gorris MT, Cambra M (2001) Single-step multiplex RT-PCR for simultaneous and colourimetric detection of six RNA viruses in olive trees. J Virol Methods 96:33–41

Bertolini E, Olmos A, Lopez MM, Cambra M (2003) Multiplex nested reverse-transcription polymerase chain reaction in a single tube for sensitive and simultaneous detection of four RNA viruses and *Pseudomonas savastanoi pv. savastanoi* in olive trees. Phytopathology 93:286–292

Biosca EG, Marco-Noales E, Ordax M, Lopez MM (2006) Long-term starvation-survival of *Erwinia amylovora* in sterile irrigation water. Acta Horticult 704:107–112

Bird RE, Hardman KD, Jacobson JW, Johnson S, Kaufman BM, Lee S, Lee T, Pope SH, Riordan GS, Whitlow M (1988) Single-chain antigen-binding proteins. Science 242:423–426

Bock KR (1982) The identification and partial characterization of plant-viruses in the tropics. Trop Pest Manag 28:399–411

Bonde MR, Micales JA, Peterson GL (1993) The use of isozyme analysis for identification of plant-pathogenic fungi. Plant Dis 77:961–968

Cahill DM, Hardham AR (1994) Exploitation of zoospore taxis in the development of a novel dipstick immunoassay for the specific detection of *Phytophthora cinnamomi*. Phytopathology 84:193–200

Cambra M, Capote N, Myrta A, Llacer G (2006) *Plum pox* virus and the estimated costs associated with sharka disease. Bull OEPP/EPPO 36:202–204

Caruso P, Bertolini E, Cambra M, Lopez MM (2003) A new and co-operational polymerase chain reaction (Co-PCR) for rapid detection of *Ralstonia solanacearum* in water. J Microbiol Methods 55(1):257–272

Chandelier A, Dubois N, Baelen F, De Leener F, Warnon S, Remacle J, Lepoivre P (2001) RT-PCR, ELISA tests on pooled sample units for the detection of virus Y in potato tubers. J Virol Methods 91:99–108

Chittaranjan S, De Boer SH (1997) Detection of *Xanthomonas campestris pv. pelargonii* in geranium and greenhouse nutrient solution by serological and PCR techniques. Eur J Plant Pathol 103:555–563

Cottyn B (1996) Bacterial diseases of rice. characterization of pathogenic bacteria associated with sheath rot complex and grain discoloration of rice in the Philippines. Plant Dis 80:438–445

Damsteegt VD (1997) *Prunus tomentosa* as a diagnostic host for detection of plum pox virus and other *Prunus* viruses. Plant Dis 81:329–332

Davey HM, Kell DB (1996) Flow cytometry and cell sorting of heterogeneous microbial populations: the importance of single cell analyses. Microbiol Rev 60:641–696

De Laat PCA, Verhoeven JTW, Janse JD (1994) Bacterial leaf rot of Aloe vera L, caused by *Erwinia chrysanthemi* biovar-3. Eur J Plant Pathol 100:81–84

Duffy BK, Weller DM (1994) A semiselective and diagnostic medium for *Gaeumannomyces graminis var tritici*. Phytopathology 84:1407–1415

Fatmi M, Schaad NW (1988) Semiselective agar medium for isolation of *Clavibacter michiganense* subsp. *michiganensis* from tomato seed. Phytopathology 78:121–126

Fatmi M, Schaad NW, Bolkan HA (1991) Seed treatments for eradicating *Clavibacter michiganensis subsp. michiganensis* from naturally infected tomato seeds. Plant Dis 75:383–385

Fillhart RC, Bachand GD, Castello JD (1998) Detection of infectious tobamoviruses in forest soils. Appl Environ Microbiol 64:1430–1435

Fraaije BA, Lovell DJ, Rohel EA, Hollomon DW (1999) Rapid detection and diagnosis of Septoria tritici epidemics in wheat using a polymerase chain reaction PicoGreen assay. J Appl Microbiol 86:701–708

Fraaije BA, Lovell D, Coelho JM, Baldwin S, Hollomon DW (2001) PCR-based assays to assess wheat varietal resistance to blotch (*Septoria tritici* and *Stagonospora nodorum*) and rust (*Puccinia striiformis* and *Puccinia recondita*) diseases. Eur J Plant Pathol 107:905–917

Fraaije BA, Butters JA, Coelho JM, Johes DR, Hollomon DW (2002) Following the dynamics of strobilurin resistance in Blumeria graminis f. sp. tritici using quantitative allele-specific realtime PCR measurements with the fluorescent dye SYBR green I. Plant Pathol 51:45–54

Gilliland G, Perrin S, Blanchard K, Bunn HF (1990) Analysis of cytokine mRNA and DNA: detection and quantization by competitive polymerase chain reaction. Proc Natl Acad Sci U S A 87:2725–2729

Gitaitis R et al (1997) Bacterial streak and bulb rot of onion.1. A diagnostic medium for the semiselective isolation and enumeration of Pseudomonas viridiflava. Plant Dis 81:897–900

Goor MC et al (1984) The use of API systems in the identification of phytopathogenic bacteria. Mededelingen Van De Faculteit Landbouwtenschappen Rijksuniversiteit. Genetics 49:499–507

Grey BE, Steck TR (2001) The viable but non culturable state of *Ralstonia solanacearum* may be involved in long-term survival and plant infection. Appl Environ Microbiol 67:3866–3872

Griep RA et al (1998) Development of specific recombinant monoclonal antibodies against the lipopolysaccharide of *Ralstonia solanacearum* race 3. Phytopathology 88:795–803

Gurtler V, Stanisich VA (1996) New approaches to typing and identification of bacteria using the 16S-23S rDNA spacer region. Microbiology 142:3–16

Henegariu O, Heerema NA, Dlouhy SR, Vance GH, Vogt PH (1997) Multiplex PCR: critical parameters and step-by step protocol. Biotechniques 23:504–511

Hodgson RAJ, Wall GC, Randles JW (1998) Specific identification of coconut tinangaja viroid for differential field diagnosis of viroids in coconut palm. Phytopathology 88:774–781

James D (1999) A simple and reliable protocol for the detection of apple stem grooving virus by RT-PCR and in a multiplex PCR assay. J Virol Methods 83:1–9

Janse JD (2007) Phytobacteriology: principles and practice. CABI Publishing, Wallingford

Janse JD, Wenneker M (2002) Possibilities of avoidance and control of bacterial plant diseases when using pathogen-tested (certified) or -treated planting material. Plant Pathol 51:523–536

Janse J, Van den Beld HE, Elphinstone J, Simpkins S, Tjou-Tam-Sin NNA, Van Vaerenbergh J (2004) Introduction to Europe of *Ralstonia solanacearum* biovar 2, race 3 in *Pelargonium zonale* cuttings. J Plant Pathol 86:147–155

Jeffers SN, Aldwinckle HS (1987) Enhancing detection of Phytophthora cactorum in naturally infested soil. Phytopathology 77:1475–1482

Jones DAC (1994) Blackleg potential of potato seed – determination of tuber contamination by *Erwinia carotovora* subsp atroseptica by immunofluorescence colony staining and stock and tuber sampling. Ann Appl Biol 124:557–568

Kittipakorn K et al (1993) Strains of peanut stripe potyvirus rapidly identified by profiling peptides of the virion proteins. J Phytopathol Phytopathologische Z 137:257–263

Knoll S, Vogel RF, Niessen L (2002) Identification of *Fusarium graminearum* in cereal samples by DNA Detection Test Strips. Lett Appl Microbiol 34:144–148

Landgraf A, Reckmann B, Pingoud A (1991) Direct analysis of polymerase chain reaction products using enzyme-linked immunosorbent assay techniques. Ann Biochem 198:86–91

Lopez MM, Bertolini E, Olmos A, Caruso P, Gorris MT, Llop P, Penyalver R, Cambra M (2003) Innovative tools for detection of plant pathogenic viruses and bacteria. Int Microbiol 6:233–243

Lopez MM, Bertolini E, Marco-Noales E, Llop P, Cambra M (2006) Update on molecular tools for detection of plant pathogenic bacteria and viruses. In: Rao JR, Fleming CC, Moore JE (eds) Molecular diagnostics: current technology and applications. Horizon Bioscience, Wymondham, pp 1–46

Louws FJ, Rademaker JLW, de Bruijn FJ (1999) The three Ds of PCR-based genomic analysis of phytobacteria: diversity, detection, and disease diagnosis. Annu Rev Phytopathol 37:81–125

Manandhar JB, Hartman GL, Wang TC (1995) Semiselective medium for *Colletotrichum gloeosporioides* and occurrence of 3 *Colletotrichum spp* on pepper plants. Plant Dis 79:376–379

Martin RR, James D, Levesque CA (2000) Impacts of molecular diagnostic technologies on plant disease management. Annu Rev Phytopathol 38:207–239

Miller SA, Martin RR (1988) Molecular diagnosis of plant disease. Annu Rev Phytopathol 26:409–432

Milne RG, Ramasso GE, Lenzi R, Masenga V, Sarindu S, Clark MF (1995) Pre- and post-embedding immunogold labeling and electron microscopy in plant host tissues of three antigenically unrelated MLOs: primula yellows, tomato big bud and bermudagrass whiteleaf. Eur J Plant Pathol 10:57–67

Mullis KB, Faloona FA (1987) Specific synthesis of DNA in vitro via a polymerase-catalyzed chain reaction. Methods Enzymol 155:335–350

Mutasa ES, Ward E, Adams MJ, Collier CR, Chwarszczynska D, Asher MJC (1993) A sensitive DNA probe for the detection of *Polymyxa betae* in sugar beet roots. Physiol Mol Plant Pathol 43:379–390

Mutasa ES, Chwarszczynska D, Adams MJ, Ward E, Asher MJC (1995) Development of PCR for the detection of *Polymyxa betae* in sugar beet roots and its application in field studies. Physiol Mol Plant Pathol 47:303–313

Mutasa ES, Chwarszczynska DM, Asher MJC (1996) Single-tube, nested PCR for the diagnosis of *Polymyxa betae* infection in sugar beet roots and colorimetric analysis of amplified products. Phytopathology 86:493–497

Nassuth A, Pollari E, Helmeczy K, Stewart S, Kofalvi SA (2000) Improved RNA extraction and one-tube RT-PCR assay for simultaneous detection of control plant RNA plus several viruses in plant extracts. J Virol Methods 90:37–49

Nicholson P, Lees A, Maurin N, Parry D, Rezanoor HN (1996) Development of a PCR assay to identify and quantify *Microdochium nivale* var. *nivale* and *Microdochium nivale* var. *majus* in wheat. Physiol Mol Plant Pathol 48:257–271

Nicholson P, Simpson DR, Weston G, Rezanoor HN, Lees AK, Parry DW, Joyce D (1998) Detection and quantification of Fusarium culmorum and Fusarium graminearum in cereals using PCR assays. Physiol Mol Plant Pathol 53:17–37

Okuda M, Hanada K (2001) RT-PCR for detecting five distinct Tospovirus species using degenerate primers and dsRNA template. J Virol Methods 96:149–156

Olmos A, Cambra M, Esteban O, Gorris MT, Terrada E (1999) New device and method for capture, reverse transcription and nested PCR in a single closed-tube. Nucleic Acids Res 27:1564–1565

Olmos A, Bertolini E, Cambra M (2002) Simultaneous and cooperational amplification (Co-PCR): a new concept for detection of plant viruses. J Virol Methods 106:51–59

Olmos A, Esteban O, Bertolini E, Cambra M (2003) Nested RT-PCR in a single closed tube. In: Bartlett JMS, Stirling D (eds) Methods in molecular biology, vol 226, 2nd edn, PCR Protocols. Humana Press, Ottowa, pp 156–161

Pernezny K (1995) An outbreak of bacterial spot of lettuce in Florida caused by *Xanthomonas campestris pv. vitians*. Plant Dis 79:359–360

Piotte C (1994) Cloning and characterization of 2 satellite DNAs in the low-C-value genome of the nematode *Meloidogyne spp*. Gene 138:175–180

Piotte C (1995) Analysis of a satellite DNA from *Meloidogyne hapla* and its use as a diagnostic probe. Phytopathology 85:458–462

Rasmussen OF, Reeves JC (1992) DNA probes for the detection of plant pathogenic bacteria. J Biotechnol 25:203–220

Rasmussen JB, Scheffer RP (1988a) Effects of selective toxin from *Helminthosporium carbonum* on chlorophyll synthesis in maize. Physiol Mol Plant Pathol 32:283–291

Rasmussen JB, Scheffer RP (1988b) Isolation and biological activities of four selective toxins from *Helminthosporium carbonum*. Plant Physiol 86:187–191

Roberts PD (1996) Survival of *Xanthomonas fragariae* on strawberry in summer nurseries in Florida detected by specific primers and nested polymerase chain reaction. Plant Dis 80:1283–1288

Roy MA (1988) Use of fatty-acids for the identification of phytopathogenic bacteria. Plant Dis 72:460

Russo P, Miller L, Singh RP, Slack SA (1999) Comparison of PLRV and PVY detection in potato seed samples tested by Florida winter field inspection and RT-PCR. Am J Potato Res 76:313–316

Saade M, Aparicio F, Sanchez-Navarro JA, Herranz MC, Di-Terlizzi AMB, Pallas V (2000) Simultaneous detection of the three ilarviruses affecting stone fruit trees by nonisotopic molecular hybridization and multiplex reverse-transcription polymerase chain reaction. Phytopathology 90:1330–1336

Schaad NW, Frederick RD (2002) Real-time PCR and its application for rapid plant disease diagnostics. Can J Plant Pathol 24:250–258

Schots A, Dewey FM, Oliver R (eds) (1994) Modern assays for plant pathogenic fungi: identification, detection and quantification. CAB International, Wallingford

Scortichini M (1995) Le malattie batteriche delle colture agrarie e delle specie forestali. Edagricole, Edizione Agricole, Bologna

Shurtleff MC, Averre CW III (1997) Glossary of plant-pathological terms. American Phytopathology Association, St. Paul

Singh RP, Kurz J, Boiteau G (1996) Detection of stylet-borne and circulative potato viruses in aphids by duplex reverse transcription polymerase chain reaction. J Virol Methods 59:189–196

Strange RN (1993) Plant disease control: towards environmentally acceptable methods, using PCR assays. Physiol Mol Plant Pathol 53:17–37

Van der Wolf JM, Van Beckhoven JRCM, Bonanats PJM, Schoen CD (2001) New technologies for sensitive and specific routine detection of plant pathogenic bacteria. In: de Boer SH (ed) Plant pathogenic bacteria. Kluwer, Dordrecht, pp 75–77

Van Dijk P, Vandermeer FA, Piron PGM (1987) Accessions of Australian Nicotiana species suitable as indicator hosts in the diagnosis of plant virus diseases. Neth J Plant Pathol 93:73–85

Van Sluys MA, Monteiro-Vitorello CB, Camargo LEA, Menck CFM, da Silva ACR, Ferro JA, Oliveira MC, Setubal JC, Kitajima JP, Sympson AJ (2002) Comparative genomic analysis of plant-associated bacteria. Ann Rev Phytopathol 40:169–190

Volkhard A, Kempf J, Trebesius K, Autenrieth IB (2000) Fluorescent in situ hybridization allows rapid identification of microorganisms in blood cultures. J Clin Microbiol 38:830–838

Wang AM, Doyle MV, Mark DF (1989) Quantitation of mRNA by the polymerase chain reaction. Proc Natl Acad Sci U S A 24:9717–9721

Ward E, Gray R (1992) Generation of a ribosomal DNA probe by PCR and its use in identification of fungi within the *Gaeumannomyces-Phialophora* complex. Plant Pathol 41:730–736

Wells JM, Vanderzwet T, Butterfield JE (1994) Differentiation of *Erwinia* species in the herbicola group by class analysis of cellular fatty-acids. J Phytopathol Phytopathol Z 140:39–48

Williams K, Blake S, Sweeney A, Singer JT, Nicholson BL (1999) Multiplex reverse transcriptase PCR assay for simultaneous detection of three fish viruses. J Clin Microbiol 37:4139–4141

Williams R, Ward E, McCartney HA (2001) Methods for integrated air sampling and DNA analysis for detection of airborne fungal spores. Appl Environ Microbiol 67:2453–2459

Winter G, Griffiths AD, Hawkins RE, Hoogenboom HR (1994) Making antibodies by phage display technology. Ann Rev Immunol 12:433–455

Wullings BA, Beuningen AR, Van Janse JD, Akkermans ADL, Van Beuningen AR (1998) Detection of *Ralstonia solanacearum*, which causes brown rot of potato, by fluorescent in situ hybridization with 23S rRNA-targeted probes. Appl Environ Microbiol 64:4546–4554

Zeringue HJ, Bhatnagar D, Cleveland TE (1993) C15H24 volatile compounds unique to aflatoxigenic strains of *Aspergillus flavus*. Appl Environ Microbiol 59:2264–2270

Plant Nutrition in the Management of Plant Diseases with Particular Reference to Wheat

20

D.P. Singh

Nutrients not only are important for the growth and development of crop plants but also can affect the microorganisms; thus, they are important factors in disease control (Agrios 2005). The nutrients can affect the disease severity in a positive or negative way (Huber and Graham 1999). For the expression of proper resistance to diseases, sufficient supply and intake of nutrients in plants are quite important. The importance of proper nutrient management for reducing losses due to diseases and getting higher yields has been demonstrated in different crops (Graham and Webb 1991; Marschner 1995; Huber and Graham 1999). Although both fertilizers and cultural practises reduce the severity of many plant diseases, the effect of fertilizers is more direct (Marschner 1995; Atkinson and McKinlay 1997; Oborn et al. 2003). Nutrients make the plants grow at a faster rate and also may change the properties of soil, thus making it unfavourable for the survival of pathogens. The resistance or susceptibility of plants to diseases, their histological or morphological structure or properties, and the virulence and survival ability of pathogens are affected by plant nutrition. The synthesis of defence compounds like phytoalexins and flavonoids may change due to nutritional supply. The resistant plants respond to optimum doses of nutrients and resist diseases, whereas susceptible ones require higher amounts of nutrients. Nutrient-deficient plants are less vigorous and more susceptible to different diseases, and all nutrients affect plant diseases in varying degrees. The disease cannot be totally eliminated by altering the of supply of a particular nutrient, but its level of severity can only be changed. Adequate supply of nutrients like calcium develops stronger cell walls and tissues in plants which inhibit the effect of enzymes released by bacteria responsible for dissolving the cell wall. Likewise, silicon reduces the feeding ability of some sucking pests like aphid on plants which are known vectors of viral pathogens and thus help reduce virus infection. Few reviews indicating the role of nutrients on plant diseases have been already published (Datnoff et al. 2007; Dordas 2008; Katan 2009; Reuveni and Reuveni 1998; Spann and Schumann 2013; Spectrum Analytic, Inc. 2013; Walters and Bingham 2007), and information taken from these sources upon preparing this chapter is thankfully acknowledged. This chapter discusses the more significant nutrients and their interactions with disease with particular reference to wheat by taking the most recent references on the subject.

The cultural practises such as cropping system, use of organic soil amendment, soil pH correction, tillage methods, and irrigation management also affect the status of nutrients and their availability to plants, therefore altering diseases.

D.P. Singh (✉)
Guyana Rice Development Board,
Georgetown, Guyana (S. America)
e-mail: dpkarnal@yahoo.com

L.P. Awasthi (ed.), *Recent Advances in the Diagnosis and Management of Plant Diseases*,
DOI 10.1007/978-81-322-2571-3_20

Likewise, conservation agriculture or use of a single herbicide may also influence the supply of plant nutrients. Macro- and microelements have long been recognized as being associated with the size, quality, and yield of crops and also with changes in the levels of the incidence of disease (Rush et al. 1997). The two major objectives of nutrient applications to crops for their protection against pathogens are to avoid plant stress, which may allow crops to better withstand pathogen attack, and to manipulate nutrients to the advantage of the plant and disadvantage of the pathogen (Palti 1981). Soil properties such as soil pH, high levels of ammonium and nitrate, moisture, etc. also affect plant diseases. The soil pH is affected by the form of nitrogen and therefore influences disease development. The potato scab severity is reduced in lower soil pH when sulphur and NH_4^+ are used. Excess nitrogen promotes thinner and weaker cell walls, increases plant density and therefore high humidity and low light conditions and delay in maturity, and therefore helps the disease development in plants. The sources of N also affect the severity of diseases, and nitrate suppresses the production of these enzymes while ammonium may increase their production (Huber and Graham 1999). The unbalanced N:K ratio affects not only crop yield but also the disease resistance. The excess supply of potassium and high K:Ca ratio might result in calcium deficiency and reduced resistance to diseases.

The incidence or severity of a disease under different doses of nutrients in both resistant and susceptible cultivars is measured to know the effect of nutrients. Factors like the stage of growth of plants, environmental conditions, and biological activity also influence the interaction of mineral nutrition and diseases. The interaction of K with N and P has been demonstrated in wheat, and disease incidence decreases in the case of sufficient supply of all three elements. In the case of deficiency of N and P, an increase in K also increases the incidence of diseases like *Fusarium* wilt of tomato and downy mildew of tobacco (Huber and Arny 1985). There are interactions between K and Mg in the case of *Fusarium* wilt of cotton and K and Ca in common scab disease (*Streptomyces scabies*) in potato (Huber 1980). Calcium application increases the resistance of cell to the extracellular macerating enzymes produced by pathogens such as *Erwinia*, *Sclerotium rolfsii*, *Pythium myriotylum*, *Rhizoctonia solani*, *Cylindrocladium crotalaria*, *Sclerotinia minor*, and *Fusarium solani*. It thus suppresses macerating diseases in plants (Bateman and Basham 1976). Magnesium in the presence of sufficient Ca reduces susceptibility to pathogen-produced macerating enzymes (Csinos and Bell 1989). Silicon along with other elements like Ca and Mn gives cell walls greater strength as physical barriers against rice blast fungus (*Magnapothe oryzae*) and powdery mildew (*Erysiphe* spp.) (Datnoff et al. 1991; Savant et al. 1997).

20.1 Mechanisms of Disease Control with Plant Nutrients

The metabolic activity of plants, responsible for developing resistance, is regulated by the supply of mineral nutrition. In resistant plants, the inhibitory compounds like phytoalexins, phenols, and flavonoids are accumulated during infection only if the nutrient supply needed for their synthesis is adequate. For example, Mn plays an important role in active defence mechanisms mediated through the shikimate pathway. The synthesis of glycoproteins (lectin) needs Mn adequately, and Mn is also required in the case of sweet potato resistance against black rot (*Ceratocystis fimbriata*) and late blight or potato blight (*Phytophthora infestans*) (Garas and Kuc 1981).

Nutrients play a major role in plants to develop strong cell walls and other tissues. The germination of fungal spores is stimulated by compounds exuded by the plants. The amount and composition of these exudates are affected by the nutrition of the plant. In nutrient-deficient plants, these exudates may contain higher amounts of compounds like sugars and amino acids, thus favouring disease development. The infection of a fungus in plants causes increased production of phenolic compounds and flavonoids, both at the site of infection and in other parts of the plant. The nutrient supply to plant directly affects the production and transport of these compounds.

For example, when there is an imbalance of N with other nutrients, production of antifungal compounds is reduced. The enzymes released by bacteria and fungi that are responsible for dissolving parts of the plant tissue are produced more under short nutrient supply. The activity of these enzymes is reduced by the calcium ion (Ca++). The oxygen radicals (O=and OH) and hydrogen peroxide (H_2O_2) are produced as a result of fungal infection in plants. These radicals may be destructive to the plant cells as well as the pathogen. Some nutrients act to detoxify oxygen radicals and hydrogen peroxide, reducing damage to plant cells (Spann and Schumann 2013).

The bacterial enzymes cause the decomposition of pectin. The production and activity of these pectolytic enzymes are affected by some nutrients. The deficiency of those nutrients leads to greater damage to host cells from these pathogens. Bacterial vascular diseases spread by way of the xylem, and the bacteria produce "slime" within the vessels, blocking them and leading to wilting and death of plants. Some plant nutrients help in blocking or reducing the bacteria's ability to form this slime. Not much information is present indicating the role of mineral nutrition of plants on viral diseases. The viruses need amino acids and nucleotides of the host cells to multiply in the host tissues. The properly nourished plants sometimes mask the effect of viral infection. For example, in the case of beet mild yellowing virus (BMYV) in sugar beet, Mn is helpful in masking the symptoms of virus. The vectors of viruses (aphids) prefer to feed on yellow reflecting surfaces, such as chlorotic leaves due to nutrient deficiency. The higher contents of amino acid in plants also favour feeding of insect vectors. The silicon application physically inhibits the feeding ability of sucking insects (Spann and Schumann 2013). Ahmad et al. (2012) conducted experiments on fertilizers and found that NPK (classic) was proved most effective against urdbean leaf crinkle virus (ULCV) and reduced the disease severity up to 65 %. Zn and B (fashion) was found effective and reduced the disease severity up to 62.57 % in Pakistan.

The role of different essential macro- and microelements on plant diseases is as follows:

20.1.1 Nitrogen (N)

It is applied most in crop production and known to increase or decrease the severity of diseases. Nitrogen is required for the synthesis of proteins, amino acids, nucleotides, coenzymes, and chlorophyll. About 70 % of total leaf nitrogen may be in the chloroplasts. The adequate N supply is helpful in reducing the severity of root rot of canola, aflatoxin in maize, root rot of cotton, stem canker of soya bean, stem rot of sunflower, and tan spot of wheat. The over-fertilization with N results in excessively weak vegetative growth increase in plant density and changes the microclimate in field, thus inviting more infection and spread of pathogens. Excess N can delay maturity and, therefore, favours higher infection and disease development. The higher N application increases the severity of diseases caused by obligate pathogens, whereas it reduces facultative pathogens. The main reason for the increased susceptibility to obligate parasites at high N rates is that it causes anatomical and biochemical changes in plants along with increase in the content of the low-molecular-weight organic nitrogen compounds which act as substrates to pathogens. On the contrary, the plants grown under low N availability are better defended against pathogens since the synthesis of defence-related compounds is increased (Bryant et al. 1983; Herms and Mattson 1992; Hoffland et al. 1999, 2000; Wilkens et al. 1996). The diseases whose severity is increased due to N are grey leaf spots of corn, rusts, and powdery mildew of wheat (Howard et al. 1994; Büschbell and Hoffmann 1992) and seed infection in the cereals (Martin 2013). Potatoes deficient in N are more susceptible to early blight (*Alternaria solani*) (Blachinski et al. 1996). Besides doses, the form of N, or the ratio of ammonium N (NH_4-N) to nitrate N (NO_3-N), have an effect on diseases (Huber and Graham 1999; Celar 2003; Harrison and Shaw 2001). The severity of diseases like stalk rot, Northern corn leaf blight, and root rot in maize; take-all in wheat; and wilt of cotton increases with application of N (NO_3-N) and decreases with N (NH_4-N). The effect of above-mentioned forms of nitrogen is just reverse in the case of black scurf (*R. solani*) in potatoes, root rots of beans and peas, and foot rot in wheat.

20.1.2 Phosphorus (P)

Phosphorus plays an important role in plant metabolism, is responsible for the formation of adenosine triphosphate (ATP) and phospholipids including those of the membranes, and plays a role in the formation of sugar phosphates and various nucleotides and coenzymes (Prabhu et al. 2007a). In seeds, phosphorus is stored in the form of phytic acid. It activates some enzymes in the plants. Phosphorus is taken up mostly as HPO_4^{2-} or $H_2PO_4^-$. It is important for energy transfer in plants, is essential in cell division and proper growth, increases tiller and yield, and creates stay-green effect in plants. The effect of P on diseases is not always consistent, and therefore, balanced soil fertility is important for reducing plant stress, improving the level of resistance, and decreasing the severity of diseases. Phosphate fertilization of wheat effectively reduces the severity of *Pythium* root rot (Huber 1980). The incidence of diseases like smut, blotch, powdery mildew, and take-all in wheat decreases, whereas scab increases due to P (Huber and Graham 1999; Kirkegaard et al. 1999; Reuveni et al. 1998, 2000). The higher rates of monoammonium phosphate fertilizer reduced root rot severity in barley by 7 % and the incidence of *Gaeumannomyces graminis* var. *tritici* in wheat by greater than 40 % (Bailey et al. 1998). In corn, P application can reduce root rot and the incidence of smut (Huber and Graham 1999; Potash and Phosphate Institute 1988). Phosphorus also reduces bacterial leaf blight and yellow and blast disease in rice, downy mildew, blue mould, tobacco leaf curl virus, and pod and stem blight in soya bean(Dordas 2008).

20.1.3 Potassium (K)

Potassium is involved in many cellular processes and therefore influences disease severity. It acts as an activator of numerous enzymes and plays a role in protein synthesis, stomatal opening and closing, and thus exchange of CO_2 in leaves (Marschner 1995). The translocation of the photosynthates from leaves to other parts of plants is also affected by K. It increases the root growth and reduces lodging. The different forms of K have no effect on diseases. The amount of the plants' natural antifungal compounds at the site of infection decreases during K deficiency. K has a direct role in the development of thick cuticles which acts as a physical barrier to infection or penetration by sucking insects. The wilt-resistant varieties of flax are able to absorb more K from soil than the susceptible ones. K interacts with Ca negatively and thus increases diseases like common scab (*S. scabies*) in potato and *Phytophthora* root rot (*P. parasitica*) in citrus. The severity of damage from root knot nematode and soya bean cyst nematode is reduced by adequate K fertilization. It has been frequently observed that K reduces the incidence of various diseases such as bacterial leaf blight, sheath blight, stem rot, and sesame leaf spot in rice; black rust in wheat; sugary disease on sorghum; bacterial leaf blight in cotton; *Cercospora* leaf spot in cassava; tikka leaf spot in peanut; red rust in tea; *Cercospora* leaf spot in mung bean; and seedling rot caused by *R. solani* (Chase 1989; Huber and Graham 1999; Sharma and Duveiller 2004; Sharma et al. 2005; Dordas 2008). In the case of wheat, increased application of K decreases the incidence of diseases like leaf blotch, powdery mildew, and stem, leaf and stripe rusts but increases root rot, take-all, tan spots, bunt, and smuts (Menzies et al. 1992; Sharma and Duveiller 2004; Sharma et al. 2005). Potassium in excess inhibits the uptake of Mg, Ca, and/or NH_4-N. In the K-deficient plant, higher amounts of compounds like sugars and amino acids are secreted in the exudates which favour the higher infection of fungal pathogens. It has been argued that potassium-deficient plants might be predisposed to diseases (Prabhu et al. 2007b), and indeed in many cases, potassium application has been shown to reduce the incidence of both foliar and soilborne diseases, while in a few cases, the opposite has been found to be true. For example, out of 165 cases of fungal, bacterial, and nematode diseases, 117 cases (71 %) resulted in a decrease in diseases while 29 % showed an increase with application of K (Prabhu et al. 2007b).

20.1.4 Sulphur (S)

Sulphur is a constituent of amino acids, such as cysteine and methionine. Thiamine, biotin, ferredoxins, and coenzyme A are examples of S compounds. Protein also contains N and S. Therefore, S deficiency in plants results in lower contents of essential proteins and carbohydrate. Sulphur mainly comes from organic sources and acts as natural biocide as well as increases resistance in plants against pathogens. The incidence of powdery mildew and stem rust is reduced under adequate supply of S to the wheat crop.

20.1.5 Calcium (Ca)

Under Ca deficiency, the permeability of cell membrane is affected adversely, and the cell membrane allows leakage of low-molecular-weight compounds like sugars and amino acids, from the cytoplasm to the apoplast, stimulating the infection by the pathogens. Again calcium is a cofactor for α-amylase and a major cation of the middle lamella of cell wall. It therefore provides mechanical strength to the tissues and resistance to fungal infections (Marschner 1995). The activity of pectolytic enzymes of fungi and bacteria in plants is inhibited by Ca ions. The contents of Ca vary in different types of soils. Its contents are low in highly weathered soils, whereas calcareous soils have high contents. The wheat root rot is reduced by application of Ca. It is also helpful in protecting peanut pods from infections by *Rhizoctonia* and *Pythium* (Huber 1980) and provides resistance against *Pythium*, *Sclerotinia*, *Botrytis*, and *Fusarium* (Graham 1983). A putative mechanism by which Ca is believed to provide protection against *Sclerotinia sclerotiorum* is by binding of oxalic acid or by strengthening the cell wall (Dordas 2008).

20.1.6 Boron (B)

Boron plays a major role in the transport of sugars in plants and also promotes the root growth. It helps the plants to restrict fungal hyphae from movement through the cell walls and synthesizes lignin which acts as a barrier to infection of pathogens. The exudates of boron-deficient plants contain higher amounts of compounds such as sugars and amino acids that promote fungal infections. Boron promotes stability and rigidity of the cell wall structure and therefore supports the shape and strength of the plant cell (Marschner 1995; Brown et al. 2002). Furthermore, B is possibly involved in the integrity of the plasma membrane (Marschner 1995; Brown et al. 2002; Dordas and Brown 2005). It decreases the severity of wheat rusts, clubroot (*Plasmodiophora brassicae*) of crucifers, bean root rot (*F. solani*), tomato and cotton wilts (*Verticillium albo-atrum*), mung bean stem rot (*R. solani*), bean mosaic virus and tobacco mosaic virus (Dordas 2008; Graham and Webb 1991), and *Blumeria graminis* on wheat (Marschner 1995).

20.1.7 Copper (Cu)

Copper is a constituent of plant proteins and plays an essential role in the oxidative process as well as electron transport in plants. The biosynthesis of lignins and activity of polyphenol oxidase are affected by Cu in plants. It acts as a direct toxic element to microbes, and higher plant forms such as crops and ornamentals can tolerate much higher Cu levels than the lower forms. It helps the plants translocate antifungal compounds to the point of infection and detoxify oxygen radicals (O=and OH-) and hydrogen peroxide (H_2O_2) produced as a response to infection limiting the damage to plants. Copper fertilization has been proved helpful in the management of diseases like powdery mildew (*Blumeria graminis* var. *tritici*) and leaf rust (*Puccinia triticina*) in wheat, ergot (*Claviceps purpurea*) of rye and barley, wilt (*Verticillium albo-atrum*) in tomato, and common scab (*S. scabies*) in potato. Zinc reduces the uptake of Cu (Dordas 2008).

20.1.8 Zinc (Zn)

Some of the first fungicide and bactericide contain Zn. It regulates enzymes that defend the plant against disease infection. Under deficient conditions of Zn, the contents of RNA and ribosomes decline sharply in cells. It is also important for the synthesis of auxins, cell division, and water and nutrient uptake. It is usually deficient in waterlogged soils having low organic matter and high pH. The changes in membrane permeability occur due to Zn deficiency, resulting in the loss of nutrients through the roots or leaf exudation to attract pathogens or enhance infection, particularly zoosporic organisms (Huber and Haneklaus 2007; Marschner 1995; Römheld and Marschner 1991). Zinc has been beneficial in the control of diseases like root rot (*Phytophthora megasperma*), wilt (*Fusarium* spp.), sheath blight (*R. solani*), cotton wilt (*Verticillium* spp.), tomato nematode (*Rotylenchulus reniformis*), peanut rot (*Rhizoctonia bataticola*), take-all (*Gaeumannomyces graminis* var. *tritici*) and head scab (*Fusarium graminearum*) in wheat (Graham and Webb 1991; Grewal et al. 1996), foot rot (*S. rolfsii*) in soya bean, and powdery mildew (*Erysiphe polygoni*) in pea.

20.1.9 Iron (Fe)

Iron plays an important role in the maintenance of chlorophyll in plants. It is carried with the help of heme proteins and has an important role as a chelating agent in plants. Iron is essential for proteins, electron transfer, and redox reactions. It is immobile in soil and plants and the most soluble iron is present as iron siderophores. These siderophores are important as agents of protection from bacteria. Fe can activate enzymes that are involved in the infection of the host by the pathogen or the defence, which is why opposite effects were found. Fe can promote the synthesis of fungal antibiotics by soil bacteria (Graham and Webb 1991). Rhizosphere microorganisms can synthesize siderophores which can lower Fe level in the soils. These siderophores can suppress germination of chlamydospores of *Fusarium oxysporum* f. sp. *cucumerinum* in vitro. However, the production of siderophores and antagonisms for Fe is not only the mechanisms to limit the growth of parasitic fungus. As far its role in disease control is concerned, it is not considered in the same category as K, Mn, Cu, or Zn. It has been reported that plant pathogens generally have a high requirement for Fe. Fe has a role in both the production and detoxification of oxygen radicals and hydrogen peroxide, thus limiting damage to plant cells. In some cases increased Fe availability or uptake also increases disease severity. For example, tolerance of tomato to *Fusarium* wilt was decreased by Fe without affecting the development of the fungus and disease incidence, and the severity of take-all in wheat and barley is also increased by Fe. The diseases like wheat leaf rust (*Puccinia recondita*), smut (*Tilletia* sp.), banana anthracnose (*Colletotrichum musae*), apple black rot (*Sphaeropsis malorum*), pear black rot (*Sphaeropsis malorum*), and cabbage virus vector (*Olpidium brassicae*) are reduced by Fe application (Graham and Webb 1991; Graham 1983). Fe reduces the severity of rust in wheat leaves and smut in wheat and *Colletotrichum musae* in banana (Graham and Webb 1991; Graham 1983; Dordas 2008). Foliar application of Fe can increase resistance of apple and pear to *Sphaeropsis malorum* and cabbage to *Olpidium brassicae* (Graham 1983).

20.1.10 Manganese (Mn)

Manganese is essential for the functioning of oxidative and carboxylating enzymes. It is involved in the electron transport in the chloroplast in photosystem II during photosynthesis. Manganese controls lignin and suberin biosynthesis (Römheld and Marschner 1991; Vidhyasekaran 1997) through the activation of several enzymes of the shikimate and phenylpropanoid pathways (Marschner 1995). Being the phenolic polymers, lignin and suberin act as important biochemical barriers to infection against fungal pathogens (Kolattukudy et al. 1994; Rioux and Biggs 1994; Hammerschmidt and Nicholson 1999; Vidhyasekaran 1997, 2004). It exists in plants as Mn^{2+}. Immobilization of Mn at the infection site due to *Gaeumannomyces graminis* var. *tritici* predisposes wheat tissues to take-all, and this is also true in the case of *Magnaporthe oryzae*, or rice blast fungus. The virulence level of pathogens and isolates of *G. graminis* and *M. grisea* directly depends on their ability to oxidize Mn (Huber and Thompson 2007). The wheat diseases like powdery mildew, rusts, and take-all are decreased by Mn application (Rovira et al. 1983; Graham and Webb 1991; Huber 1996; Krauss 1999). In general, the severity of fungal and bacterial diseases decreases, whereas viral diseases increase with increasing doses of Mn. The host plants require much more Mn than fungi and bacteria. It may be the case that by supplying high levels of Mn to some plants through fungicides or nutrition, we are causing a simple Mn toxicity for fungi and bacteria. The soil applications of Mn reduce common scab of potato (Keinath and Loria 1996), *Fusarium* spp. infections in cotton, and *Sclerotinia sclerotiorum* in squash (Graham and Webb 1991; Agrios 2005).

20.1.11 Magnesium (Mg)

Magnesium is a constituent of chlorophyll. It is a cofactor of almost all enzymes which act on phosphorylated substrates and therefore plays role in energy metabolism. The deficiency of Mg thus affects the metabolism of the plants adversely. The smut disease in wheat increases with application of Mg.

20.1.12 Chlorine (Cl)

Chlorine in the form of electrically charged ion (Cl^-) is needed for oxygen evolution in photosystem II during photosynthesis. The plants tend to accumulate free amino acids under deficient conditions of Cl. It had synergistic effect with K and combine to form KCl which is quite effective in reducing the severity of plant diseases and drought stress. It is known to inhibit nitrification and regulate osmosis. The Cl has positive effect on the control of diseases of wheat like root rot, take-all, powdery mildew, glume blotch, tan spots, stripe rust, and leaf rust. It also helps reduce the severity of diseases like stalk rot (*Gibberella zeae*) and leaf blight (*Helminthosporium* sp.) in corn, sudden death syndrome in soya bean (*F. solani*), and physiological disorders of potato like hollow heart and brown centre in potato (Dordas 2008). Thomason et al. (2001) reported inconsistent results on the severity of take-all diseases with Cl application, even following the application of lime where increased soil pH can increase disease severity. Cl is reported to help in the control of a number of diseases such as stalk rot of corn, Northern corn leaf blight, and downy mildew of millet (Graham and Webb 1991; Mann et al. 2004).

20.1.13 Silicon

It is quite abundant in soil (about 28 %). It creates a physical barrier in the cuticle of plant leaves and sheaths and helpful in the control of blast and brown spots in rice. It also reduces the severity of powdery mildew of barley, grapes, and cucumbers and bean rust. Silicon is helpful in managing the diseases of wheat like foot rot, spot blotch, powdery mildew, and eye spots. Increased Si in plants has been shown to increase the difficulty of sucking insects, like aphids or leaf hoppers thus preventing the spread of viruses. Higher amounts of accumulation of Si around the site of infection in plants is related to the tolerance of these against diseases. Si application reduces the severity of diseases like brown spot (*Cochliobolus miyabeanus*) and sheath blight (*Thanatephorus cucumeris*) in rice, blast (*Magnaporthe oryzae*) in St. Augustine grass, and *R. solani*, *Pythium* spp., *M. oryzae*, and *Blumeria graminis* in various turf grasses (Carver et al. 1998; Savant et al. 1997; Alvarez and Datnoff 2001; Seebold et al. 2000, 2004; Zhang et al. 2006).

20.2 Effect of Balanced Fertilizers on Wheat Diseases

The metabolism of the crop plants is affected by the amount, proportion, time, and mode of application of fertilizers which in turn influence the occurrence and severity of diseases. The nutrients supplied to the wheat crop affect the severity of diseases in many ways and thus are important components of integrated disease management. Both P and K significantly reduce the severity of leaf rust and powdery mildew. Among mineral nutrients, K plays a particular role in the tolerance of crop plants to stress conditions. It is essential in photosynthesis, translocation of photosynthates into sink organs, maintenance of turgescence, activation of enzymes, and reduction of excess uptake of ions (Na and Fe) in saline and flooded soils. Grey leaf spot incited by *Cercospora zeae-maydis* is a new devastating foliar disease of maize in East Africa. The experiments conducted by Patrick et al. (2004) showed the highest disease development occurring in nitrogen-augmented plots. Exclusive phosphorus application had no clear effect on grey leaf spot epidemics, but combined application with nitrogen significantly reduced the predisposition effects of nitrogen to the disease.

The diseases are affected by the time and amount of N fertilizer application in wheat. The severity of stripe rust, powdery mildew, and *Septoria* leaf spot in wheat increased with increase in the amount of N particularly in cases of early application of N. Therefore, early N demands for an early disease control also. The nitrogen status of a wheat crop affects the level of stripe rust (*P. striiformis*). It may be due to increased biomass production which creates humidity within the canopy that is more conducive to the development of disease. The adult plant resistance (APR) seems to be delayed in crops with high nitrogen status, whereas in deficient crop, it triggers early (Daniel and Parlevliet 1995). The genotype × N × year effect was also significant. In the case of leaf rust, also N application increased the severity (up to 180 kg/ha), whereas K application up to 80 kg/ha reduced the rust incidence (Sinha 1998). Higher incidence of powdery mildew (*Erysiphe graminis*) in wheat was also reported by Keblikas (1997) and Pokacka and Blonska (1973) with application of higher N fertilizers (120 kg/ha). Sisterna and Sarandon (1996) however observed no significant difference as far as the numbers of infected grains are concerned due to *Bipolaris sorokiniana* in wheat supplied with 0 and 90 kg N/ha in Argentina. For foliar pathogens, the severity of mildew and *S. tritici* was increased by early nitrogen (Jordan et al. 1988). Heier et al. (2005) recorded increased level of mycotoxins such as deoxynivalenol (DON) and zearalenone (ZEA) of *Fusarium* head blight under immoderate N fertilizers in Germany.

The studies were also conducted to assess the effect of micronutrients on wheat diseases. The spraying of spring wheat (*Triticum* spp. var. Sarotovskaya 36) with nickel sulphate (1000 l/ha) at tillering, shooting, earing, and after flowering decreased leaf rust infection by 3.3 times and with nickel sulphate + manganese, zinc, or boric acid by 2.5 times. A single spray of nickel sulphate decreased powdery mildew severity by 4.3 times and one of the mixtures by 1.8–2.3 times. Microelements increased yield and improved the quality of grain in wheat (Romanova and Gibadullin 1980). Fernandez et al. (1998) reported that reduced P application helped in reducing the leaf spot severity. Odorfer and Pommer (1997) observed increased incidence of powdery mildew with increase in the N supply of the different systems in Germany. Derecha et al. (1980) reported that the incidence of root rot in sandy loam soil was at a minimum (19.7–22.7 %) with application of N60-90, P60-90, and K60 kg/h. The combined application of N60P60K60 gave high disease control than with N60K60 in the case of *Septoria* spp. in wheat. The effect of foliar fertilizers such as Ekolist PK 1, Ekolist Mg, Mikrosol Zn, and urea was seen on *E. graminis*, *Phaeosphaeria nodorum*, *P. tritici-repentis*, and *P. recondita* in wheat in Poland. The fertilizers improved the green leaf area (Maczynska et al. 2005). Sharma et al. (2006) reported that N alone reduced the severity of spot blotch (*B. sorokiniana*) by 8 %, whereas P had no effect in Nepal. The NPK-balanced dose reduced disease by 15–22 %. The KCl and $CaCl_2$ reduced the spot blotch. Goos et al. (1994) reported increased incidence of root rot caused by *C. sativus* with increased P fertilizers. The effect of S on powdery mildew was observed in the UK by Hussain and Leitch (2005) and application of S delayed senescence of upper leaves and ears. Thongbai et al. (1993) observed the reduced incidence of *Rhizoctonia* root rot in wheat by application of Zn than in check. Prashar et al. (1995) observed reduced incidence of flag smut (*Urocystis agropyri*) and increase in contents of phenols and sugars with application of K fertilizer. High levels of P (90 kg/ha) increased the incidence of smut, whereas optimum dose of 60 kg/ha of P was beneficial in controlling the disease. The recommended doses of P + K reduced the incidence but higher doses of N + P along with K increased the disease in pot conditions.

In most of the cases, application of fertilizers is helpful in decreasing the incidence of disease in crop plants. It may be due the involvement of these elements in the resistance mechanisms of the host plant. The nutrients in excess may result in the toxicity in plants, and therefore, it might have caused higher disease incidence. A balanced nutrition is important in integrative crop protection programme and helps in containing the diseases in a cost-effective and environmentally friendly manner and therefore may be a component of integrated disease management approach. It can be achieved through multidisciplinary research involving plant pathologists, soil scientists, and agronomists.

References

Agrios NG (2005) Plant pathology, 5th edn. Elsevier, Amsterdam, p 635

Ahmad M, Ali Z, Safdar KMA, Sahi ST (2012) Role of nutrients and naphthalene acetic acid in the management of urdbean leaf crinkle virus. Pak J Phytopathol 24:79–81

Alvarez J, Datnoff LE (2001) The economic potential of silicon for integrated management and sustainable rice production. Crop Prot 20:43–48

Atkinson D, McKinlay RG (1997) Crop protection and its integration within sustainable farming systems. Agric Ecosyst Environ 64:87–93

Bailey KL, Lafond GP, Domitruk D (1998) Effects of row spacing, seeding rate and seed-placed phosphorus on root diseases of spring wheat and barley under zero tillage. Can J Plant Sci 78:145–150

Bateman DF, Basham HG (1976) Degradation of plant cell walls and membranes by microbial enzymes. In: Heitefuss R, Williams PH (eds) Encyclopedia of plant physiology, physiological plant pathology, vol 4. Springer, New York, pp 316–355

Blachinski D, Shtienberg D, Dinoor A, Kafkafi U, Sujkowski LS, Zitter TA, Fry WE (1996) Influence of foliar application of nitrogen and potassium on Alternaria diseases in potato, tomato and cotton. Phytoparasitica 24:281–292

Brown PH, Bellaloui N, Wimmer MA, Bassil ES, Ruiz J, Hu H, Pfeffe H, Dannel F, Romheld V (2002) Boron in plant biology. Plant Biol 4:205–223

Bryant JP, Chapin FS, Klein DR (1983) Carbon/nutrient balance of boreal plants in relation to vertebrate herbivory. Oikos 40:357–368

Büschbell T, Hoffmann GM (1992) The effects of different nitrogen regimes on the epidemiologic development of pathogens on winter-wheat and their control. J Plant Dis Prot 99:381–403

Carver TLW, Thomas BJ, Robbins MP, Zeyen RJ (1998) Phenylalanine ammonia-lyase inhibition, autofluorescence and localized accumulation of silicon, calcium and manganese in oat epidermis attacked by the powdery mildew fungus *Blumeria graminis* (DC) Speer. Physiol Mol Plant Pathol 52:223–243

Celar F (2003) Competition for ammonium and nitrate forms of nitrogen between some phytopathogenic and antagonistic soil fungi. Biol Control 28:19–24

Chase AR (1989) Effect of nitrogen and potassium fertilizer rates on severity of Xanthomonas blight of *Syngonium podophyllum*. Plant Dis 73:972–975

Csinos AS, Bell DK (1989) Pathology and nutrition in the peanut pod rot complex. In: Engelhard AW (ed) Soilborne plant pathogens: management of diseases with macro- and microelements. APS Press, St. Paul, pp 124–136

Daniel DL, Parlevliet JE (1995) Effects of nitrogen fertilization on disease severity and infection type of yellow rust on wheat genotypes varying in quantitative resistance. J Phytopathol 143:679–681

Datnoff LE, Raid RN, Snyder G, Jones DB (1991) Effect of calcium silicate on blast and brown spot intensities and yields of rice. Plant Dis 75:729–732

Datnoff LE, Elmer WH, Huber DM (2007) Mineral nutrition and plant disease. American Phytopathological Society, APS Press, St. Paul, 278p

Derecha AA, Khanin MD, Mikhailova RG (1980) Effect of increased doses of mineral fertilizers on wheat infection in the Poles'e region of the Ukraine. Khimiya-v-Sel'skom-Khozyaistve 7:21–23

Dordas C (2008) Role of nutrients in controlling plant diseases in sustainable agriculture: a review. Agron Sustain Dev 28:33–46

Dordas C, Brown PH (2005) Boron deficiency affects cell viability, phenolic leakage and oxidative burst in rose cell cultures. Plant Soil 268:293–301

Fernandez MR, Zentner RP, McConkey BG, Campbell CA (1998) Effect of crop rotations and fertilizer management on leaf spotting diseases of spring wheat in southwestern Saskatchewan. Can J Plant Sci 78:489–496

Garas NA, Kuc J (1981) Potato lectin lyses zoospores of Phytophthora infestans and precipitates elicitors of terpenoid accumulation produced by the fungus. Physiol Plant Pathol 18:227–238

Goos RJ, Johnson BE, Stack RW (1994) *Penicillium billaie* and P fertilization effect on the growth, development, yield and common root rot severity of spring wheat. Fertil Res 39:97–103

Graham DR (1983) Effects of nutrients stress on susceptibility of plants to disease with particular reference to the trace elements. Adv Bot Res 10:221–276

Graham DR, Webb MJ (1991) Micronutrients and disease resistance and tolerance in plants. In: Mortvedt JJ, Cox FR, Shuman LM, Welch RM (eds) Micronutrients in agriculture, 2nd edn. Soil Science Society of America Inc., Madison, pp 329–370

Grewal HS, Graham RD, Rangel Z (1996) Genotypic variation in zinc efficiency and resistance to crown rot disease (*Fusarium graminearum* Schwa. Group 1) in wheat. Plant Soil 186:219–226

Hammerschmidt R, Nicholson RL (1999) A survey of plant defense responses to pathogens. In: Agrawal AA, Tauzin S, Bent E (eds) Induced plant defenses against pathogens and herbivores. APS Press, St. Paul, pp 55–71.

Harrison UJ, Shaw HD (2001) Effects of soil pH and nitrogen fertility on the population dynamics of *Thielaviopsis basicola*. Plant Soil 228:147–155

Heier T, Jain SK, Kogel KH, Pons Kuhnemann J (2005) Influence of N fertilization and fungicide strategies on Fusarium head blight severity and mycotoxin content in winter wheat. J Phytopathol 153:551–557

Herms DA, Mattson WJ (1992) The dilemma of plants – to grow or defend. Q Rev Biol 67:283–335

Hoffland E, van Beusichem ML, Jegger MJ (1999) Nitrogen availability and susceptibility of tomato leaves to *Botrytis cinerea*. Plant Soil 210:263–272

Hoffland E, Jegger MJ, van Beusichem ML (2000) Effect of nitrogen supply rate on disease resistance in tomato depends on the pathogen. Plant Soil 218:239–247

Howard DD, Chambers AY, Logan J (1994) Nitrogen and fungicide effects on yield components and disease severity in wheat. J Prod Agric 7:448–454

Huber DM (1980) The role of mineral nutrition in defense. In: Horsfall JG, Cowling EB (eds) Plant disease, an advanced treatise, volume 5, how plants defend themselves. Academic, New York, pp 381–406

Huber MD (1996) The role of nutrition in the take-all disease of wheat and other small grains. In: Engelhard WA (ed) Management of diseases with macro- and microelements. APS Press, Minneapolis, pp 46–74

Huber DM, Arny DC (1985) Interactions of potassium with plant disease. In: Munson RD, Bishop WD (eds) Potassium in agriculture: proceedings of an international symposium, American Society of Agronomy, Madison, pp 467–488

Huber DM, Graham RD (1999) The role of nutrition in crop resistance and tolerance to disease. In: Rengel Z (ed) Mineral nutrition of crops fundamental mechanisms and implications. Food Product Press, New York, pp 205–226

Huber DM, Haneklaus S (2007) Managing nutrition to control plant disease. Landbauforschung Völkenrode 4(57):313–322

Huber DM, Thompson IA (2007) Nitrogen and plant disease. In: Datnoff LE, Elmer WH, Huber DM (eds) Mineral nutrition and plant disease. APS Press, St. Paul, pp 31–44

Hussain Z, Leitch MH (2005) The effect of applied sulphur on growth, grain yield and control of powdery mildew in spring wheat. Ann Appl Biol 147:49–56

Jordan VWL, Hunter T, Fielding EC (1988) Interactions of fungicides with nitrogen and plant growth regulators for cost-effective control of wheat diseases. Brighton crop protection conference. Pests Dis 3:873–880

Katan J (2009) Mineral nutrient management and plant disease. e-IFC 21(9):6–8. Source: http://www.ipipotash.org/udocs/eifc_no21-rf2.pdf

Keblikas D (1997) Incidence of powdery mildew of different spring wheat varieties at various fertilization with nitrogen fertilizers. In: Integrated plant protection: achievements and problems. Proceedings of scientific conference devoted to the 70th anniversary of plant protection science in Lithuania, 7–9 September 1997, pp 76–79

Keinath PA, Loria R (1996) Management of common scab of potato with plant nutrients. In: Engelhard WA (ed) Management of diseases with macro- and microelements. APS Press, Minneapolis, pp 152–166

Kirkegaard JA, Munns R, James RA, Neate SM (1999) Does water and phosphorus uptake limit leaf growth of Rhizoctonia-infected wheat seedlings? Plant Soil 209:157–166

Kolattukudy EP, Kämper J, Kämper U, González-Candelas L, Guo W (1994) Fungus-induced degradation and reinforcement of defensive barriers of plants. In: Petrini O, Guellete GB (eds) Host wall alterations by parasitic fungi. APS Press, Minneapolis, pp 67–79

Krauss A (1999) Balanced nutrition and biotic stress. In: IFA agricultural conference on managing plant nutrition, 29 June–2 July 1999, Barcelona, Spain

Maczynska A, Glazek M, Krezyzinska B (2005) Effect of foliar fertilizers and fungicidal protection against leaf spot diseases on winter wheat. Acta Agrobot 58:51–58

Mann RL, Kettlewell PS, Jenkinson P (2004) Effect of foliar applied potassium chloride on septoria leaf blotch of winter wheat. Plant Pathol 53:653–659

Marschner H (1995) Mineral nutrition of higher plants, 2nd edn. Academic, London. 889 p

Martin KL (2013) Nutrient interactions with plant diseases. Source: www.southwest.ksu.edu/DesktopModules/ViewDocument.aspx?Docum

Menzies J, Bowen P, Ehret D, Glass ADM (1992) Foliar applications of potassium silicate reduce severity of powdery mildew on cucumber, muskmelon and zucchini squash. J Am Soc Hortic Sci 117:902–905

Oborn I, Edwards AC, Witter E, Oenema O, Ivarsson K, Withers PJA, Nilsson SI, Richert Stinzing A (2003) Element balances as a toll for sustainable nutrient management: a critical appraisal of their merits and limitations within an agronomic and environmental context. Eur J Agron 20:211–225

Odorfer A, Pommer G (1997) Correlation between growing intensity and infection with wheat diseases. In: 4th scientific meeting on ecological agriculture, held on 3–4 March 1997, at Rheinische Friedrich-Wilhelms-Universitat Bonn. Schriftenreihe Institut fur Organischen Landbau 4:302–307

Palti J (1981) Cultural practices and plant diseases. Springer, Berlin. 243p

Patrick O, Rubaihayo P, Ekwamu Adipala R, Dixelius C (2004) Interactive effects of host, pathogen and mineral nutrition on grey leaf spot epidemics in Uganda. Eur J Plant Pathol 110:119–128

Pokacka Z, Blonska Pawlak A (1973) The effect of high nitrogen dosage on the field resistance of winter-wheat varieties to powdery mildew (*Erysiphe graminis* D.C.) and brown rust (*Puccinia triticina* Erikss.). Biuletyn Instytutu Ochrony Roslin 56:153–180

Potash and Phosphate Institute (PPI) (1988) Phosphorus nutrition improves plant disease resistance. In: PPI (ed) Better crops with plant food, Fall 1988. Atlanta, pp 22–23

Prabhu AS, Fageria ND, Berni RF, Rodrigues FA (2007a) Phosphorous and plant disease. In: Datnoff LE, Elmer WH, Huber DM (eds) Mineral nutrition and plant disease. APS Press, St. Paul, pp 45–55

Prabhu AS, Fageria ND, Huber DM, Rodrigues FA (2007b) Potassium and plant disease. In: Datnoff LE, Elmer WH, Huber DM (eds) Mineral nutrition and plant disease. APS Press, St. Paul, pp 57–78

Prashar RD, Sindhan GS, Hooda I (1995) Effect of fertilizers in relation to phenol and carbohydrate metabolism and susceptibility of wheat plants to flag smut disease. Plant Dis Res 10:43–46

Reuveni R, Reuveni M (1998) Foliar-fertilizer therapy-a concept in integrated pest management. Crop Prot 17:111–118

Reuveni M, Oppernheim D, Reuveni R (1998) Integrated control of powdery mildew on apple trees by foliar sprays of mono-potassium phosphate fertilizer and sterol inhibiting fungicides. Crop Prot 17:563–568

Reuveni R, Dor G, Raviv M, Reuveni M, Tuzun S (2000) Systemic resistance against *Sphaerotheca fuliginea* in cucumber plants exposed to phosphate in hydroponics system and its control by foliar spray of mono-potassium phosphate. Crop Prot 19:355–361

Rioux D, Biggs AR (1994) Cell wall changes and nonhost systems: microscopic aspects. In: Pertini O, Quellette B (eds) Host wall alterations by parasitic fungi. APS Press, Minneapolis, pp 31–44

Romanova AM, Gibadullin MG (1980) Results of treatment with microelements. Zashchita-Rastenii 1:23

Römheld V, Marschner H (1991) Function of micronutrients in plants. In: Mortvedt JJ, Cox FR, Shuman LM, Welch RM (eds) Micronutrients in agriculture. Soil Science Society of America, Inc. Madison, pp 297–328

Rovira AD, Graham RD, Ascher JS (1983) Reduction in infection of wheat roots by *Gaeumannomyces graminis* var. *tritici* with application of manganese to soil. In: Parker CA et al (eds) Ecology and management of soil borne pathogens. APS, Minneapolis

Rush CM, Piccinni G, Harveson RM (1997) Agronomic measures. In: Rechcigel NA, Rechcigel JE (eds) Environmentally safe approaches to crop disease control. CRC Publications, Boca Raton

Savant NK, Snyder GH, Datnoff LE (1997) Silicon management and sustainable rice production. Adv Agron 58:151–199

Seebold KW, Datnoff LE, Correa-Victoria FJ, Kucharek TA, Snyder GH (2000) Effect of silicon rate and host resistance on blast, scald and yield of upland rice. Plant Dis 84:871–876

Seebold KW, Datnoff LE, Correa-Victoria FJ, Kucharek TA, Snyder GH (2004) Effect of silicon and fungicides on the control of leaf and neck blast in upland rice. Plant Dis 88:253–258

Sharma RC, Duveiller E (2004) Effect of helminthosporium leaf blight on performance of timely and late-seeded wheat under optimal and stressed levels of soil fertility and moisture. Field Crop Res 89:205–218

Sharma S, Duveiller E, Basnet R, Karki CB, Sharma RC (2005) Effect of potash fertilization on helminthosporium leaf blight severity in wheat, and associated increases in grain yield and kernel weight. Field Crop Res 93:142–150

Sharma P, Duveiller E, Sharma RC (2006) Effect of mineral nutrients on spot blotch severity in wheat and associated increases in grain yield. Field Crop Res 95:426–430

Sinha AP (1998) Effect of potash and nitrogenous fertilizers on brown rust of wheat. Indian Phytopathol 51:100–101

Sisterna MN, Sarandon SJ (1996) Black point of wheat (*Bipolaris sorokiniana* Sacc. Shoem.) influenced by N fertilization under no till and conventional tillage. Cereal Res Commun 24:217–221

Spann, TM, Schumann AW (2013) Mineral nutrition contributes to plant disease and pest resistance. Source: http://edis.ifas.ufl.edu/pdffiles/HS/HS118100.pdf. pp 1–4

Spectrum Analytic, Inc. (2013) The relationship between nutrients and other elements to plant disease. Source: http://www.spectrumanalytic.com/support/library/pdf/relationship_between_nutrients_and_other_elements_to_plant_diseases.pdf. pp 1–24

Thomason WE, Wynn KJ, Freeman KW, Lukina EV, Mullen RW, Johnson GV, Westerman RL, Raun WR (2001) Effect of chloride fertilizers and lime on wheat grain yield and take-all disease. J Plant Nutr 24:683–692

Thongbai P, Graham RD, Neate SM, Webb MJ (1993) Interaction between Zn nutritional status of cereals and Rhizoctonia root rot severity II. Effect of Zn on disease severity of wheat under controlled conditions. Plant Soil 153:207–214

Vidhyasekaran P (1997) Fungal pathogenesis in plants and crops. In: Molecular biology and host defense mechanisms. Marcel Dekker, New York, p 568

Vidhyasekaran P (2004) Concise encyclopedia of plant pathology. Food Products Press, New York, p 619

Walters DR, Bingham IJ (2007) Influence of nutrition on disease development caused by fungal pathogens: implications for plant disease control. Ann Appl Biol 151:307–324

Wilkens RT, Spoerke JM, Stamp NE (1996) Differential response of growth and two soluble phenolics of tomato to resource availability. Ecology 77:247–258

Zhang Q, Fry J, Lowe K, Tisserat N (2006) Evaluation of calcium silicate for brown patch and dollar spot suppression on turf grasses. Crop Sci 46:1635–1643

Nutritional Disorders of Citrus and Their Management

21

J.N. Srivastava, P.K. Sharma, Upma Dutta, A.K. Srivastava, and Rakesh Kumar

Fruit trees and crop plants suffer from nutritional disorders. Nutritional disorders are a result of inadequate supply or excess of desired minerals/fertilizer. The term macro- and micronutrients are used to denote collectively group of mineral/nutritional chemical elements, which are indispensable for optimal growth and which plants absorb primarily through roots. Nitrogen, phosphorous and potassium requiring larger quantity are known as macronutrients, while calcium, magnesium, iron, manganese, zinc, boron, copper, molybdenum, sulphur, etc., are known as micronutrients and require in smaller amounts. Deficiency of minerals, viz. macronutrients and micronutrients, results in disorders in plant metabolism, and plants express hunger signs, viz. chlorosis, leaf spot, leaf blotch, leaf blight, dieback, reduced growth of tree, poor fruit quality and decreased number of fruits in citrus tree. Excess of mineral disturbs nutritional balance which is most necessary for the proper metabolism in citrus tree. If the supply of minerals is high, trees show toxicity symptoms. Deficiencies and excesses of these minerals also reduce resistance of plants to fungal, bacterial and other diseases (Singh 1983).

Apart from visual diagnosis, the analysis of plant and leaf samples is also helpful in identification of nutritional disorders in plant which can be supplemented through soil analysis. Twelve such minerals and disorders caused by them in citrus are discussed as under.

21.1 Nitrogen (N)

Being an important constituent of protein amino acid, enzymes, hormones, vitamins and chlorophyll, N is more important than other minerals. It is an integral part of plant tissues (0.2–4.1 % N on dry matter basis) which is required for proper photosynthesis by green tissue. Nitrogen greatly influences important tree functions such as growth, leaf protection, flower initiation, fruits setting and fruit development and quality (Mooney et al. 1991).

The main cause of N deficiency is lack of available N in soil, which can be due to many factors. Nitrogen leaching is caused by the combination of heavy summer rain fall or over-irrigation in highly porous soil. Water logging in soil can cause N loss through denitrification that may lead

J.N. Srivastava (✉) • U. Dutta • R. Kumar
Department of Plant Pathology, Bihar Agricultural University, Sabour, Bhagalpur, Bihar, India
e-mail: j.n.srivastava1971@gmail.com

P.K. Sharma
National Bureau of Agriculturally Important Microorganisms, Mau Nath Bhanjan, U. P., India

A.K. Srivastava
Programme Coordinator R.D.S.KVK, Pratapgarh District, Uttar Pradesh, India

L.P. Awasthi (ed.), *Recent Advances in the Diagnosis and Management of Plant Diseases*, DOI 10.1007/978-81-322-2571-3_21

to a temporary N deficiency that can be relieved by dry weather (Zekri 1995a).

21.1.1 Deficiency Symptoms

In citrus tree N deficiency symptoms include yellowing of the foliage that begins with older leaves and then appears on younger leaf flush. Leaves become progressively more yellow, with no distinct pattern, but sometimes mature green leaves slowly bleach to a mottled irregular green and yellow pattern, become entirely yellow, and are shed (Zekri and Obreza 2003). N deficiency often occurs in winter or early spring because of low tree N reverse, low soil temperature and/or lack of root activity. The N-deficient trees are stunted with thin canopy and no fruit load and can be highly erratic in bearing habit. They bloom sparsely and flushes emerge irregularly and produce limited twig and leaf growth (Zekri 1995a, b).

The colour of citrus fruit peel tends to be pale and smooth, and the juice has lower soluble solids and acid concentration. If N is deficient during summer and fall seasons, when the fruit is expending and maturing, some of the green leaves will turn yellow and may shed. Trees that are constantly short of N are stunted with irregular and very short twig growth, twig dieback can occur, and crop production is greatly reduced (Zekri and Obreza 2004).

21.1.2 Management

Citrus trees deficient in N can be improved by applying supplementary N fertilizer in frequent application according to the age of plants. The use of low biuret urea as foliar spray is a very effective and rapid way to correct N deficiency (Zekri 2003).

21.2 Phosphorus (P)

Phosphorus is important for normal plant growth and reproduction. It is a constituent of phospholipids, nucleic acid and many proteins. Phosphorus is involved in carbohydrate and fat transformation and protein metabolism and also respiration. It plays important role in high energy bonding (i.e. ATP ⇔ ADP) and in respiration. It is essential for timely differentiation and maturation of plant tissues (Mehrotra 1999).

The cause of phosphorus deficiency is the poor availability of P in soil. Phosphorus deficiency may occur in area of high rainfall due to leaching and erosion. In strongly acid soils, P becomes quickly unavailable. Phosphorus availability is also reduced in calcareous soil.

21.2.1 Deficiency Symptoms

Phosphorus deficiency symptoms appear first on older leaves and then on younger tissues, which lose their deep green colour (Zekri 1995a). Citrus trees deficient in P have reduced growth and leaves are small and narrow with purplish or bronze discolouration. This type of chlorosis spread inwards from the midrib, some time leaving areas of healthy green tissues. Necrosis of tissue leads to withering of leaves and breaking petioles at the pseudostem. Some leaves may later develop necrotic areas and young leaves will show reduced growth rate. Leaves shed prematurely and fruit drop before normal harvesting time (Zekri and Obreza 2003). Citrus trees show limited flower development with reduced fruit set and yield. The fruit will be coarse and rough in texture and have thick rind and a hollow core. The fruit will also have a high acidity in proportion to total soluble solids. In such trees, fruit maturity is delayed. Usually roots are stunted and poorly branched (Zekri and Obreza 2004).

21.2.2 Management

Phosphorus deficiency can be improved by applying water soluble P fertilizer to soil after confirmation of P deficiency by leaf and soil analysis (Zekri and Obreza 2003). Foliar concentration ranges between 0.10 and 0.60 % because healthy citrus leaves are capable of tolerating wide variation in P content (Zekri 1999).

21.3 Potassium (K)

Potassium is important for normal growth and development of plant; in addition it also influences the ghost reaction (tolerance) to various pests/pathogens. Potassium is needed by plant in large amount. Potassium plays important role in carbohydrate and protein synthesis, osmoregulation and also stomatal movement. It is essential as a catalyst of many reactions. But it is not effective without co-nutrients such as N and P (Mehrotra 1999).

Potassium deficiency symptoms usually result from an insufficient K supply in the soil. Lack of soil moisture also reduces K uptake and may lead to K deficiency. If the supply of N and P is high relative to that of K, growth may be rapid at first, but the K concentration in the plant may be ultimately decreased to K deficiency (Zekri 1995a). K deficiency may occur in sandy acid soils where leaching takes place. The supply of K to plants may be decreased in soil that has very high concentration of Ca and Mg or by heavy application of N. Decreased K uptake occurs typically in some calcareous soils (Zekri and Obreza 2003).

21.3.1 Deficiency Symptoms

Deficiency of K in citrus tree causes a general reduction in growth and dropping of leaves without visual deficiency symptoms. The K deficiency results in premature yellow-brown discolouration of leaves from the tips and margins, which then gets border. Necrotic areas and spotting can develop on leaves. Purplish-brown patches may appear at the base of petiole, and in severe case water-soaked areas may be seen (Zekri 1996).

Symptoms appear first on older leaves because K tends to concentrate in the rapidly growing tissues. Potassium deficiency causes compact tree appearance, slow growth, small leaves, smaller fruit size with inferior flavour, very thin peel and smooth texture, premature shedding of fruit, lower acid concentration in the fruit and an increase in susceptibility to drought and cold (Camp and Fudge 1939).

21.3.2 Management

Potassium deficiency can be improved by applying K fertilizer, viz. potassium chloride or potassium sulphate, to the soil. Foliar application of potassium nitrate or monopotassium phosphate can be very effective and is a rapid control measure to improve K deficiency (Zekri and Obreza 2003).

21.4 Calcium (Ca)

Calcium is involved in cell wall (calcium pectate) formation which gives turgidity of cell and indirectly in cell division, highly required in telophase for cell plate formation. It is essential to activate the growing point especially root tip. The activity of several enzymes is influenced by this mineral (Mehrotra 1999). In acid soil of hill regions calcium becomes unavailable and its deficiency takes place (Saha 2002).

21.4.1 Deficiency Symptoms

In affected citrus trees, young leaves become distorted; tips of leaves hook back and their margins become curled. Leaves may be irregular in shape, sometimes brown scorching, or spots may appear on leaves. Terminal buds are also affected or die (Zekri 1995b). Trees are extremely stunted having inadequate root system (Bhargava et al. 2000).

21.4.2 Management

Quicklime (bujha chuna) is major source for calcium. The quicklime should be powdered and mixed with soil around the trunk in a radius of 1.5–2 m. Apply lime at 4–6 Kg/tree/year to manage the deficiency of this mineral (Saha 2002).

21.5 Magnesium (Mg)

Magnesium is the constituent of chlorophyll and plays an important role in the structural material of certain enzymes involved in carbohydrates synthesis (Mehrotra 1999). Magnesium deficiency is the result of imbalanced availability of calcium or imbalanced use of potassium fertilizer or calcium-containing fertilizers.

21.5.1 Deficiency Symptoms

The symptoms occur first on mature/older leaves, with younger leaves becoming mottled or chlorotic; then reddish sometimes necrotic spots appear. Leaves yellowing are on both sides of the midrib and extend from the base to the apex of the leaf (Saha 2002). The green portion tapers towards the tip of the leaf so that inverted "V" shape is formed. Affected leaves fall prematurely in late summer and autumn (Camp and Fudge 1939).

21.5.2 Management

- Drench tree basin with magnesium sulphate at 100 g + quicklime at 200 g/100 L water (Saha 2002).
- Spraying magnesium nitrate at 1 g/L water also controls the disorder (Glendining 1999).

21.6 Iron (Fe)

Iron is a constituent of certain enzymes and proteins. It is essential for synthesis of chlorophyll and seems to play a catalytic role. It is an electron carrier in oxidation-reduction during respiration of plants (Mehrotra 1999). Iron deficiency often appears in winter due to low soil temperature and root inactivity. High soil pH can cause iron deficiency, especially in trees on trifoliate hybrid rootstocks or trifoliate root stocks. Iron deficiency can also occur in poorly drained soil and also in alkaline soil (Brown 1956).

21.6.1 Deficiency Symptoms

Young leaves are worst affected and show the symptoms, while older leaves may remain green. Leaves of deficient plants become chlorotic and cream coloured nearly too white with the main veins remaining green. In severe cases, even the veins may turn yellow (Zekri 1995c). The affected plants show poor growth and have small fruit. Dieback of braches takes place. In a tree single branch or in orchard, a single or few trees may be affected (Wallace and Lunt 1960).

21.6.2 Management

- Spraying with the mixture of ferrous sulphate 400 g and lime 400 g in 100 L of water gives the best remedy of deficiency of iron (Saha 2002).
- Sparing ferrous sulphate at 0.5–0.9 % alleviates the deficiency symptoms of iron (Alvs and Tucker 1992).
- Apply ferrous sulphate to the acidic soil at 20 g/tree and alkaline soil at 50 g/tree in a year (Tisdale and Nelsomn 1975).

21.7 Zinc (Zn)

Zinc is a component of many enzymes involved in auxin and carbohydrates synthesis. It plays an important role in chlorophyll formation and photosynthesis activity (Mehrotra 1999).

Zinc deficiency in citrus is described as "little leaf", "mottle leaf" and "resetting leaf" because of the distinctive leaf pattern produced on most citrus species. In case of severe deficiency, it is also called "mottling". Excessive phosphate or nitrogen has been shown to induce zinc deficiency (Saha 2002). It is most acute in alkaline or acid coastal soils (Buckman and Brady 1969).

21.7.1 Deficiency Symptoms

Symptoms of zinc deficiency are more noticeable on the north side of tree. Leaves show interveinal chlorosis. Leaf symptoms include small, narrow leaves (little leaf) and whitish-yellow areas between the veins (mottle leaf). Leaves are few and small, internodes are short, and shoots form rosettes. The leaves especially the terminal growth develop "mottle leaf" symptoms, and the growth becomes yellowish and bright creamy and unthrifty (Saha 2002).

"Mottling" becomes more pronounced with severe deficiency. Leaves are reduced in size and have pale colour. Necrosis may occur beginning at leaf tip and margins, and terminal growth is affected. Affected twigs are erect and bushy. Chlorotic leaves drop off early leading to dieback of twigs. Trees remain stunted. Zinc deficiency is most severe in spring growth in citrus (Zekri 1995c).

Reduced vigour, low fruit production, smaller fruit size and poor fruit quality result from deficiency of zinc in the citrus tree. In lemon, the yield may be reduced even through without evident symptoms. In case of orange, mild deficiency may have little effect on yield.

21.7.2 Management

- Zinc deficiency can be corrected by foliar spray of 0.5–1.2 % zinc sulphate twice at weekly intervals on tender foliage followed by soil applications of zinc sulphate at 100–500 g/tree, according age of tree (Zekri 2003).
- Spraying by of ziram or Dithane Z-78 at 2 g/L water can also correct the deficiency of zinc (Saha 2002).

21.8 Manganese (Mn)

Manganese is a cofactor of enzymes of cellular respiration, nitrogen metabolism and photosynthesis (Mehrotra 1999). Manganese deficiency is particularly evident in the spring after a cold winter, because it frequently occurs in combination with deficiencies of zinc or iron or both, and its symptoms may be over shadowed.

21.8.1 Symptoms

Symptoms of manganese deficiency are usually more noticeable on the south side of the tree and are more pronounced in the spring growth flush. Symptoms are evident on both young and mature leaves (Sprague 1964).

Leaves exhibit interveinal yellowing with a darker green band along the midrib and veins and interveinal chlorosis on the new foliage. Necrotic spot may appear scattered on leaf, and severely affected leaves turn brown. Leaf size is not reduced due to the Mn deficiency. In case of mild deficiency in citrus tree, a slight reduction of vigour and yield occurs. While severe manganese deficiency can induce defoliation, loss of vigour and lower fruit yield. In extreme cases, symptoms are accompanied by premature leaf drop (Zekri 1995c).

21.8.2 Management

Manganese deficiency can be corrected by the spraying 0.2–0.5 % manganese sulphate twice at weekly interval or manganese sulphate at 5–10 kg/ha (Zekri 2002a, b).

21.9 Boron (B)

Boron plays an important role in many activities of plants, viz. cell division, protein synthesis, pollination of flower, flower formation, fruit setting and seed production. Boron is essential for plant growth too (Mehrotra 1999). Boron deficiency usually occurs in calcareous soil or irrigation with alkaline water. Hard fruit disease of citrus is caused by boron deficiency (Zekri 2004).

21.9.1 Deficiency Symptoms

Symptoms of boron deficiency are seen on young leaves and fruit. The first visual symptoms of B deficiency are generally the death of the terminal growing points of the main stem (Smith and Reuther 1949).

In leaf, characteristic symptom is the discolouration and downward curling of leaves. The margins of leaves turn brown and the veins turn yellow in colour. The veins on the upper leaf surface are enlarged, become thickened and corky and split (Zekri 1995d).

Fruits are brownish in colour and mis-shaped and become hard and dry due to lumps in the rind caused by gum impregnations. So boron deficiency is also known as "hard fruit". Cracking of fruits can be also seen (Haas 1945; Zekri and Obreza 2003).

Boron deficiency causes growing point to die, growth is reduced, and twigs may split. Stems are brittle, internodes are shortened, and roots are thick and stunted (Zekri and obreza 2003).

21.9.2 Management

- Boron deficiency can be corrected by soil application of borax at 250 g per tree or 10 kg/ha in a year.
- Spraying 0.2–0.5 % solution of borax or boric acid on the foliage once or twice a year after the formation of new flushes is helpful in alleviating boron deficiency (Saha 2002).

21.10 Copper (Cu)

Copper is a cofactor of several oxidation enzymes. It acts as catalyst in certain reaction of respiration (Mehrotra 1999).

Copper deficiency is also known as "dieback", "ammonization" and "exanthema". These names are derived from the dying back of twigs, frequent association with heavy application of N (ammonia), and gum exudes/excrescences on the surface of twigs and fruit (Floyd 1977; Zekri and Obreza 2003).

21.10.1 Deficiency Symptoms

Copper deficiency is characterized by dark green young leaves, development of multiple buds, general stunting and bushy appearance. The leaves become mottled, narrowed and reduced in size (Saha 2002). Thin cut strip appears on bark of stem and gum exudes. Gum exudation can be seen on the rind also (Dickey et al. 1948). In severe cases, shoots become dry leading to dieback of twigs and become distorted in irregular shapes generally "S" shaped. The fruits are small in size and discoloured. Sometimes fruits split and drop before maturity. Fruit shows gum pockets around central pith (Zekri 1995d).

21.10.2 Management

- One spraying of Bordeaux mixture (1 %) is recommended for correcting the deficiency of copper.
- Spray of copper sulphate at 0.5–0.9 is recommended (Saha 2002).
- Copper sulphate is also recommended at 4–6 Kg/tree in a year for copper deficiency.

21.11 Molybdenum (Mo)

Molybdenum is a constituent of enzymes that are responsible for the reduction of nitrates to nitrites. It is essential for protein synthesis. Molybdenum deficiency is most common in acidic soil (Mehrotra 1999).

21.11.1 Deficiency Symptoms

The symptoms appear on mature leaves, while young leaves appear normal. Leaves become pale yellow with marginal chlorosis. The leaf lamina becomes thin and dry. Yellow spots can be seen on leaves; these spots initially appear as water-soaked areas and gradually develop into yellow spots. Gum formation occurs on the underside of the leaf. In severe cases, the necrotic yellow spots enlarge and extend to margins. Affected tree becomes almost defoliated during the winter (Zekri 1995d). Large irregular brown spots surrounded with yellow may develop on the fruit (Sprague 1964).

21.11.2 Management

Foliar spray of sodium molybdate at 0.2 % is useful for correcting molybdenum deficiency in the trees (Zekri 2002a, b).

21.12 Sulphur (S)

The amount of sulphur in plant ranges between 0.1 and 1.0 % on dry weight basis. A portion of total S in plants is mobile and it moves through phloem. Sulphur is a component of amino acids, tripeptides, protein, vitamins (thiamine and biotin) and enzymes. Mostly soil organic matter is held and sulphur deficiency is not common. Sulphur availability is reduced by high soil pH (Knorr 1973; Walter et al. 1958).

21.12.1 Deficiency Symptoms

Symptoms always start with the newest foliage. Visual symptoms are small pale green or light yellow on leaves with lighter veins without any spots. The symptoms resemble those of nitrogen deficiency (Zekri 2002a, b).

21.12.2 Management

Additional dose is applied with recommended dose of organic fertilizer.

21.12.3 General Management of Deficiency Disorders

In view of overlapping symptoms of different nutritional orders, combination of following nutrients can be sprayed on flushes of the trees.

Zinc sulphate	500 g
Ferrous sulphate	200 g
Copper sulphate	500 g
Manganese sulphate	200 g
Borax	100 g
Urea	450 g
Lime	400 g
Water	100 L

21.13 Preparation of Solution with Above Minerals

The salts should be dissolved in small quantity of water separately, the quicklime is mixed with half quantity of water in a separate container, and then the salt solution is added to quicklime solution for better mixing (Saha 2002; Embleton et al. 1967a, b).

21.14 Nutrients/Fertilizer Requirements of Young Citrus Trees

Fertilizer application should be divided in two parts; first part is applied in the beginning of September and second part by the end of February. The approximate quantities of nutrients required by young tree are listed in Table 21.1.

21.15 Foliar Sprays for Treating Micronutrient Deficiencies in Citrus (Table 21.2)

Table 21.1 Estimated nutrient/fertilizer requirement (g/tree) for young citrus trees

Tree age (years)	Fertilizer requirement			
	FYM (kg/tree)	Urea (g/tree)	Diammonium phosphate (g/tree)	Muriate of potash (g/tree)
1	5	95	50	35
2	5	190	105	65
3	10	285	155	100
4	15	380	205	135
5	20	475	260	165
6	25	570	310	200
7	30	715	360	235
8	35	865	415	2265
9	40	1010	465	300
10	45	1175	515	335

Source: Package and practices for horticultural crops SKUAST-Jammu

Table 21.2 Nutritional foliar sprays for correcting deficiencies in young citrus tree

Nutrient	Treatment	Application rate	Comments	Timing
Magnesium	Magnesium	1 kg/100 L		When spring flush leaves are ½ or 2/3 expanded
	Magnesium sulphate + calcium nitrate	1 kg/100 L 1 kg/100 L	Mix magnesium sulphate in half full vat, then add calcium nitrate separately while agitator is running, and then fill vat	
Zinc	Zinc sulphate heptahydrate (23 % Zn)	150 g/100 L	–	As above
	Zinc sulphate (23 % (Zn) + hydrate line	500 g/100 L 250 g/100 L	–	As above
Manganese	Manganese sulphate	100 g/100 L	500 g of urea can be added to improve uptake of manganese	As above
Iron	Ferrous sulphate 0.5–0.9 %	–	–	As above
Zinc and manganese	Zinc sulphate + manganese sulphate	150 g/100 L 100 g/100 L	500–750 g urea can be added to improve uptake	As above
	Zinc sulphate 23 % (Zn) + manganese sulphate + hydrate lime	500 g/100 L 300 g/100 L 250 g/100 L	Zinc and manganese deficiencies often occur together. A combined spray correcting both	
Copper	Copper sulphate	As per label	–	Spring or autumn to suit fungicide programme
	Copper oxychloride	As per label	–	
	Copper hydroxide	As per label	–	

Source: Koo (1983), Embleton et al. (1973), and Weir and Sarooshi (1991)

21.16 Nutrient Recommendation for Citrus Trees

Observation of visual symptoms for nutrient deficiency and their correction is adequate for average production of citrus fruit. For optimum production and quality of citrus fruit, developing a range of tools such as soil analysis and leaf analysis can be used to assess the nutrient requirements of trees. Nutrient analysis is essential to correct potential problem before it becomes a limiting factor in production (Rajput and Haribabu 1985; Zekri 2002a, b).

21.17 Soil Analysis

Analytical procedures used in soil testing vary considerably among labs, as do the results they obtain. None are capable of reporting available nutrient levels in a sample; they can only report the chemically extractable levels. Moreover, no soil testing extraction procedure has yet been calibrated to correlate the extractable value of any nutrient element with citrus production levels of citrus fruit quality (Embleton et al. 1967a,b; Zekri and Obreza 2003).

It is simple to verify pH and available P and certain exchangeable cations notably Ca and mg

Table 21.3 Leaf analysis standards for citrus as percentage of dry matter of leaf

Minerals	Symbol	Unit	Deficient	Low	Optimum	High	Excess
Nitrogen	N	%	<2.20	2.20–2.40	2.50–2.70	2.80–3.00	>3.00
Phosphorus	P	%	<0.09	0.09–0.11	0.12–0.16	0.17–0.29	>0.30
Potassium	K	%	<0.70	0.70–1.10	1.20–1.70	1.80–2.30	>2.40
Calcium	Ca	%	<1.50	1.50–2.90	3.00–4.90	5.00–7.00	>7.00
Magnesium	Mg	%	<0.20	0.20–0.29	0.30–0.49	0.50–0.70	>0.80
Sulphur	S	%	<0.14	0.14–0.19	0.20–0.39	0.40–0.60	>0.60
Iron	Fe	ppm	<35	36–59	60–120	121–200	>200
Zinc	Zn	ppm	<17	18–24	25–100	101–300	>500
Manganese	Mn	ppm	<17	18–24	25–100	101–300	>500
Boron	Br	ppm	<20	21–35	36–100	101–200	>250
Copper	Cu	ppm	<3	3–4	5–16	17–20	>20
Molybdenum	Mo	ppm	<0.05	0.06–0.09	0.10–1.0	2.0–5.0	>5.0

Source: Smith (1966), Koo (1983), Malavolta and Netto (1989)

by soil analysis. It is usually more difficult to assess the N and K status in the soil because both these elements are subject to leaching in humid regions (Embleton 1967a,b).

21.18 Leaf Analysis

Leaf analysis is an effective technique for monitoring the nutrient status of citrus trees. It has been the most extensively researched tool for determining the needs of citrus trees (Smith 1966). The leaf sample consists of at least 100 leaves that are 4–6 months old, taken from non-fruiting twigs or terminals of the previous spring's growth flush. Thus, sampling should be conducted from July to September. The orchard to be represented by the leaf sample should consist of only one rootstock/scion combination of uniform-aged trees, within a single soil type (Smith 1966; Zekri 1994).

Select 15–20 trees randomly across the orchard from which to collect five or six leaves each. Leaves should be free of damage from insects or disease. Leaves with obvious chlorosis should be excluded, unless the sample is being taken specially to ascertain a potential cause of the chlorosis. While still fresh, the leaves should be washed to remove soil or dust, then air-dried, packaged and submitted to the laboratory for analysis (Jorgensen and Price 1978; Embleton et al. 1967a, b).

When the result are obtained from the laboratory, compare them with the standard shown in Table 21.3. The laboratory will report the levels of major element as percentage of dry weight, while microelements will be reported as parts per million.

References

Alvs AK, Tucker DPH (1992) Foliar application of various sources of iron, manganese and zinc to citrus. Proc Fla State Hortic Soc 105:70–74

Bhargava BS, Raghupathi HB, Dohroo NP, Bharat NK, Nayar S (2000) Nutrient deficiency and toxicity disorders in tropical and subtropical fruits. In: Gupta VK, Sharma SK (eds) Diseases of fruit crops. Kalyani Publishers, New Delhi, pp 283–286

Brown JC (1956) Iron chlorosis. Ann Rev Plant Physiol 7:171–190

Buckman HO, Brady NC (1969) The nature and properties of soil. MacMillan Co., New York, pp 395–397, 504–523

Camp AP, Fudge BR (1939) Some symptoms of citrus malnutrition in Florida. Florida Agricultural. Stn Bull: 335

Dickey RD, Drosdoff M, Hamilton J (1948) Copper deficiency of tung in Florida. Fla Agric Exp Stn Bull 447:1–32

Embleton T, Jones W, Reiz H (1967a) Citrus fertilization. In: Reuther W, Webber HJ, Batchelor LD (eds) Citrus industry, vol 3. University of California, Berkeley, pp 122–182

Embleton T, Jones W, Labanauskas C, Reuther W (1967b) Leaf analysis as a diagnostic tool and guide to fertilization. In: Reather W, Webber HJ, Batchelo LD (eds) Citrus industry, vol 3. University of California, Berkeley, pp 183–210

Embleton TWH, Reitz HJ, Jones WW (1973) Citrus fertilization. In: Recther W (ed) Citrus industry, vol 3. University of California Division of Agricultural Science, Berkeley

Floyd BF (1977) Dieback or exanthema of citrus tree. Proc.Fla Fr Gr Assoc (Jan) as cited by H.H. Hume (1900). Fla Agric Exp Stn Bull 53:157

Glendining JS (1999) Australian soil fertility manual. CSIRO, Collingwood

Haas ARC (1945) Boron in citrus tree. Plant Physiol 20:323–343

Jorgensen KR, Price GH (1978) The citrus leaf and soil analysis system in queens land. In: Proceeding of the international society of Citriculture, Singapore, p 297

Knorr LC (1973) Citrus diseases and disorders. University press of Florida, Gainesville, pp 26–27

Koo RCJ (1983) Recommended fertilizers and nutritional spray for citrus. Fla Agric Exp Stn Bull, 536 D, USA

Malavolta E, Netto AV (1989) Nutricao mineral, calagem, cessageme adubcao dos citros. Associacoa Brasileria para pesquisa da potasa do fosfato, Piracicada, S.P. Brazil

Mehrotra RS (1999) Plant pathology. Tata McGraw-Hill Publishing Co. Ltd, New Delhi, pp 750–753

Mooney PA, Richardson A, Harty AR (1991) Citrus nitrogen nutrition – a fundamental approach. N.Z. kerikeri Horticultural Research Station, Citrus Research Seminar, June 1991, pp 69–88

Rajput CBS, Haribabu R (1985) Citriculture. Kalyani Publishers, New Delhi, pp 273–355

Saha LR (2002) Hand book of plant diseases. Kalyani Publishers, Ludhiana, pp 290–292

Singh RS (1983) Plant diseases. Oxford & IBH Publishing Co. Pvt. Ltd, New Delhi, pp 552–553

Smith PF (1966) Leaf analysis of citrus. In: Childers NF (ed) Temperate and tropical fruit nutrition. Horticulture Publication, Rutgers-The state university, New Brunswick, pp 208–228

Smith PF, Reuther W (1949) Observation on boron deficiency in citrus. Proc Fla State Hortic Soc 62:31–37

Sprague HB (1964) Hunger signs in crops, 3rd ed. Darid McKay Co. Inc, New York, 61 pp

Tisdale SL, Nelsomn WL (1975) Soil fertility and fertilizer. Mac Million Publishing Co., New York, pp 300–313

Wallace A, Lunt OR (1960) Iron chlorosis in horticulture plants – a review. Proc Am Soc Hortic Sci 75: 819–841

Walter R, Embleton TW, Jones WW (1958) Mineral nutrition in tree crops. Ann Rev Plant Physiol. University of California, Citrus experiment station Bull, 9, p 6

Weir RG, Sarooshi R (1991) Citrus nutrition. Agric Fact H.2.3.11: N SWDPI

Zekri M (1994) Assessment of fertilizer needs. Citrus Ind 75(9):30

Zekri M (1995a) Nutritional deficiencies in citrus trees. I. Nitrogen, phosphorus and potassium. Citrus Ind 76(8):58–61

Zekri M (1995b) Nutritional deficiencies in citrus trees. II. Calcium, magnesium and sulphur. Citrus Ind 76(9):19–20

Zekri M (1995c) Nutritional deficiencies in citrus trees. III. Iron, Zinc and manganese. Citrus Ind 76(10): 16–17

Zekri M (1995d) Nutritional deficiencies in citrus trees. IV. Boron, copper, molybdenum and chlorine. Citrus Ind 76(11):34–36

Zekri M (1996) Several disorders in citrus trees. Citrus Ind 77(2):18–20, 53

Zekri M (1999) Citrus fertilizer management. Citrus Veg Mag 63(9):28–30

Zekri M (2002a) Fertilizer management and nutrition of citrus trees. Citrus Ind 83(12):29–30

Zekri M (2002b) Micronutrients in citrus nutrition. Citrus Ind 83(40):21–23, 32

Zekri M (2003) Guidelines for fertilizer programme in citrus. Citrus Veg Mag 67(8):56–60

Zekri M (2004) Citrus nutrition in relation to soil acidity and calcareous soils. Citrus Ind 85(40):21–24

Zekri M, Obreza TA (2003) Essential micronutrients for citrus. Functions, deficiency. Symptoms and correction. In: Futch SH (ed) Nutrient management for optimum citrus tree growth and yield short courses. University of Florida, Lake Alfred, pp 17–24

Zekri M, Obreza TA (2004) Plant nutrients for citrus trees. Citrus Veg Mag 68(10):17–20

The manufacturer's authorised representative in the EU is Springer Nature Customer Service Centre GmbH, Europaplatz 3, 69115 Heidelberg, Germany. If you have any concerns regarding our products, please contact ProductSafety@springernature.com

Printed and bound by CPI Group (UK) Ltd, Croydon, CR0 4YY
02/06/2026
02125540-0016